The Illustrated Directory of

MODERN WEAPONS

Warplanes, tanks, missiles, warships, artillery, small arms

<aside>Published by
CRESCENT BOOKS
New York</aside>

The Illustrated Directory of
MODERN WEAPONS

Warplanes, tanks, missiles, warships, artillery, small arms

Edited by Ray Bonds

A Salamander Book

First English edition published by
Salamander Books Ltd.,
Salamander House,
27 Old Gloucester Street,
London WC1N 3AF,
United Kingdom.

This edition is published by Crescent Books,
distributed by Crown Publishers, Inc.,
One Park Avenue,
New York,
New York 10016, United States of America.

h g f e d c b a

Library of Congress Cataloging in Publication Data
Main entry under title:

Modern weapons.

1. Weapons systems.   I. Bonds, Ray.
UF500.M63   1985   355.8'2   85-7849
ISBN 0-517-46959-6

Credits

Editor: Ray Bonds
Designer: Philip Gorton
Filmset: The Old Mill, England
Color reproduction: Melbourne Graphics, England
Printed in Belgium: Henri Proost et Cie, Turnhout

Acknowledgments

I would like to thank wholeheartedly everyone who has supplied
information and illustrations for this book, including all the
manufacturers of the weapons systems, many branches of various
armed services throughout the world, official government archives
and many private individuals. A full list of illustration credits is
given at the back of the book, but the following people and
organizations merit special mention: the US Department of
Defense Audio Visual Agencies, the British Ministry of Defence
and the NATO Press Department Photographic Service.
In particular, I thank Bernard Fitzsimons for his tremendous
editorial help throughout the project.

Ray Bonds

Contributors

Christopher F. Foss, Editor of *Jane's Armoured Fighting Vehicles,
Armour and Artillery* and *Military Vehicles and Ground Support
Equipment,* and author of Salamander's *Encyclopedia of Tanks and
Fighting Vehicles.*

John Jordan, a contributor to many important defense journals, a
consultant for the Soviet section of *Jane's Fighting Ships* 1980-81,
co-author of Salamander's *Balance of Military Power,* and author
of four of Salamander's illustrated "Guides": *The Modern US
Navy, The Modern Soviet Navy, Modern Naval Aviation* and
Battleships and Battlecruisers.

David Miller, a contributor to numerous technical defense
journals, author of the Salamander "Guides" to *Modern
Submarines* and *Modern Sub Hunters,* and co-author of their
Balance of Military Power and *Advanced Technology Warfare.*

Bill Gunston, author of many Salamander technical reference
books, in particular *Modern Air Combat* and *Modern Fighting
Aircraft,* author of many of their illustrated military "Guides" and
co-author of *Advanced Technology Warfare.* He is former
Technical Editor of *Flight International* magazine and an advisor to
several international aerospace companies. He is also an Assistant
Compiler of *Jane's All the World's Aircraft.*

Contents

Introduction

As this directory shows, there is no shortage of customers for, or suppliers of, military hardware. Clearly it is a self-perpetuating situation, since the more powerful and sophisticated the offensive weapon is, the more effective and sophisticated the defensive weapon has to be to counter it. And the more effective the counter or defensive system is, then the more effective the offensive weapon has to be to overcome it. This applies right across the military spectrum, from strategic nuclear weapons to small arms, on land, at sea and in the air — and now of course in space.

The effectiveness of a weapon is also increasingly more dependent upon miniaturization of guidance and control systems rather than the amount of explosive material in the warhead. For it is such micro-electronic devices that are enabling weapons to find their targets with more deadly accuracy, and with more predictability. Such is the sophistication of modern weapons technology that shells or missiles can be launched at a target, such as a slow-moving or stationary warship, from out of sight but with the absolute certainty that they are on course for their target — and yet, paradoxically, they may never get there because of the accuracy and timeliness of a defensive gun or missile system that can detect, track and destroy a sea-skimming shell or missile with equal "certainty"!

Economics, as always, is a most important factor in the development of weapons. Armies find themselves equipped with the weapons their nation can afford — even though in truth

nations cannot really afford the weapons they field. This, however, is a matter for individual governments' sense of national priorities. But without doubt there are many nations (including the superpowers) whose "defensive" armoury outstrips by far the level required to do the job. Often, too, the weapons developed appear to be more sophisticated than they need to be: perhaps it is a case of "We have the technology, and we shall use it". Of course, this can lead to a situation where certain weapons are too valuable to use, lest they themselves should be destroyed! Certainly, many are too expensive to be used extensively merely in training, so that if the time comes when they are needed in earnest they might be found to malfunction, or generally not to perform "as advertised".

But we have not concentrated solely on the high-technology weapons in this directory; they certainly predominate, reflecting the march of progress in one of the most strident industries in the world, but we have also included very many of the "simple" systems, like artillery pieces, mortars and small arms.

No single volume or even series of volumes in the public domain could hope to include descriptions and illustrations of every weapon in service today. However, we have attempted to cover the most important weapon systems in service or coming through; often we have selected a single system as representative of a type. In this way we have been able to provide technical details and a description of some 400 weapons and platforms, with reference to many more as variants within entries. Almost all of the weapons are illustrated, and mostly in colour.

Furthermore, the Appendix gives concise information on another 200 weapons, and includes those systems that fill in many of the unavoidable gaps in the main directory; some of these weapons, such as the Soviet AK-47 rifle and T-54/55 are obsolescent but are important nonetheless since they are still in widespread service.

The entries are presented in main sections relative to their theatre of operations (Strategic, Air, Land and Sea Weapons), and then in subsections relative to their specific type. Within subsections, the entries are arranged alphabetically or numerically, as appropriate.

Strategic
Weapons

Strategic nuclear weapons systems split conveniently into three categories — land, sea and air — constituting the "triad", so often referred to by Pentagon planners. Land-based inter-continental ballistic missiles (ICBMs) are currently the primary counter-force (ie targetted against hostile strategic nuclear forces) because they possess not only good range and throw-weight, but also are extremely accurate. The latest US re-entry vehicle (RV) has a circular area probable (CEP, that is, accuracy) of 656ft (200m) while the Soviet SS-19 has one of 853ft (260m). Current US intentions are to replace 100 Minutemen ICBMs with the new Peacekeeper, while the elderly Titan is in the last stages of wasting-out. On the Soviet side deployment of the SS-18 and SS-19 is virtually complete, and the next generation is already under development. China and France both have small ICBM forces.

Submarine-launched ballistic missiles (SLBMs) continue to provide a virtually guaranteed second-strike counter-value capability (ie, against cities). The launch vehicles (SSBN) remain difficult to detect, despite a great increase in size, with the Soviet Typhoon being the largest submarine vehicle ever constructed. SLBMs are becoming much more accurate and if Manoeuvrable Re-entry Vehicles (MaRVs) are perfected then SLBMs could develop a counter-force role. Submarine-launched cruise missiles (SLCMs) add a new capability, although currently confined to the US and Soviet Navies. The UK, France and China also have SSBN forces.

Manned bombers comprise the third leg of the triad; the US has a substantial force while the Soviets retain some very elderly aircraft in the role, but with the new "Blackjack" about to enter service. The US is about to put the B-1B into service, but a super-secret "stealth" bomber is known to be in development. Both the US and USSR have air-launched cruise missiles (ALCMs) in service. Neither bombers nor the ALCMs (and GLCMs) are first-strike weapons, but would be used in a second-strike retaliatory role.

Above: An ominous message on the side of a Minuteman ICBM in its silo at Ellsworth AFB, South Dakota.

ABM/ASAT SYSTEMS

ABM-1B Galosh

Origin: Soviet Union.
Type: Anti-ballistic missile.
Dimensions: Length about 65ft (19·8m); diameter (fins folded) 8·4ft (2·57m).
Launch weight: About 72,000lb (32,700kg).
Range: About 200 miles (322km) reported, probably several times greater, but accurate figures have not been revealed.

Development: Originally called SA-7 by the DoD, a designation later applied to the shoulder-fired Grail, this system is the only ABM in operational use in the world. It uses a large conical multi-stage missile, first seen (or rather not seen) in its tubular container in the Red Square parade of November 1964. This was the first showing of the MAZ-543 eight-wheel tug, which has since become standard for towing many ICBMs on separate trailers, the tug accommodating numerous crew-members.

With Galosh the missile travels in a ribbed tubular container pivoted above the rear wheels and with its own powered truck further back. With this missile no attempt is made to carry extra crew in the tug. The missile travels base-first, the open front of the container revealing four first-stage nozzles. The other end of the container was covered by a fabric shroud attached over a light framework until 1969, when a light rigid convex end-closure apparently of plastic material was substituted.

There are three propulsion stages and a thermonuclear warhead with a yield of several megatons (one report says 2-3 MT). The SALT-I treaty allowed the Soviet Union to deploy 100 ABM launchers, and in the late 1960s four ABM complexes were started in areas around Moscow, clearly intended for the protection of the Party leadership.

The Galosh sites themselves each contain two large battle-management radars of the Dog House or Cat House phased-array type, four Triad (Try Add) engagement radars, including Chekhov target/missile

trackers, and 16 launch silos. Dog House became operational around 1968 and has a range of some 1,750 miles (2,816km). Each complex also has large computer installations and other supporting services. In the event only four complexes (16 launchers) have been completed, but ABM research continues "on a massive scale". Flight testing of an improved SH-4 Galosh missile was reported in 1976, with a manoeuvrable bus which can "loiter" while incoming warheads are separated from chaff and decoys, restarting its propulsion several times and then homing for the kill.

It was reported in 1984 that this system is being increased from 64 to the full 100 allowed under the ABM Treaty. In the two-layer system silo-based modified ABM-1B missiles will undertake exoatmospheric interceptions backed up by a new high acceleration ABM for second-layer endoatmospheric missions.

Left: US DoD artist's impression of one of the 64 ABM-1B Galosh surface launchers deployed in defence of Moscow. Current upgrading of the ABM system will relocate modified Galoshes in silo-based launchers.

BOMBERS

Boeing B-52 Stratofortress

Origin: USA.
Type: Heavy bomber and missile platform.
Engines: (D) Eight 12,100lb (5,489kg) thrust P&WA J57-19W or 29W turbojets, (G) eight 13,750lb (6,237kg) thrust P&WA J57-43W or -43WB turbojets, (H) eight 17,000lb (7,711kg) thrust P&WA TF33-1 or -3 turbofans.
Dimensions: Span 185ft 0in (56·39m); length (D, and G/H as built) 157ft 7in (48·0m), (G/H modified) 160ft 11in (49·05m); height (D) 48ft 4^{1}/$_{2}$in (14·7m), (G/H) 40ft 8in (12·4m); wing area 4,000sq ft (371·6m^2).
Weights: Empty (D) about 175,000lb (79,380kg), (G/H) about 195,000lb (88,450kg). loaded (D) about 470,000lb (213,200kg), (G) 505,000lb (229,000kg), (H) 505,000 at takeoff, inflight refuel to 566,000lb (256,738kg).
Performance: Max speed (true airspeed, clean), (D) 575mph (925km/h), (G/H) 595mph (957km/h); penetration speed at low altitude (all) about 405mph (652km/h), Mach 0·53; service ceiling (D) 45,000ft (13·7km), (G) 46,000ft (14·0km), (H) 47,000ft (14·3km); range (max fuel, no external bombs/missiles, optimum hi-alt cruise), (D) 7,370 miles (11,861km), (G) 8,406 miles (13,528km), (H) 10,130 miles (16,303km); takeoff run, (D) 11,100ft (3,383m), (G) 10,000ft (3,050m), (H) 9,500ft (2,895m).
Armament: (D) Four 0·5in (12·7mm) guns in occupied tail turret, MD-9 system, plus 84 bombs of nominal 500lb (227kg) in

bomb bay plus 24 of nominal 750lb (340kg) on wing pylons, total 60,000lb (27,215kg); (G) four 0·5in (12·7mm) guns in remote-control tail turret, ASG-15 systems, plus eight nuclear bombs, or up to 20 SRAM (eight on internal dispenser plus 12 on wing pylons), or 12 AGM-86B ALCMs on wing pylons; (H) single 20mm six-barrel gun in remote-control tail turret, ASG-21 system, plus bombload as G (later to have AGM-86B internal dispenser).

Development: A legend in its own time, the B-52 was designed to the very limits of the state of the art in 1948-49 to meet the demands of SAC for a long-range bomber and yet achieve the high performance possible with jet propulsion. The two prototypes had tandem pilot positions and were notable for their great size and fuel capacity, four

double engine pods and four twin-wheel landing trucks which could be slewed to crab the aircraft on to the runway in a crosswind landing. The B-52A changed to a side-by-side pilot cockpit in the nose and entered service in August 1954, becoming operational in June 1955. Subsequently 744 aircraft were built in eight major types, all of which have been withdrawn except the B-52D, G and H.

The B-52D fleet numbered 170 (55-068/-117, 56-580/-630 built at Seattle and 55-049/-067, 55-673/-680 and 56-657/-698 built at Wichita) delivered at 20 per month alongside the same rate for KC-135 tankers in support. The B-52G was the most numerous variant, 193 being delivered from early 1959 (57-6468/-6520, 58-158/-258 and 59-2564/-2602, all from Wichita), introducing a wet (integral-tank) wing which increased internal fuel from 35,550 to 46,575 US gal and also featured shaft-drive generators, roll control by spoilers only, powered tail controls, injection water in the leading edge, a short vertical tail, rear gunner moved to the main pressurized crew compartment, and an inner wing stressed

Below: A modified B-52H with its rarely-used 32ft (9·75m) braking parachute deployed. ALQ-153 pulse-Doppler blisters for missile warning are apparent on the fin.

ALMV (ASAT)

For many years the US Department of Defense, and particularly the USAF and the US Army, have been studying Anti-Satellite (ASAT) weapons systems. The Air Force studies led in 1979 to a programme for operational hardware, with a USAF contract for $78.2 million being awarded to the Vought Corporation for an ASAT system to be deployed by about 1985. This was to comprise an advanced interceptor vehicle with guidance so accurate it would destroy targets by physical impact, no explosive warhead being necessary.

This weapon, designated the Air-Launched Miniature Vehicle, is now in the testing stage; it is launched by an F-15 fighter and then boosted into orbital height by two rocket stages. The first stage is based on the Short Range Attack Missile (SRAM), a Boeing Aerospace product, and Boeing has developed this stage, as well as providing integration services and managing development of the mission control centre. The second stage is the Altair III, the Thiokol motor which for many years has been the fourth stage of Vought's Scout vehicle. McDonnell-Douglas has modified the F-15 to serve as the launch platform. An $82.3 million contract was voted in 1980 and a further $268 million in 1981 to continue development through to flight testing in 1984.

In February 1985 US Secretary of State for Defense Caspar Weinberger stated that the ASAT programme was in the test and evaluation stage and that he was requesting funds in FY85 to start procurement.

Above: Captive flight test of a Vought ASAT aboard an F-15, two squadrons of which will use the missile from 1987.

Left: Chin bulges house the LLTV and FLIR sensors installed as part of the EVS update applied to these SAC B-52Hs.

Above: Unlike the cockpits of any earlier models, the flight decks of the B-52G and H are dominated by the EVS displays.

for a large pylon on each side of the fuselage. The final model, the B-52H, numbered 102 (60-001/-062 and 61-001/-040), and was essentially a G with the TF33 fan engine and a new tail gun.

During the Vietnam war the B-52D was structurally rebuilt for HDB (high-density bombing) with conventional bombs, never considered in the original design. The wings were given inboard pylons of great length for four tandem triplets of bombs on each side, and as noted in the data 108 bombs could be carried in all with a true weight not the "book value" given but closer to 89,100lb (40,400kg). Another far-reaching and costly series of structural modifications was needed on all models to permit sustained operations at low level, to keep as far as possible under hostile radars.

The newest models, the G and H, were given a stability augmentation system from 1969 to improve comfort and airframe life in turbulent dense air. From 1972 these aircraft were outfitted to carry the SRAM (Short-Range Attack Missile), some 1,300 of which are still with the SAC Bomb Wings. Next came the EVS (Electro-optical Viewing System) which added twin chin bulges. The Phase VI ECM (electronic countermeasures) cost $362·5 million from 1973. Quick Start added cartridge engine starters to the G and H for a quick getaway to escape missile attack. Next came a new threat-warning system, a satellite link and "smart noise" jammers to thwart enemy radars. From 1980 the venerable D-force was updated by $126·3 million digital nav/bombing system. Further major changes to the G and H include the OAS (offensive avionics system) which is now in progress costing $1,662 million.

The equally big CMI (cruise-missile interface) will eventually fit the G-force for 12 AGM-86B missiles on the pylons in tandem triplets, the internal bay being the wrong size and remaining available for a SRAM dispenser.

Under current plans 99 B-52Gs are being modified for cruise missiles, each also having a curved wing root fairing to identify it to Soviet satellites as a missile carrier (to comply with the SALT II treaty articles, even though these were never ratified). The 416th BW at Griffiiss AFB became fully operational with AGM-86B in December 1982, and other wings have converted, or are doing so, at Blytheville, Grand Forks, Fairchild and Barksdale AFBs. Some of these have the B-52H, 96 of which are also being converted, and later in the 1980s the H will be more radically modified to accommodate the AGM-86B rotary launcher in the internal bay.

As 168 B-52Gs were rebuilt with OAS, 69 are left over from the CMI conversion, and these are replacing Ds in the non-nuclear maritime support role. Here one of their new weapons is the AGM-84 Harpoon cruise missile, clusters of which can be carried in the long-range anti-ship role. In 1983 one of the numerous test launches was from 30,000ft (9,144m), but all were successful and two squadrons of Harpoon-Gs are now operational.

Yet another new weapon is the GD AGM-109H MRASM (medium-range ASM), a variant of the Tomahawk cruise missile. With a range of 285 miles (459km), this dispenses large conventional submunitions, the usual design being for cratering runways. Four missiles are carried on each wing pylon.

In 1984/85 many B-52s were being repainted in European 1 camouflage, a considerable task.

Dassault-Breguet Mirage IVA, IVP and IVR

Origin: France.
Type: (A) Strategic bomber, (P) missile carrier, (R) strategic reconnaissance.
Engines: Two 15,432lb (7,000kg) thrust (max afterburner) SNECMA Atar 9K augmented turbojets.
Dimensions: Span 38ft 10^1/2in (11·85m); length 77ft 1in (23·5m); height 17ft 8^1/2 (5·4m).
Weights: Empty about 33,000lb (15,000kg); loaded 73,800lb (33,475kg).
Performance: Max speed (dash) 1,454mph (2,340km/h, Mach 2·2) at 40,000ft, (sustained) 1,222mph (1,966km/h, Mach 1·7) at 60,000ft (19,680m); time to climb to 36,090ft (11,000m), 4min 15sec; service ceiling 65,620ft (20,000m) tactical radius (dash to target, hi-subsonic return) 770 miles (1,240km); ferry range 2,485 miles (4,000km).
Armament: (A) One AN22 free-fall nuclear bomb, usually of 60kT yield, or up to 16,000lb (7,258kg) of conventional weapons on external attachments, (P) one ASMP nuclear cruise missile, (R) none.

Development: In the Gaullist era the creation of a national French nuclear deterrent was an overriding priority, and one of the three delivery systems chosen was the Mirage IVA manned supersonic bomber: It was eventually planned around two Atar engines, and this

so reduced the size in comparison with the first schemes as to render the aircraft incapable of flying round-trip missions from French territory to likely targets. Thus, heavy reliance has always been placed on the 14 (later reduced by a crash to 13) C-135F tankers, and on "buddy" techniques in which one Mirage IVA refuels its companion.

Even so, the initial planning of the Commandement des Forces Aériennes Stratégique presupposed that most missions would be one-way (or at least would not return to France). Dispersal was not maximized, with the force divided into three Escadres (91 at Mont de Marsan, 93 at Istres and 94 at Avord), which in turn were subdivided into three four-aircraft groups, two of which were always dispersed away from Escadrille HQ. Despite being a heavy and "hot" aircraft, the IVA has also been rocket-blasted out of short unpaved strips hardened by quick setting chemicals sprayed on the soil.

Of the original total of 62 Mirage IVs, only 36 are currently at readiness in the bomber role, but another 12 have been largely rebuilt as dedicated reconnaissance aircraft, designated IVR. Losing their ability to carry a nuclear store, they have large, classified installations which include optical cameras, an IR linescan and, it is believed, a SLAR (Side Looking Airborne Radar), all installed in the underside of the fuselage. There is a possibility that

an SAR (Synthetic-Aperture Radar) may become available, although none is known in France.

The Mirage IVAs currently in the strategic bomber role carry a single 70kT AN22, free-fall nuclear bomb semi-recessed in the underside of the fuselage.

In June 1983 a Mirage IVA trials aircraft carried the first ASMP (air/sol moyenne portée, air/ground long range) cruise missile to have operative ramjet propulsion. This Aérospatiale weapon has a range of about 62 miles (100km) at highly supersonic speed, with a warhead of 100 or 150kT yield, and programmed to make violent course changes and other diversionary tactics. A batch of 18 IVA bombers are going through the Atelier Industrielle de L'Air at Aulnat (which converted the IVRs) being rebuilt as IVP missile carriers. They have new inertial navigation systems, computers and cockpit displays, with digital position information transfer to the

Above: Rocket-assisted takeoff of Mirage IVP (modified IVA) carrying an ASMP nuclear missile. Eventually, France hopes to have 18 IVP bombers as well as dual-role IVAs.

missile.

Unlike the IVAs, which can be either strategic bombers or reconnaissance machines, but not both, the aircraft being converted to ASMP-carrying IVPs will have useful flexibility in that they can be quite quickly transformed from bomber into reconnaissance aircraft by removing both the cruise missile and its pylon, and replacing them with a camera pod.

The IVP (Pénétration) will equip EB1/91 Gascogne and 2/91 Bretagne at Mont de Marsan, with IOC in 1986. Meanwhile the rundown in the IVA force continues, with 3/91 having been withdrawn in late 1983 followed in 1984 by all escadrons of EB93.

Rockwell B-1

Origin: USA.
Type: Strategic bomber.
Engines: Four 30,000lb (13,608kg) thrust class General Electric F101-102 augmented turbofans.
Dimensions: Span (15^{0}LE angle) 136·7ft (41·67m); length 147·0ft (44·81m); height overall 34·0ft (10·36m); wing area (gross) 1,950sq ft (181·2m²).
Weights: Empty about 172,000lb (78,000kg); max 477,000lb (216,367kg).
Performance: Lo penetration speed 600mph (966km/h); max speed (hi, clean) 825mph (1,330km/h, Mach 1·25); 7,455 miles (12,000km); field length, under 4,500ft (1,372m).
Armament: Eight ALCM internal plus 14 (later mofified to 22) external; 24 SRAM internal plus 14 (later 22) external; 12 B28 or B43 internal plus 8/14 external; 84 Mk 82 internal plus 44 external (80,000lb, 35,288kg).

Development: No air weapon in history has ever taken so long to replace as the B-52; and this is largely because what looks like at last becoming its replacement, the B-1B, has already taken more than 22 years to proceed from inception to first flight. Inevitably this means that the B-1B is already obsolescent. In particular it was designed before what are called "stealth" concepts were worked out.

Back in the 1960s the emphasis was still on supersonic speed, so the four B-1 prototypes (today designated as B-1As) were built with maximum swing sweep of 67º/30ft and were planned to have variable engine inlets and ejectable crew capsules of extremely advanced design. The latter feature was abandoned to save costs, and though the second aircraft reached Mach 2·22 in October 1978 this end of the speed spectrum steadily became of small importance.

By 1978 the emphasis was totally on low-level penetration at subsonic speeds, with protection deriving entirely from defensive electronics and primitive "stealth" characteristics. Not much could be done to reduce radar cross-section, but actual radar signature could be substantially modified, and the effort applied to research and development of bomber defensive electronic systems grew rapidly.

The B-1 features a blended wing/body shape with the four engines in paired nacelles under the fixed inboard wing immediately outboard of the bogie main gears. Though designed more than 15 years ago, the aerodynamics and structure of the B-1 remain competitive, and the extremely large and comprehensive defensive electronics systems (managed by AIL Division of Cutler-Hammer under the overall avionics integration of

Boeing Aerospace) far surpass those designed into any other known aircraft, and could not reasonably have been added as post-flight modifications. During prototype construction it was decided to save further costs by dropping the variable engine inlets, which were redesigned to be optimized at the high-subsonic cruise regime.

Another problem, as with the B-52, was the increased length of the chosen ALCM, which meant that the SRAM-size rotary launcher was no longer compatible. The original B-1 was designed with three tandem weapon bays, each able to house many free-fall bombs or one eight-round launcher. Provision was also made for external loads.

A particular feature was the LARC (Low-Altitude Ride Control), an active-control modification which, by sensing vertical accelerations due to atmospheric gusts at low level and countering these by deflecting small foreplanes and the bottom rudder section, greatly reduced fatigue of crew and airframe during low-level penetration.

All four prototypes exceeded planned qualities. The third was fitted with the ECM system and DBS (doppler beam-sharpening) of the

Right: The refined contours of the B-1B are accentuated by the Europe 1 camouflage scheme applied to the first example. First flight (October 1984) was six months ahead of schedule.

Tupolev "Blackjack"

Origin: Soviet Union.
Type: Bomber, missile carrier and strategic recon aircraft.
Engines: Two large augmented turbofans, believed to be Kuznetsov NK-144 derivatives each rated at about 44,090lb (20,000kg) thrust.
Dimensions: (estimated) Span (spread, 20°) 113·0ft (34·45m), (max sweep, 65°) 86·0ft (26·21m); length overall 137·8ft (42m), (excl probe) 132·0ft (40·23m); height overall 33·0ft (10·06m); wing area (20°) 1,830sq ft (170m²).
Weights: (estimated) Empty 120,000lb (54,400kg); internal fuel 125,500lb (57,000kg); max 270,000lb (122,500kg).
Performance: (estimated) Max speed (36,000ft/11,000m and above) 1,320mph (2,125km/h, Mach 2), (SL) 680mph (1,100km/h, Mach 0·9); high-speed cruise (med/hi level) Mach 0·9; normal cruise 560mph (900km/h); service ceiling (afterburner) 62,300ft (19,000m), (dry) 55,000ft (17,000m); unrefuelled combat radius (DoD figure) 3,420 miles (5,500km); max range 7,500 miles (12,000km); endurance on internal fuel over 10 hours.
Armament: Twin 23mm guns in remotely aimed tail barbette; either one "AS-4 Kitchen" or "AS-6 Kingfish" missile recessed under fuselage, or two similar missiles on glove pylons, or load of up to about 26,455lb (12,000kg) in weapon bay and on external racks under inlet ducts.

Development: Using the same philosophy and aerodynamics as applied to turn the Su-7 into the Su-17, the Tupolev OKB turned the fixed-wing Tu-22 into the swing-wing Tu-22M, thereby conferring a major improvement in mission radius and payload. A small number thought to equip "one DA squadron", were delivered of the initial version, called "Backfire-A" by NATO. The main production model, "Backfire-B", has a totally redesigned rear fuselage and new landing gears retracting inwards into the fuselage. All models have two very powerful engines in the rear fuselage fed by giant lateral inlet ducts.

The result is a formidable aircraft of great value to the DA and AVMF in projecting power over substantial areas of the globe. During the SALT 2 talks it was stressed that the Tu-22M is not a strategic aircraft and is therefore not subject to any numerical limitations; at the same time the flight-fuelling probes were removed, though as these could be replaced in minutes the point was obscure. The fact that this aircraft was not designed primarily for missions directly against the USA for a time clouded the real significance of the Tu-22M in diverting attention from its enormous influence on theatre operations.

The initial reports all mentioned

Below: One of the estimated 120 "Backfire-B" models of the Tu-22M in service with western theatre strategic nuclear forces.

missiles, notably the then-new "AS-6 Kingfish", hung on pylons at the wing pivots. Available photographs all show either a single missile recessed under the fuselage or external racks (of a strangely high-drag form) under the inlet ducts for rows of conventional bombs. The DoD figure of 3,420 miles (5,500km) combat radius presupposes a hi-lo-hi mission with max weapon load, though it would probably have to be revised downwards with the external racks loaded.

Unlike the Tu-22 the crew compartment seats a crew of four, two by two, the rear men being electronic systems officers who also look after the rear guns. From 28 to 35 different avionics installations have been located on these aircraft, covering almost half the external surface, though few have been positively identified. EW/ECM installations are likely to be particularly comprehensive and are probably being continually updated.

Since 1980 there has been repeated reference to a version in service with "wedge type inlets" resembling those of the MiG-25; the reporting name "Backfire-C" has been allotted to this.

Reports of the number of "Backfires" in service vary, but the US DoD stated in April 1985 that there were about 250 (including 120 in Soviet Naval Aviation), and that production of at least 30 a year was expected to be maintained through the end of the decade. The front-line force is expected to remain at about 230, the spare aircraft kept in reserve.

main radar, while the fourth had complete offensive and defensive electronics and was almost a production B-1A. The Carter administration decided not to build the B-1 for the inventory, and the four aircraft were stored after completing 1,985·2h in 247 missions.

After further prolonged evaluation the Reagan administration decided in 1981 in favour of a derived B-1B, and announced the intention to put 100 into the SAC inventory from 1986, with IOC the following year.

The B-1B dispenses with further high-altitude dash features. As well as refined engines it can carry much more fuel; a detailed weight-reduction programme reduced empty weight, while gross weight is raised by over 37 tonnes. Other changes include: main gears stronger, wing gloves and engine inlets totally redesigned, many parts (ride-control fins, flaps and bomb door for example) made of composite material, pneumatic starters with cross-bleed fitted, offensive avionics completely updated, the ALQ-161 defensive avionics subsystem fitted, RAM (radar-absorbent material) fitted at some 85 locations throughout the airframe, and the whole aircraft nuclear-hardened and given Multiplex wiring. LARC has become SMCS (Structural-Mode Control System), and many parts of the airframe and systems have been refined.

Radar cross-section is less than one-hundredth that of a B-52. The offensive radar is based on the small APG-66 of the F-16, but includes a "low-observable" phased-array subsystem for precision navigation and terrain-following. Most important of all, of course, is the vast ALQ-161 defensive subsystem, which is expected to enable this large pre-stealth bomber to penetrate defended airspace "well into the 1990s".

B-1B development, mainly using B-1A No 2, has gone well despite a hold-up caused by loss of this prototype in 1984. The first production B-1B was rolled out some six months ahead of schedule, in September 1984, and made its first flight on 18 October. Deliveries were planned to begin to Dyess AFB (now home of the 96th BW) in mid-1985, with the first squadron of 15 aircraft becoming operational in September 1986. Ellsworth AFB will receive 32 aircraft from late 1986, replacing 19 B-52Hs, followed by Grand Forks, where the 16 B-1Bs replace B-52Gs. The 100th and last aircraft will enter service at McConnel AFB in 1988. It is planned to have compatibility with a second-generation ALCM and also to provide external launchers for 22 ALCMs or other large stores. Programme cost is put at $28·4 billion (on budget) or about $40 billion including all supporting services.

Many USAF planners believe 200 aircraft will be needed, and the extra 100 would cost, it is said, $10 billion. There is pressure to fund these as B-1Cs, with more rounded fuselages and more complete RAM coatings.

Tupolev "Blackjack"

Origin: Soviet Union.
Type: Strategic bomber and missile carrier.
Engines: Four large augmented turbofans or turbojets (possibly of Koliesov type) each in the 48,500lb (22,000kg) thrust class.
Dimensions: (estimated) Span (spread, 20° sweep) 177ft (54m), (max sweep, 57°) 135ft (41·2m); length 171ft (52·1m); height 47·5ft (14·5m); wing area 4,000sq ft (370m²).
Weights: (estimated) Empty 275,600lb (125,000kg); internal fuel 368,000lb (167,000kg); max 683,400lb (310,000kg).
Performance: (estimated) Max speed (hi, clean) 1,200kt (1,382mph, 2,224km/h, mach 2·1); max combat radius with full weapon load 4,536 miles (7,300km); endurance 14h.
Armament: In US DoD statement carries cruise missiles, bombs or a combination of both. Unofficial Western estimate of internal bombload 36,000lb (16,330kg).

Development: By far the biggest bomber in the world today, this giant swing-wing machine was observed on 25 November 1981 on film transmitted by a US recon satellite. The prototype was parked next to Tu-44 or 144D SSTs on the test airfield at Ramenskoye, and comparison with the SSTs enabled fairly accurate assessment of the new bomber's size and capability to be made — or so it was thought, until 1984 when the figures were dramatically revised.

The latest figures are given above, most of them coming from the DoD. The previous estimate of span at max sweep was only 95ft (29m) and max weight a mere 575,000lb (260,800kg), whereas the length has been revised down from 180ft (55m). There now seems little doubt that the configuration is almost precisely the same as that of the USAF B-1, though the Tupolev bomber is much larger. According to estimates it is also faster, though probably this is less important than its defensive electronic systems,

"stealth" qualities and ability to follow terrain in all weathers at high speed.

Certainly "Blackjack" has nothing whatever in common with the much smaller Tu-22M, and it may not even come from the Tupolev OKB (though it is hard to imagine any other source). The giant fixed wing glove, which extends almost to the nose, and the two pairs of engines mounted underneath its rear portion, have much in common with the Tu-144D SST, though it is extremely unlikely that the similarity is more than superficial.

Instead of cruising at Mach 2-plus at extreme altitude the bomber will spend almost all of each mission at subsonic speed just above the Earth's surface. It has been surmised that its weapons are carried in a box between the engine group on each side. With or without flight refuelling this aircraft clearly has the ability to reach virtually any target on Earth from Soviet or client-state bases, and its nature (unlike the Tu-22M) is deeply strategic. Thus it is unlikely to bother with anti-ship missiles, but it is expected to cary the AS-X-15 cruise weapon in multiple.

The gigantic new production factory being swiftly completed next to the existing Tupolev plant at Kazan is seen as the source of the production machines, which are predicted to become operational in 1987. Initially they are expected to replace M-4 "Bison" and Tu-20 "Bear-A".

Previous Soviet attempts to create an intercontinental-range bomber have been disappointing. The Tupolev "Bear" has the range capability, but is hopelessly slow for use on strategic missions into US airspace.

The Tu-22 "Blinder" and Tu-22M "Backfire" were both medium-range designs, and "Blackjack" marks the Tupolev organization's return to the strategic bomber field. The Tu-144 airliner had adequate range, but as mentioned above, "Blackjack" will not be required to cruise supersonically. As a result, it will have a better change of meeting its operational requirement.

Right: US DoD drawing of the Tupolev "Blackjack", showing the relatively large fixed area (by Western standards) of the variable-geometry wing.

NUCLEAR ASMs

ALCM, AGM-86B

Origin: USA.
Type: ALCM.
Dimensions: With wings/tailplane extended, length 20·7ft (6·32m); body diameter 24·5ft (6·22mm) span 12ft (3·66m).
Launch weight: 2,825lb (1,282kg).
Propulsion: One Williams F107-101 turbofan with sea-level rating of 600lb (272kg) static thrust.
Range: Max without belly tank, 760 miles (1,200km).
Flight speed: Cruise, about Mach 0·65; terminal phase, possibly Mach 0·8.
Warhead: W-80 thermonuclear as originally developed for SRAM-B.

Development: One of the most important weapons in the West's inventory, ALCM (Air-Launched Cruise Missile) was presented by President Carter as a new idea when he terminated B-1 as a bomber; he even said B-1 had been developed "in absence of the cruise missile factor", whose presence in 1976 made the bomber unnecessary. This is simply not true. The cruise missile never ceased to be studied from 1943, and — apart form such USAF examples as Mace and Snark — it was cruise-missile studies in 1963-6 that led to AGM-86 SCAD (Subsonic Cruise Armed Decoy) approved by DoD in July 1970. This was to be a miniature aircraft powered by a Williams WR19 turbofan, launched by a B-52 when some hundreds of miles short of major targets.

Like Quail, SCAD was to confuse and dilute hostile defences; but the fact that some or all would carry nuclear warheads — by 1963 small enough to fit such vehicles — meant that SCAD could do far better than

Quail. No longer could the enemy ignore decoys and wait and see which were the bombers. Every SCAD had to be engaged, thus revealing the locations and operating frequencies of the defence sites, which could be hit by surviving SCADSs, SRAMs or ARMs. SCAD was to be installationally interchangeable with SRAM, with a maximum range of around 750 miles (1,207km).

SCAD ran into tough Congresional opposition, but the USAF knew what it was about and in 1972 recast the project as ALCM, retaining the designation AGM-86A. SCAD had had only a secondary attack function, but ALCM is totally a nuclear delivery vehicle, and like SRAM has the ability to multiply each bomber's targets and increase defence problems by approaching from any direction along any kind of profile. Compared with SRAM it is much easier to intercept, being larger and much slower, but it has considerably greater range and allows the bomber to stand off at distances of at least 1,000 miles (1,609km).

The original AGM-86A ALCM was interchangeable with SRAM, so that a B-52G or H could carry eight on the internal rotary launcher plus 12 externally, and an FB-111A four externally plus two internally (though the latter aircraft has never been named as an ALCM carrier). This influenced the shape, though not to the missile's detriment, and necessitated folding or retracting wings, tail and engine air-inlet duct. Boeing, who won SCAD and carried across to ALCM without further competition, based ALCM very closely on SCAD but increased the fuel capacity and the sophistication of the guidance, with

a Litton inertial platform (finally chosen as the P-1000). and computer (4516C), updated progressively when over hostile territory by McDonnell Douglas Tercom.

In 1976 the decision was taken to aim at maximum commonality with AGM-109 Tomahawk, but the guidance packages are not identical. The engine in both missiles is the Williams F107 of approximately 600lb (272kg) thrust, but in totally different versions; the ALCM engine is the F107-101, with accessories underneath and different starting system from the Dash-400 of AGM-109. The warhead is W-80, from SRAM-B.

Above: Test launch of an AGM-86A from a B-52: designed to fit SRAM launchers, this version of the ALCM was replaced by AGM-86B.

AGM-86A first flew at WSMR on 5 March 1976. Many of the early flights failed — one undershot its target by a mile because its tankage had been underfilled! — but by the sixth shot most objectives had been attained and 1977 was spent chiefly in improving commonality with Navy AGM-109, in preparation for something unforeseen until that year: a fly-off against AGM-109 Tomahawk in 1979

to decide which to buy for the B-52 force. It was commonly said Boeing was told to make AGM-86A short on range to avoid competing with the B-1. In fact no more fuel could be accommodated and still retain compatibility with SRAM launchers, and in 1976 Boeing proposed an underbelly auxiliary fuel tank for missiles carried externally.

A better answer was to throw away dimensional compatibility with SRAM and develop a considerably stretched missile, called AGM-86B. This has a fuselage more than 30 per cent longer, housing fuel for double the range with a given warhead. Other changes include wing sweep reduced to 25°, thermal batteries for on-board electrical power, all-welded sealed tankage, improved avionics cooling and 10 year shelf life. President Carter's decision to cancel the B-1 in June 1977 opened the way for Boeing to promote this longer missile, which could still be

carried externally under the wings of a B-52 but would not have fitted into the weapon bays of a B-1.

From July 1979 Boeing's AGM-86B was engaged in a fly-off against GD's AGM-109. Results were hardly impressive, each missile losing four out of ten in crashes, quite apart from other mission-related failures, but after a long delay the USAF announced choice of Boeing on 25 March 1980. A month later it was announced that the USAF/Navy joint management was dissolved and that the USAF Systems Command would solely manage 19 follow-on test flights in 1980 and subsequent production of 3,418 missiles by 1987. The first two rounds assigned to SAC joined the 416th BW at Griffiss AFB in January 1981. Since then about half the 169 operational B-52G bombers have been converted to carry up to 12 rounds each, in two tandem triplets, and in 1982 President Reagan increased the buy to

3,780 missiles by 1990 to permit 96 B-52H bombers to be equipped also. From 1986 the internal bomb bays are to be rebuilt by Boeing-Wichita to permit each aircraft to carry a further eight rounds on an internal rotary launcher. Each B-52, after conversion, will have a permanently attached wing-root "strakelet", visible in satellite pictures, as demanded by SALT II provisions. The pre-loaded

Above: AGM-86Bs, with increased range and improved survivability, are loaded on to a B-52.

wing pylons will be carried only in time of emergency.

The production B-1B will carry the same eight-barrel rotary launcher as the rebuilt B-52, and except for the first few aircraft will also carry a further 14 on eight external racks.

ASMP

Origin: France.
Type: Stand-off nuclear missile.
Propulsion: Integrated rocket/ramjet, probably using SNPE rocket with Statolite smokeless filling and Aérospatiale advanced kerosene-fuelled ramjet.
Dimensions: Length about 16ft 5in (5m), body diameter about 16·5in (420mm); width across inlet ducts about 32·2in (820mm).
Weight: At launch about 2,000lb (900kg).
Performance: Speed about Mach 4, range variable up to 186 miles (300km).
Warhead: CEA nuclear, 150 kilotonne.

Development: It is difficult to know whether to class this weapon as tactical or strategic, and the French are

not sure themselves. Though it has a fair range, for the initials signify Air/Sol Moyenne Portée, it will have a nuclear warhead. ASMP was initiated in 1971 to arm whatever emerged as the next-generation Armée de l'Air deep-penetration aircraft, successively the Mirage G, ACF (Avion de Combat Futur) and Super Mirage.

Cancellation of the latter in 1976 reduced the pace of development, and no deep-penetration platform is now in prospect. Development was initially competitive between Matra with turbojet propulsion, and Aérospatiale with a ram-rocket or ramjet. In March 1978 the go-ahead was given to Aérospatiale, with liquid-fuel ramjet propulsion. Today an integrated hybrid system has been chosen; France has only limited ex-

perience with such propulsion, and may licence technology from CSD, Vought, Marquardt, MBB or other company.

Range specified for the original (January 1974) ASMP was 50-93 miles (80-150km). This has since been more than doubled, because of the short range of the only available carrier aircraft (Mirage IVA. 2000 and Super Etendard) and the chief puzzle now is how the Antilope 5 radar can

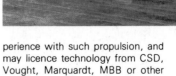

Above: Twin side inlets provide air for ASMP's ramjet after boost to supersonic speed.

acquire targets at over 186 miles (300km). Mid-course guidance is pre-programmed, with Sagem playing a major role in the main inertial guidance. About FF4,000 million is to be spent on 100 missiles, the first of which are now in service with Mirage IVA aircraft of the French Air Force.

SS-NX-21/AS-X-15/SS-C-X4

Origin: Soviet Union.
Type: SLCM/ALCM/GLCM.
Dimensions: Length 22·6ft (6·9m); wingspan 11·3ft (3·44m).
Propulsion: Small turbojet engine.
Weight: Not known.
Range: 1,865 miles (3,000km).
Speed: Subsonic.
Warhead: One nuclear.

Development: Following the lead given by the USA, the Soviet armed forces have developed a cruise missile capable of being launched from a submarine (SS-NX-21), an aircraft (AS-X-15) or from a ground launcher (SS-C-X4). The sea-based variant is launched from a standard 21in (533mm) torpedo tube and may well equip both the Mike and Sierra class SSNs which are just entering service (qv). SS-NX-21 was expected to enter service in 1984. It can, of course, be deployed in submarines patrolling off the coast of the USA.

The AS-X-15 was also expected to become operational in 1984 on board the Bear-H ALCM carrier aircraft, a new version of a somewhat elderly machine, as well as the new Black-

jack bomber, when that reaches IOC. These combinations will give the Soviet air force a real intercontinental capability for the first time.

Third version is the SS-C-X4 GLCM, which will be deployed on mobile ground launchers, probably similar to those used for the SS-20 IRBM (qv). It is expected to reach IOC in 1985.

An even newer cruise missile, about double the size of the SS-NX-21, is in a late stage of development and is expected to enter service in 1986–7. Some 43ft (63m) long and with swept wings, this system will have an intercontinental range. Ground and submarine-launched versions have been reported.

Below: US DoD artist's impression of an SS-NX-21 attack profile after firing from a torpedo tube.

Tomahawk (Sea), BGM-109A, B, C and E

Origin: USA.
Type: SLCM.
Dimensions: Length, loosely given as 21ft (6·4m) with boost motor, but varies type to type; length in flight, typically 18·25ft (5·56m); body diameter 21in (533mm); span 8·3ft (2·54m) or 8·6ft (2·61m).
Launch weight: All naval versions, typically 4,190lb (1,900kg).
Propulsion: One Williams F107-WR-400 turbojet rated at 600lb (272kg); launch by 7,000lb (3,175kg) Atlantic Research solid rocket motor.
Guidance: (A) INS(Tercom, (B/E) strapdown AHRS/active radar, (C) INS(Tercom + Dsmac).
Range: (A, C) 1,553 miles (2,500km), (B/E) 280 miles (450km).
Flight speed: 550mph (885km/h).
Warhead: (A) W-80 thermonuclear, (B/E, C) conventional, Bullpup type.

Development: Since it was first developed in December 1972 the Tomahawk family has diversified, so that, despite losing the big USAF buy of an air-launched strategic version to Boeing's AGM-86B, four variants remain active programmes: three for the US Navy described here, and one for the USAF for land mobile deployment described next. The three Navy versions are all basically similar and differ in fuel capacity, guidance and warhead. All have conventional light-alloy aircraft-type airframes, with a tubular body, pivoted wings which unfold to zero sweep after launch, four powered tail fins for trajectory control and a ventral inlet. In naval versions of the missile the tail fins and

inlet also deploy after launch.

The BGM-109A version is the TLAM-N (Tactical Land Attack Missile, Nuclear). It has inertial midcourse guidance updated by Tercom (terrain-contour-matching) guidance in the terminal phase, giving outstanding accuracy despite the fact that it has an extremely powerful warhead (thus, it can be used against hardened targets). It is deployed aboard submarines and surface warships. In the submarine mode it is delivered encapsulated in a stainless-steel container which provides environmental protection even at great ocean depths as well as control, communication and launch facilities, including gas-drive cold launch. In the SSN-688 class attack submarines 12 vertical launch tubes are built into the bow section outside the pressure hull (the original plan was to fire from existing torpedo tubes). Retired Polaris-type SSBNs are being studied as CMC (cruise-missile carrier) boats with up to 80 rounds.

BGM-109B/E is the TASM (Tactical Anti-Ship Missile), with conventional Bullpup-type warhead, much less fuel and active radar homing based on that of Harpoon. It follows a pre-programmed flight profile and can be fired from submarine capsules or surface ships, vertical launch tubes being planned for CG-47, DD-963 and DDG-X class ships. Funding continues to support development of this variant, the ship installations and the required OTH (over the horizon) command/control systems.

BGM-109C is TLAM-C, the conventional counterpart to BGM-109A and like other versions planned for both sea and land warfare, and its guidance will have the super-

accurate Dsmac (Digital scene-matching area correlation), in which scenes ahead of the missiles are scanned and analysed into digital information which is compared with information on the target approach terrain stored in the guidance system.

Above: The submarine-launched versions of Tomahawk were designed to fit in a standard torpedo tube, but the SSN-688 Los Angeles class boats carry their 12 rounds in vertical launch tubes in the bow section.

Tomahawk (Land), BGM-109G

Origin: USA.
Type: GLCM.
Dimensions: Length, 19·6ft (600m) body diameter 20·5in (520mm); span 8·2ft (2·5m).
Launch weight: 3,200lb (1,451kg).
Propulsion: One Williams F107-WR-400 turbofan rated at 600lb (272kg); launch by 7,000lb (3,175kg) Atlantic Research solid rocket motor.
Guidance: INS/Tercom, + Dsmac.
Range: 1,550 miles (2,500km).
Flight speed: 550mph (885km/h).
Warhead: W-84 nuclear or thermonuclear (classified).

Development: In most essentials BGM-109G is identical to the long-range naval versions of Tomahawk, but it has Dsmac (digital scene-matching area-correlation) guidance and the new W-84 warhead. Its mission designation is GLCM (Ground-Launched Cruise Missile), and it is assigned to the Air Force for deployment relatively close to potential trouble-spots, notably Europe. Under the desperately needed NATO TNF (theatre nuclear force) modernization programme a total of 464 GLCMs are to be stationed in NATO European nations under the management of TAC on behalf of SAC.

Each missile is packaged with aerodynamic surfaces and engine inlet retracted inside an aluminium drum. Tanks are filled and guidance pre-targeted. The BGM-109G may then be left for many months without attention. For use the canister is loaded aboard the TEL (transporter/erector/launcher) vehicle, a GD product weighing 33 tons which carries four missile tubes in a single box which can be elevated to the desired launch angle. Each Combat Flight Group comprises four TELs (16 rounds) and two LCCs (launch control centres), the latter also being built into a vehicle, in this case weighing 36 tons largely because of the CBR protection for the crew. Thus the total force will comprise 29 Flights totalling 116 TEL vehicles. These are divided between sites in Britain, West Germany, Belgium, Italy and (possibly) the Netherlands.

Deployment of what the so-called peace movement calls "cruise" is subject to a huge campaign of opposition by European protesters who appear not to understand that, in any time of political crisis, the GLCMs would be driven many miles from their peacetime base to hidden locations in other parts of the host country. Thus, unlike the nuclear bombers which have been parked on NATO airbases for years, these missiles cannot invite retaliation because an enemy could not know their future locations. The scale of the Communist propaganda reaction is a fair measure of this missile's deterrent power.

Below: March 1983 test launch of a BGM-109G Tomahawk at Dugway Proving Grounds, Utah.

Bottom: Initiation of a GLCM launch from the mobile transporter/erector/launcher (TEL).

ICBMs

CSS-3

Origin: People's Republic of China.
Type: ICBM.
Dimensions: Length 83·7ft (25·5m); diameter 8ft (2·43m).
Launch weight: Not known.
Propulsion: 2-stage liquid-propellant rocket motors. Hot launch.
Guidance: Intertial.
Range: 4,350 miles (7,000km).
Warhead: One 2MT.

Development: The development of Chinese strategic weapons has been steady but not spectacular, and the limited deployment so far indicates that the only perceived target is the USSR. The CSS-1, a medium-range ballistic missile (MRBM), entered service in 1966 and about 50 are still deployed in north-west China. The CSS-2 single-stage Intermediate Range Ballistic Missile (IRBM) began deployment in the 1970s and some 80 are now in service, possibly increasing at a rate of some five to ten per year.

The first Chinese ICBM — the CSS-3 — has been deployed at a very slow rate, and only some 10 to 12 are now in service. Another ICBM — the CSS-4 — is under test, and this appears to have greater range and payload.

Above: Only a handful of CSS-3s have been deployed since 1975, the emphasis in China apparently being on weapon development rather than deployment.

Minuteman, LGM-30

Origin: USA.
Type: ICBM.
Dimensions: Length 59·7ft (18·2m); body diameter (first stage) 6ft (183cm).
Launch weight: 76,015lb (34,475kg).
Propulsion: First stage, Thiokol TU-120 (M55E) solid rocket, 200,000lb (91,000kg) thrust for 60 sec; second stage, Aerojet SR19 solid rocket, with liquid-injection thrust-vector control, 60,000lb (27,200kg) thrust; third stage Aerojet/Thiokol solid rocket, 34,876lb (16,000kg) thrust, plus post-boost control system.
Guidance: Inertial.

Range: Over 8,000 miles (12,875km).
Warhead: Three (sometimes two) General Electric Mk 12 MIRVs.

Development: Minuteman was designed in 1958-60 as a smaller and simpler second-generation ICBM using solid propellant. Originally envisaged as a mobile weapon launched from trains, it was actually deployed (probably mistakenly) in fixed hardened silos. Minuteman I (LGM-30B) became operational in 1963 but

Below: June 1980 test launch of an LGM-30G Minuteman III ICBM from Vandenberg AFB.

19

is no longer in use. Minuteman II (LGM-30F) became operational from December 1966, and today 450 are still in use, and replacement of expired component-lifetimes has recently started. By 1978 all the re-entry vehicles were of the Mk 11C type hardened against EMP; consideration has been given to prolonging missile life, and improving accuracy, by retrofitting the NS-20 guidance used on Minuteman III (LGM-30G). The latter, operational since 1970, has a new third stage and completely new re-entry vehicles forming a fourth stage with its own propulsion, guidance package and pitch-roll motors; as well as several warheads, individually targetable, it houses chaff, decoys and possibly other penaids.

Production of LGM-30G Minuteman III ended in late 1977, but the force is continually being updated, with improved silos, better guidance software, and the Command Data Buffer System, which, with other

add-ons, reduces retargeting time per missile from around 24 hours to about half an hour, and allows it to be done remotely from the Wing's Launch Control Centre or from an ALCS (Airborne launch Control System) aircraft or an NEACP (National Emergency Airborne Command Post). The latter comprise the E-4B while the ALCS authority is vested in nine EC-135C aircraft which can monitor 200 Minuteman missiles and, via improved satellite communications links, retarget and launch any of these missiles even if their ground LCC (Launch Control Centre) is destroyed. Installation of the Mk 12A RV began in 1979. This has various advantages and carries the 330kT W-78 warhead, but it is about 16kg heavier and this slightly reduces range and MIRV footprint.

One hundred Minuteman silos are due to be converted to take the Peacekeeper ICRM (qv), and an unspecified number of Minuteman IIs are allocated to the Emergency

Rocket Communications System. The elderly Titan II ICBM, of which 52 were in service in 1980, is being phased out, the process being due for completion in 1987.

Above: Minuteman III missile in a silo at Vandenberg AFB. A total of 1,000 Minuteman II and III ICBMs are operationally deployed at six SAC bases.

Peacekeeper

Origin: USA.
Type: ICBM.
Dimensions: Length 70·8ft (21·6m); diameter 7·6ft (2·34m).
Launch weight: 195,015lb (8,845kg).
Propulsion: Stages 1, 2 and 3 use HTBP propellant, upper stages having extendable-skirt exit cones; stage 4 has hypergolic liquid propellants feeding a vectoring main chamber and eight small altitude-control engines.
Guidance: High-precision inertial.
Range: Over 6,900 miles (11,100km).
Warhead: Ten Mk 12A MIRVs each of 330kT yield.

Development: Like the B-1 bomber the Peacekeeper (formerly MX) weapon system has consumed

Below: First stage ignition during an October 1983 test launch of a Peacekeeper ICBM from Vandenberg AFB, California.

money at a prodigious rate for many years without making the slightest contribution to Western defence or deterrence. Though the need has been self-evident for many years, and there are no problems in producing the missile, arguments raged for many years on how to base it. In 1974 a packaged Minuteman was pulled by parachute from a C-5A, and prolonged research has established the feasibility of an air basing concept.

Early in MX development (pre-1978) all interest centred on mobile deployment using road (curiously, not railway) cars or various forms of transporter/launcher driven around underground rail networks, in most variations with the missile erected so that it would break through the surface in virgin terrain. The Carter administration favoured the MPS (multiple protective shelter) scheme, but the Reagan administration found this faulty and announced in December 1981: ''. . . initial deployment will be in existing Minuteman

silos. At least 40 MXs will be deployed, with the first unit of ten missiles operational in late 1986. The Specific location . . . will be determined in spring 1982 . . . In addition, the Air Force has initiated R&D to find the best long-term option . . . The options include the following: ballistic-missile defence of silo-based or deceptively based missiles; DBS, deep-basing system, in underground citadels; and air-mobile basing . . . Congress approved $1,900m for Fiscal 1982 and some $2,500m has been spent to date. The cost to produce 226 missiles and deploy 40 in Minuteman silos is estimated at 1,800-1,900m in 1982 dollars.''

Next, the Senate Armed Services Committee rejected Minuteman-silo-basing and asked for a permanent basing plan by 1 December 1982. This resulted in other suggestions, notably DUB (deep underground basing) about 3,280ft (1,000m) down in rock, with a self-contained tunnelling machine for each launch capsule and crew, and CSB (closely spaced basing). The latter, called ''dense pack'', relies on the so-called

fratricide effect in that debris from each nuclear warhead is supposed to disable those following behind it (it apparently being assumed that hostile warheads would be spaced only a second or two apart). President Reagan approved CSB in May 1982, for 100 missiles spaced at 1,640ft (500m) intervals over a region almost 4 miles (6km+) across, but the validity of CSB was later doubted by several authorities. The first cold-launch pop-up test took place in January 1982, and the first MX flight test in January 1983.

The definitive plan is that 100 Peacekeeper ICBMs will be deployed, (starting in 1986 and completing in 1990) in former Minuteman silos near the FE Warren AFB in Wyoming. Better C³ systems will be installed and more sophisticated shock isolation devices fitted, but so far as is planned to date the silos will not be further hardened.

Below: A Peacekeeper's 10 MIRV warheads are caught by time exposure during a test over the South Pacific in March 1984.

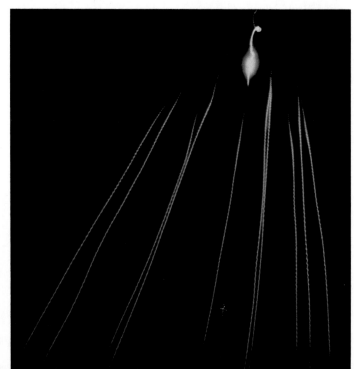

SSBS S-3

Origin: France.
Type: IRBM/ICBM.
Dimensions: Length 45·93ft (14m); diameter 4·92ft (1·5m).
Launch weight: 56,879lb (25,800kg).
Propulsion: 2-stage solid-propellant. First stage controlled by gimballed nozzles; second stage by thrust vector control.
Guidance: Inertial.
Range: Approx 1,864 miles (3,000km).
Warhead: Thermonuclear 1·2mT.

Development: Although technically an Intermediate Range Ballistic Missile (IRBM) the Sol-Sol Ballistique Stratégique (SSBS) S-3 is, in European terms, an ICBM. The first service missile was the S-2, which entered service in 1971, and since then France has maintained a force of 18 SSBS in two separate fields of nine silos each. The second generation S-3 missile began development in 1973. This has the same first stage as the S-2 but with the P-6 second stage developed for the MSBS (SLBM), and a completely new 1·2MT warhead, incorporating penetration aids. The whole missile is hardened against EMP.

The first flight of the S-3 came in 1976, and IOC was achieved in 1980, when the Plateau d'Albion site became operational with the new missile. The second site followed in 1982. It seems unlikely that France will incur the expense of developing a third generation IRBM/ICBM, although the technological and industrial capability clearly exists.

Right: An S-3 IRBM lifts off from the Landes test centre on a flight that will end in a splashdown near the Azores.

SS-17 Mod 3

Origin: Soviet Union.
Type: ICBM.
Dimensions: Length 69ft (21m); diameter 6·9ft (2·1m).
Launch weight: 143,300lb (65,000kg).
Propulsion: 2-stage liquid-propellant rocket motors. Cold launch.
Guidance: Inertial.
Range: 6,800 miles (11,000km).
Warhead: Four MIRVs.

Development: The SS-17 was developed in competition with the SS-19 as a successor to the SS-11. Slightly larger than the SS-11, the SS-17 uses liquid propellants, but utilises cold-launch techniques, i.e., the missile is expelled from the silo by a gas generator before ignition, thus making it possible to re-use the silo.

SS-17 Mod 1 carries 4 to 6 MIRV warheads, each of not less than 200kT yield, while the Mod 2 carries a single high-yield RV. The principal service version, however, is the Mod 3, of which the 150 deployed are split between two sites in north-western USSR: Yedrovo and Kostroma. It is described by the USA as a "somewhat less capable ICBM than the SS-19, but (with) similar targetting flexibility."

SS-18 Mod 4

Origin: Soviet Union.
Type: ICBM.
Dimensions: Length 108·25ft (33m); diameter 9·8ft (3m).
Launch weight: 171,958lb (78,000kg).
Propulsion: 2-stage liquid-propellant rocket motors. Cold launch.
Guidance: Inertial.
Warhead: Ten 500kT MIRVs.

Development: The SS-18 was developed in the late 1960s as the successor to the SS-9, and is by far the largest of the current (fourth) generation of Soviet ICBMs. It was first deployed in 1974 in former SS-9 silos which were modified and upgraded to take the new missile. Deployment continued until 1982 and the numbers are now steady at 308, with the six launch complexes in a large semi-circle around the Tyuratam Missile Centre in Central USSR.

The first version (Mod 1) has a single large warhead, as has the Mod 3 (20MT). Mod 2 has 8 to 10 MIRVs, while Mod 4 has ten 500kT MIRVs, and for SALT II accounting purposes all operationally deployed missiles are assumed to be of this type. According to the USA, the SS-18 Mod 4 was developed specifically to attack and destroy US ICBM silos and other hardened counterforce targets. The same source claims that, by using two warheads per target, the SS-18 Mod 4 force could destroy more than 80 per cent of US ICBM silos in a first-strike attack.

Above: Comparative scale drawings of the Soviet SS-17, -18 and -19 ICBMs and the SS-20 IRBM. Estimated totals of 150, 308, 360b and 378 respectively are operationally deployed, and continued production of the SS-20 is one of the main obstacles to SALT agreement.

SS-19 Mod 3

Origin: Soviet Union.
Type: ICBM.
Dimensions: Length 78·7ft (24m); diameter 7·72ft (2·35m).
Launch weight: Not known.
Propulsion: 2-stage liquid-propellant rocket motors. Hot launch.
Guidance: Inertial.
Range: 6,213 miles (10,000km).
Warhead: Six 500kT MIRVs.

Development: Developed in parallel with the SS-17 (qv) as a replacement for the SS-11, the SS-19 has undoubtedly proved the more successful of the two, and with 360 deployed is the most widely used Soviet ICBM. Unlike the cold-launched SS-17 and SS-18, the SS-19 uses hot-launch techniques, although the provision of a launch canister prevents too much damage to the silo.

Current SS-19 sites all lie in the Western USSR, although the missile might be deployed later in the Soviet Far East as a replacement for the remaining SS-11s. The Mod 1 has 4 to 6 MIRVs of about 200kT each, and the Mod 2 a large single RV. The Mod 3 has six 500kT MIRVs and is said by the USA to be a highly capable system, and probably more accurate than the US Minuteman III.

21

SS-20

Origin: Soviet Union.
Type: IRBM/ICBM.
Dimensions: Length 55·8ft (17m); diameter 8·2ft (2·5m).
Launch weight: 28,659lb (13,000kg).
Propulsion: 2-stage solid-propellant rocket motors. Hot launch.
Guidance: Inertial.
Range: 3,100 miles (5,000km).
Warhead: Three MIRVs.

Development: The SS-20 is one of the best known of all Soviet missiles, owing to its repeated appearance in political debates. Carried on a mobile, tracked launch vehicle, there are now some 378 in service, of which 243 face NATO, but the numbers could increase to as many as 600 by 1990. The problem with SS-20 is its range — 3,100 miles (5,000km) — which, as far as the USA is concerned, puts it into the ICBM category. The USSR hotly disputes this, because to agree would bring the missile into the SALT process.

The Circular Error Probable (CEP) is stated to be 1,312ft (400m), although this must depend to a certain extent upon the accuracy of the launch site survey. Retargeting can be carried out in seconds if the new target is within 1° to 2° of the original azimuth, but increases with the divergence, up to a maximum of about 30 minutes. Each battery comprises some nine vehicles, including one reload for each vehicle.

Right: SS-20 can be launched from unprepared sites as well as the pre-surveyed base shown in this DoD artist's impression.

SLBMs

MSBS M-4

Origin: France.
Type: SLBM.
Dimensions: Length 36·8ft (11m); diameter 6·29ft (192cm).
Launch weight: 79,365lb (36,000kg).
Propulsion: 3-stage solid fuel; first and second stage have mechanical steering, third has thrust vector control.
Guidance: Intertial.
Range: 2,485 miles (4,000km).
Warhead: Six MIRVs, each of 150kT yield.

Development: Mer-Sol Ballistique Stratégique (MSBS) is the French national strategic nuclear deterrent. It was broadly based on the Polaris concept, but has been achieved with little non-French help other than the licensing of essential technology. The early test vehicles flew between 1967 and 1970 and led to the MSBS M-1, which entered service in *Le Redoutable* in December 1971. The M-1 had a single warhead with a yield of 500kT and a range of 1,491 miles (2,400km). This was followed by the M-2 which had a new second stage and a range increased to 1,926 miles (3,100km). The M-2 was installed in *Le Foudroyant* during construction and was retrofitted to the earlier boats during the course of refits.

The current missile system is the M-20, standard on the five operational Sousmarin Nucléaire Lanceur d'Engins (SNLE). This is generally similar to M-2, but has a more powerful warhead with advanced penetration aids.

For the second half of the 1980s a completely new M-4 missile is being developed, which has very little in common with earlier French MSBS. It can, however, be fitted into the existing submarines after considerable modifications. It has a three-stage propulsion system and will mount six MIRVs, each with a yield of approximately 150kT. Flight testing has already begun and an IOC around 1985 is envisaged. Like the SSBN (SNLE) programme, the development and fielding of an entirely French strategic nuclear missile system is a remarkable achievement.

Below: An MSBS M-4 is loaded aboard the experimental missile submarine Gymnote.

Poseidon C-3

Origin: USA.
Type: SLBM.
Dimensions: Length 34ft (10·36m); diameter 74in 6188cm).
Launch Weight: Approx (5,000lb (29,500kg).
Propulsion: First stage, Thiokol solid-fuel motor with gas-pressurized gimballed nozzle; second stage, Hercules motor with similar type nozzle.
Guidance: Inertial.
Range See *Warhead*.
Warhead: MIRV system carrying maximum of 14 MIRVs for 2,485 miles (4,000km), or 10 MIRVs for the maximum range of 3,230 miles (5,198km). Each warhead of 50KT yeld.

Development: The result of lengthy studies into the benefits of later technology, Poseidon C-3 SLBMs were first installed in Franklin and Lafayette class SSBNs, starting with USS *James Madison* (SSBN-627). This boat carried out the first Poseidon deployment on 31 March 1971. Compared with the Polaris A-3 (now only serving on the four British SSBNs), Poseidon has at least equal range, carries double the payload, and has twice the accuracy (ie, half the circular area probable/CEP) as

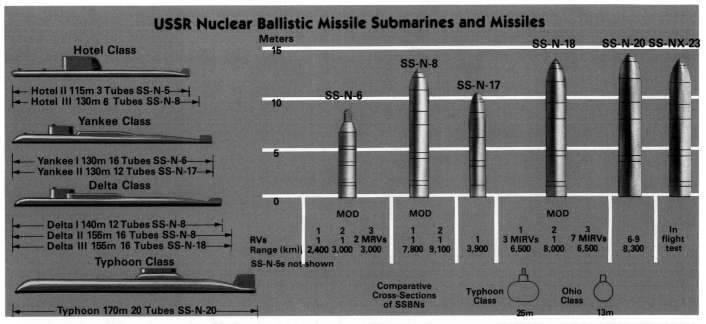

USSR Nuclear Ballistic Missile Submarines and Missiles

Hotel Class

Hotel II 115m 3 Tubes SS-N-5
Hotel III 130m 6 Tubes SS-N-8

Yankee Class

Yankee I 130m 16 Tubes SS-N-6
Yankee II 130m 12 Tubes SS-N-17

Delta Class

Delta I 140m 12 Tubes SS-N-8
Delta II 155m 16 Tubes SS-N-8
Delta III 155m 16 Tubes SS-N-18

Typhoon Class

Typhoon 170m 20 Tubes SS-N-20

SS-N-6   SS-N-8   SS-N-17   SS-N-18   SS-N-20   SS-NX-23

	SS-N-6			SS-N-8		SS-N-17	SS-N-18			SS-N-20	SS-NX-23
MOD	1	2	3	1	2	1	1	2	3	6-9	In flight test
RVs	1	1	2 MRVs	1	1	1	3 MIRVs	1	7 MIRVs		
Range (km)	2,400	3,000	3,000	7,800	9,100	3,900	6,500	8,000	6,500	8,300	

SS-N-5s not shown

Comparative Cross-Sections of SSBNs

Typhoon Class 25m   Ohio Class 13m

US Nuclear Ballistic Missile Submarines and Missiles

Trident (Ohio Class) SSBN

Trident 170.7m 24 Tubes

Poseidon SSBN

Poseidon 129.5m 16 Tubes

Poseidon SLBM C-3   Trident SLBM C-4

	Poseidon SLBM C-3	Trident SLBM C-4
RVs	10	8
Range (km)	4,000	7,400

---

well as much improved MIRV and penetration aid (penaid) capability. A modification programme, started in 1973 to rectify deficiencies which showed up after the 1970 IOC, was completed in 1978. More than 40 missiles, which were withdrawn from submarines after operational patrols, have been fired with excellent results.

Left: Pad 25C at Cape Canaveral is the site of one of the last launches of a Poseidon development missile in May 1970. First deployment aboard USS *James Madison* followed in March 1971, and the missile continues in service on 19 SSBNs.

Poseidon C-3 was installed on 31 Franklin and Lafayette SSBNs, and 12 of the latter have since been retrofitted with Trident missiles. The remaining 19 Poseidon boats will remain in service well into the 1990s, and there are no plans to convert any more to take Trident, nor are there likely to be any more than routine updates to the missiles themselves.

At maximum range ten Mark 3 MIRVs (each of 50KT yield) can be flown, together with numerous penaids. At such ranges CEP is of the order of ½nm (0·8km). At shorter ranges more MIRVs are carried.

Compared with over 1,000 Soviet SLBMs, the US fields 640 including 496 Poseidon C-3/C-4.

SS-N-20

Origin: Soviet Union.
Type: SLBM.
Dimensions: Length 49·2ft (15m); diameter 85·56ft (2·17m).
Launch weight: Approx 46,300lb (21,000kg).
Propulsion: 2-stage solid propellant, with post-boost vehicle.
Guidance: High precision Inertial.
Range: 5,157 miles (8,300km).
Warhead: Six to nine MIRVs.

Development: The SS-N-20 is the SLBM developed for service in the new giant Typhoon class SSBN (qv), and is the largest missile ever to be used at sea. First tests of the SS-NX-20 were notified in the West in January 1980, and it was reported that the initial four flights all ended in failure. At least two successful flights were made in 1981, and on 14 October 1982, in a unique test, a salvo of four missiles was fired simultaneously from a submerged Typhoon SSBN, which carries 20 SS-N-20 MIRVed SLBMs.

Testing of yet another new missile — SS-N-23 — has already started. This is expected to be deployed in 1985-86 in Delta-III SSBNs in place of their current SS-N-18s.

SS-N-18 Mod 1

Origin: Soviet Union.
Type: SLBM.
Dimensions: Length 46·25ft (14·1m); diameter 5·9ft (180cm).
Launch weight: 44,000lb (19,958kg).
Propulsion: 2-stage liquid fuel, with Post Boost Vehicle.
Guidance: Stellar-Inertial.
Range: 4,040 miles (6,500km).
Warhead: Three 200kT MIRV.

Development: The SS-N-18 was first seen on test flights some weeks after testing had begun on the SS-N-17. It is now fully deployed on Delta III submarines and has sufficient range for the launch vehicles to be able to operate from havens in the Barents Sea and the Sea of Okhotsk. Two developments of the original type have been identified, but whether they are yet deployed is uncertain. The Mod 2 has one 450kT warhead and a range of 4,970 miles (8,000km), while the Mod 3 has no less than seven MIRVs and a range of 4,040 miles (6,500km). Circular Error Probable (CEP) is of the order of 0·76nm (1,410m). This is a most formidable strategic nuclear missile by any standard.

Trident C-4

Origin: USA.
Type: SLBM.
Dimensions: Length 34ft (10·36m); diameter 6·1ft (188cm).
Launch weight: 70,000lb (32,000kg).
Propulsion: Three tandem stages; advanced solid motors with thrust vectoring.
Guidance: Inertial.
Range: 4,400 miles (7,100km).
Warhead: Eight Mark 4 100kT MIRV. Mark 500 Evader MaRV may be installed later.

Development: Some 24 extremely large submarines of the Ohio class, each boat mounting 24 launch tubes, are projected to carry this large, long-range SLBM system. Flight testing of the Trident I (C-4) missile improved after a somewhat shaky start and the system is now operational. Trident I (C-4) has been installed in 12 units of the Franklin/Lafayette class and the first such conversion, USS *Francis Scott Key* (SSBN-657), joined the fleet as the first operational Trident-armed submarine on 20 October 1979. Trident II (D-5), a longer missile which has much improved throw-weight and/or range, is currently under development. It will fit the C-4 launch tubes for diameter, but, being somewhat longer, will not only need new tubes but also a new submarine launch platform.

The British Government has announced plans to purchase Trident missiles, which will be installed in a new class of British-built SSBNs. As with Polaris, an entirely British front-end will be fitted, thus ensuring ultimate national control over deployment and targetting. After initially deciding to order the C-4, the British Government changed its mind and ordered the D-5, but stated that the total number of warheads per missile would never exceed that planned for the C-4.

Right: First stage ignition of a Trident 1 SLBM after ejection by gas pressure from the launch tube of USS Ohio SSBN.

Far right: Re-entry of C-4 Trident 1 photographed by time exposure from a height of 20,000ft (6,100m) and at 15nm (28km) from impact point.

SSBNs

Delta III Class

Origin: Soviet Union.
Type: SSBN.
Displacement: 11,000 tons (surfaced); 13,250 tons (submerged).
Dimensions: Length 509ft (155·1m); beam 39·3ft (12m); draught 33·4ft (10·2m).
Missiles: Sixteen SS-N-18.
Propulsion: Nuclear (60,000shp).
Shafts: Two.
Speed: 24 kt (44·5km/h) (submerged).
Complement: 132.

Development: Until 1973 the US Navy had a considerable advantage in the quality and performance of its SLBMs, but in that year the Soviet Navy introduced the SS-N-8 missile with a range of more than 4,800 miles (7,720km) and a CEP of only 0·84nm (1,550m). This outranges not only Poseidon but Trident 1 as well. Initial

trials were conducted in a Hotel III class SSBN and the missile system was then installed in the Delta I and Delta II class. The Delta I (18 built) carries 12 missiles in two rows of six abaft the fin; the Delta II (4 built) carries four more, matching the Western SSBNs with their 16 missiles. A second slipway was built at the Severodvinsk shipyard in 1975 to enable these boats to be built more quickly.

The Delta III is somewhat larger than the Delta II, and carries 16 SS-N-18 missiles, which have a range of some 4,040 miles (6,500km) and mount MIRVed warheads. These submarines are a great threat to the USA because they can hit North America from launching areas in the Sea of Okhotsk and the Barents Sea, well out of reach of any known countermeasures.

Above: An SS-N-18-armed Delta III class Soviet SSBN running on the surface.

Below: DoD artist's impression of the Severodvinsk depot with moored Delta III and Oscar boats.

Ohio Class

Origin: USA.
Type: SSBN.
Displacement: 16,600 tons (surfaced), 18,700 tons (submerged).
Dimensions: Length 560ft (170·7m); beam 42ft (12·8m); draught 35·4ft (10·8m).
Missiles: 24 Trident 1 (C-4).
Torpedo tubes: Four 21in (533mm).
Propulsion: Nuclear (60,000shp); 1 shaft.
Speed: Not known.
Complement: 133.

Development: While the programme of upgrading the later Polaris SLBM submarines to carry Poseidon was under way in the early

1970s, development of an entirely new missile was started. This was to have a much longer range — 4,400 miles (7,100km) — which in turn necessitated a new and much larger submarine to carry it. The missile, Trident 1, is now in service on converted Lafayette class SSBNs, while the first five submarines purpose-built for Trident, the Ohio class, have also joined the fleet. Initially Congress baulked at the immense cost of the new system — but then the Soviet Navy introduced its own long-range SLBM, the 4,200-mile (6,760km) SS-N-8, in the Delta class. This was followed in 1976 by the firing of the first of the increased-range SS-N-18s (4,846 miles, 7,800km). US reaction was to speed up the Tri-

Le Redoutable Class

Origin: France.
Type: SSBN.
Displacement: 8,045 tons (surfaced), 8,940 tons (submerged).
Dimensions: Length 422·2ft (128·7m); beam 34·7ft (10·6m); draught 32·8ft (10m).
Missiles: 16 MSBS M-20.
Torpedo tubes: Four 21 in (533mm).
Propulsion: Nuclear (16,000shp); reserve diesel (1,306shp); 1 shaft.
Speed: 25kt (46km/h) (submerged).
Complement: 135.

Development: Like Britain, France decided to build nuclear ballistic missile submarines to ensure a viable national nuclear deterrent; but unlike the British Polaris and Trident submarines, the French *"Force de dissuasion"* has been developed completely independently of the USA. This has resulted in a much greater effort spread over a much longer timescale, and in heavier missiles carrying a smaller warhead over a shorter range. The Mer-Sol Ballistique Stratégique (MSBS) M-1 SLBM fitted in the first two boats had a range of only 1,500 miles (2,400km), but this has been progressively increased in succcessive missile systems. The first four boats were all modified to take the MSBS M-2, but have subsequently been modified again to take the M-20. The fifth boat was constructed from the outset to take the latter missile. The sixth boat was constructed to take the even newer M-4 and this will be retrofitted to all except *Le Redoutable*.

The M-4, with a MIRVed warhead, was tested in the experimental diesel-electric submarine *Gymnote* in 1978–9. This submarine has been used to develop the Le Redoutable boats and their missiles. Built between 1963 and 1966, it has two SLBM launching tubes and a laboratory suite.

The Le Redoutable boats resemble the American SSBNs in that they have two rows of eight vertical SLBM launch tubes abaft the fin, which carries the forward hydroplanes. The French SSNBs have turbo-electric propulsion, but also have an auxiliary diesel that can be cut in to provide power should the nuclear reactor fail.

An order was placed in late 1978 for the sixth SSBN, mentioned above, and this boat, *L'Inflexible,* entered service in 1985. She has similar dimensions to the earlier boats, but has improved propulsive machinery and incorporates updated electronics. She is armed with the MSBS M-4 SLBM, as we have said, and can also launch SM39 missiles from her torpedo tubes.

On 13 November 1981, following a two-day visit to *Le Tonnant* on an operational patrol, the French Prime Minister announced an order for a seventh SSBN, also of the interim class, for delivery in the mid-1980s — a clear endorsement by the Socialist government of France's independent nuclear deterrent. It is planned to replace all five first-generation boats with a totally new class in 1990-2000, with the sixth interim design being replaced later.

The French SSBN/SLBM programme is a major national achievement, although it is also one which has been achieved at a great price.

Below: *Le Redoutable,* **the first French ballistic missile-armed nuclear submarine. Six of the class are now in service.**

---

dent programme, and the first of the Ohio class submarines was laid down on 10 April 1976. In addition to the five now in commission, six are building and at least 13 more will be ordered.

The eventual number of Trident-carrying SSBNs depends on two principal factors. The first is the outcome of any talks between the US and the USSR. Any agreement would presumably include, as in SALT-II proposals, the maximum numbers of SLBMs and launch platforms that each super-power was prepared to permit the other to possess. The other factor is the development of new types of long-range cruise missiles, some of which can be used in a strategic role even when launched from a standard 21in (533mm) submerged torpedo tube. This, and similar progress in other fields, may restrict the need for large numbers of SLBMs in huge and very expensive SSBNs. The great advantage, however, of the current generation of very long-range SLBMs is that they can be launched from American or Soviet home waters, thus making detection of the launch platform and destruction of either the submarine or the missiles launched from it extremely difficult, if not impossible.

Left. A Trident I C-4 SLBM is lowered into a launch tube of USS *Ohio*, first of a projected class of 24 SSBNs, though the eventual total remains in doubt.

Right: View along the deck of USS *Ohio* with the caps of the 24 Trident I tubes open; by 1985 unit cost of these boats had risen to $1,755m.

Resolution Class

Origin: UK.
Type: SSBN.
Displacement: 7,500 tons (surfaced), 8,400 tons (submerged).
Dimensions: Length 425ft (129·5m); beam 33ft (10·1m); draught 30ft (9·1m).
Missiles: 16 Polaris A-3 (Chevaline).
Torpedo tubes: Six 21in (533mm) bow.
Propulsion: Nuclear (15,000shp).
Shafts: One.
Speed: 20kt (37km/h) (surfaced): 25kt (46km/h) (submerged).
Complement: 143.

Development: In the late 1950s it was planned that the RAF would provide the British strategic deterrent in the 1960s and 1970s, using V-bombers armed with the Skybolt missile. But at the Nassau Conference in 1962, President Kennedy told Britain that the USA was abandoning Skybolt because of apparently insuperable development problems. It was then agreed that Britain would build her own SSBNs: the USA was to provide Polaris SLBMs which would, however, have an entirely British front end.

Four submarines were built of a planned total of five, the last boat being cancelled in the Labour government's defence review in 1965. Much

technical assistance was obtained from the USA and the Resolution class is generally similar to the American Lafayette class SSBNs, although the Resolutions' actual design is based on that of the Valiant SSN, but with a missile compartment between the control centre and the reactor room.

In the longer term the UK intends to purchase Trident missiles from the USA, but as with Polaris, an entirely British warhead will be installed. These missiles will be deployed in new British SSBNs; four are currently planned, and a decision on a fifth was to have been made in 1982-3. Nothing was announced but the rapidly escalating costs of the programme suggest that it is highly improbable that the fifth will be ordered. Construction of the SSBNs will begin in 1987 with an IOC in the early 1990s and a life expectancy through to 2020 at least.

Above right: HMS *Resolution*, the Royal Navy's first nuclear-powered ballistic missile submarine, was commissioned on October 2, 1967.

Right: Four boats of the class were built, a fifth being cancelled in 1965; each of them can carry 16 Polaris A-3 SLBMs.

Typhoon Class

Origin: Soviet Union.
Type: SSBN.
Displacement: 25,000 tons (submerged).
Dimensions: Length 558ft (170m); beam 82ft (25m); draught not known.
Missiles: 20 SS-N-20.
Torpedo tubes: Six or eight 21in (533mm).
Propulsion: Nuclear (80,000shp).
Speed: 24kt (44km/h) (submerged).
Complement: 150.

Development: Persistent and growing rumours in the West were confirmed by the NATO announcement in November 1980 that the USSR had launched the first of the Typhoon class SSBNs. This enormous craft has a submerged displacement of 25,000 tons and an overall length of 558ft (170m), making it by far the largest submarine ever built. Among the many unusual features of the design is the 82ft (25m) beam; the normal length:beam ratio in SSBNs is in the region of 13:1, but the extraordinary girth of the Typhoon reduces the ratio to 7:1. This may indicate a considerable degree of separation (up to 14-15ft, 4·3-4·57m) between outer and inner hulls, or simply a huge inner hull.

Another departure from previous practice is that the 20 missile tubes are forward of the fin. The reason for this is not yet clear, although one possibility is that the necessarily powerful propulsion machinery for this giant is so large and heavy that the missile compartment has had to

be moved forward to compensate.

If it were to venture out into the open oceans, this submarine would be relatively easy for NATO to follow: its very size would facilitate detection by virtually every type of ASW sensor. It seems possible, therefore, that it is simply intended as a relatively invulnerable missile launching platform, required only to move a short distance out into the Barents Sea

and to loiter there, its time on station limited only by the endurance of the crew. For the latter, conditions can be assumed to be more spacious and comfortable than in any previous type of SSBN.

The Typhoon has been observed in tests which involved firing up to four missiles simultaneously: a unique capability, as all previous SSBNs must perforce launch in a ripple. It is now believed that one Soviet intention is for the Typhoon to shelter under the Arctic icecap and then

force its way through a thin section of ice to the surface from where it would launch its missiles.

It is worth noting that the Russians have frequently exhibited a fascination with sheer size and their latest ships — *Kirov* (battle-cruiser), *Kiev* (aircraft carrier), *Oscar* (SSGN) — seem to fit in with such a general pattern.

Below: Typhoon class SSBN, the world's largest submarine, first seen publicly in 1985.

Xia Class

Origin: People's Republic of China.
Type: SSBN.
Displacement: 8,000 tons (submerged).
Dimensions: Length 393·6ft (120m); beam 33ft (10m); draught 26·2ft (8m).
Missiles: 12 or 16 CSS-NX-3.
Propulsion: Nuclear.

Development: The PRC has long had at least one Type 200 submarine, similar in design to the Soviet Golf class. One was lost in an underwater missile test in 1981, but a second test "from Number 200" took place successfully in 1982. As far as is known, no non-nuclear-powered missile submarine has become operational, so it must be assumed that the Type 200 played the same development role for the Chinese SLBMs as did the *Gymnote* for the French Navy.

The first PRC SSNB class is the Xia, the lead boat having been laid down in 1978 and launched in 1981. Present plans are for six boats, giving the PRC a capability on a par with the UK and France, although one problem for the Chinese is that the Soviet heartland is out of range of current SLBMs.

Above right: Probable configuration of the Xia class SSBNs.

Right: Launch of a CSS-NX3. Each Xia class sub will carry 16.

Yankee Class

Origin: Soviet Union.
Type: Yankee I/II SSBN; Yankee III, SSN.
Displacement: Yankee I/II: 7,800 tons (surfaced), 9,300 tons (submerged), Yankee III: c4,600 tons (surfaced), c5,600 tons (submerged).
Dimensions: Yankee I/II: length 424·6ft (129·4m); beam, 38·0ft (11·6m); draught, 25·6ft (7·8m). Yankee III: length 329·6ft (100·5m); beam 38·05ft (11·6m); draught, 24·93ft (7·6m).
Missiles: Yankeee I: 16 SS-N-6; Yankee II: 12 SS-N-17; Yankee III: nil.
Torpedo tubes: 6 x 21in (533mm).
Propulsion: Nuclear (40,000shp).
Shafts: Two.
Speed: 20kt (surfaced); 30kt (submerged).
Complement: Yankee I/II: 120; Yankee III: c90.

Development: The Yankee class were the first Soviet purpose-built designed nuclear ballistic missile submarines to enter service, and (a decade after the Americans) they were the first Soviet submarines to mount SLBMs within the hull. The 16 missiles are arranged in two vertical rows of eight abaft the fin in a similar fashion to the US Polaris boats. The first 20 boats were armed with the SS-N-6 Mod I (Sawfly) SLBM which has one 1 to 2MT warhead and a range of some 1,495 miles (2,405km). The next 13 boats mounted the longer-ranged SS-N-6 Mod 3 which has two MRV warheads, while the last of the class (the single Yankee II) mounted 12 SS-N-17 in a trial installation.

Like all Soviet boats, the Yankees are noisier than their Western counterparts and are correspondingly easier to detect. The relatively short range of the SS-N-6, even in its later versions, means that the Yankees must approach the American coast before launching their missiles, although they are less vulnerable than the earlier boats which had to surface to launch their missiles.

Since 1978 10 Yankees have had their missile section removed, thus shortening them by 95ft (28.9m), and they are now employed as SSNs. This conversion is intended to keep the Soviet strategic missile force within the SALT limits, especially in view of the impending entry into service of the Typhoon class, each carrying 20 missiles. The remaining 23 Yankee Is remain in front-line service since their deployment near the US coast adds greatly to the threat to the Continental USA as well as exacerbating the ASW problem.

Above right: The Yankee class were the first purpose-designed Soviet SSBNs, and the first of 3 entered service in the late 1960s.

Right: The sixteen launch tubes abaft the fin originally housed SS-N-6 Mod 1 SLBMs, later replaced by the longer-ranged dual-warhead Mod 3.

Below: Five of the Yankee class have had their missile sections removed, shortening the hull by 95ft (28.9m) for service as SSN attack submarines

Air Weapons

Air power remains one of the most easily identifiable and high-profile measures of military power; it is also currently one of the most expensive and subject to the most rapid escalation of costs. The past two years have seen the entry into service of many new types, especially in the fighter/attack aircraft. The US F-16 and F/A-18 are now in the inventories of numerous air forces and naval air arms, while in Europe the Tornado IDS and ADV are also now in the front line. In the Warsaw Pact a whole new range of MiG and Sukhoi designs have entered service or are about to, although as usual only rather vague "artists' impressions" are available of some of them to give an idea of their appearance. In the PRC new aircraft are also appearing and the Shenyang J-8 is now certainly with the squadrons.

One area of major development is that of helicopters which are now firmly in the arena as weapon systems. ASW helicopters are now fully autonomous, no longer acting as relay stations for their parent ships, while for land operations the anti-tank gunship is now appearing in ever greater numbers. Indeed, the helicopter is now such a threat that helicopter "fighters" are now on the drawing-boards, designed to find and shoot-down enemy rotary-wing aircraft.

At sea US fixed-wing aircraft remain in a virtual monopoly, with only the French Super Etendard and the British Sea Harrier in competition. The forthcoming Soviet 75,000-ton aircraft carrier will, however, give rise to a series of navalised fighter and early-warning aircraft to supplement the only current Soviet type, the Yak-36 "Forger".

A minor trend is the expansion of nations capable of designing and producing their own aircraft. Yugoslavia and Romania have got together to produce the transonic Orao, while Brazil has joined Italy in the flight-test phase with their very neat and capable AMX.

Just around the corner are some revolutionary designs utilising "stealth" technology, or featuring forward-swept wings, canard controls or totally new concepts of control.

Above: F-14 Tomcat and Phoenix missiles, the world's most potent air defence combination.

FIGHTER AND ATTACK/ STRIKE AIRCRAFT

AM-X

Origin: Italy/Brazil.
Type: Tactical attack and reconnaissance aircraft.
Engine: One 11,030lb (5,003kg) Rolls-Royce Spey 807 turbofan.
Dimensions: Span (over AAMs), 32ft 9·75in (10·0m); length 44ft 6in (13·575m); height 15ft 0·25in (4·576m); wing area 226sq ft (21·0m²).
Weights: Empty 13,228b (6,000kg); max loaded 25,353lb (11,500kg).
Performance: Max speed with full external mission load at sea level 723mph (1,163km/h, Mach 0·95); cruising speed in bracket Mach 0·75 to 0·8; takeoff run at max weight 3,120ft (950m); attack radius with 5min combat and 10 per cent reserves with 6,000lb (2,722kg) of external ordnance (hi-lo-hi) 320 miles (570km), (lo-lo-lo) 230 miles (370km).
Armament: Total external load of 7,716lb (3,500kg) carried on centreline pylon, four underwing pylons and AAM wingtip rails; internal gun(s) (Italy) one 20mm M61A-1 with 350 rounds, (Brazil) two 30mm DEFA 5-54 with 125 rounds each.

Development: This machine's designation stems from Aeritalia/Mac-

chi Xperimental, and it was started in 1977 as an Italian study project for a replacement for the G91 and F-104G. The eminently sensible decision was taken to aim at the subsonic limiting Mach number, the result being an aircraft that promises to be light, compact, relatively cheap, possessed of good short-field performance, versatile in operation and capable of carrying a wide assortment of equipment and weapons. The participation of the Brazilian partner has not only broadened the market and manufacturing base but also expanded the variety of equipment and weapon fits.

The basic AM-X will be used chiefly in tactical roles such as close air support and battlefield interdiction, operatinng with full fuel and weapons from unpaved strips less than 3,280ft (1,000m) in length. The design has attempted to maximize reliability and the ability to withstand battle damage, and to an exceptional extent everything on board is modular and quickly replaceable. The seat is a Martin-Baker 10L, fuel is divided between the fuselage and integral wings and flight controls are dual hydraulic with manual back-up.

The two original customers have

specified a simple range-only radar derived in Italy from the Israeli Elta M-2021. The Italian aircraft, of which 187 are planned to be delivered to the AMI to equip eight squadrons, will have a high standard of avionics, with a HUD, INS and Tacan, digital data highways and processing, an advanced cockpit display and a very comprehensive ECM installation. Any of three different photo-recon modules can be installed in a large bay in the lower right side of the fuselage, while an external IR/optronics recon pod can be carried on the centreline pylon.

The Brazilian FAB expects to acquire 79, with VOR/ILS but no INS, different guns and other avionic variations. The original FAB force

Above: Second prototype of the Italian-Brazilian AM-X after roll-out from Aermacchi's Venegono plant in July 1984. Production deliveries are scheduled to start in 1986.

requirement was for 144, and it is possible that this may be restored.

An unuual feature of the AM-X is that it has no direct competitor, other than refurbished used A-4 Skyhawks or A-7 Corsair IIs, which are appreciably larger and much heavier aircraft.

The first prototype flew in May 1984 and crashed a few weeks later, but the second prototype flew in September 1984, as planned. Deliveries will start in late 1986.

Aermacchi M.B.339

Origin: Italy.
Type: (339A) Light attack and trainer; (339K) single-seat light attack and operational trainer.
Engine: One Rolls-Royce Viper turbojet, (339A) 4,000lb (1,814kg) Viper 632-43; (339K) 4,450lb (2,019kg) Viper 680.
Dimensions: Span (over tip tanks) (A) 35ft 7·48in (10·858m), (K) 36ft 2·84in (11·045m); length (A) 35ft 11·96in (10·972m), (K) 35ft 4·88in (10·792m); height (both) 13ft 1·24in (3·994m); wing area 207·7sq ft (19·3m²).
Weights: Empty (A) 6,889lb (3,125kg), (K) 7,066lb (3,205kg); max (A) 12,996lb (5,895kg), (K) 13,558lb (6,150kg).
Performance: Max speed (clean, S/L) (A) 558mph (898km/h), (K) 564mph (907km/h); takeoff run at max wt (A) 3,000ft (915m), (K) 2,986ft (910m); combat radius (both, with four Mk82 bombs, lo-lo-lo) 230 miles (371km).
Armament: (A) Wide range of external stores to maximum weight of 4000lb (1,814kg) carried on six underwing pylons, the four inners being stressed to 1,000lb (454kg) and the centre station on each wing being plumbed for a 71·5gal (325lit) tank, stores including AIM-9 or Magic AAMs and two pods each containing either a 30mm DEFA 5-53 gun and 120 rounds or a 12·7mm M3 with 350

rounds; (K)internal installation of two 30mm DEFA 5-53 guns each with 120 rounds, and maximum of 4,270lb (1,937kg) of external weapons carried on same six underwing hardpoints.

Development: The M.B.339 is a modernized development of the best-selling M.B.326 tandem jet trainer, hundreds of which are in use all over the world. Compared with the 326 the 339 has a revised airframe, the most significant change being the provision of a raised rear cockpit for the instructor. The cockpit is pressurized and fitted with Martin-Baker Mk 10 zero/zero seats, and standard avionics include Tacan, VOR, DME and ILS and a Marconi dead-reckoning computer.

Altogether the 339A is frontrunner in the market slot immediately below the top layer as exemplified by the Hawk and Alpha Jet. The latter have been widely sold not only as trainers but also as dedicated frontline attack machines. In 1982 the Italian Air Force received some camouflaged M.B. 339AS, and these are intended for use as part of an emergency close air support force. The lower powered 339A, however, cannot really succeed in such missions except in third-world visual environments (where its low price is attractive). The IndAer company of Peru hopes to build a large

number under licence and may be permitted to export in certain markets.

The Italian Air Force aerobatic team, Frecce Tricolori, uses M.B. 339PANs.

Aermacchi has upgraded the performance with a more powerful single-seat model the 339K Veltro 2 (Veltro, Greyhound, was the name

of a famed Macchi fighter of World War II). This is offered with such customer options as a HUD, TV type display and integral ECM installation, but no customer had signed as this book went to press in 1984.

A number of M.B.339A were deployed by Argentina to the Falkland Islands (Malvinas) in the South Atlantic War of 1982.

Aero L-39 Albatros

Origin: Czechoslovakia.
Type: Attack reconnaissance and target towing; trainer.
Engine: One 3,792lb (1,720kg) Ivchyenko AI-25TL turbofan.
Dimensions: Span 31ft 0·5in (9·46m); length 39ft 9·56in (12·13m); height 15ft 7·8in (4·77m); wing area 202·36sq ft (18·8m²).
Weights: Empty 7,859lb (3,565kg), (ZA,8,060lb, 3,656kg); maximum 10,141lb (4,600kg) (ZA, 12,346lb, 5,600kg).
Performance: Max speed (clean, tip tanks empty) 485mph (780km/h) at 19,685ft (6,000m), (ZA, same height) 469mph (755km/h); takeoff run at 9,480lb (4,300kg) 1,575ft (480m); time to 16,400ft (5,000m) at same weight, 5min; range at same height, max internal fuel, clean, 621 miles (1,000km).
Armament: (ZO, ZA, only) Four wing hardpoints (inboard 1,102lb, 500kg, each, and outers 551lb, 250kg, each) for total maximum load of 2,425lb (1,100kg) made up of various bombs, rocket pods, K-13A AAMs, recon pod or tanks; (ZA only) underfuselage pod housing GSh-23 gun with up to 150 rounds of 23mm ammunition.

Development: Czechoslovakia is the source for military trainers for the Warsaw Pact nations other than Poland. The Aero L-29 Delfin remains in service throughout WP air forces and many other countries, some 3,600 being built. The completely new L-39 was designed as a more fuel-efficient replacement with a modern stepped cockpit arrangement, and it has proved to be a very satisfactory aircraft. The cockpit is pressurized and fitted with rocket assisted seats usable down to zero height but not below a speed of 94mph (150km/h). Fuel is housed in a group of fuselage bags and the permanently attached tip tanks. The regular L-39 is unarmed and used for training basic and advanced pilots (including those designated to fly helicopters in air forces).

Overseas acceptance of the L-39, assisted by large bulk deals for political ends, has been on an exceptional scale. Several countries have upwards of 24, and Iraq and Libya (the only announced recipients of the ZO armed version) have well over 100 each. Even the Sandinista regime in Nicaragua has received almost 100, and eyewitnesses state that these also are of the armed version.

The L-39ZA, with the powerful Soviet gun, has been in production since early 1983, though recipients has not been identified. The L-39VO is a variant fitted with a winch for towing KT-04 targets for AA gunnery. By mid-1984 the number of all versions was estimated to exceed 2,000. A completely revised prototype was expected to have flown in 1985, and it is possible this may supplant the L-39 in production later in the decade.

Above: Company demonstrator of the L-39 Albatros, with which Aero Vodochodny has achieved considerable success.

Below: The second prototype L-39 shows a rather odd humpbacked aspect not really borne out in the side view.

Above: Camouflaged M.B.339As are used by the Aeronautica Militare Italiano in the emergency close-support role.

Right: The M.B.339PANs used by the Frecce Tricolori can be used in a military role, as demonstrated by this example with gun pods and bombs.

British Aerospace Buccaneer

Origin: UK.
Type: Two-seat attack and reconnaissance.
Engines: Two 11,030lb (5,003kg) Rolls-Royce Spey 101 turbofans.
Dimensions: Span 44ft (13·41m); length 63ft 5in (19·33m). height 16ft 3in (4·95). wing area 514·7sq ft (47·82m²).
Weights: Empty about 30,000lb (13,610kg); maximum loaded 62,000lb (27,123kg).
Performance: Max speed 690mph (1,110km/h) at sea level. range on typical hi-lo-hi strike mission with weapon load 2,300 miles (3,700km).
Armament: Rotating bomb door carries four 1,000lb (454kg) bombs or multisensor reconnaissance pack or 440gal tank; four wing pylons each stressed to 3,000lb (1,361kg), compatible with very wide range of guided and/or free-fall missiles. Total internal and external stores load 16,000lb (7,257kg).

Development: In April 1957 the notorious "Defence White Paper" proclaimed manned combat aircraft obsolete. Subsequently the Blackburn N.103, built to meet the naval attack specification NA.39, was the only new British military aircraft that was not cancelled. Designed for carrier operation, its wing and tail were dramatically reduced in size as a result of powerful tip-to-tip supercirculation (BLC, boundary-layer control) achieved by blasting hot compressed air bled from the engines from narrow slits. The S1 (strike Mk 1) was marginal on power. but the greatly improved S.2 was a reliable and formidable aircraft.

The first 84 were ordered by the Royal Navy but, when the government ordered the phase-out of Britain's conventional carrier force, most were transferred to RAF Strike Command, designated S.2B when converted to launch Martel missiles. The RAF signed in 1968 for 43 new S.2Bs with new avionics and probe.

Within the limits of crippling budgets the RAF Buccaneers have been updated by a few improved avionics, and have gradually been recognized as among the world's best long-range interdiction aircraft. When carrying a 4,000-lb (1,814kg) bombload a "Bucc" at full power is faster than a Mirage, Phantom or F-16 at low level, and burns less fuel per mile. Many Red Flag exercises have demonstrated that a well-flown example is among the most difficult of all today's aircraft to shoot down. On most occasions an intercepting aircraft has failed to get within missile- or gun-firing parameters before having to abandon the chase because of low fuel state. Almost universally the Buccaneer aircrews consider that "the only replacement for a Buccaneer in the 1990s will be another Buccaneer, with updated avionics".

Buccaneers equipped 15 and 16 Sqns of RAF Germany (2 ATAF) at Laarbruch in the land attack role, but these were replaced by Tornado GR.1s in 1983-5 (with replacement of Jaguar squadrons as well, RAF Germany will in effect gain one extra squadron). Tornados will also (despite the wishes of their crews, so popular is the present aircraft) replace 208 Sqn, No 1 Group, at Honington.

The other two UK Buccaneer units, 12 and 216, are tasked with maritime patrol and will go on well into the 1990s. All they want is bet-

Above: 208 Sqn's Buccaneer S.2s are based at Lossiemouth for anti-shipping strikes: these S.2As are seen off Britain's south coast during training.

Right: Buccaneer S.2B of 208 Sqn, showing wing hinges, stores pylon and the tracks of the leading edge vortex generators.

ter avionics, not only internal (and a new nav/attack system due in 1984) but also better ECM than the external ALQ-101 pod. Their main anti-ship weapon will be Sea Eagle.

British Aerospace Harrier

Origin: UK.
Type: Single-seat STOVL tactical attack and reconnaissance; (T.4) dual trainer or special missions.
Engine: One 21,500lb (9,752kg) thrust Rolls-Royce Pegasus 103 vectored-thrust turbofan.
Dimensions: Span 25ft 3in (7·7m), (with bolt-on tips, 29ft 8in); length (GR.3) 47ft 2in (14·38m), (T.4) 57ft 3in (17·45m); height (GR.3) 11ft 3in (3·43m); (T.4) 13ft 8in (4·17m); wing area 201·1sq ft (18·68m²).
Weights: Empty (GR.3) 12,200lb (5,533kg); (T.4) 13,600lb (6,168kg); max (non-VTOL) 26,000lb (11,793kg).
Performance: Max speed over 737mph (1,186km/h, Mach 0·972) at low level; max dive Mach number 1·3; initial climb (VTOL weight) 50,000ft (15,240m)/min; service ceiling, over 50,000ft (15,240m); tactical radius on strike mission without drop tanks (hi-lo-hi) 260 miles (418km); ferry range 2,070 miles (3,300km).
Armament: All external, with many options. Under-fuseable strakes each replaceable by pod containing one 30mm Aden or similar gun, with 150 rounds. Five or seven stores pylons, centre and two inboard each rated at 2,000lb (907kg), outers at 650lb (295kg) and tips (if used) at 220lb (100kg) for Sidewinder AAMs, first fitted during the Falklands crisis. Normal load 5,300lb (2,400kg), but 8,000lb (3,630kg) has been flown.

Left: US Marine Corps AV-8A Harriers fitted with refuelling probes practise operational deployment during an exercise.

Above: In 1971 concepts of operational deployment envisaged more elaborate hides than would be used in a real war.

Development: Until May 1982 the Harrier was generally regarded (except by those familiar with it) as a quaint toy of an experimental nature. Since then it has been a battle-proven weapon which sustained intensive operations in conditions which would have kept other aircraft grounded.

When the experimental P.1127 got daylight under its wheels in 1960 the RAF showed not the slightest interest (in any case, British combat aircraft had been officially pronounced obsolete). Gradually the RAF did show interest in a much more powerful Mach 2 aircraft, the P.1154, but in 1964 this was cancelled. The Government did, however, permit the development of a much smaller subsonic aircraft, and this became the Harrier, basically a machine of classic simplicity which pioneered the entire concept of STOVL (short takeoff, vertical landing) combat operations, and the sustained mounting of close-support and recon missions from dispersed sites in many parts of Europe.

Though the Harrier is small it has better range and weapon load than a Hunter, and it has also rather surprisingly emerged as an air-combat adversary of extreme difficulty. Though not designed as a fighter, its combination of small size, unusual shape, lack of visible smoke and unique agility conferred by the ability to vector the engine thrust direction (to make "impossible" square turns, violent deceleration or unexpected vertical movements without change

Above: Gutersloh-based RAF Harrier GR.3s of 4 Sqn carrying centreline multisensor pods on a reconnaissance mission.

of attitude) make even the original Harrier a most unpopular opponent for any modern interceptor.

The RAF Harrier GR.3 has an inertial nav/attack system, laser ranger and marked-target seeker and fin-mounted passive warning receivers; it is planned to install internal ECM (unlikely before 1985). RAF Germany has two squadrons (3 and 4) at Gutersloh, while in the UK in 1

Group are No 1 Sqn and 233 OCU, both at Wittering.

All these units are vastly experienced, No 1 having played a central role in the recovery of the Falklands and many RAF Harrier pilots having fought with RN Sea Harrier units.

Pilots of No 1 Sqn performed an outstanding feat of airmanship in ferrying their Harriers not only from

the UK to Ascension Island, but also — in the case of four aircraft — onwards to the flight deck of HMS *Hermes* in the South Atlantic, a total air-refuelled trip of over 8,000 miles (12,800km).

British Aerospace Hawk

Origin: UK.
Type: Light interceptor, light attack, and trainer.
Engine: One Rolls-Royce Turboméca Adour turbofan, (T.1) 5,200lb (2,359kg) Mk 151, (50-53 and T-45A) 5,340lb (2,422kg) Mk 851, (Mks 60 onwards) 5,700lb (2,586kg) Mk 860.
Dimensions: Span 30ft 10in (9·4m); length (over probe) 39ft 2¹/₂in (11·95m); height 13ft 5in (4·09m); wing area 179·54sq ft (16·69m²).
Weights: Empty 7,450lb (3,379kg); loaded (trainer, clean) 12,000lb (5,443kg), (attack mission) 16,260lb (7,375kg).
Performance: Max speed 630mph (1,014km/h) at low level; Mach number in shallow dive 1·1; initial climb, 6,000ft (1,830m)/min; service ceiling 50,000ft (15,240m); range on internal fuel 750 miles (1,270km); endurance with external fuel 3hr.
Armament: Three or five hardpoints (two outboard optional) each rated at 1,000lb (454kg), (export Hawk 6,800lb/3,085kg weapon load); centreline point normally equipped with 30mm gun pod and ammunition; intercept role, two AIM-9L Sidewinder.

Development: Though this is the only new All-British military aircraft for 15 years, the Hawk serves as a model of the speed and success that can be achieved when an experi-enced team is allowed to get on with the job. The RAF ordered 175, all of which were delivered by 1982, equipping No 4 FTS at Valley in the advanced pilot training role (replac-ing the Gnat and Hunter) and also with No 1 TWU (Tac Weapons Unit) at Brawdy, and No 2 TWU at Chivenor, in the weapon training role. RAF Hawks normally do not have the outer pylons fitted but these could be added in hours.

By late 1982 RAF Hawks had flown 170,000 hours, with the lowest accident record for any known military jet in history. It cut defect rates by 70 per cent whilst halving maintenance man-hours per flight hour. Despite the aircraft's greater size and power, fuel burn has been dramatically reduced com-pared with the Gnat. Hawks also equip the Red Arrows aerobatic display team, again establishing an unprecedented record for trouble-free operation.

In the weapon-training role air-craft are routinely turned around between sorties in 15 minutes by teams of four armourers.

In 1981 it was announced that, to back up RAF Strike Command's very limited fighter defence forces, about 90 Hawks would be equipped-to fire AIM-9L Sidewinders in the light interception role. Under current planning about 72 are actually armed with the missiles.

In addition the Hawk was selected in 1981 as the future undergraduate pilot trainer of the US Navy, as the T-45A with full carrier equipment and major airframe changes in-cluding a nose gear with two wheels strengthened for nose-tow catapulting, two side airbrakes in-stead of one belly speed brake, and various items in carbon-fibre com-posites. Up to 300 are being supplied from 1987, but the plan by the RAF to lend 12 Hawk T.1s for five years from 1984 without charge was not taken up by the US Navy.

Most of the export customers other than the US Navy use their Hawks in at least a weapons training role, and some task them with com-bat missions. BAe is marketing a Series 100 dedicated attack version of the Hawk with digital avionics, in-ternal navigation, optional laser ranging, HUDs in both cockpits and various features based on the F-16A. The Series 200 will be a single seat version with even greater combat capabilities; a full-scale mock-up of this formidable aircraft was unveiled at the Farnborough airshow in September 1984.

Above: Hawk T1A modified to carry Sidewinder AAMs as well as the 30mm centreline gun pod for airfield defence duties.

British Aerospace Lightning

Origin: UK.
Type: All-weather interceptor.
Engines: Two 15,680lb (7,112kg) thrust Rolls-Royce Avon 302 after-burning turbojets.
Dimensions: Span 34ft 10in (10·6m); length 53ft 3in (16·25m); height 19ft 7in (5·95m); wing area 380·1sq ft (35·31m²).
Weights: Empty about 28,000lb (12,700kg); loaded 50,000lb (22 680kg).
Performance: Max speed 1,500mph (2,415km/h) at 40,000ft (12,000m); initial climb 50,000ft (15,240m)/min; service ceiling over 60,000ft (18,290m); range without overwing tanks 800 miles (1,290km).
Armament: Interchangeable packs for two all-altitude Red Top or stern-chase Firestreak guided missiles; option of two 30mm Aden cannon in forward part of belly tank; export versions up to 6,000lb (2,722kg) bombs or other offensive stores above and below wings.

Development: Britain's only nation-ally developed supersonic military aircraft, 338 Lightnings were built, and despite extreme disinterest by the RAF and political dislike by the Government (because it was a man-ned aircraft), they were eventually allowed to grow in power, fuel capa-city and weapon capability. In the RAF, however, it has always been a pure local-defence interceptor, and even the definitive F.6 variant has no air/ground capability. Indeed, even the overwing ferry tanks are no longer fitted, restricting the aircraft to 1,200gal (5,455 litres), which would be consumed in six minutes in full afterburner.

Primary armament of Red Tops remains fairly effective, and can be used from any firing angle including head-on. The two cannon in the front half of the belly tank are a good installation, causing no visible flash at night, and pilots have always und-ergone an intensive air/air gunnery course at an annual Armament Prac-tice Camp at Akrotiri, Cyprus.

No longer in service with RAF Germany, the Lightning remains an operational interceptor with Nos 5 and 11 Sqns, No 11 Group, Strike Command (No 11 Group was merged into No 1 Group in 1984).

The last F.3 single-seaters are now stored, together with about 40 Lightnings of various marks which in 1979-82 had been expected to form an additional air-defence squadron. The F.6 and a few T.5 two-seaters will now remain as local-defence in-terceptors until replaced by the Tor-nado F.2 by 1986.

As this book went to press the MoD and RAF had not taken a deci-sion on whether to convert surplus Lightnings as RPVs and targets or to buy Sperry QF-100 Super Sabres from the USA.

Below: An elderly but smart Lightning T.5 of the RAF's Lightning Training Flight (note initials on the tailfin), with Red Top training missiles and flight refuelling probe.

British Aerospace Sea Harrier

Origin: UK.
Type: Multirole STOVL naval combat aircraft.
Engine: One 21,500lb (9,752kg) thrust Rolls-Royce Pegasus 104 vectored-thrust turbofan.
Dimensions: Span 25ft 3in (7·7m); length 47ft 7in (14·5in); height 12ft 2in (3·71m); wing area 201·1sq ft (18·68m²).
Weights: 12,960lb (5,878kg); max (non-VTOL) probably 26,200lb (11,880kg).
Performance: Max speed over 737mph (1,186km/h); typical lo attack speed 690mph (1,110km/h); hi intercept radius (3min combat plus reserves and vertical landing) 460 miles (750km); lo strike radius 288 miles (463km).
Armament: Normally fitted with two 30mm Aden Mk 4 each with 150 rounds;. five hardpoints for max weapon load of 8,000lb (3,630kg) including Sea Eagle or Harpoon ASMs, four Sidewinder AAMs and wide range of other stores.

Development: This STOVL (short takeoff, vertical landing) aircraft was most successfully developed from the Harrier chiefly by redesigning the forward fuselage. The deeper structure provides for a versatile and compact Ferranti Blue Fox radar, which folds 180º for shipboard stowage, and a new cockpit with the seat raised to provide space for a much-enhanced nav/attack/combat system, and to give an all-round view.

The Royal Navy purchased 24, plus a further 10, FRS.1s, the designation meaning ''fighter, recon, strike'' (strike normally means nuclear, but the Fleet Air Arm has not confirmed this capability). In the NATO context the main task is air defence at all heights, normally with direction from surface vessels, either as DLI (deck-launched intercept) or CAP (combat air patrol).

In the Falklands fighting, in which almost all the RN Sea Harriers (28 out of 32) took part, these aircraft repeatedly demonstrated their ability to fly six sorties a day in extremely severe weather, with maintenance by torchlight at night often in hail blizzards. Serviceability was consistently around 95 per cent each morning. CAPs were flown at 10,000ft (3km) at 290mph (463kmH), and within a few seconds it was possible to be closing on an enemy at 690mph just above the sea; 24 Argentine aircraft were claimed by AIM-9L Sidewinders and seven by guns. In air-ground missions main stores were 1,000lb (454kg) bombs, Paveway ''smart'' bombs and BL.755 clusters. Many new techniques were demonstrated including 4,000-mile (6,440km) flights to land on ships (sometimes by pilots who had never landed on a ship) and operations from quickly added sheet laid on the top row of containers in a merchant ship.

From the harsh self-sufficient campaign it is a major step to the more sophisticated European environment of greater density and diversity of forces, and especially of emitters (though Sea Harriers did use jammer pods in the South Atlantic). The E-2C and other aircraft would normally be available for direction, and the Sea Harrier is envisaged as filling the fleet defensive band between ship-to-air missiles and long-range F-14s with Phoenix AAMs. Its ESM fit is more advanced than that of Harriers and is used as a primary aid to intercept emitting aircraft (or, it is expected, sea-skimming missiles). Pilots normally operate as individuals, flying any mission for which they are qualified.

By the middle of 1984 23 additional aircraft were ordered to replace losses from all causes (6) and increase establishment of the three combat squadrons (800, 801 and 809, normally embarked aboard *Invincible*, *Illustrious* and *Hermes* (later *Ark Royal*) and the training unit 899 Sqn at Yeovilton.

Under a mid-life improvement programme, due to be implemented later in the decade, the Royal Navy FRS.1 aircraft are to be updated to have lookdown-shootdown capability with a new radar of pulse-doppler type, named Blue Vixen. This will considerably upgrade all-round capability, and in particular will

Above: A Sea Harrier FRS.1 leaves the 7º ski-jump of HMS *Invincible* with two tanks and three practice bomb carriers.

match the range of the new Sea Eagle anti-ship missile. It is expected that Zeus active ECM will be installed, together with pairs of AIM-120A (Amraam) AAMs, and the cockpit will be updated. British Aerospace have published an artist's impression showing Lerx (leading-edge root extensions) and wingtip rails for Sidewinder or Asraam close-range AAMs, but the future of these is uncertain.

In 1984 the only export customer was the Indian Navy. This service purchased six FRS.51 and two T.60 two-seat trainers, followed by a repeat order for ten MK 51s and two trainers. They operate with No. 300 Sqn from shore bases and from INS *Vikrant*.

Below: A Sea Harrier T.4N trainer (top) and an 899 Sqn FRS.1 which destroyed three enemy aircraft in the Falklands.

CASA C-101 Aviojet

Origin: Spain.
Type: Light attack aircraft, and trainer.
Engine: One Garrett TFE731 turbofan, (E-25) TFE731-2-2J rated at 3,500lb (1,588kg) thrust, (BB) TFE631-3-1J, 3,700lb (1,678kg), (CC) TFE731-5, 4,300lb (1,950kg).
Dimensions: Span 34ft 9in (10·6m); length 41ft 0in (12·5m). height 13ft 11·3in (4·25m); wing area 215·3sq ft (20·0m²).
Weights: (E-25) Empty 7,606lb (3,450kg); max (clean) 10,692lb (4,850kg), (attack) 12,345lb (5,600kg).
Performance: (E-25) Max speed (SL) 430mph (691km/h), (25,000ft, 7,620m) Mach 0·71, 495mph (797km/h); time to 25,000ft (7,620m) 8·5min; endurance 7h; combat radius (lo-lo-hi, four 250kg bombs and 30mm gun, 3h over target plus 30min reserve) 236 miles (380km); ferry range (30min reserve) 2,303 miles (3,706km).
Armament: Six wing hardpoints for total load of 4,960lb (2,250kg) made up of any tactical store up to 1,102lb (500kg) individual weight;. internal bay below rear cockpit for 30mm gun pack, twin-12·7mm gun pack, laser designator, ECM pack, recon camera or similar tailored installation.

Development: Designed and developed by CASA in collaboration with MBB and Northrop (who provided design and test facilities), the C-101 was planned to meet the needs of Spain's EdA, which uses 88 in the training role at the Academia del Aire and the 41⁰ Grupo. The basic EdA model is the C-101EB, with EdA designation E-25. The C-101BB is the armed export version, subject of a major deal with Chile which imported four aircraft and assembled eight more supplied in kit form. The Chilean deal also includes an option for 23 more aircraft. In the service of *Fuerza Aérea de Chile* the C-101 is called T-36 Halcón. Four plus an option on four more aircraft have been ordered by Honduras.

The C-101CC is a further uprated attack version, the first of which (the 90th production Aviojet) is being supplied to Chile together with three production machines. In the final phase of the Chilean programme, the CC will be assembled by FAC/Ind-Aer.

The C-101 is relatively fuel-efficient, and combines good all-round performance, and quite exceptional range and endurance, with more than the normal spectrum of attack weapons and devices. The 101DD will have Ferranti HUD and INS/weapon-aiming systems.

Above: The second prototype C-101 demonstrates its ordnance capacity with six bombs and a twin 12.7mm machine gun pod.

Below: The first prototype on a test flight with a practice bomb carrier mounted on the inboard wing pylon.

Cessna A-37B Dragonfly

Origin: USA.
Type: A-37, light attack.
Engines: Two 2,850lb (1,293kg) thrust General Electric J85-17A turbojets.
Dimensions: Span (over tanks) 35ft 10·5in (10·93m); length (excl refuelling probe) 28ft 3·25in (8·62m); wing area 183·9sq ft (17·09m²).
Weights: Empty 6,211lb (2,817kg); loaded 14,000lb (6,350kg).
Performance: Max speed 507mph (816km/h); normal cruising speed (clean) 489mph (787km/h); initial climb 6,900ft (2,130m)/min; service ceiling 41,765ft (12,730m); range (max fuel, four drop tanks) 1,012 miles (1,628km), (max payload including 4,100lb/1,860kg ordnance) 460 miles (740km).
Armament: GAU-2B/A 7·62mm Minigun in fuselage, eight underwing pylons (four inners 870lb/394kg each, next 600lb/272kg and outers 500lb/227kg) for large number of weapons, pods, dispensers, clusters, launchers or recon/EW equipment.

Development: For almost 30 years the basic pilot trainer of the USAF, the T-37 seats an instructor and pupil in side-by-side ejection seats. The much more powerful A-37 has a strengthened airframe to carry heavy underwing and wingtip loads and still fly uninhibited manoeuvres. Surprisingly, ejection seats were not fitted to this version, which was

originally created for "brushfire" wars where ground defences are primitive. Many Dragonflies have inflight-refuelling probes; about 600 were built. USAF-operated A-37B Dragonfly ground-support aircraft have been adapted for forward air control duty with some Air National Guard Groups. Others are serving in South Korea.

Right: The A-37B Dragonfly, a much more powerful derivative of the T-37 trainer, was built as a light attack platform.

Below: Among the export customers for the A-37 has been the Peruvian Air Force.

Dassault-Breguet Mirage III and 5

Origin: France.
Type: Single-seat or two-seat interceptor, tactical aircraft (depending on sub-type).
Engine: (IIIC) 13,225lb (6,000kg) thrust (max afterburner) SNECMA Atar 9B turbojet; (most other III and some 5) 13,670lb (6,200kg) Atar 9C; (IIIR2Z, NG and 50) 15,873lb (7,200kg) Atar 9K50.
Dimensions: Span 27ft (8·22m); length (excl probe) (IIIC) 48ft 5in (14·75m), (IIIE) 49ft 3^1/2in (15·03m), (5) 51ft 0^1/4in (15·55m); height 13ft 11^1/2in (4·25m); wing area 375sq ft (35·0m^2).
Weights: Empty (IIIC) 13,570lb (6,156kg); (IIIE) 15,540lb (7,050kg); (IIIR) 14,550lb (6,600kg); (IIIB) 13,820lb (6,270kg); (5) 14,550lb (6,600kg); loaded (IIIC) 19,700lb (8,936kg); (IIIE, IIIR, 5) 29,760lb (13,500kg), (IIIB) 26,455lb (12,000kg).
Performance: Max speed (all models, clean) 863mph (1,390km/h, Mach 1·14) at sea level, 1,460mph (2,350km/h, Mach 2·2) at altitude; initial climb, over 16,400ft (5,000m)/min (time to 36,070ft/11,000m, 3min); service ceiling (Mach 1·8) 55,775ft (17,000m); range (clean) at altitude about 1,000 miles (1,601km); combat radius in attack mission with two bombs and tanks (hi-altitude) 745 miles (1,200km); ferry range with three external tanks 2,485 miles (4,000km).
Armament: Two 30mm DEFA 5-52 cannon, each with 125 rounds (normally fitted to all versions except when IIIC carries rocket-boost pack); three 1,000lb (454kg) external pylons for bombs, missile or tanks (Mirage 5, seven external pylons with max capacity of 9,260lb/4,200kg).

Development: By far the most commercially successful fighter ever built in Western Europe, the tailless delta Mirage has had little local competition, but has had to contend in the world market with the F-104, MiG-21 and F-5. It had the advantage of proven combat success in

1967 with Israel, the first export customer, and this catapulted it into the limelight and brought sales of 1,410 to 20 air forces.

The initial Mirage IIIC, now being withdrawn from l'Armée de l'Air but still operating in South Africa, has an early Cyrano radar giving limited all-weather and night interception capability. Original armament comprised the Matra R530 missile, with IR or radar homing head, and an alternative of two 30mm guns or a booster rocket pack whose tankage occupied the gun ammunition bay. This rocket is still available, but with the advent of more powerful Atar engines has not been adopted by other customers.

Most of today's aircraft are variants either of the IIIE, a slightly lengthened fighter-bomber with radar and more comprehensive navigation and weapon-delivery avionics, or the 5, a simplified model without radar and able to carry additional fuel or bombs within the same gross weight. This has appealed strongly to third-world countries because of its lower price, and if flown in good weather it has few serious limitations apart from the basic ones that afflict all early tailless deltas: the need for a long runway and the inability to make sustained tight turns without speed bleeding off rapidly. The Mirage III was, in fact, planned to have low-pressure

tyres and to operate from rough strips, but the actual tyre pressure combined with the extremely high takeoff and landing speed make this a very rare occurrence.

Some customers have bought tandem dual trainers, others the IIIR with a camera-filled nose (South Africa's R2Z having the uprated 9K50 engine, and several users the chin bulge showing installation of navigational doppler). Chile brought the latest production model, the 50.

At the time of writing a customer has yet to emerge for the IIING (Nouvelle Génération) development which adds a fly-by-wire flight control system, wing-root strakes, fixed canards, inertial navigation, HUD, Cyrano IV radar, laser ranger and additional weapon stations to take advantage of the increased max weight of 32,400lb (14,700kg). Some customers, such as Peru, have up-

dated their Mirages using Dassault-Breguet kits for a HUD, inertial navigation system, laser ranger and Magic AAMs.

The Mirage III/5 story is far from over however. In Switzerland and in France, canard foreplanes are being fitted to add more manoeuvrability in combat. A modification programme is underway in Swiss IIIs.

Above: Mirage 5 with a heavy load of 12·250kg (551lb) and two 125kg (276lb) Matra general-purpose bombs.

Below: The prototype Mirage IIING, with canard foreplanes and fly-by-wire for enhanced dogfight manoeuvrability.

Bottom: One of the 12 Mirage 5AD interceptors supplied to the Abu Dhabi Air Force in 1974.

Dassault-Breguet Mirage F1

Origin: France.
Type: Single-seat multimission fighter; (E) all-weather strike, (R) recon, (B) dual trainer.
Engine: 15,873lb (7,200kg) thrust (max afterburner(SNECMA Atar 9K-50 augmented turbojet.
Dimensions: Span 27ft 6³/₄in (8·4m); length (F1.C) 49ft 2¹/₂in (15m); (F1.E) 50ft 11in (15·53m); height (F.1C) 14ft 9in (4·5m); (F1.E) 14ft 10¹/₂in (4·56m); wing area 269·sq ft (25·0m²).
Weights: Empty (F1.C) 16,314lb (7,400kg); (F1.E) 17,857lb (8,100kg); loaded (clean) (F1.C) 24,030lb (10,900kg), (F1.E) 25,450lb (11,540kg); (max) (F1.C) 32,850lb (14,900kg); (F1.E) 33,510lb (15,200kg).
Performance: Max speed (clean, both versions) 915mph (1,472km/h, Mach 1·2) at sea level, 1,450mph (2,335km/h, Mach 2·2) at altitude (with modification to cockpit transparency and airframe leading edges F1.E capable of 2·5); rate of climb (sustained to Mach 2 at 33,000ft) (F1.C) 41,930-47,835ft (12,780-14,580m)/min, (F1.E) above 59,000ft (18,000m)/min; service ceiling (F1.C) 65,600ft (20,000m), (F1.E) 69,750ft (21,250m); range with max weapons (hi-lo-hi) (F1.C) 560 miles (900km), (F1.E) 621 miles (1,000km); ferry range (F1.C) 2,050 miles (3,300km), (F1.E) 2,340 miles (3,765km).

Armament: (both versions) Two 300mm DEFA 5-53 cannon, each with 135 rounds; five pylons, rated at 4,500lb (2,000kg) on centreline 2,800lb (1,350kg) inners and 1,100lb (500kg) outers; launch rails on tips rated at 280lb (120kg) for air-to-air missiles; total weapon loaded 8,820lb (4,000kg). Typical air combat weapons, two Matra 550 Magic for close combat, one/two Matra Super 530 for long-range homing. Optional reconnaissance pod containing cameras, SAT Cyclope infrared linescan and EMI side-looking radar.

Development: In the early 1960s Dassault thought the Mirage III would soon need replacing and studied various configurations before deciding on a high fixed wing of conventional form, plus a horizontal tail. The F1 has a wing much smaller than the deltas but so much more efficient that the F1 has much shorter field length, slower landing, and (with 40 per cent greater internal fuel) three times the supersonic endurance, or twice the tactical radius at low level, with superior all-round manoeuvrability. With long-stroke twin-wheel main gears and a landing speed of 143mph (230km/h) the F1 is also more genuinely able to use short un-paved airstrips.
  The Armée de l'Air achieved operational capability with the F1.C

Above: Mirage F1.C in typical air defence configuration, with two wingtip Magics and underwing Super 530 missiles.

Below: Launch of an R.550 Magic dogfight missile from the wingtip rail of an early Mirage F1.C during trials.

Dassault-Breguet Mirage 2000

Origin: France.
Type Multirole fighter with emphasis on interception and air superiority combat.
Engine: One SNECMA M53-5 afterburning by-pass turbojet (low-ratio turbofan) with max thrust of 12,350lb (5,602kg) dry and 19,840lb (9,000kg) with afterburner.
Dimensions: Span 29ft 6in (9·0m); length (2000) 47ft 1in (14·35m), (2000B) 47ft 9in (14·55m); height 17ft 6in (5·3m); wing area 441sq ft (41m²).
Weights: Empty 16,315lb (7,400kg); normal takeoff, air-intercept mission 33,000lb

(14,969kg); max 36,375lb (16,500kg).
Performance: Max continuous speed at 36,000ft (11,000m) Mach 2·2, 1,320mph (2,124km/h); maximum attack speed at low level 690mph (1,110km/h). range with two tanks, over 1,118 miles (1,800km).
Armament: Two 30mm DEFA 5-53 cannon; normal air-intercept load two Matra Super 530 and two Matra 550 Magic air-to-air missiles; intention is to develop ground-attack version with max overload of 13,225lb (6,000kg) of weapons and/or tanks and ECM pods on nine external hardpoints.

Development: Superficially almost identical to earlier Mirage deltas, the 2000 is in fact a totally new aircraft, with dramatically enhanced capabilities. Strongly biased to the air superiority mission, the 2000 was made possible only by the emergency of CCV (control configured vehicle) technology in the USA, which by combining FBW (fly-by-wire) signalling with instant-action computer control and multi-lane channel redundancy allows a fighter to be made basically unstable. A further great advantage, compared with the Mirage III, is that the wing has full-span leading-edge flaps which provide controlled camber for use in different flight regimes. For slow speeds and landing the fully drooped leading edge allows the

trailing-edge elevons to act as flaps, in contrast to the elevons of older Mirages which have to be deflected upwards, in effect greatly adding to weight, not lift.
  The M53 engine is almost a turbo-jet, its bypass ratio being very low, and the overall pressure ratio of only 9·3 showing the degree to which it is optimized to supersonic speed at very high altitude. Thus it is at a great disadvantage in fuel burn in the normal subsonic regime which accounts for more than 95 per cent of even a Mirage 2000's flight time.

Below: The Super 530D missile, seen here being test launched from the second prototype, gives the Mirage 2000 an impressive snap-up capability.

at Reims (30e Escadre), followed by 5e Escadre at Orange (whose three squadrons include 25 of the F1.C-200 type with permanent FR probes to permit non-stop deployment to Djibouti, 3,100 miles, 5,000km, and similar distant points) and EC 12 at Cambrai.

Equipped with Cyrano IV radar and the excellent combinations of Magic and Super 530 AAMs, the F1.C is one of the best interceptors in Western Europe, though its relatively small size and power inevitably place it at a disadvantage in either the long-range patrol mission or close air combat.

All models have the uprated 9K50 engine, and most customers have bought a comprehensive EW suite including the Thomson-CSF Type BF RWR. In a few air forces, including the Armée de l'Air, the same supplier's Remora self-protection jammer pod is carried, and in the dedicated ECM role the usual jammer is the powerful Caiman. On the other hand very few customers have bothered about the very large increase in stand-off interception cability conferred by the Super 530F radar-guided AAM, even though this is a capability available from no other aircraft outside the two super-powers apart from the Viggen JA37 and Tornado ADV.

Every export customer has, it is believed, matched the F1 with the relatively cheap Magic dogfight AAM. In the second half of the decade the completely new Mle 30-791B gun will become available, firing higher-performance 30mm

Above: A Mirage F1.C lets fly with a salvo of 68mm (2.68in) SNEB rockets from its four 18-round underwing launchers.

ammunition at 2,500rds/min.

There are four basic sub-variants of F1. Most have been derived from the F1.C with radar, and this family includes the F1.E multirole model with inertial navigation, central computer and HUD. The F1.A is a simplified model with a slim non-radar nose, configured mainly for ground attack. The F1.B tandem-seat dual-pilot model is a trainer with full combat capability but reduced fuel. The F1.R is a dedicated reconnaissance version with cameras and sensors.

The F1 is operated by the air forces of France, Ecuador, Greece, Iraq, Jordan, Kuwait, Libya, Morocco, Qatar, South Africa and Spain.

In the attack mission the aircraft is further penalized by its large delta wing, which puts it very high in the "unacceptable" regime in the plot of indicated airspeed at low level against number of 0·5g bumps experienced per minute. This does not worry some customers, who see their Mirage 2000s purely as air combat aircraft, in which role the combination of guns, Magic dogfight AAMs and radar-guided Super 530D AAMs promises to be a good one, though the 530D matched to the RDI pulse-doppler radar has yet to be exported and the radar itself has been seriously delayed. Thus, the RDM radar, a simplified earlier set, is fitted to all current Mirage 2000s instead of only those which are equipped for export.

Dassault has managed to conclude important exports sales despite unit prices which in some cases have — according to published figures — run to $38 million. Egypt's price has been announced as only $20 million, with increasing Egyptian content in the aircraft assembled at Helwan from 1985. The first batch of 16 single-seaters and four two-seat 2000Bs were ordered in 1983, of a total of 80. India expects to purchase 150; the first 36 2000H and four 2000TH two-seaters are all-French, but the final 110 are to be assembled in India with increasing local content. Peru's buy is 26 and Abu Dhabi 18, but China decided this small aircraft was "too expensive" after a prolonged study. France's own total is expected

eventually to reach 243, of which the first group, Escadre de Chasse 2, were being equipped in 1983, with 19 aircraft delivered in that year. A further 87 are due by the end of 1988, including several of the 2000N nuclear interdiction model with two seats, improved navigation and ECM, and the ASMP stand-off missile. An indication of the true price of the 2000 is provided by the Fr8,000 million voted in 1984 for 12 2000DA fighters and 16 2000Ns, a unit price of £26·64 million.

Above: Third prototype of the Mirage 2000, a very different aircraft from the Mirage III, despite similar overall appearance. As with the Mirage III, Dassault expect to build more 2000s for export than for home use, by a factor of 2 to 1.

Dassault-Breguet Super Etendard

Origin: France.
Type: Single-seat strike fighter.
Engine: 11,265lb (5,110kg) thrust SNECMA Atar 8K-50 turbojet.
Dimensions: Span 31ft 5³/₄in (9·6m). length 46ft 11¹/₂in (14·31m); height 12ft 8in (3·85m). wing area 305·7sq ft (28·4m²).
Weights: Empty 14,200lb (6,450kg); loaded 25,350lb (11,500kg).
Performance: Max speed 745mph (1,200km/h) at sea level, Mach 1 at altitude; initial climb 24,600ft (7,500m)/min; service ceiling 45,000ft (13,700m); radius (hi-lo-hi, one AM 39, one tank) 403 miles (650km).
Armament: Two 30mm DEFA cannon, each with 125 rounds; five pylons for weapon load with full internal fuel of 4,630lb (2,100kg); one AM 39 Exocet can be carried (right wing) with one tank (left).

Development: The French Aéronavale still uses the original Etendard IVN attack aircraft and IVP photo-reconnaissance machine, in each case often as a "buddy" air refuelling tanker carrying a hose-reel pod. The replacement for the former in four combat units, as described later, is the Super Etendard. This is a very much updated aircraft, though with the advantage of some commonality with the earlier machine. Though called a strike fighter the Super Etendard has little air-combat capability against enemy high-performance aircraft and is used almost wholly in an air/surface role.

Equipment includes an Agave multi-mode radar which is fully adequate for most attacks on surface ships, a Sagem (Kearfott licence) inertial nav/attack system, BF radar warning system and DB-3141 ECM jammer pod. Free-fall bombs of 250 and 400kg sizes can be carried, but the chief anti-ship weapon is the AM 39 Exocet. Super Etendards of the Argentine navy destroyed HMS *Sheffield* and the *Atlantic Conveyor* with these missiles.

The Aéronavale planned to buy 100 Super Etendards but inflation reduced the total to 71 in 1978-82. These equip Flotilles 11F, 12F and 14F at Landvisiau, and 17F at Hyères; 12F is in the intercept role replacing the Mach 2 Crusader. The IVP remains in use but a reconnaissance version of the Super has long been projected. Super Etendard flotilles go to sea aboard the small and aged *Clemenceau* and *Foch,* to replace which a 33,000-tonne nuclear carrier is planned for the end of the century, the keel for the *Charles de Gaulle* being due to be laid in 1986. By this time, however, the basic obsolescence of the Super Etendard will probably have begun to show, and it is hoped to replace it by a carrier-based version of the proposed ACX.

Since 1982 Iraq has been trying to purchase Super Etendards armed with AM 39 missiles to bolster its capability against oil terminals in the war against Iran. There has been intense international pressure on France not to sell, and reopening the production line would be uneconomic, the last of 14 for Argentina being delivered in early 1983. Nevertheless some 30 Iraqi personnel completed training on

Above: Steam-catapult launch of the eighth Super Etendard with the usual underwing tanks. Note the tailplane angle.

both the aircraft and missile in August 1983, at Landivisiau, and the Aéronavale loaned five missile-armed Super Etendards from reserve stocks. These have been used by Iraq to attack shipping in the Gulf, much of it belonging to third parties not involved in the conflict.

Below: Super Etendards of the Argentinian Navy's 3rd Escuadra in formation over France during a training flight prior to delivery of the aircraft.

Dassault-Breguet/Dornier Alpha Jet

Origin: Jointly France and West Germany.

Type: Two-seat light strike/reconnaissance aircraft, and trainer.

Engines: Two 2,976lb (1,350kg) thrust SNECMA/Turboméca Larzac C5 turbofans; (NGEA) 3,175lb (1,440kg) Larzac C20.

Dimensions: Span 20ft 10$\frac{3}{4}$in (9·11m); length (excluding any probe) 40ft 3$\frac{3}{4}$in (12·29m); height 13ft 9in (4·2m); wing area 188·4sq ft (17·5m^2).

Weights: Empty (trainer) 7,374lb (3,345kg); loaded (clean) 11,023lb (5,000kg) (max) 16,535lb (7,500kg).

Performance: Max speed (clean) 576mph (927km/h) at sea level, 560mph (900km/h) (Mach 0·85) at altitude; climb to 39,370ft (12,000m), less than 10min; service ceiling 48,000ft (14,630m); typical mission endurance 2hr 30min; ferry range with two external tanks 1,827 miles (2,940km).

Armament: Optional for weapon training or combat missions, detachable belly fairing housing one 30mm DEFA or 27min Mauser cannon, with 125 rounds; same centreline hardpoint and either one or two under each wing (to max of five) can be provided with pylons for max external load of 5,511lb (2,500kg), made up of tanks, weapons, reconnaissance pod, ECM or other devices.

Development: Though France and Britain were already collaborating on a trainer and light attack aircraft, in 1969 France and West Germany announced a collaborative programme for a less-powerful machine in this class, and after an industry competition the Alpha Jet was selected in 1970. Production was seriously delayed, but eventually a multinational manufacturing group achieved a high rate of output.

The design was specially arranged with a high wing to give plenty of clearance for underwing stores, though this resulted in a narrow main landing gear track with the units folding into the fuselage. The basic Alpha Jet E, for training and light attack, has tandem staggered Martin-Baker seats, those for the Armée de l'Air being Mk 4s usable at zero height but not below 104mph (167km/h) airspeed, although Egypt, Belgium and Qatar have Mk 10 seats with zero/zero capability.

This model serves with the Armée de l'Air (200 total) to equip the entire Groupement-Ecole 314 "Christian Martel" at Tours, the Patrouille de France aerobatic team at Salon, the Centre d'Entrainement au Vol Sans Visibilité and the 8e Escadre de Transformation at Cazaux. It is also used (33 supplied) by Belgium's 7, 9 and 11 Sqns at St Truiden (St Trond). All these are pure training or display units, but the Federal German Luftwaffe uses a different version in the close-support and reconnaissance roles.

The Alpha Jet A has the Mauser gun, a pointed nose with pitot probe (aircraft length 43ft 5in, 13·23m) and MBB-built Stencel seats instead of Martin-Baker. A total of 153 was supplied to three fighter/bomber wings: JaboG 49 at Fürstenfeldbruck, JaboG 43 at Oldenburg and JaboG 41 at Husum, each with 51 aircraft on strength. They are austerely equipped for attack missions in the European environment, though navigation systems are good and a HUD (head-up display) is provided. The LaCroix BOZ-10 chaff pod has been developed jointly by France and Germany and is expected to appear with these JaboGs. In the recon role a Super Cyclope pod can be carried with optical cameras, IR linescan and a decoy launcher. Combat missions are expected to be strongly supported by AWACS (E-3A Sentry) coverage to make up for deficiencies in the Alpha Jet's defensive avionics. The Luftwaffe has 18 Alpha jets in the weapon-training role at Beja, Portugal, the German total being 175.

The alternative close-support version has inertial navigation, a HUD, laser ranger and a radar altimeter, and is co-produced at Helwan, Egypt, as the MS2, the original trainer being known as the MS1. Dassault-Breguet has itself developed this model further into the NGEA (Nouvelle Génération Ecolé et Appui, new generation trainer and attack), with the more powerful Larzac C20 engine and enhanced pylon capability for tanks or Magic AAMs. In 1984 Helwan was producing both the MS1 and MS2 at the rate of two every three months, with the rear fuselage and all control or movable surfaces manufactured locally. Altogether 499 had then been ordered, with 467 delivered.

Below: Luftwaffe Alpha Jet A with centreline 27mm cannon, wing-mounted fuel tanks and four 250kg (551lb) bombs.

Above: Alpha Jet E trainers of the "Christian Martel" GE 314 school at Tours St Symphorien.

Below: Alpha Jet NGEA with underwing fuel tanks and Magic air-to-air missiles.

Fairchild Republic A-10A Thunderbolt II

Origin: USA.
Type: Close-support attack aircraft.
Engines: Two 9,065lb (4,112kg) thrust General Electric TF34-100 turbofans.
Dimensions: Span 57ft 6in (17·53m); length 53ft 4in (16·26m). height (regular) 14ft 8in (4·47m), (NAW) 15ft 4in (4·67m); wing area 506sq ft (47m²).
Weights: Empty 21,519lb (9,761kg); forward airstrip weight (no fuel but four Mk 82 bombs and 750 rounds) 32,730lb (14,846kg); max 50,000lb (22,680kg). Operating weight empty 24,918lb (11,302kg), (NAW) 28,630lb (12,986kg).
Performance: Max speed (max weight, A-10A) 423mph (681km/h), (NAW) 420mph (676km/h); cruising speed at sea level (both) 345mph (555km/h); stabilized speed below 8,000ft (2,440m) in 45° dive at weight 35,125lb (15,932kg) 299mph (481km/h); max climb at basic design weight of 31,790lb (14,420kg), 6,000ft (1,828m)/min; service ceiling not stated; takeoff run to 50ft (15m) at max weight 4,000ft (1,220m); operating radius in CAS mission with 1·8h loiter and reserves 288 miles (463km); radius or single deep strike penetration 620 miles (1,000km); ferry range with allowances 2,542 miles (4,091km).

Armament: One GAU-8/A Avenger 30mm seven-barrel gun with 1,174 rounds, total external ordnance load of 16,000lb (7,257kg) hung on 11 pylons, three side-by-side on body and four under each wing; several hundred combinations of stores up to individual weight of 5,000lb (2,268kg) with max total weight 14,638lb (6,640kg) with full internal fuel.

Development: The concept of a relatively slow tactical weapons platform, able to survive over a battlefield by flying very low and carrying special armour and duplicated systems, has always been a matter for prolonged argument. What has never been in doubt is the devastating punch that the A-10A can deliver, at least in conditions of reasonable visibility. Lingering doubts about the cost/effectiveness of this unique aircraft may have been dispelled by the appearance of the Soviet Su-25, which is an exact counterpart (but modelled more on the A-10's unsuccessful rival, the Northrop A-9A).

Until 1967 the USAF had never bothered to procure a close-support aircraft, instead flying CAS missions with fighters and attack machines. With the A-10, emphasis was placed on the ability to operate from short unpaved front-line airstrips, to carry an exceptional load of weapons — in particular a very powerful high-velocity gun — and to withstand prolonged exposure to gunfire from the ground. Avionics were left to a minimum, the official description for the fit being "austere", but a few extra items are now being added.

The original A-10A was a basically simple single-seater, larger than most tactical attack aircraft and carefully designed as a compromise between capability and low cost. As an example of the latter many of the major parts, including flaps, main landing gears and movable tail surfaces, are interchangeable left/right, and systems and engineering features were designed with duplication and redundancy to survive parts being shot away. The unusual engine location minimizes infra-red signature and makes it almost simple to fly with one engine inoperative or even shot off.

Above: The distinctive Warthog nose markings — reflecting the unofficial name bestowed on the A-10 — are those of the 917th TFG, AFRES; common to all Thunderbolts is the aircraft's principal weapon, the massive GAU-8/A cannon.

FMA IA 58 and IA 66 Pucará

Origin: Argentina.
Type: Close-support attack and reconnaissance aircraft.
Engines: Two turboprops, (58) 988shp (Turboméca Astazou XVIG, (66) 1,000shp Garrett TPE331-11-601W.
Dimensions: Span 47ft 6·9in (14·5m); length 46ft 9·15in (15·253m); height overall 17ft 7·1in (5·362m); wing area 326·1sq ft (30·3m²).
Weights: Empty (58A) 8,900lb (4,037kg), (66) 8,862lb (4,020kg); max 14,991lb (6,800kg).
Performance: (both) Max speed (9,840ft/3,000m) 310mph (500km/h); econ cruise 267mph (430km/h); takeoff to 50ft/(15m) at 12,125lb (5,500kg) 2,313ft (705m); landing from 50ft (15m) 1,978ft (603m); attack radius with max external weapons, 10 per cent reserve fuel (hi-lo-hi) 155 miles (250km), (1,764lb/800kg weapons and external fuel) 559 miles (900km); ferry range 1,890 miles (3,042km).
Armament: Two 20mm Hispano HS804 each with 270 rounds and four 7·62mm FN Browning each with 900 rounds all firing ahead; up to 3,307lb (1,500kg) of wide range of stores carried on three pylons, with individual stores up to 2,205lb (1,000kg), examples including 12 bombs of 276lb (125kg), 12 large napalm tanks, three 1,102lb (500kg) DA bombs, seven 19 x 2·75in (70mm) rocket pods or a cannon pod and two 72·5gal (330lit) drop tanks.

Development: The Pucará, named after an early hilltop type of stone fortress, was influenced by the US interest in light turboprop Co-In aircraft in the early 1960s. Intended for use against unsophisticated forces, and in fact ordered by the FAA (Argentine air force) for suppressing internal disorders, it was planned to have considerable firepower yet operate from austere airstrips with the minimum of ground support.

Features include pilot and copilot in staggered Martin-Baker "zero/zero" seats, carefully disposed armour and equipment for operation by night but not in adverse weather. There is good avionic provision for communications and navigation, a ILS is standard, but weapon aiming is visual. Weather radar is an option.

In the South Atlantic war in Spring 1982 Pucarás played a major role in the Falklands, being able to use various airstrips through the islands. Despite their good weapon load and inflight agility the 20 island-based machines accomplished little beyond shooting down a British Army Scout AH.1 on 28 May. None survived; six were brought to Britain and one was carefully evaluated at Boscombe Down.

Meanwhile, production of 100 for the FAA was completed in 1984. Continuing production is for export, and a Libyan mission evaluated the IA 66 export model in spring 1983. Most of the missions flown over the Falklands were by single pilots, and a single-seat IA 66 is seen as the most likely definitive model. FMA flew an IA 58B with 30mm guns and upgraded avionics, and planned a turbofan version, but dropped both.

Below: Pucará with typical Co-In warload of six general purpose bombs and four rocket pods.

jammer pods are always hung externally.

In 1979 Fairchild flew a company-funded NAW (night/adverse weather) demonstrator with agumented avionics and a rear-cockpit for a WSO seated at a higher level and with good forward view. Both the regular and NAW aircraft carry a Pave Penny laser seeker pod under the nose, vital for laser-guided munitions, and the NAW also has a Ferranti laser ranger, Texas Instruments FLIR (forward-looking infra-red), GE low-light TV and many other items including a Westinghouse multi-mode radar with WSO display. It is probable that during the rest of the decade A-10As will be brought at least close to the NAW standard, with the LANTIRN pod, though the two-seat NAW itself was never funded. A-10A funding was abruptly terminated in 1982 at a total of 707 aircraft (not including the six RDT&E [research, development, test and engineering] prototypes), the last one in 1983.

In service with the 57th TTW, four regular TFWs (the 23rd, 81st, 354th and 355th) and the 66th FWS of the USAF, and the 174th TFW and four TFGs (103rd, 104th, 128th and 175th) of the ANG, the A-10A has proved generally popular and unquestionably effective with many weapons. Serviceability and manpower burden have been as predicted, and the only cause for worry is the sustained attrition rate, rather higher than normal, caused by hitting the ground during operations at low level.

Weapon pylons were added from tip to tip, but the chief tank-killing ordnance is the gun, the most powerful (in terms of muzzle horsepower) ever mounted in an aircraft, firing milk-bottle-size rounds at rates hydraulically controlled at 2,100 or 4,200 shots/min. The gun is mounted 2° nose-down and offset to the left so that the firing barrel is always on the centreline (the nose landing gear being offset to the right).

The basic aircraft has a HUD (head-up display), good communications fit and both Tacan and inertial navigation. RHAWS and ECM have been internal from the beginning of the programme, but

Above: Manoeuvrability is the key to the A-10's survival: stores are Pave Penny and ALQ-119 pods plus four Mavericks.

General Dynamics F-16 Fighting Falcon

Origin: USA.
Type: (A,C) Multirole fighter, (B,D) operational fighter/trainer.
Engine: (A, B, C, D) One 23,840lb (10,814kg) thrust Pratt & Whitney F100-200 afterburning turbofan, (/79) one 18,000lb (8,165kg) General Electric J79-119 afterburning turbojet, (F) F100-220 engine, same rating as -200, (G) General Electric F110-100 afterburning turbofan "in 30,000lb class".
Dimensions: Span 31ft 0in (9·449m) (32ft 10in/10·1m over missile fins), length (both versions, excl probe) 47ft 7in (14·52m); wing area 300·0sq ft (27·87m²).
Weights: Empty (A) 15,137lb (6,866kg), (B) 15,778lb (7,157kg); loaded (AAMs only) (A) 23,357lb (10,594kg), (B) 22,814lb (10,348kg), (max external load) (both) 35,400lb (16,057kg), (Block 25 on) 37,500lb (17,010kg).
Performance: Max speed (both, AAMs only) 1,350mph (2,173km/h, Mach 2,05) at 40,000ft (12·19km); max at SL, 915mph (1,472km/h, Mach 1·2), initial climb (AAMs only) 50,000ft (15·24km)/min; service ceiling, over 50,000ft (15·24km); tactical radius (A, six Mk 82, internal fuel, Hi-Lo-Hi) 340 miles (547km); ferry range 2,415 miles (3,890km).
Armament: One M61A-1 20mm

gun with 500/515 rounds, centreline pylon for 250gal (1,136lit) drop tank or 2,200lb (998kg) bomb, inboard wing pylons for 4,500lb (2,041kg) each, middle wing pylons for 3,500lb (1,587kg) each, outer wing pylons for 700lb (318kg) each (being uprated under MSIP-1 to 3,500lb), wingtip pylons for 425lb (193kg), all ratings being at 9g. Normal max load 11,950lb (5,420kg) for 9g, 20, 450lb (9,276kg) at reduced load factor.

Development: Starting as a small technology demonstrator, the F-16 swiftly matured into a brilliantly capable multirole fighter, which, in the eyes of many observers, is No. 1 in the Western world. Basic features include a fixed wing tapered on the leading edge, with automatic variable camber from hinged leading and trailing edges, a slab horizontal tail, a single large engine of the type already used in the F-15, fed by a plain ventral inlet without any variable geometry, a modern cockpit with a reclining seat, sidestick force-transducer controller linked to FBW flight controls and an overall

Right: The F-16's single F100 has plenty of poke, as demonstrated by this pair of F-16As in a perfect formation loop.

concept of relaxed static stability which even today represents an exceptional application of the CCV concept.

At the time of its design the F-16's ability to sustain 9g in prolonged turns was unique, and turns at 5·5g can be made with a theoretical external stores load of 20,450lb (9,276kg), roughly the same as the original clean gross weight!

It was in January 1975 that the F-16 was selected as a major type for the USAF inventory, suddenly transforming it from a mere demo programme (generally regarded as having no place in the inventory of a service which regarded the F-15 as sacrosanct) into a 1,388-aircraft programme for worldwide deployment. This total included 204 F-16B two-seaters, with full avionics and weapons for 17 per cent less fuel.

In June 1975 the F-15 was selected for Belgium, Denmark, Netherlands and Norway, mainly to replace the F-104. These European countries insisted on substantial industrial offsets, and with remarkable speed a multinational manufacturing programme was set up to build production aircraft. All major aircraft, engine, avionics and accessory firms in the four European countries participate and there are assembly lines in Belgium (SABCA/SONACA) and the Netherlands, (Fokker) as well as at Fort Worth. Though this has put up the costs, and still not achieved the potential output of which Forth Worth alone would be capable if working at maximum rate, the multinational programme has worked quite well and by end-1984 had delivered 1,500 aircraft to nine countries.

By far the largest user is the USAF. This chose the 388th TFW at Hill AFB, Utah, as lead operator, and Hill has also served as a principal logistics centre and training base for the international F-16 programme. Other F-16 units in the USAF include the 56th and 58th TTWs, the 8th, 50th, 363rd and 474th TFWs, the Thunderbirds display team, the 169th TFG of the ANG and the 466th TFS of the AFRes.

The FAB (Belgium) has 116 aircraft, with a follow-on order for 44, with two squadrons each at Beauvechain and Kleine Brogel. Egypt has 40 equipping 23 Fighter Brigade at Inchas, with plans for two similar brigades to be based at Al Mansurah and Abu Hammad (aircraft not yet under contract in mid-1984). Israel has 75 F-16s which on 7 June 1981 made a combat mission from Etzion AB with eight aircraft, each with two 2,000lb (907kg) bombs, which flew a very long dogleg mission at very low level to destroy the Iraqi nuclear reactor at Osirak with pinpoint accuracy. President Reagan immediately embargoed a second batch of 75 aircraft, but these have now been delivered.

South Korea will receive 36 aircraft from 1986. The Netherlands has an initial batch of 102, equipping two squadrons at Leeuwarden, two at Volkel, and a third (No 306) at Volkel tasked in the recon role carrying a multisensor pod on the centreline. This completed conversion from the F-104G, and further batches have since been ordered of 22, 18, 12 and finally 57, to re-equip NF-5 units, the final delivery being in early 1992. Norway bought 72 F-16s; the US has offered to sell Norway 24 co-produced aircraft at $460m as attrition reserve. Pakistan is receiving 40, due to be completed in January 1986, and has been negotiating for a follow-on of 60 more.

Singapore is the only country to accept the downgraded F-16/79, with the old-technology J79 engine; the price of $280m for the first eight must cover considerable support and training costs. Thailand has been assigned one squadron initially, with more to follow, but the regular aircraft rather than the F-16/79. Turkey finally picked the F-16 as the basis not only for a revitalized air force with new (not secondhand) aircraft but also for the long-planned TUSAS, the national aircraft industry. At a programme cost of $4b, $25m per aircraft, 160 F-16s are to be acquired, the first 40 from GD and the rest assembled by TUSAS with growing local content. All will be F-16Cs and Ds as described later. Venezuela's FAV received 18As and six Bs between September 1983 and late 1985.

In 1980 the USAF launched the MSIP (multistaged improvement programme), to extend multirole all-weather capability of the basic aircraft. Since late 1981 all aircraft have wiring and avionic system architecture for later updating with LAN-TIRN night attack pods, one on each side of the inlet duct, the ASPJ (airborne self-protection jammer) EW system and the AIM-120A Amraam AAM. The latter rectifies the curious lack of a radar-guided AAM which cost the F-16 several export sales to the F-18 Hornet. The upgraded aircraft, now operational with the USAF, is the F-16C, the F-16D being the two-seat version.

General Dynamics built an F-16/J79 demonstrator which the Department of Defence is offering to nations supposedly not qualified for the regular aircraft. GD also flew an F-16 with the F101DFE engine, as a result of which the F110 engine is now in production to power future F-16s, the initial 1984 buy being 120 engines. The resulting more powerful (and in other ways superior) aircraft may be designated F-16F, because F-16E was applied to the proposed production model of the redesigned F-16XL, two prototypes of which (one with F110 engine) flew in 1982. These aircraft have a giant cranked-arrow wing of 663sq ft (61·59m^2) area, with no hoizontal tail, and the fuselage is lengthened to 54ft 1·86in (16·51m).

The XI was a contender in the USAF Enhanced Tactical Fighter competition, and showed its ability to carry 29 weapons on 17 stations and not only beat the regular F-16 in field length but also carry double the weapon load 45 per cent further. Though not accepted, the F-16E was judged so outstanding that development is continuing for USAF use in the 1990 period.

GD also build a research

Top: Formation of 388th TFW F-16As on a training mission over a Utah bombing range.

Above: An F-16B releases a Maverick ASM in a medium-altitude diving attack.

Below: Even with free-fall iron bombs the F-16 offers exceptional accuracy: during trials, the diameter of the weapon impact area has been no more than one third of the specified figure.

F-16/AFTI (Advanced Fighter Technology Integration) with full CCV capability and large canted canard controls. Together with modified flight controls these enable the AFTI to manoeuvre instantly in any direction, without rotation of the fuselage axis. Features include a wide-field Marconi HUD, foot pedals for pointing the aircraft in any direction and a throttle to control direct lift, pitch and vertical translation. Phase I testing was completed in July 1983, after which combat tactics and weapon delivery have been explored progressively.

General Dynamics F-111

Origin: USA.

Type: A, C, D, E, F, all-weather attack; FB, strategic attack.

Engines: Two Pratt & Whitney TF30 afterburning turbofans, as follows, (A/C) 18,500lb (8,390kg) TF30-3, (D, E) 19,600lb (8,891kg) TF30-9, (FB) 20,350lb (9,231kg) TF30-7, (F) 25,100lb (11,385kg) TF30-100.

Dimensions: Span (fully spread) A, D, E, F) 63ft 0in (19·2m), (C, FB) 70ft 0in (21·34m), (fully swept) (A, D, E, F) 31ft 11¹/₂in (9·74m) (C, FB) 33ft 11in (10·34m); length 73ft 6in (22·4m), wing area (A, D, E, F, gross, 16°) 525sq ft (48·77m²).

Weights: Empty (A) 46,172lb (20,943kg), (C) 47,300lb (21,,455kg), (D) 49,090lb (22,267kg), (E) about 47,000lb (21,319kg), (F) 47,481lb (21,537kg), (FB) close to 50,000lb (22,680kg); loaded (A) 91,500lb (41,500kg), (D, E) 92,500lb (41,954kg), (C, F) 100,000lb (45,360kg), (FB) 114,300lb (51,846kg).

Performance: Max speed at 36,000ft (11,000m), clean and with max afterburner, (A, C, D, E) Mach 2·2, 1,450mph (2,335km/h), (FB) Mach 2, 1,320mph (2,124km/h), (F) Mach 2·5, 1,653mph (2,660km/h); 1,160mph (1,865km/h); cruising speed penetration, 571mph (919km/h); service ceiling at combat weight, max afterburner, (A) 51,000ft

(15,500m), (F) 60,000ft (18,290m); range with max internal fuel (A, D) 3,165 miles (5,093km), (F) 2,925 miles (4,707km); takeoff run (A) 4,000ft (1,219m), (F) under 3,000ft (914m), (FB) 4,700ft (1,433m).

Armament: Internal weapon bay for two B43 bombs or (D, F) one B43 and one M61 gun; three pylons under each wing (four inboard swivelling with wing, outers being fixed and usable only at 16°, otherwise being jettisoned) for max external load 31,500lb (14,288kg), (FB only) provision for up to six SRAM, two internal.

Development: In its early years the F-111, originally known as the TFX (tactical fighter experimental), suffered from the most severe problems ever to afflict a basically good aircraft. These problems were political, financial, technical (mainly associated with structure and propulsion) and concerned with flight performance, so that the popular image of the F-111 was that of a scandalous failure. This has thankfully receded, leaving the USAF and RAAF with an outstanding long-range aircraft which is not a fighter and should never have been called a fighter but instead has for 17 years been the world's best all-weather long-range attack aircraft.

Basic features of the F-111 include a variable sweep "swing wing" (the first in production in the world) with limits of 16° and 72·5°, with exceptional high-lift devices, side-by-side seating for the pilot and right-seat navigator (usually also a pilot) or (EO) electronic-warfare officer, large main gears with low-pressure tyres for no-flare landings on soft strips (these prevent the carriage of ordnance on fuselage pylons), a small internal weapon bay, very great internal fuel capacity (typically 5,022 US gal, 19,010 litres), and emergency escape by jettisoning the entire crew compartment, which has its own parachutes and can serve as a survival shelter or boat.

General Dynamics cleared the original aircraft for service in 2¹/₂ years, and built 141 of this F-111A

Above: Pave Tack pod and Paveway laser-guided bombs on a 48th TFW F-111F.

version, which equips 366TFW at Mountain Home AFB, Idaho (others have been reserved for conversion into the EF-111A). It is planned to update the A by fitting a digital computer to the original analog-type AJQ-20A nav/bomb system, together with the Air Force standard INS and a new control/display set. The F-111E was similar but had larger inlet ducts and engines of slightly greater power; 94 were

Below: An unusual near-vertical view of two 474th TFW F-111As operating out of Nellis over the dry watercourses of Nevada.

delivered and survivors equip the 20th TFW at Upper Heyford, England. These are to receive the same updates as the A.

Next came the F-111D, which at great cost was fitted with an almost completely different avionic system of a basically digital nature including the APQ-30 attack radar, APN-189 doppler and HUDs for both crewmembers. This aircraft had great potential but caused severe technical and manpower problems in service and never fully realized its capabilities, though it remains a major advance on the A and E. The 96 built have always equipped the 27th TFW at Cannon AFB, New Mexico.

The F-111F is by far the best of all tactical F-111 versions, almost entirely because Pratt & Whitney at last produced a really powerful TF30 which incorporated many other advanced features giving enhanced life with fewer problems. With much greater performance than any other model the F could if necessary double in an air-control role though it has no air-to-air weapons except the gun and if necessary AIM-9. The 106 of this model served at Mountain Home until transfer to the 48th TFW in England, at Lakenheath.

The most important of all F-111 post-delivery modifications has been the conversion of the F force to use the Pave Tack pod, normally stowed in the weapon bay but rotated out on a cradle for use. This complex-package provides a day/night all-weather capability to acquire, track, designate and hit surface targets using EO, IR or laser guided weapons. The first squadron to convert was the 48th TFW's 494th TFS in September 1981. Their operations officer, Maj Bob Rudiger, has said: "Important targets that once required several aircraft can now be disabled with a single Pave Tack aircraft; the radar tells the pod where to look, and the laser allows us to put the weapon precisely on target."

The long-span FB-111A was bought to replace the B-58 and early models of B-52 in SAC, though the rising price resulted in a cut in procurement from 210 to 76, entering service in October 1969. It has so-called MK IIB avionics, derived from those of the D but configured for SAC missions using nuclear bombs or SRAMs. With strengthened structure and landing gear the FB has a capability of carrying 41,250lb (18,711kg) of bombs, made up of 50 bombs of 825lb (nominal 750lb size) each. This is not normally used, and the outer pylons associated with this load are not normally installed. The FB equips SAC's 380th BW at Plattsburgh AFB, NY, and the 509th at Pease, New Hampshire.

The RAAF purchased 24 F-111Cs in 1963, and after prolonged technical problems finally received them in 1973. These have the long-span wing and strengthened landing gear of the FB, but are basically otherwise F-111As with the original engine and inlet duct. Back in 1968 General Dynamics developed and flew a comprehensively equipped RF-111A reconnaissance version for the USAF, but this was not adopted. The RAAF, however, decided to have its reconnaissance pallet developed by GD, and in 1978 an F-111C was returned to Forth Worth where the pallet was installed, with cameras, IR linescan, TV, optical sights and sensor control and displays in the right-hand cockpit. This aircraft and three more converted in Australia serve with RAAF No 6 Sqn at Amberley, the other F-111 unit being No 1 Sqn at the same base. Four F-111Cs lost have

Above: An F-111D releases 12 Mk 82 1,000lb (454kg) bombs over Nevada in July 1972.

been replaced by F-111As bought secondhand. The RAAF has evaluated various sensors and missiles for use in the maritime and anti-ship role.

In October 1986 GD will begin updating the avionics of all 381 aircraft in the USAF inventory, under a basic S1.1 billion six-year contract. GE will provide a new attack radars, TI new TFRs (terrain-following radas) and other suppliers new (but mainly off-the-shelf) navigation, communications, IFF and EW subsystems.

Grumman A-6 Intruder and EA-6 Prowler

Origin: USA.

Type: (A-6A, B, C, E) two-seat carrier-based all-weather attack; (EA-6A) two-seat ECM/attack; (EA-6B) four-seat ECM; (KA-6D) two-seat air-refuelling tanker.

Engines: (Except EA-6B) two 9,300lb (4,218kg) thrust Pratt & Whitney J52-8A turbojets; (EA-6B) two 11,200lb (5,080kg) J52-408.

Dimensions: Span 53ft (16·15m); length (except EA-6B) 54ft 7in (16·64m); (EA-6B) 59ft 5in (18·11m); height (A-6A, A-6C, KA-6D) 15ft 7in (4·75m); (A-6E, EA-6A and B) 16ft 3in (4·95m); wing area 528·9sq ft (49·1m^2).

Weights: Empty (A-6A), 25,684lb (11,650kg); (EA-6A) 27,769lb (12,596kg); (EA-6B) 34,5871lb (15,686kg); (A-6E) 25,630lb (11,625kg); max loaded (A-6A and E) 60,400lb (27,397kg); (EA-6B) 58,500lb (26,535kg).

Performance; Max speed (clean A-6A) 685mph (1,102km/h) at sea level or 625mph (1,006km/h, Mach 0·94) at height, (EA-6A) over 630mph, (EA-6B) 599mph (964km/h) at sea level, (A-6E) 648mph (1,043km/h) at sea level; initial climb (A-6E, clean) 8,600ft (2,621m)/min; service ceiling (A-6A) 41,660ft (12,700m), (A-6E) 44,600ft (13,595m), (EA-6B) 39,000ft (11,582m); range with full combat load (A-6E) 1,077 miles (1,733km); ferry range with external fuel (all) about 3,100 miles (4,890km).

Armament: All attack versions, including EA-6A, five stores locations each rated at 3,600lb (1,633kg) with max total load of 15,000lb (6,804kg); typical load thirty 500lb (227kg) bombs; (EA-6B, KA-6D) none.

Development: Despite its seemingly outdated concept, the A-6 Intruder will remain in low-rate production throughout the foreseeable future as the standard equipment of all the heavy attack squadrons of the US Navy and Marine Corps. The design was formulated during the later part of the Korean war, in 1953, when the need for a truly all-weather attack aircraft was first recognized. After much refinement of the requirement an industry competition was held in 1957, Grumman's G-128 design being chosen late in that year. The prototype was notable for its downward-tilting engine jetpipes, intended to give STOL performance, but these were not featured in the production machine.

Basic characteristics of all aircraft of the family include a conventional long-span wing with almost full-span flaps on both the leading and trailing edges. Ahead of the trailing-edge flaps are "flaperons" used as lift spoilers and ailerons, while the tips contain split airbrakes which are

Above: An EA-6B Prowler from CVW-17 carrying fuel tanks and three self-powered pods housing two jamming transmitters each.

Below: An A-6E of CVW-6 with drop tanks, 12 general-purpose bombs and TRAM turret just visible under the nose.

fully opened on each carrier landing. Plain turbojets were used, and these have remained in all successive production versions. The nose is occupied by a giant radar array, with a fixed inflight-refuelling probe above on the centreline in front of the side-by-side cockpit with Martin-Baker seats (slightly inclined and staggered) which can be tilted back to reduce fatique.

Like all Grumman products the A-6 soon gained a reputation for unbreakable strength as well as crisp handling and (except for the avionics) reliability. The internal fuel load of 15,939lb (7,230kg), with the option of 8,020lb (3,638kg) in four drop tanks, has proved adequate to counteract the basically poor fuel economy of the engines.

Gruman delivered 482 of the original A-6A model, ending in December 1969, and 62 of these were converted into KA-6D air-refuelling tankers which can transfer over 21,000lb (9,526kg) of fuel through its hosereel. This remains the standard tanker of the 14 carrier air wings, with limited attack capability and equipment for use as an air/sea rescue control platform. The A-6A, B and C are no longer in use, the standard attack model being the A-6E. This has totally new radar, the Norden APQ-148 replacing two radars in earlier versions as well as an IBM/Fairchild computer-based attack and weapon-delivery system.

In 1974 an A-6E was fitted with the TRAM (target recognition and attack multisensor) package, comprising a stabilized chin turret containing a FLIR and a laser interlinked with the radar for detection, identification and weapon-guidance at greater ranges in adverse conditions. Other updates with TRAM include the Litton ASN-92 CAINS (carrier aircraft inertial navigation system), a new CNI suite and automatic carrier landing.

It was planned to deploy 318 A-6Es, but production has been continued with new airframes being produced at six a year until 1987. In that year it is planned to switch to the A-6F, whose specification was

still being finalized in 1984. Meanwhile a major update programme has fitted TRAM to all the 250 A-6Es in frontline carrier air wings, to be completed in early 1985, and to the similar aircraft equipping the five all-weather VMA(AW) attack squadrons of the Marines. Since 1981 new and updated A-6Es have been equipped to launch Harpoon.

The EA-6B Prowler is the standard electronic warfare platform of the Navy carrier air wings and Marine Corps. Though it is based on the airframe of the A-6E, with local reinforcement to cater for the increased weights, fatigue life and 5·5g load factor, it is gutted of attack avionics and instead houses the AIL ALQ-99 tactical jamming system, which covers all anticipated hostile emitter frequency bands. Surveillance receivers are grouped in the fairing on the fin, and the active jammers are mounted in up to five external pods, each energized by a nose windmill generator and containing two (fore/aft) transmitters covering one of seven selected frequency bands.

To manage the equipment a crew of four is carried, comprising pilot (left front), an ECM officer in the right front seat to manage navigation, communications, defensive ECM and chaff/flare dispensing, and two more ECMOs in the rear seats who can each detect, assign, adjust and monitor the jammers.

All VAQ (fixed-wing EW) squadrons of the Navy fly the EA-6B, as do the three Marine EW squadrons. Production at six per year is continuing to 1990. All in service have been updated with new avionics in two stages of ICAP (increased capability), and Norden is supplying about 100 advanced APS-130 navigation radars for retrofit to existing Prowlers.

During the rest of the decade Grumman has to define a further update stage for the EA-6B, with most of the A-6E upgrade features.

Below: An A-6E TRAM of test and evaluation squadron VX-5 armed with four Harpoon anti-ship missiles.

Grumman F-14 Tomcat

Origin: USA.
Type: Two-seat carrier-based multirole fighter.
Engines: (F-14A) two 20,900lb (9,480kg) thrust Pratt & Whitney TF30-412A afterburning turbofans; (C) two 20,900lb (9,480kg) thrust Pratt & Whitney TF30-414A afterburning turbofans; (D) two 29,000lb (13,154kg) thrust General Electric F110-401 afterburning turbofans.
Dimensions: Span (68° sweep) 38ft 2in (11·63m), (20° sweep) 64ft 1·5in (19·54m); length 62ft 8in (19·1m); height 16ft (4·88m); wing area (spread) 575sq ft (52·49m²).
Weights: Empty 37,500lb (17,010kg); loaded (fighter mission) 55,000lb (24,948kg), (max) 72,000lb (32,658kg).
Performance: Max speed 1,564mph (2,157km/h, Mach 2·34) at height, 910mph (1,470km/h, Mach 1·2) at sea level; initial climb at normal gross weight over 30,000ft (9,144m)/min; service ceiling over 56,000ft (17·070m); range (fighter with external fuel) about 2,000 miles (3,200km).
Armament: One 20mm M61A-1 multi-barrel cannon in fuselage; four AIM-7 Sparrow and four or eight AIM-9 Sidewinder air-to-air missiles, or up to six AIM-54 Phoenix and two AIM-9; maximum external weapon load in surface attack role 14,500lb (6,577kg).

Development: Designed and developed with outstanding speed and skill to fill the gap left by the cancelled F-111B, the Grumman G-303 won an industry competition for the VFX (experimental fighter) in January 1969. An aircraft of extraordinary refinement, complexity and capability, it has had a long career with virtually no major modification, and will remain in production until the 1990s with updates to the avio-

nics and a new engine. Throughout its career the engine has been the only continual source of worry over what is otherwise an aircraft without direct rivals outside the Soviet Union. In particular, the range and multiple target capability of the radar and AAMs carried puts it in a class of its own. It also has great potential in the surface attack role, though this has never been put to use.

Features include pilot and naval flight officer in unstaggered tandem Martin-Baker GRU-7A zero/zero seats, a retractable refuelling probe on the right of the nose, widely spaced engines fed by fully variable long inlet ducts, swing wings with widely separated pivots at the extremities of an almost flat aircraft upper surface of enormous area, main gears retracting forwards with the wheels rotating to lie in the fixed wing gloves, full-span powered leading-edge slats, trailing-edge flaps and roll-control spoilers (the latter augmented by the slab stabiliators or horizontal tails), slightly canted twin vertical tails, and retractable canards called glove vanes which with increasing wing sweep pivot outwards from the centre of the leading edge of each glove.

Avionics are exceptional, and stem in part from the Bendix stand-off AAM system planned for the defunct F6D Missileer of 1960. The Hughes AWG-9 main radar has a 36in (0·914m) planar aerial and operates in many modes usually in pulse/doppler to give clear indication of airborne targets out to 195 miles (315km). It can track 24 targets while simultaneously engaging any selected six and guiding AAMs on to those targets.

The F-14 has always been the only aircraft in service able to fire the AIM-54 Phoenix missile, also a Hughes product, which has repeatedly demonstrated its ability to kill at ranges in excess of 125 miles (201km). Four of these large AAMs can be carried on special under-fuselage pallets which serve as aerodynamic fairings; two more

Below: Full fleet defence load of six Phoenix long range missiles aboard an F-14 Tomcat of VF-32 Swordsmen.

can be mounted on pylons 1B and 8B under the wing gloves, which are hung well outboard on sloping struts to clear the landing gears. These pylons normally have two launch adapters, both able to carry an AIM-9 Sidewinder and the main shoe able to take an AIM-7 Sparrow as alternative to an AIM-54. Specially configured drop tanks, of 222gal (1,101lit) capacity, can be hung under the engine ducts. The gun is mounted at the bottom of the left side of the fuselage, fed by a horizontal drum with 675 rounds.

By mid-1984 Grumman had delivered 480 F-14s, almost all of them F-14As. These have worked their way through various improved TF30 engines finishing with the Dash-414A with new first-stage fan blades and a surrounding containment ring to prevent catastrophic damage from broken blades. The nozzle has also been improved, with 18 sliding petals giving smooth variation of nozzle profile over a ratio of areas greater than 2:1.

One Tomcat (F-14A No. 157986) was flown as the prototype F-14B in September 1973 with F401 engines giving more thrust for less weight and reduced fuel burn, but the expected production (from aircraft No 68) was cancelled. Between July and September 1981 the same aircraft was tested with GE F101DFE engines, the results being so good that in FY87 (fiscal year 1987) the production line will switch to the F110-powered F-14D. Before that, the F-14C designation will apply to existing aircraft updated with new digital avionics, new cockpit displays, improved radar, ASPJ, JTIDS and the AIM-120A missile. The F-14D is to fly in 1988, and by mid-1984 GE was planning for 700 F110-401 engines for 300 aircraft.

Of the 500 aircraft built by the time this book appears, 80 were supplied to Iran. Here their advanced nature has probably kept them out of the war with Iraq (although

there have been confirming reports of Tomcats operating together with F-4 Phantoms, the F-14s "drawing-in" Iraqi fighters like decoys, and the F-4s taking over for the kill). Support from the USA has been withheld, and Grumman's opinion is that Iran cannot use its F-14 missiles without support.

The rest of the F-14s serve with US Navy VFs 1, 2, 11, 14, 21, 24, 31, 32, 33, 41, 51, 74, 84, 101, 102, 103, 111, 114, 124, 142, 143, 154, 211 and 213. Two aircraft from VF-41, saw brief combat over the Gulf of Sidra on 19 August 1981 when they were attacked by two Libyan Su-22s. Both the latter were destroyed, by a Sidewinder AAM apiece, and the engagement merely confirmed the superior tactics of the Navy crews.

In close combat the F-14 is one of

the few aircraft with an auto wingsweep programmer, to adjust wing angle continuously to the demands of Mach number and manoeuvres. F-14 missions are almost always governed by E-2C control aircraft, and comprise three basic CAP (combat air patrol tasks): Forcap (cover for the task force), Barcap (barrier against an oncoming air attack), and Tarcap (target cover for friendly attack aircraft in hostile airspace).

Most F-14As have a chin fairing housing a FLIR, APN-154 radar beacon and ACLS (auto carrier landing system) aerial, but since 1982 Northrop has been supplying the TCS (TV Camera Set) as a replacement. This has a large forward-facing lens through which the crew can see a wide-angle TV picture of

Above: Four of VF-32's Tomcats in formation with wings at maximum 68° sweep angle.

Right: The former F-14B prototype in "Super Tomcat" F101DFE-powered configuration.

targets for acquisition, followed by a narrow magnified image for identification, at beyond visual distance. Another optional store is TARPS (tactical air reconnaissance pod system), normally carried on the right rear body pylon. This has a forward looking KS-87B frame camera, KA-99 low altitude lateral oblique panoramic camera and AAD-5 IR linescan. So far 49 aircraft carry TARPS, and these serve as the Navy's interim reconnaissance aircraft.

HAL Ajeet

Origin: India.
Type: Air-combat fighter and light attack.
Engine: HAL-built Rolls-Royce Orpheus turbojet, 4,670lb (2,118kg) thrust Mk 701E.
Dimensions: Span 22ft 1in (6·73m); length 29ft 8in (9·04m), height overall 8ft 10in (2·69m), wing area 136·12sq (12·646m^2).
Weights: Empty 4,850lb (2,200kg), max (clean) 6,650lb (3,016kg), (external stores) 8,885lb (4,030kg).
Performance: Max speed (SL) 714mph (1,150km/h), initial climb 20,000ft (6,100m)/min; service ceiling 52,000ft (15,850m), range (max fuel with external tanks) 1,180 miles (1,900km).
Armament: Two 30mm Aden Mk 4 guns each with 115 rounds; four wing pylons for all usual gun pods, rocket launchers and other loads, a typical combination with the trainer being two 55lb (250kg) bombs inboard and two 30gal (136·5lit) tanks outboard.

Development: Unlike Britain, India recognized the cost/effectiveness of Teddy Potter's miniature fighter, the Folland Gnat, and in various wars found that its air combat performance fully lived up to expectations. A licence agreement was signed in 1956, as a result of which Folland supplied 25 complete Gnats, 25 sets of parts and then (as Hawker Siddeley) provided support while HAL built 213 Gnats by fitting integral-tank wings accommodating the same quantity of fuel as previously carried in drop tanks, thus freeing the pylons for weapons. The resulting Ajeet ("Unconquerable") also has the Martin-Baker seat), (instead of the small Folland seat), improved tail control and avionics, and 80 were delivered from 1976.

Right: One of the 80 Ajeets built by HAL to a design based on that of the original Gnat. These equip six IAF squadrons, but are due for replacement by the Soviet MiG-23 and MiG-29.

IAI Kfir and Nesher (Dagger)

Origin: Israel.
Type: Multirole fighter and attack, (TC-2) trainer and EW.
Engine: (except Nesher): One 17,900lb (8,119kg) General Electric J79-J1E afterburning turbojet.
Dimensions: (Nesher) as Mirage CJ, (Kfir) span 26ft 11·5in (8·22m); length (C1, C2) 51ft 4·4in (15·65m), (C2 with Elta radar) 53ft 11·6in) (16·45m); (TC2) 54ft 1·12in (16·49m); height 14ft 11in (4·55m); wing area 374·6sq ft (34·8m²); foreplane 17·87sq ft (1·66m²).
Weights: (Kfir) Empty (C2, interceptor) 16,060lb (7,285kg); loaded (C2, half internal fuel plus two Shafrir) 20,700lb (9,390kg), (C2, max with full internal fuel, two tanks, seven 500lb/227kg bombs and two Shafrir 32,340lb (14,670kg).
Performance: (C2) Max speed (clean) 863mph (1,389km/h) at sea level, over Mach 2·3 (1,516mph, 2,440km/h) above 36,090ft (11,000m); initial climb

45,950ft/min (233m/s); service ceiling 58,000ft (17,680m); combat radius (20min reserve), (interceptor, two 110gal, 500lit tanks plus two Shafrir) 215 miles (346km), (attack, three 330gal, 1,500lit tanks plus seven bombs and two Shafrir, hi-lo-hi) 477 miles (768km).
Armament: Two IAI-built DEFA 5-52 guns each with 140 rounds; seven hardpoints for total of 9,469lb (4,295kg) of various stores, always including two Shafrir 2 AAMs (outer wings) plus tanks, bombs (ten 500lb, 227kg), Gabriel III, Maverick or Hobos missiles, rocket pods, Matra Durandal or other anti-runway weapons, napalm, ECM pods and tanks.

Development: It is often thought that development of an Israeli version of the Dassault Mirage, was triggered by Gen de Gaulle's curt refusal to permit the first batch of Mirage 5s — developed specifically

Above: IAI has delivered both TC2 and TC7 long-nose two-seat versions of the Kfir for pilot and weapons training and EW.

for Israel and paid for largely in advance — to leave France in 1973, as a result of the Yom Kippur war (in which, incidentally, Israel was the innocent victim). In fact, IAI had begun as early as 1969 to study the

prospects both for making the Mirage itself and for developing an improved model powered by the J79 engine.

The first programme results in the IAI Nesher (Eagle), first flown in September 1969. This was essentially the same as the Mirage 5, powered by an Atar 9C, but equipped with Israeli avionics. Substantial numbers were built, but after clearance for service of the later Kfir these were replaced in Heyl Ha'Avir service. All have been sold to Argentina, in batches both before and after the Falklands war in 1982. In 1984 survivors were being equipped with flight-refuelling probes, with IAI engineering support.

The Kfir (Lion Cub) is a largely redesigned aircraft whose main difference was at first the J79 engine. Among 270 engineering changes, compared with the Nesher, are a modified inlet system, enlarged ducts, a new engine bay of reduced length, a new dorsal fin with ram inlet, four further engine-bay cooling inlets, completely revised cockpit, new nose with basically triangular section, new fuel system with greater capacity than the Mirage 5, strengthened longer-stroke main landing gears and totally new avionics of mainly Israeli origin.

The Kfir-C1 entered production in 1974 as a multirole air superiority and ground-attack fighter. The first two aircraft were delivered in June 1975. Further development led to the Kfir-C2, with major aerodynamic and avionics improvements. Adding small strakes along the sides of the nose and removable (but fixed-incidence) foreplanes above the inlets dramatically improved low-speed and combat-manoeuvre capability, which was further enhanced by extending the outer wing leading edge with a sharp dogtooth.

The C2 has shorter takeoff and landing, steeper landing approach in a flatter attitude and markedly better combat agility, including reduced gust reponse in low-level attack. The C2 was revealed at Hatzerim in July 1976, when it was announced as available for export at some S5m.

The standard C2 has an extended nose housing the Elta EL/M-2001B ranging radar. Standard C2 equipment includes inertial navigation, very comprehensive flight-control and weapon-delivery systems and a high standard of EW/ECM installations. Further subsystems are in the rear cockpit of the two-seat TC2 version, first flown in 1981, with a distinctive down-sloping extended nose. The two-seater has an almost identical flight performance to the single-seat Kfir, and as well as

conversion training the two-pilot version is used as a dedicated EW aircraft.

In 1983 production switched to the Kfir-C7, with uprated engine, increased internal fuel, a new Hotas cockpit, increased gross weight and a major improvement in avionics. The latter include a digital computer for stores management, the WDNS 341 navigation and weapon-aiming system and, as an option, the big M 2021 pulse-doppler radar in place of the little ranging set.

By mid-1984 IAI had delivered about 250 Kfirs, and was also offering update kits for export users of the regular Mirage (Colombia having bought these as well as the Kfir-C2 itself). The US State Department, which had earlier vetoed various proposed Kfir exports, was reported to have approved sales to Austria, Mexico and Taiwan. More interestingly, the US Navy has borrowed 12 of the early Kfir-C1 (non-canard) fighters from Israel, with contract support by IAI, to try them out as "hostile" opponents at NAS Miramar, as more powerful replacements for the Northrop F-5E in the Top Gun fighter pilot school. A leasing deal was announced in September 1984.

Below: Heyl Ha'Avir armourers hustle to load LAU-68B rocket launchers and Shafrir 2 AAMs on a Kfir-C2 during a late-1970s exercise. It is believed that C2s have been upgraded to C7 standard, with uprated engines and Hotas cockpits.

Above: Externally difficult to distinguish from a Kfir-C2, the C7, shown here with bomb triplets, tanks and Sidewinders, has a J79-IAI-J1E uprated by Bedek Division's "Combat Plus" system to 18,750lb (8,500kg).

Below: Markedly different from those of earlier Mirage IIIs, the Kfir-C2's cockpit has one programmable display, an Electro-Optics HUD and Hotas (hands on throttle and stick) arrangement of switches.

Lockheed F-104 Starfighter

Origin: USA.
Type: (G) Multimission strike fighter, (CF) strike-reconnaissance; (TF) dual trainer; (QF) drone RPV (F-104S) all-weather interceptor; (RF and RTF) reconnaissance.
Engine: One General Electric J79 turbojet with afterburner; (G, RF/RTF, CF) 15,800lb (7,167kg) J79-11A; (S) 17,900lb (8,120kg) J79-19 or J1Q.
Dimensions: Span (without tip tanks) 21ft 11in (6·68m); length 54ft 9in (16·69m); height 13ft 6in (4·11m); wing area 1961 sq ft (18·22m²).
Weights: Empty 14,082lb (6,387kg), (F-104S, 14,900lb, 6,760kg); max 28,779lb (13,054kg), (F-104S, 31,000lb (14,060kg).
Performance: Max speed 1,450mph (2,334km/h, Mach 2·2); initial climb 50,000ft (15,250m)/min; service ceiling 58,000ft (17,680m) (zoom ceiling over 90,000ft, 27,400m); range with max weapons about 300 miles (483km); range with four drop tanks (high altitude, subsonic) 1,815 miles (2,920km).
Armament: In most versions, centreline rack rated at 2,000lb (907kg) and two underwing pylons each rated at 1,000lb (454kg); additional racks for small missiles (eg Sidewinder) on fuselage, under wings or on tips; certain versions have reduced fuel and one 20mm M61 Vulcan multi-barrel gun in fuselage; (S) M61 gun, two Sparrow or Aspide and two Sidewinder.

Development: Originally developed as a pure air-combat fighter for USAF Tactical Air Command, the F-104 proved to have all the performance needed, but to be disappointing in some aspects of agility, range and weapons capability. The programme was rescued by Lockheed's company-funded development of the F-104G, with a revised and strengthened airframe and totally new avionic fit with the first INS ever to enter service in a fighter, Nasarr multimode radar, manoeuvring flaps and many other new items. This was selected as its chief new combat type by the Federal German Luftwaffe and Marineflieger, mainly because (though the F-104G had not been built) it promised to have unrivalled low-level penetrability either with a nuclear store or in a tactical reconnaissance role.

A total of 1,266 were built in many sub-types including two-seaters used for training and EW missions. Most have been replaced in their original air forces, the main recipient of retired examples being Turkey.

Turkey had already purchased 40 new F-104S from Italy, this being the final variant developed by Lockheed and Aeritalia as an all-weather interceptor, with R21G radar and primarily air-to-air weapons including radar-guided Sparrow or Aspide AAMs. The Italian AF bought 205 of this type which equip three fighter gruppi operating under Nadge control. They have proved effective, and have not suffered anything like the severe attrition which gave the F-104G a bad name in the early 1960s (the many well-publicized losses proved it is not a type for inexperienced pilots).

Some F-104Gs and CF-104s have run out of hours, but the CAF 1st Air Group in Europe are managing to

Above: A formation of Italian Aeronautica Militare F-104S Starfighters: Italy is the chief user of this, the most important remaining variant.

keep up to strength while waiting for CF-18s. Italy, instead of passing on F-104Gs replaced by Tornados, has used them to form a new 18° Gruppo Difesa Aerea which became operational in mid-1984 at Trapani-Birgi in Sicily (where no fighters had been based since 1945).

McDonnell Douglas A-4 Skyhawk

Origin: USA.
Type: Single-seat attack bomber; OA, two-seat FAC; TA, dual-control trainer.
Engine: (E,J) One 8,500lb (3,856kg) Pratt & Whitney J52-6 turbojet; (F, G, H, K) 9,300lb (4,218kg) J52-8A; (M, N, Y) 11,200lb (5,080kg) J52-408A.
Dimensions: Span 27·5ft (8·38m); length (E, F, G, H, K, L, P, Q, S) 40·1ft (12·22m); (M, N, Y) 40·3ft (12·27m), (OA, and TA, excluding probe) 42·6ft (12·98m); height 15ft (4·57m), (TA series 15·25ft); wing area 250sq ft (24·17m²).
Weights: Empty (E) 9,284lb, (typical single-seat, eg Y) 10,465lb (4,747kg), (TA-4F) 10,602lb (4,809kg); max (shipboard) 24,500lb (11,113kg); (land-based) 27,420lb (12,437kg).
Performance: Max speed (clean) (E) 685mph, (Y) 670mph (1,078km/h), (TA-4F) 675mph; max speed (4,000lb/1,814kg bomb load) (Y) 645mph; initial climb (Y) 8,440ft (2,572m)/min; service ceiling (all, clean) about 49,000ft (14,935m); range (clean, or with 4,000lb weapons and max fuel, all late versions) about 920 miles (1,480km); maximum range (Y) 2,055 miles (3,307km).
Armament: Standard on most versions, two 20mm Mk 12 cannon, each with 200 rounds; (H, N, and optional on other export versions) two 30mm DEFA 553, each with 150 rounds. Pylons under fuselage and wings for total ordnance load of (E, F, G, H, K, L, P, Q, S) 8,200lb (3,720kg); (M, N, Y) 9,155lb (4,153kg).

Development: The ultimate expression of designer Ed Heinemann's art in meeting a challenging requirement (for a carrier-based attack aircraft) with an aircraft which weighted just half the suggested figure and flew 100 knots (185km/h) faster, the A-4 remained in production 26 years, from 1954 to 1979 inclusive. The final total was 2,980, and thanks to continual updating many are in service today, and third-world airpowers are taking careful note of each used Skyhawk that comes on to the market.

The US Marine Corps currently operates a large number of A-4s in various roles. The single-seat combat aircraft is the A-4Y, some built new and others reworked A-4Ms with Marconi HUD, an advanced ARBS similar to that of the Harrier II and enhanced EW systems. The OA-4M tandem-seat FAC platform is a rebuild of the TA-4F trainer, and the latter together with the simple and agile TA-4J continue as advanced pilot trainers, especially for teaching air combat tactics.

Singapore's fleet includes unique TA-4S trainers with separate canopies over the staggered tandem cockpits. Biggest of the export customers was Israel, which purchased 279 single-seaters of various types and 27 two-seaters. Today Israel's IAI is on to good business with its offer of retrofitting A-4s for export customers.

The IAI-modified A-4s have extended nose for various new avionic options, extended jetpipe to reduce IR signature, underwing spoilers, steerable nosewheel, dual disc mainwheel brakes, braking parachute, 30mm guns, two extra wing pylons, updated EW and chaff/flare dispensers, a new weapon-delivery avionic system, and complete structural life-extension and rewiring. Skyhawks featured prominently in the 1982 South Atlantic War, when Argentine air force and navy pilots flew them with great verve against the British forces. Argentina continues to be a good Israeli customer, and it was reported in 1984 that IAI had begun to supply three Gabriel III/AS missiles for each A-4.

Below: One of 40 A-4L Skyhawks (Cs with F avionics) which Grumman is refurbishing for the Royal Malaysian Air Force.

McDonnell Douglas F-4 Phantom II

Origin: USA.

Type: All-weather multirole fighter for ship or land operation; (F-4G) EW defence suppression; (RF) all-weather multisensor reconnaissance.

Engines: (C, D, RF) two 17,000lb (7,711kg) General Electric J79-15 turbojets with afterburner; (E, F, G) 17,900lb (8,120kg) J79-17; (J, N, S) 17,900lb J79-10; (K, M) 20,515lb (9,305kg) Rolls-Royce Spey 202/203 augmented turbofans.

Dimensions: Span 38·4ft (11·7m); length (C, D, J, N, S) 58·2ft (17·76m), (E, G, F and all RF versions) 63ft (19·2m), (K, M) 57·6ft (17·55m); height (all) 16·2ft (4·96m); wing area 530sq ft (49·2m²).

Weights: Empty (C, D, J, N) 28,000lb (12,700kg), (E, F and RF) 29,000lb (13,150kg), (G, K, M) 31,000lb (14,060kg); max (C, D, J, K, M, N, RF) 58,000lb (26,308kg), (E, G, F) 60,630lb (27,502kg).

Performance: Max speed with Sparrow missiles only (low) 910mph (1,464km/h), Mach 1·19 with J79 engines, 920mph with Spey, (high) 1,500mph (2,414km/h), Mach, 2,27 with J79, 1,386mph with Spey; initial climb, typically 28,000ft (8,534m)/min with J79, 32,000ft/min with Spey; service ceiling over 60,000ft (19,685m) with J79, 60,000ft with Spey; range on internal fuel (no weapons) about 1,750 miles (2,817km); ferry range with external fuel, typically 2,300 miles (3,700km), (E and variants), 2,600 miles (4,184km).

Armament: (all versions except RF models which have no armament) Four AIM-7 Sparrow or Sky Flash (later AMRAAM) air-to-air missiles recessed under fuselage; inner wing pylons can carry two more AIM-7 or four AIM-9 Sidewinder missiles; in addition E versions except RF have internal 20mm M61 multi-barrel gun, and virtually all versions can carry the same gun in external centreline pod; all except RF have centreline and four wing pylons for tanks, bombs or other stores to total weight of 16,000lb (7,257kg).

Development: By far the most important fighter in the non-Communist world during the past 20 years, the F-4 has an evergreen quality of sheer capability that from time to time is recognized. Egypt, for example, was about to sell its 34 F-4Es to Turkey, but then realized this would leave them with no stand-off kill capability, using radar-guided AAMs (a strange shortcoming of the otherwise much superior F-16) and the sale did not go through.

Even today the F-4 combines modest cost, maturity, toughness, long range, heavy weapon loads of almost every type of store in the inventory, and a flight performance that is still impressive, in a way that no other aircraft can do on either side of the Iron Curtain.

There were 5,195 F-4s constructed, of all models — including aircraft assembled in Japan and airframes built but not delivered to Iran — far surpassing all other Western fighters since 1960.

The F-4A and F-4B for the US Navy introduced blown flaps and dropped leading edges, a broad fuselage with four radar-guided Sparrow AAMs recessed into the underside, wing and centreline pylons for tanks, Sidewinders, bombs or other stores, tremendous internal fuel capacity, tandem seats for pilot and RIO (radar intercept officer) and a big and powerful Westinghouse nose radar. This sub-family was continued via the F-4J of 1965 to today's F-4N and F-4S rebuilds with more fuel, revised structures and avionics, slatted wings and tailplanes, and many other updates, which continue in service with the Navy and Marines.

The original USAF version was a minimum-change variant, the F-4C of 1963, with minor changes to wheels and brakes, cockpits (usually configured for two pilots) and with an FR receptacle instead of a retractable probe. This proved so satisfactory that the Air Force was allowed to have its own F-4D, with ground attack avionics. This was followed during the Vietnam War by the F-4E with uprated engines, an extra rear-fuselage tank, a new and smaller radar, an M61 gun recess under the nose and, in the course of the production run, powerful leading-edge slats instead of the blown droops, to improve the previously very poor agility when heavily laden with weapons near the ground.

The F-4E became the definitive fighter version, bought by several export customers, though West Germany's Luftwaffe chose a simpler F-4F model without provision for Sparrow medium-range AAMs or various EW subsystems. Mitsubishi assembled 138 F-4E(J) Phantoms in Japan, with increasing local content, the only Phantoms not built at St Louis. In 1964-5 Britain cancelled its own fighters and bought the F-4K (Phantom FG.1) for the Navy and F-4M (FGR.2) for the RAF. Both were largely redesigned with Spey engines, and this proved to be a very expensive mistake

Above: One of at least 180 F-4E Phantoms equipped to carry the centreline Pave Tack pod.

which at high altitude and high speed actually degraded the performance by a considerable margin. In normal low-level use the big fan engines do a good job, and other features include special carrier provisions on the FG.1 (which have all been passed to the airfield-based RAF) and on the FGR.2 a Ferranti INS, AWG/11/12 radar fire control, strike camera in one AAM recess and fin-cap RWR.

Tornadoes are steadily replacing RAF Phantoms, but in the air-defence role Nos 111 and 43 at Leuchars (43 with the FG.1), 29 at Coningsby and 56 at Wattisham are still operational, with 29 divided between Coningsby and RAF Stanley in the Falklands, the latter force being redesignated as 23 Sqn.

To make up the loss of the Falklands aircraft 15 ex-US F-4J fighters have been purchased at the high price of ⅔125m and, after complete refurbishment at NAS North Island (San Diego), have formed 74 Sqn at Wattisham. It had been expected these aircraft would be designated Phantom F.3, but this has yet to be announed. Utterly unlike any other RAF aircraft or engines, these Phantoms may be maintained under civilian contract.

The US Navy did not need a dedicated reconnaissance Phantom in the early 1960s, but the USAF quickly contracted for the RF-4C,

Below: A 3rd TFW F-4E displays the uniquely angular — and widely derided — form of the versatile Phantom.

which set a new standard in fighter-type aircraft used for multisensor reconnaissance without armament. It introduced a new nose, longer and slimmer, with no main radar but a small APQ-99 forward-looking radar and a main bay occupied by forward oblique, vertical and lateral oblique cameras. In the fuselage is an APQ-102 SLAR and an ASS-18A IR linescan. Various new communications include HF radio with a shunt aerial flush with the skin on each side of the fin. Ahead of the fin are two photoflash installations firing the cartridges upwards. The Marines bought RF-4Bs with similar equipment but also including cameras on rotating mountings, aimed by the pilot, and an inertial navigation system. Newest and best of the recon variants is the RF-4, all built for export, based on the F-4E.

Latest version is the F-4G Avanced Wild Weasal EW platform, whose mission is to sense, locate and destroy hostile ground air-defence emitters. Used only by the USAF, this is a rebuild of late-model F-4E fighters with the APR-38 EW system whose 52 special aerials (antennae) include large pods facing forwards under the nose and to the rear above the rudder. The system is governed by a Texas Instruments computer with reprogrammable software to keep up to date on all known hostile emitters.

This Phantom carries such weapons as triplets of the AGM-65 EO-guided Maverick precision attack weapon, Shrike ARM (anti-radar missile) and HARM (high-speed ARM). Like almost all Phantoms the left front fuselage recess often carries an ECM jammer pod (usually an ALQ-119), leaving the other three available for Sparrow AAMs if necessary; or Sidewinders can be carried under the wings.

In 1985 Germany's F-4Fs and Japan's F-4E(J)s are subjects of major update programmes with lookdown pulse-doppler radars. Meanwhile, Boeing and Pratt & Whitney have had a positive response to their joint proposal to re-engine as many F-4s as possible with the PW1120 engine (as used in the IAI Lavi). This would improve the Phantom, but obviously might damage market prospects for follow-on American fighters, so McDonnell appear to be cool to the idea. Instead McDonnell is pushing a detailed update of the avionics, equipment, windscreen and fuel system, leaving the engines alone.

McDonnell Douglas F-15 Eagle

Origin: USA.
Type: Air superiority fighter with attack capability, (E) dual-role fighter/attack.
Engines: Two 23,930lb (10,855kg) thrust Pratt & Whitney F100-100 afterburning turbofans, (E) two F100-200, same rating.
Dimensions: Span 42·8ft (13·05m); length 63·75ft (19·43m); height overall 18·46ft (5·63m); wing area 608sq ft (56·5m²).
Weights: Empty (A) 27,381lb (12,420kg); takeoff (intercept mission, A) 42,206lb (19,145kg); max (A) 56,000lb (25,401kg), (C, FAST packs) 68,000lb (30,845kg), (E) 81,000lb (36,742kg).
Performance: Max speed (clean, over 45,000ft/13,716m) 1,650mph (2,655km/h, Mach 2·5), (clean, SL) 912mph (1,468km/h, Mach 1·2); combat ceiling (A, clean) 63,000ft (19,200m); time to 50,000ft (15,240m) (intercept configuration) 2·5min; mission radius, no data; ferry range (C) over 3,450 miles (5,560km).
Armament: One 20mm M61A1 gun with 940 rounds; four AIM-7 Sparrow AAMs or eight AIM-120A (Amraam), Plus four AIM-9 Sidewinders; three attack weapon stations (five with FAST packs) for external load of up to 16,000lb (7,258kg), or (E) 24,500lb (11,113kg).

Development: Most observers in the Western World regard the F-15 as the natural successor to its ancestor, the F-4, as the best fighter in the world. To a considerable degree its qualities rest on the giant fixed-geometry wing, F100 engine and Hughes APG-63 pulse-doppler radar.

Inevitably the F-15 emerged as a large aircraft. Two of the extremely powerful engines were needed to achieve the desired ratio of thrust/weight, which near sea level in the clean condition exceeds unity. The lower edge of the fuselage is tailored to snug fitting of four medium-range AAMs. The gun is in the bulged strake at the root of the right wing, drawing ammunition from a tank inboard of the duct. There is no fuel between the engines but abundant room in the integral-tank inner wing and between the ducts for 11,600lb (5,260kg,

1,448gal, 6,592lit), and three 500gal (2,270lit) drop tanks can be carried, each stressed to 5g manoeuvres when full. Roll is by ailerons only at low speeds, the dogtoothed slab tailplanes taking over entirely at over Mach 1, together with the vertical twin rudders.

Avionics and flight/weapon control systems are typical of the 1970 period, with a flat-plate scanner pulse-doppler radar, vertical situation display presenting ADI (attitude/director indicator), radar and EO information in one picture, a HUD, INS and central digital computer. In its integral ECM/IFF subsystems the F-15 was far better than most Western fighters, with Loral radar warning (with front/rear aerials on the left fin tip), Northrop ALQ-135 internal counter measures system, Magnavoc EW warning set and Hazeltine APX-76 IFF with Litton reply-evaluator. High-power jammers, however, must still be hung externally, Westinghouse pods normally occupying an outer wing pylon. The APG-63 offered excellent capability to track low-level targets with cockpit switches giving a Hotas (Hands on throttle and stick) capability which dramatically improved dogfight performance.

The original plan was to procure 729 F-15s, but this number has risen to 1,488, of which well over 840 had been delivered to the USAF and other customers by early 1985. Current production is centred on the F-15C and two-seat F-15D, which provide substantial updates in mission capability.

Other air forces which have ordered the F-15 are: Israel (40); Saudi Arabia (62); and Japan (88 F-15J, 12 F-150J). All are being produced in the USA, except for 86 of the JASDF order, which are being built under licence by Mitsubishi.

A programmable signal processor gives the ability to switch from one locked-on target to another, to switch between air and ground targets and keep searching whilst already locked-on to one or more targets. Increase in memory capacity from 24 to 96K gives a new high-

Below: A formation of F-15As from Alaskan Air Command's Elmendorf-based 21st TFW.

resolution radar mode which can pick one target from a large group at extreme range. Internal fuel is increased by 2,000lb (907kg) and conformal pallets, called FAST (Fuel And Sensor, Tactical) fit snugly on each side of the fuselage to increase total fuel by 9,750lb (4,422kg).

In 1983 an improved so-called tangential carriage arrangement was introduced which enables 12 bombs of 1,000lb (454kg), or four of twice this size, to be hung on short stub pylons along the lower edges of the FAST packs, giving reduced drag and leaving existing pylons free (thus an F-15C can carry 12 bombs, four AAMs, two IR sensor pods and still carry three tanks).

In the late 1970s the USAF began studying an ETF (enhanced tactical fighter), configured equally for surface attack and air superiority roles. McDonnell modified an early F-15B as the Enhanced Eagle and after pro-

Below: Development aircraft for the F-15E dual-role fighter, 71-0291 with 22 Rockeye bombs.

Above: Two F-15A Eagles from the 1st TFW, based at TAC's Langley AFB, Virginia, HQ.

longed tests this carried the day over the rival F-16XL. 372 of the resulting F-15Es are to be procured between 1985 and the mid-1990s. Completing these aircraft to F-15E standard will add an estimated [1,500m to the bill, but the result will be an aircraft of awesome capability. It will stick with the F100 engine, but in an improved Dash-200 form.

A tandem-seater, the F-15E will have the new APG-70 high-resolution radar, with DBS (doppler beam sharpening), totally new computer and programmable armament control system, wide-field HUD, internal ASPJ ECM, LANTIRN all-weather nav/targeting pod, and multifunction displays in the rear cockpit for managing the complete mission. Remarkably few structural changes apart from the landing gears are needed to handle the gross weight, which is almost double that of a B-17 Flying Fortress.

McDonnell Douglas F-18 Hornet

Origin: USA.
Type: Originally carrier-based strike fighter; most now land-based dual-role fighter and attack, with reconnaissance capability (RF-18).
Engines: Two General Electric F404-400 afterburning turbofans, each "in 16,000lb (7,258kg) thrust class"
Dimensions: Span (with missiles) 40·4ft (12·31m), (without missiles) 37·5ft (11·42m); length 56ft (17·07m); height 15·3ft (4·66m); wing area 400sq ft (36·16m²).
Weights: (provisional) Empty 20,583lb (9,336kg); loaded (clean) 33,642lb (15,260kg); loaded (attack mission) 49,200lb (22,317kg); max loaded (catapult limit) 50,064lb (22,710kg).
Performance: Max speed (clean, at altitude) 1,190mph (1,915km/h, Mach 1·8), (max weight, sea level) subsonic; sustained combat manoeuvre ceiling, over 49,000ft (14,935m); combat radius (air-to-air mission, high, no external fuel) 461 miles (741km); ferry range, more than 2,300 miles (3,700km).
Armament: One 20mm M61 gun with 570 rounds in upper part of the forward fuselage; nine external weapon stations for max load (catapult launch) of 13,400lb (6,080kg) or (land takeoff) of 17,000lb (7,711kg), including bombs, sensor pods, missiles (including Sparrow) and other stores, with tip-mounted Sidewinders.

Development: It is remarkable that this twin-engined machine designed for carrier operation should have won three major export sales almost entirely because of its ability to kill hostile aircraft at stand-off distance using radar-guided AAMs (which with a few trivial changes could be done by its losing rival, the F-16). From the outset in 1974 the Hornet was designed to be equally good at both fighter and attack roles, replacing the F-4 in the first and the A-7 in the second. It was also hoped that the Hornet — strictly the F-18 but usually designated F/A-18 by the Navy to emphasize its dual role — would prove a "cheaper alternative" to the F-14. Predictably the long and not wholly successful development process has resulted in an aircraft priced at well over [20 million; indeed the initial Spanish contract for 72 aircraft is priced at [3,000 million, though this includes spares and training. Spain has 12 more on option.

Where the Hornet is unquestionably superior is in the engineering of the aircraft itself, which is probably the best yet achieved in any production combat type, and in particular in the detail design for easy routine maintenance and sustained reliability. Though not a large aircraft, with dimensions between those of the compact Tornado and the F-4, and significantly smaller than the F-15, the F-18 combines the Tornado's advantage of small afterburning engines and large internal fuel capacity with avionics and weapons configured from the start

for both F and A missions. Of course, in the low-level attack role it cannot equal Tornado because it has a large wide-span fixed wing, giving severe gust response, and suffers from relatively low maximum speed and lack of terrain-following radar; but in the typical Navy/Marines scenario with a mission mainly over the sea and a dive on target these shortcomings are less important.

In weapon carriage the Hornet is first-class, with plenty of pylons and payload capability, and clearance for a wide spectrum of stores. In the fighter mission it is excellent, now that the wing has been redesigned to meet the specified rates of roll; and unlike its most immediate rival, the F-16, it has from the start carried a high-power liquid-cooled radar, the Hughes APG-65, matched with radar-guided AAMs.

It would be easy to criticize the armament as the only old part of a new aircraft, but in fact the M61 gun is still hard to beat, and the AIM-7 Sparrow and -9 Sidewinder AAMs have been so updated over the years that both remain competitive. The fundamental shortcoming of the Sparrow in requiring continuous illumination of the target — which means the fighter must keep flying towards the enemy long after it has fired its missile — is a drawback to all Western fighters, and will remain so until AIM-120 (AMRAAM) is in service in 1986. In most air-combat situations the Hornet can hold its own, and compared with previous offensive (attack) aircraft is in a different class; it is in the long-range interdiction role that Hornet has significant shortcomings.

These centre chiefly on radius with a given weapon load, though it has been pointed out this can be rectified to some extent by using larger external tanks and by air refuelling, the aircraft being equipped with a British-style retractable probe, permanently installed in the upper right side of the nose. Forward vision is good, though unlike the F-16 there is a transverse frame, and the cockpit will always be a major "plus" for this aircraft, with Hotas controls (the stick being conventional instead of a sidestick), upfront CNI controls and three excellent MFDs (multifunction displays) which replace virtually all the traditional instruments. This cockpit goes further than anything previously achieved in enabling one man to handle the whole of a defensive or an offensive mission. But a second crewmember would ease the workload, especially in hostile airspace, and there have been studies of a two-seat version since the start of the programme, though escalating costs have made its go-ahead increasingly unlikely.

There is, of course, a two-seat dual-pilot version for conversion training; this retains weapons capability and the APG-65 radar, but has about 6 per cent less internal fuel. There is also a prototype of a dedicated reconnaisssance RF-18, testing of which began in 1984. This has a new nose, with the gun and

ammunition replaced by a recon package which would normally include optical cameras and/or AAD-5 IR linescan. It is stated that this model could "overnight" be converted into the fighter/attack configuration.

The first Navy/Marines training squadron, VFA-125, commissioned at NAS Lemoore in November 1980. Three Marine Corps squadrons, VMFA-314, -323 and -531, were equipped by mid-1984, with three others following, while Navy squadrons are now also converting at lower priority. Until 1984 the Hornet was cleared operationally only for the fighter mission, because the Marines needed to replace the F-4 in this role more urgently. The attack mission depends to some degree on adding the laser spot tracker and FLIR.

In the summer of 1984 inspection of aircraft based at the Naval Air Test Centre at Patuxent River showed evidence of a structural problem — cracking of the tips of the fins and the fin/fuselage attachment. Investigation showed that this was due to the effect of side forces on the fins in conjunction with vortices. As an interim measure, angle of attack was limited to 25° below 30,000ft when the aircraft was flying between 30-400kts. Modification kits were quickly devised and installed in the field, while modified fins entered production later that year.

The CF-18 differs in small items such as having a spotlight for visual identification of aircraft at night.

Top: Medium-altitude bomb release by a Hornet of VFA-113 Stingers over the Leach Lake range in California.

Above: F-18 with Sidewinders, four Mk 84 bombs, fuel tanks and FLIR and LST/SCAM pods.

Above right: The distinctive wing planform and leading edge extensions of the F-18 Hornet.

Canada has only a small manufacturing offset (Canadair makes nose sections), despite the size of the order; 138, including 24 TF-18s. Deliveries began in October 1982, the first units to convert being 410, followed by 409, 416 and 425 Sqns previously equipped with the CF-101. Deliveries have also begun to 439 in the Canadian Air Group at Baden-Söllingen, followed by the group's other squadrons, 421 and 441.

Australia's buy of 57, plus 18 TFs, has triggered a vast and complex involvement of local industry. The first two TFs were delivered from St Louis soon after completion in October 1984. The rest will all be assembled by Government Aircraft Factories, with major Australian content. GAF expect to assemle seven TFs in 1985, with single-seaters following in 1986, to re-equip Nos 3, 75 and 77 Sqns. Australian offsets are 25 per cent, but Spain's total $1,800 million (60 per cent), includes local manufacture of various movable surfaces, panels and pylons.

McDonnell Douglas/BAe
AV-8B Harrier II

Origin: UK/USA.
Type: Multirole close-support attack fighter and (RAF) reconnaissance.
Engine: One 21,700lb (9,843kg) thrust Rolls-Royce F402-406 Pegasus 105 vectored-thrust turbofan.
Dimensions: Span 30·35ft (9·25m); length 46·32ft (14·12m); height 11·7ft (3·56m); wing area 230sq ft (21·37m²).
Weights: Empty 12,750lb (5,783kg); max (VTO) 19,185lb (8,720kg), (STO) 29,750lb (1,349kg).
Performance: Max speed (clean, SL) 668mph (1,075km/h); dive Mach limit 0·93; combat radius (STO, seven Mk 82 bombs plus tanks, lo profile, no loiter) 748 miles (1,204km); ferry range 2,879 miles (4,633km).
Armament: Seven external pylons, centreline rated at 1,000lb (454kg) inboard wing 2,000lb (907kg), centre wing 1,000lb (454kg) and outboard 630lb (286kg), for total external load of 7,000lb (3,175kg) for vertical takeoff, or 17,000lb (7,711kg) for short takeoff, (GR. 5) two additional AAM wing pylons; in addition ventral gun pods for (US) one 25mm GAU-12/U gun and 300 rounds or (RAF) two 25mm Aden each with 130 rounds.

Development: Though developed directly from the original British Aerospace Harrier, the Harrier II is a totally new aircraft showing quite remarkable improvement and refinement in almost every part. This is especially the case in radius of action with any given weapon load, but it also extends to the scope and variety of possible loads, and to the general comfort and pleasure of flying.

The original Harrier required a lot of attention, especially during accelerating or decelerating transitions, and suffered from poor allround view and a distinctly traditional cockpit, whereas the Harrier II offers a completely new experience which makes full use of experience gained with the F-15 and F-18. At the same time, apart from the wing, which is a completely new long-span structure made almost entirely from graphite composites, the new aircraft is a joint effort with inputs from both partners.

The wing is the most obvious visible difference, compared with earlier Harriers. Apart from giving extra drag, it houses much more fuel, so that total internal fuel capacity is 50 per cent greater. With eight sinewave spars and composite construction it is virtually unbreakable, with essentially limitless fatigue life, and the curved LERX (leading-edge

root extensions) developed by BAC greatly enhanced combat manoeuvrability and bring turn radius closer to what the RAF required (The only real fault of the Harrier II, in British eyes, is that it was planned as a "superior bomb truck" for the US Marine Corps, whereas the RAF were more interested in air-combat agility and speed.)

Under the wing are six stores pylons (eight in the RAF GR.5), four of them plumbed for tanks which in the Marine Corps AV-8B are normally of 250gal (1,136lit) size. The extra pylons on the GR.5 are in line with the outrigger landing gears and will normally carry AIM-9L Sidewinders. The under-fuselage gun pods are specially configured to serve as LIDS (lift-improvement devices) which, joined across the front by a retractable dam, provide a cushion of high-pressure air under the air-

Above: The USMC ordered the AV-8B as a V/STOL "bomb truck", to provide short-notice support for front-line positions.

craft which counters the suck-down effect of rising air columns around the fuselage. In the AV-8B a 25mm gun is housed in the left pod with its ammunition fed from the right pod; the GR.5 has two individual gun/ammunition pods, but the gun is the new 25mm calibre Aden.

In the matter of avionics and EW the Harrier II is dramatically updated, the basic kit including INS (Litton ASN-130A or, in the GR.5, a Ferranti set), digital air-data and weapons computers, large field of view HUD, fibre-optic data highways and comprehensive RWR and ECM systems. The primary weapon-delivery system is the Hughes ARBS (angle/rate bombing system), with dual-wavelength

TV/laser target acquisition and tracking. A multimode radar has been studied, and in October 1984 McDonnell Douglas was given a Marine Corps go-ahead on an all-weather version with radar among several other avionic improvements.

The GR.5 is to differ in many avionic items, though the RWR will have its forward-looking receivers in the wingtips as on the AV-8B, rather than in a fin-tip fairing as in early Harriers. The RAF may use a different chaff/flare payload dispenser, though it will be installed in the same bay behind the airbrake. The AV-8B is likely to carry the ALQ-164 jammer pod on the centreline pylon. Both variants will have a bolt-on FR probe pack, added above the left inlet duct, the probe being extended hydraulically when required.

All 60 RAF GR.5s are to be frontline aircraft for RAF Germany. They will have Martin-Baker Mk 12 seats, stronger leading edges, nose and windshield to meet a severe birdstrike requirement, and considerable internal mission equipment including Marconi/Northrop Zeus active ECM, and BAe Dynamics MIRLS (miniature IR linescan), both of which result in small fuselage blister fiarings. Another non-standard feature is the Ferranti moving-map display, though both cockpits have a single MFD (multi-function display) similar to those in the F-18. The RAF will have no difficulty converting pilots on the existing T.4s, but the Marines will buy a planned 28 two-seat TAV-8Bs, of a new design not yet funded but ex-

pected to fly in 1986.

The Marines expect to receive a total of 300 AV-8Bs, not including four FSD (full-scale development) machines or the two-seaters. Deliveries are now taking place and the first squadron should achieve IOC in

late 1985, by which time there will be 33 AV-8Bs at Cherry Point. The first GR.5 squadron should form in 1987, roughly in parallel with the Spanish Navy squadron which will equip the new Spanish V/STOL carrier, *Principe de Asuritias*.

Above: A plan view emphasises the more obvious new features of the AV-8B, especially the new carbon-fibre composite wing. The low visibility marking scheme highlights the APU inlet and exhaust apertures.

MiG-19, J-6 "Farmer"

Origin: Soviet Union.
Type: Single-seat fighter/attack/recon; (JJ) dual trainer.
Engines: Two Tumanskii R-9BF-811 turbojets each rated at 7,167lb (3,250kg) thrust with afterburner, (J, JJ) two WP-6, same engine.
Dimensions:. Span 30·18ft (9·2m); length (excl probe) 41·33ft (12·6m), (JJ) 44·1ft (13·4m); height overall 12·73ft (3·88m); wing area 269·1sq ft (25·0m²).
Weights: (J-6) Empty 12,700lb (5,760kg); loaded (two tanks, two AIM-9B) 19,764lb (8,965kg); max 22,046lb (10,000kg).
Performance: Max speed (hi, clean) 957mph (1,540km/h, Mach 1·45), (SL,clean) 832mph (1,340km/h, Mach 1·09); cruising speed 590mph (950km/h); takeoff/landing over 82ft (25m) with afterburner/drag chute about 2,200ft (670m); combat radius (hi, two tanks) 426 miles (685km); max range 1,366 miles (2,200km).
Armament: (most) Three 30mm NR-30 guns, each usually with 80 rounds; (JZ-6) wing root guns only; (JJ-6) fuselage gun only; (J-6B) no guns but equipped to launch and guide four AA-1 Alkali AAMs. In addition four wing pylons, the outers normally carrying tanks (167 or 251gal, 760 or 1,140lit) and the inners bombs of up to 500lb (227kg), rocket launchers or single rockets.

Development: First supersonic fighter in the world, the MiG-19 remains to this day a formidable foe in close combat, because of its good turn radius, fair thrust/weight ratio when lightly loaded, and devastating power of its guns (far greater than Western guns of 30mm calibre). Nevertheless, the age of this design makes it useful chiefly as an advanced weapons trainer, and in general it is restricted to simple visual missions in good weather.

The J-6B, a copy of the MiG-19PM, is a night and all-weather

fighter with very early radar and AAMs, used only by the People's Republic alongside the Chinese-developed J-6Xin with a small gunsight radar in a pointed radome centred in the nose inlet and armed with guns. Most J-6 and JJ-6 have a breaking parachute below the rudder, replacing the original ventral location and enabling it to be streamed on the approach.

The tandem-seat JJ-6 is a Chinese design, with the semi-automatic seats slightly staggered in height and enclosed by flat-topped canopies hinged to the right. The JJ is stressed to the same 8g factor as the single-seater, and removing the

wing guns enables internal fuel to be almost unchanged, though training sorties at low level are held to 45min. This model continues in production at two factories, as does the JZ-6 fighter-recon version with two cameras between the nose inlet ducts replacing the fuselage gun.

The MiG-19 is in service with some 23 countries including all Warsaw Pact nations, Afghanistan, Angola, Cuba, North Korea, Bangladesh, China, Egypt, Iran, Iraq, Kampuchea, Pakistan, Tanzania, Vietnam and Zimbabwe.

Below: A Pakistan Air Force Shenyang J-6 (MiG-19SF).

MiG-21 J-7 "Fishbed", "Mongol"

Origin: Soviet Union.
Type: (most) Fighter, (some) fighter/bomber or reconnaissance.
Engine: (21) one 11,243lb (5,100kg) Tumanskii R-11 afterburning turbojet, (21F) 12,677lb (5,750kg) R-11F, (21PF) 13,120lb (5,950kg) R-11F2 (21FL, PFS, PFM, US) 13,688lb (6,200kg) R-11-300, (PFMA, M, R) R-11F2S-300, same rating (MF, RF, SMT, UM, early 21bis) 14,550lb (6,600kg), R-13-300, (21bis) 16,535lb (7,500kg) R-25.
Dimensions: Span 23·46ft (7·15m); length (almost all versions, including instrumentation boom) 51·71ft (15·76m), (excluding boom and inlet centrebody) 44·16ft (13·46m); height overall (typical) 13·45ft (4·1m); wing area 247·57sq ft (23m²).
Weights: Empty (F) 12,440lb (5,634kg), (MF) about 12,300lb (5,580kg), (bis) 12,600lb (5,715kg); loaded (typical, half internal fuel and two K-13A) 15,000lb (6,800kg), (full internal fuel and four K-13A) 18,078lb (8,200kg); max (bis, two K-13A and three drop tanks) 20,725lb (9,400kg).
Performance: Max speed (typical of all, SL) 800mph (1,290km/h, Mach 1·05), (36,000ft/11,000m, clean) 1,385mph (2,230km/h, Mach 2·1); initial climb (F) about 30,000ft (9,144m)/min (bis) 58,000ft (17,680m)/min; service ceiling (bis, max) 59,055ft (18,000m); practical ceiling (all), rarely above 50,000ft (15,240m); range with internal fuel (F) 395 miles (635km), (bis) 683 miles (1,100km); max range with three tanks (bis 1,118km).
Armament: (21 and F) One (rarely two) NR-30 guns, two pylons for K-13A AAMs or UV-16-57 rockets, (FL and all subsequent) one GP-9 belly pack containing GSh-23 gun with 200 rounds and four wing pylons for K-13A, R-60 or (possibly) AS-7 missiles, or up to 3,307lb (1,500kg) other loads including bombs, rockets or two 108gal (490lit) drop tanks; (HAL-built) carries Magic AAM; (J-7) carries CAA-1 and -2 AAMs.

Development: The basic design was settled in 1955 after exhaustive aerodynamic research as a 57° tailed delta with large constant-chord flaps inboard of large ailerons, and with a fixed leading edge. Fully powered controls gave superb agility, and even the first versions (plain MiG-21) were judged adequate to handle Frontal Aviation air-combat duties, releasing the bigger Su-7 for use in the attack role. Like its partner, the small MiG suffered from high fuel burn and very limited internal capacity, and the need always to carry a 108gal (490lit) centreline tank virtually eliminated the possibility of flying attack missions.

In 1958 the American Sidewinder AAM was copied to provide the K-13 (AA-2 Atoll) AAM, small pylons for two of these being added to the first major production MiG-21, the 21F, which also introduced a broader fin, more powerful engine and the option of carrying three tanks or various light loads of bombs or rocket pods. Usually the left gun was omitted. Large numbers of these continue in use as advanced trainers, mainly for aerobatics, supersonic techniques and weapon delivery.

By 1960 production had switched to the PF with a further increase in power and a larger engine inlet duct made possible by a new forward fuselage of almost constant diameter, the resultant larger inlet (36in/910mm instead of 27in/690mm) providing room for a larger three-position centrebody housing the scanner of RIL ('Spin Scan') radar. The dorsal spine was enlarged to raise internal fuel to 616gal (2,800lit), the gun(s) were removed (simplifying the forward airbrakes), main tyres were increased in size (causing bulges above and below the wing root to accommodate them in an enlarged bay) to give good flotation on unpaved surfaces, the instrument boom was moved to the top of the nose, and avionics were changed and upgraded with repositioned aerials.

By 1961 PF and PFL variants had introduced a further broadened fin (twice the area of the original), a drag-chute compartment at the base of the rudder, small lips above and below the jet nozzle, and provision for ATO (assisted-takeoff) rockets. Most aircraft by late 1961 had seven fuel tanks for 627gal (2,850lit), and a conventional side-hinged canopy with fixed screen, and a proportion switched to blown flaps in an important modification called SPS which considerably reduced landing speed and the required field length (ATO rockets having similarly shortened the takeoff run). The SPS flaps became standard and by 1965 further sub-types (suffix PFS and PFM) introduced the F2S-300 series engine, a further increase in chord and the vertical tail, more powerful R2L radar and the new GP-9 belly gunpack incorporating the twin-barrel GSh-23.

All these are still in service in various air forces, some having local or subcontracted modifications. More important are the later models derived from the Ye-9 prototype (1965), such as the MiG-21PFMA fighter bomber with Jay Bird radar, four pylons and an enlarged dorsal spine giving a straight line from canopy to fin.

Indian-built aircraft use type numbers based on the Soviet family, the Ye-9 series giving rise to the HAL Type 96; earlier HAL versions were the Type 74 (21F), 76 (PF) and 77 (FL). Up to this point all models also had tandem dual trainer versions, though most MiG-21U trainers are based on early series such as the HAL Type 66-400 and 66-600, the latter with broad fin. So far as is known, all Chinese production, at Xian, has been of early 66-series 21Fs, though many have been updated with (mainly Western) avionics and equipment.

By 1968 basic Soviet production featured over 20 different fits of EW (electronic warfare) equipment, reconnaissance cameras and flash cartridges, some aircraft being dedicated 21R and RF reconnaissance versions with pods which include SLAR. The KM-1 rocket-assisted seat, angle-of-attack sensor on the left of the nose, very large fin-top dielectric (mainly for VHF/UHF communications aerials) and internal GP-9 pack with only the barrels projecting progressively became standard, and often were retro-fitted to aircraft in service. Maximum external load rose to 3,307lb (1,500kg).

In 1970 the 21MF entered service with the R-13 engine which offered more thrust for less weight, as well as suck-in auxiliary inlet doors under the leading edge with debris deflectors and a rear-view mirror above the windshield. An even larger dorsal fairing from canopy to drag-chute box characterized the SMT of about 1972, many of which have ECM pods on the wingtips. A fourth enlargement in the saddle fairing resulted in what is believed to be the current production family, called 21bis, produced in numerous sub-variants. The R-25 engine is considerably more powerful and economical, and the avionics have also been further augmented by Jay Bird radar which is believed to be compatible with the AA-8 AAM.

While HAL production of the 21bis continues at about 30 aircraft annually, probably to be replaced by the MiG-27 in late 1986, more than 1,600 MiG-21s of assorted vintages remain in front-line use in some 32 non-WP air forces. These represent a significant market for updating, and Ferranti, Smiths, Dunlop and at least 16 US companies have already begun to pick up business here. To take one example, Egypt's total of some 350 need a great deal of support before all can be restored to flight status. These now have the AIM-9P3 AAM.

Above: MiG-21bis of the Indian Air Force, manufactured under licence by HAL in India.

Below: The small but efficient Jugoslav Air Force has 110 MiG-21s of various sub-types.

MiG-23 and -27 "Flogger"

Origin: Soviet Union.
Type: (most) Multirole fighter and attack, (23BN and 27) dedicated close-support and attack, (U) operational trainer.
Engine: (most) One 27,560lb (12,500kg) thrust Tumanskii R-29 afterburning turbojet, (BN and -27) 25,353lb (11,500kg) Tumanskii R-29B, (S, SM and U) 22,485lb (10,200kg) Tumanskii R-27.
Dimensions: Span (72° sweep) 26·8ft (8·17m), (16°) 46·75ft (14·25m); length (most, excl probe) 55·1ft (16·8m), (BN, 27) 52·5ft (16·0m); height overall 14·33ft (4·35m); wing area (16°) 301·4sq ft (28·0m²).
Weights: Empty (MF) 24,250lb (11,000kg), (27) 23,787lb (10,790kg); loaded (clean) (MF) 34,390lb (15,600kg), (27) 34,170lb (15,500kg); max (BM, two tanks and six FAB-500) 44,312lb (20,100kg), (27 with 16 FAB-250) 43,210lb (19,600kg).
Performance: Max speed (hi, clean) (MF) 1,553mph (2,500km/h, Mach 2·35), (BN, 27) 1,124mph (1,800km/h, Mach 1·7); speed at SL (clean) (MF) 912mph (1,468km/h, Mach 1·2), (BN, 27) 836mph (1,345km/h, mach 1·1); service ceiling (MF) 61,000ft (18,600m), (27) 50,850ft (15,500m); takeoff to 50ft (15m) at clean gross (27) 2,625 (800m); combat radius (MF, fighter mission) 560-805 miles (900-1,300km), (27, four FAB-500 and AAMs, lo-lo-lo) 240 miles (390km); ferry range (27) 1,500 miles (2,500km).
Armament: (23MF interceptor) One 23mm GSh-23 gun with 200 rounds, two R-23R (AA-7 Apex) and four R-60 (AA-8 Aphid) AAMs; (BN and 27) various bombloads such as 16 FAB-250 (tandem pairs of glove pylons and tandem quads under fuselage) or (overload) six FAB-1000 or cluster dispensers, or (glove pylons) two GSh-23 pods with pivoted guns depressed down for attacking surface targets, or two AS-7 Kerry or AS-10 missiles.

Development: Built at a higher rate than any other combat aircraft throughout the past ten years, the MiG-23 and -27 together constitute the most important single type in the WP air forces, and are also being exported in substantial quantities.

The basic design has the same aerodynamics as the Su-24, but is smaller and half as powerful. The VG wing, which appears to have gone out of fashion in Western nations, gives great lift for takeoff and loiter with heavy loads of fuel and weapons, and in the MiG-23MF interceptor roughly doubles the patrol endurance to a maximum of almost 4h. With the wings at 72° supersonic drag is greatly reduced and the aircraft ideally configured for either an air-to-air interception with stand-off kill by missile or for a lo attack on a surface target.

All versions have more or less the same airframe, designed to a load factor of 8g and for operation from rough airstrips. The first production series was the MiG-23MF, in various sub-types called "Flogger-B" by NATO. This usually carries the J-band radar called "High Lark", with a large nose radome carrying a downward-sloping pitot boom at its tip; yaw and angle of attack sensors are on the forward fuselage together with doppler radar, IFF, ILS and, in a ventral fairing, an IR sensor. Standard Sirena 3 radar warning is fitted, together with comprehensive ECM threat-analysis and jamming (though for penetrating high-threat areas jammer/flare/chaff dispensers must occupy at least one pylon). The radar was likened by the USA to that of the F-4J and described as the first Soviet type to have a significant

Above: The massive under-carriage enables the MiG-23 to operate from rough airstrips.

capability against low-level targets; later the 23MF demonstrated an impressive ability to engage targets at far above its own altitude using the large R-23R missile.

The trainer version has a substantially redesigned forward fuselage with a slimmer nose housing R2L-series radar, two stepped cockpits with separate hinged canopies, a periscope for the instructor at the rear and a sloping sill extending to the front of the windscreen, and a larger dorsal spine fairing covering the much larger air-conditioning system. Internal fuel is naturally reduced, and the engine is the R-27. No M-series NATO name has been assigned.

In 1978 a variant of the MF, called "Flogger-G", visited Finland and

---

MiG-25 "Foxbat"

Origin: Soviet Union.
Type: High-altitude interceptor, (R) multisensor recon.
Engines: (most) Two 27,010lb (12,250kg) thrust Tumanskii R-31 afterburning turbojnets, (M) two 30,865lb (14,000kg) R-31F.
Dimensions: Span 45·75ft (13·95m), (25R) 44·0ft (13·40m); length (all known variants) (overall), 78·15ft (23·82m), (fuselage only) 63·65ft (19·40m); height 20·0ft (6·10m); wing area, gross, 611.7sq ft (56·83m²), (25R) 603 sq ft 56·0m²).
Weights: (typical) Empty equipped (25) just over 44,090lb (20,000kg), (25R) 43,200lb (19,600kg); max loaded (25) 79,800lb (36,200kg), (25R) 73,635lb (33,400kg).
Performance: Max speed (low level) about 650mph (1,050km/h, Mach 0·85), (36,000ft/11,000m and above, MiG-25 clean) 2,115mph (3,400km/h, Mach 3·2), (11,000m and above, 4 AAMs) 1,850mph (2,987km/h, Mach 2·8); max rate of climb 40,950ft (12,480m)/min; time to 36,090ft (11,000m) with sustained afterburner, 2·5min; service ceiling (25) 80,000ft (24,400m), (both 25R versions) 88,580ft (27,000m); combat radius (all) 900 miles (1,450km); takeoff run (25, max weight) 4,525ft (1,380m); landing (25) touchdown 168mph (270km/h), run 7,150ft (2,180m).
Armament: Four wing pylons each equipped to launch either version of AA-6 Acrid AAM (usually two radar, two IR); alternatively various combinations of AA-6, R-23R (AA-7) and R-60 (AA-8), one pylon carrying an R-23R with an R-60 on each side; (R) none seen; (U) usually none, but recent photographs show missile pylons.

Development: Fastest combat aircraft ever put into service, the MiG-25 was originally designed to intercept the USAF RS-70 Valkyrie (which was cancelled). Far from being cancelled when the threat disappeared, the MiG-25 went into production both as a long-range stand-off interceptor and as a multisensor strategic reconnaissance aircraft.

From the start it was clear that,

Above: The Tumanskii turbojets of this MiG-25 "Foxbat-E" are the most powerful of their type in service in any air force.

Below: The defection to Japan of Victor Belenko in this MiG-25A gave Western analysts their first chance to inspect the type.

France. At first these were regarded merely as simplified short-series aircraft to make goodwill foreign visits without compromising the security of the advanced avionics, but later this variant was recognized as a regular type in FA regiments. It has a smaller dorsal fin and has been seen with a new undernose sensor pod.

Dedicated attack members of the MiG-23 family are styled MiG-23BN and exist in many variants, some export models with simpler avionics. One, called "Flogger-H", is virtually a MiG-27 with the MiG-23 engine installation with fully variable inlets and nozzle.

In contrast, the basic MiG-27 is a dedicated subsonic low-level attack aircraft with a simplified propulsion system making no attempt to fly fast at high altitude; it also has a new nose with a down-sloping broad profile — resulting in the pilot nickname "Ducknose" — not only giving better forward view but also accommodating every desired avionic item for the surface attack mission, in place of a radar. Even the cockpit is repositioned at a higher level with a deeper hinged hood and windshield to give the best possible pilot view over the terrain ahead (the resulting extra drag is not important).

A further change is the fitting of tyres of greater size and reduced inflation pressure for operations from rough unpaved airstrips. The engine is basically an R-29 but has a smaller afterburner with simple nozzle matched to maximum thrust at takeoff and in low-level missions (the nozzle is noticeably shorter than that of the MiG-23 family).

The oblique forward "chisel" window covers a laser ranger and marked-target seeker. The small radome at the tip of the nose is for air/air ranging in conjunction with the gun. Under the nose is a doppler navigation radar and further aft on each side are small blisters over CW target-illuminating radars. Aerials on the nose include the matched trio for SRO-2M "Odd Rods" IFF and the forward-pointing "Swift Rod" ILS, matched by a similar probe aerial on the fin facing aft. Sirena 3 radar homing and warning uses aerials on the leading edges and tail in the usual way, while forward-pointing pods on the fixed wing glove leading edges are thought to be an ASM guidance transmitter (left) and an active ECM jammer transmitter (right), though both beliefs are as yet unconfirmed.

Certainly the MiG-27 is designed to carry all available tactical missiles, fuel/air explosives, cluster munitions and laser-guided "smart" weapons.

There are various sub-types, and since 1980 the main production version has had a kinked tailplane trailing edge, long leading-edge root extension strakes, a complete revision of the nose sensors with changes in external appearance, and no bullet avionics fairings on the wing gloves. Most MiG-27s have thick armour plates attached on the outside of the fuselage on each side of the cockpit.

As noted earlier, some of this family (the MiG-23BN series, Flogger-F and -H) have the propulsion system of the MiG-23 fighter. This mixed model equips Nos 10, 220 and 221 Sqns of the Indian AF, but the version expected to be licence-built by HAL is the MiG-27 Flogger-J, a pure attack variant with the root strakes. Despite this, it is reported that both the R-60 dogfight AAM and the large stand-off R-23R AAM are being supplied to India, though the latter is almost certainly incompatible with the attack MiG-27.

A carrier-based variant is expected to be the first aircraft carried aboard the large new fixed-wing carrier the Soviets are now building.

Below: MiG-23MF "Flogger-G" during a "showing-the-flag" visit to Sweden in August 1981.

The massive undercarriage would seem ideally suited to the high sink rate of carrier landings, but the presence of a folding ventral fin seems to rule out the fitting of the tail hook needed for deck landing.

Recent reports, however, suggest that the definitive carrier home fighter will be a navalised version of the Su-27 (Flanker).

Current users of the MiG-23 include all Warsaw Pact nations, plus Algeria, Cuba, Egypt, Ethiopia, Iraq, Libya, Sudan, Syria, Vietnam, and India (BN). The MiG-27 is operated only by the USSR and East Germany, but further exports are highly probable. Apparently there has been customer criticism that export models are not equipped to Soviet Air Force standards.

speed and agility being incompatible, the MiG-25 would be designed as a "straight line" aircraft. It takes time to work up to full speed, burning fuel at a prodigious rate, and once at Mach 2·8 (the limit with AAMs) the aircraft has a turn radius of many miles. At low levels all versions prior to the M are restricted to below Mach 1 and to below 5g.

The basic aircraft is in many respects unique. The wing is unswept, but tapered on the leading edge and set at 4° anhedral, and with thickness/chord ratio of well under 4 per cent. The slim oval fuselage merges into giant flanking air ducts with enormous sloping inlets with powered doors above and below, large bleed outlets, variable roof profile and variable transverse control shutters and high-rate water/alcohol sprays. About 3,850gal (17,500lit) of special T-6 fuel is housed in nine welded-steel tanks in the fuselage and a large integral tank forming an inner box of each wing. The basic airframe material is steel, with leading edges of titanium (except for the left inclined fin whose leading edge is a dielectric aerial).

The only movable surfaces comprise powered ailerons well inboard from the tips, powered slab tailplanes, twin powered rudders and plain flaps (apparently not blown). Twin ventral fins incorporate tail bumpers, with an airbrake between them, and twin braking parachutes can be streamed from the rear of the dorsal spine. The landing gears have high-pressure tyres, single on the forward-retracting main legs and twin on the steerable nose unit, and there has been no attempt to fit the MiG-25 to anything but a long paved runway.

The original radar, called Fox Fire by NATO, was a typical 1959 set, of great size and weight and using vacuum tubes. Its rated output of 600kW was used to burn through hostile jamming, but even the earliest MiG-25 interceptors were well equipped with EW systems including the Sirena 3 RWR with receivers in the right fin tip and in each wingtip anti-flutter rod, with comprehensive internal jammers and

Above: MiG-25M "Foxbat-E" looks similar to the earlier version but has uprated engines and IR seeker under the nose.

dispensers. CW illuminating transmitters occupy the front of each wingtip fairing in this version.

Usually one pair of radar missiles

and one pair of IR-homing missiles are carried, the AA-6 Acrid packing such a giant punch as to destroy the largest aircraft even with a near miss. Today later AAMs are carried (see data), but despite repeated rumours there is no internal gun even in the current interceptor, the MiG-25M (Foxbat-E).

This variant went into production in about 1977, a prototype (Ye-266M) having set various world records in 1975 including times to various heights; it still holds the absolute height record at 123,523ft (37,650m). Apart from the uprated engine the 25M has a completely new radar and many other avionics improvements, though full details are not yet known.

According to the US DoD many of these aircraft are converted MiG-25s, which originally had no lookdown or low-level capability. Though the airframe is still severely limited at low altitudes (according to one report, to Mach 0·8), the 25M is said to have low-level interception capability "somewhat comparable to Flogger" (the MiG-23F is clearly meant). The only visible distinguishing feature of the 25M is an undernose sensor similar to the IR fairing carried by the 23MF. The new-generation MiG-31 Foxhound is described separately.

Compared with the MiG-25 interceptor, the reconnaissance versions have a wing of reduced area, with slightly less span and constant sweep from root to tip. The nose radar is removed, giving a conical nose offering reduced drag; inside this (in the basic version known to NATO as Foxbat-B) are five very large vertical, forward oblique, lateral and panoramic cameras and a SLAR "looking" through a dielectric panel on the left side of the nose. Doppler radar is believed to be fitted, as on many of the interceptor version. Foxbat-D is a less common variant with a much larger SLAR installation and probably IR linescan

but no cameras; this is used by the Soviet Union and Libya only. About 160 MiG-25Rs of both models are estimated to be in FA service, plus about 45 with foreign customers, including eight with No 106 Sqn, Indian AF. These were purchased to replace obsolete Canberra PR.7s.

The MiG-25U trainer has the instructor cockpit in front of and below the original cockpit (occupied by the pupil), which is not only the reverse of normal procedure but means that the extra cockpit displaces the radar and other sensors but not fuel. The MiG-25U is usually used in small numbers for plain pilot conversion, but recent photographs of Soviet PVO examples show missile pylons in place, so it may at least be possible to run through simulated interceptions with training missiles and perhaps with simulated radar displays.

Algeria, Libya and Syria operate both MiG-25 and -25R, while India has MiG-25R and -25U. The Soviet Union has all three versions and is the only one (so far) to have the MiG-25M.

Top: MiG-25M "Foxbat-E", with the older "Foxbat-A", makes up about a quarter of the Soviet strategic fighter force. All "Foxbat-As" are being upgraded to -Es, giving them limited lookdown/shoot-down capability.

Above: MiG-25 "Foxbat-A", a strategic interceptor armed with AA-7 and AA-8 missiles. The MiG-25 was designed to counter the USAF's B-70 high-altitude strategic bomber, which did not enter service.

MiG-29 "Fulcrum"

Origin: Soviet Union.
Type: Multirole fighter and attack.
Engines: Two engines, believed to have an afterburning rating of about 16,535lb (7,500kg) thrust.
Dimensions: (1984 DoD estimates) Span 39·3ft (12·0m); length (excl probe) 50·83ft (15·5m); height 16·0ft (4·9m); wing area 380sq ft (35·5m²).
Weights: Empty about 18,000lb (8,200kg) (a DoD estimate of 28,000lb (12,700kg) in November 1983 must be wildly inaccurate); loaded (clean) 28,000lb (12,700kg); max (assuming 4,000kg weapon load) 36,800lb (16,700kg).
Performance: Max speed (hi, clean) 1,520mph (2,450km/h, Mach 2·3), (clean, SL) 913mph (1,470km/h, Mach 1·2); turn rate at 15,000 (4,572m) (sustained) 16°/s at Mach 0·9, (instantaneous) 21°/s pulling 9g; combat radius (hi-lo-hi) (clean) 438 miles (705km), (four AAMs) 415 miles (668km), (four FAB-500) 374 miles

(602km). Note: mid-1984 DoD combat radius figure is "800km" (almost 500 miles).
Armament: Six AAMs. These are likely to be of the new type called AA-X-10 by NATO, with a range of some 30 miles (48km) and equipped with an active terminal sensor. Older AAMs such as the R-32R and R-60 (AA-7 and AA-8) could presumably also be carried, though at a loss in capability. In the attack role a weapon load of 8,820lb (4,000kg) is estimated.

Development: In appearance this very important new combat aircraft resembles a scaled-down MiG-25 but with a subtle softening of the former's harsh lines (which were configured for high supersonic flight regimes that the MiG-29 will seldom reach). Alternatively, it has been pointed out how in some respects the new fighter resembles such Western types as the F-15 (but on a reduced scale), F-16 and F-18. A prototype was photographed by US

reconnaissance satellite on the test airfield at Ramenskoye in 1979, and for a while the type was simply dubbed Ram-L (as one of a number seen on test at that location). Seen from above it was clearly twin-engined, with widely separated swept tailerons well behind the engine nozzles. The almost unswept wing could be seen to have long root extensions, but no inlets could positively be indentified.

Since then the vague Ram-L has been refined into the MiG-29, along with a profusion of US-originated data which may be good guesstimates but in many cases are so far apart as to be meaningless. For example, "Washington sources" stated in 1982 that a prototype had achieved an instantaneous turn rate of 16·8°/s and a sustained turn rate of 8·26°/s. It was not disclosed how this information was obtained, though at the same time it was widely rumoured that the West was in possession of a flight test cine film showing impressive agility. Yet in November 1983 "Washington sources" came out with the turn rate figures given in the data, which are

almost twice as good! Certainly it is universally agreed that nothing in the sky can out-turn the new MiG, unless it is the much bigger Su-27 whose data are even more conflicting than those of the MiG-29.

The MiG-29 is a real air-combat aircraft. It exists in substantial numbers, a Moscow-area plant delivering over 200 in 1983, and according to the USAF TAC Deputy Chief of Staff/Intelligence it was declared operational within Frontal Aviation in October 1983. Probably the nearest Western parallel to it would be the stillborn Northrop F-18L, with two engines giving a thrust/weight ratio well beyond unity at low levels, tremendous agility and look-down shoot-down capability with BVR missiles (something still absent from the F-16);

As a replacement (in the first instance) for the MiG-21 there was some reason to think the MiG-29 would be not only agile but also small. But even the 1983 estimates of span fluctuated between 10·25 and 10·5m (say, 34ft), with wing area put at 380sq ft, compared with 300 for the F-16. The DoD figures,

MiG-31 "Foxhound"

Origin: Soviet Union.
Type: Long-range interceptor.
Engines: Two large afterburning turbojets or turbofans (not necessarily 30,865lb/14,000kg thrust R-31Fs).
Dimensions: (estimate) Span 45·87ft (14m); length 83·89ft (25·57m); height about 20ft (6·1m); wing area about 615sq ft (57m²).
Weights: (estimate) Empty 47,400lb (21,500kg); max 90,400lb (41,000kg).
Performance: (estimate) Max speed (hi, 8 AAM) 1,586mph (2,553km/h, Mach 2·4), (SL) 900mph (1,450km/h, Mach 1·18); service ceiling 75,500ft (23,000m); combat radius (hi intercept) 932 miles (1,500km).
Armament: Eight AA-9 AAMs.

Development: Though based closely on the MiG-25, the MiG-31 is a completely new aircraft. Indeed it is no small tribute to the now very old MiG-25 to note that the aerodynamics have hardly changed at all, beyond adding wingroot strakes. Structurally the airframe has been restressed for manoeuvrability at all altitudes, and instead of being a very fast straight-line interceptor at great heights the MiG-31 is a versatile and formidable aircraft right down to sea level. It can outrun almost anything in the sky, and has complete capability against multiple targets at all heights including ground-hugging lo attack.

Among the major differences compared with the MiG-25 are the addition of a second crew-member, the elimination of the wingtip anti-flutter bodies and the ability to carry eight of the formidable AA-9 missiles. The adjective "formidable" is required because, while these AAMs are much smaller than the monster AA-6 Acrid, they have a similar punch and require no inflight guidance from the carrier aircraft. They possibly use strapdown inertial midcourse guidance, but the terminal homing is autonomous. Presumably US satellite coverage has verified that an active terminal seeker is used, rather than IR or other passive homing. According to *Aviation Week,* quoting official Washington sources in 1983, MiG-31s have demonstrated very great radar range and TWS capability while intercepting RPV targets including simulated cruise missiles at Vladimirovka, a test range on the Caspian. Another report specifically states that radar-guided missiles fired from high-flying MiG-31s have intercepted targets flying at about 200ft (90m).

There has been speculation that, as the MiG-31 is a flexible aircraft able to fly various missions, it no longer uses the low-pressure-ratio R-31 type engine. Instead it is surmised to have high-pressure engines burning much less fuel as a result of higher cycle efficiency at subsonic speeds. DoD estimates of MiG-31 speed are lower than for the MiG-25, though in fact the longer fuselage — said to be stretched 39in (1m) ahead of the wing, including the inlet ducts, and 30in (0·75m) aft — should actually reduce drag as well as increase fuel capacity.

There are apparently four wing pylons, the outers being shown carrying 440gal (2,000lit) drop tanks with large horizontal fins. It would be logical to try to carry some of the AAMs conformally, as the F-15 carries Sparrows. With the MiG-25 this has been impossible but the MiG-31 may have enough body length

Above: MiG-31 "Foxhound". Over 70 of these strategic interceptors are in service.

ahead of or behind the landing gear for AAMs. An alternative arrangement would be F-4 style recesses underneath. Even the AA-9 is still a substantial missile, considerably bigger than Sparrow, and eight would impose a serious drag penalty.

Early reports of the MiG-31 invariably mentioned an internal gun, usually said to be "heavy calibre" (presumably meaning 30mm). This gun is not mentioned in any of the Pentagon-based reports published since November 1983. What is emphasized, however, is that four regiments of the MiG-31s were already operational by November 1983, with output from Gorkii building up all the time.

given in the April 1984 *Soviet Military Power,* jumped to nearly 40ft (12m), putting the MiG-29 almost in the F-15 class.

As for the "six AAMs", there is no indication that these are carried, or whether any are recessed or carried conformally — as would probably be needed to achieve the alleged speed of 1,500mph (2,500km/h). The drawing merely showed four wing pylons each with an AAM, those on the inboard pylons having canards. A gun would seem a natural for this aircraft.

Another feature of the official April 1984 drawings is that the vertical and horizontal tails are shown attached to the rear fuselage as in the MiG-25. All 1983 artwork showed the tails attached to cantilevered beams, as in the F-14 and F-15, with the engines well inboard. All recent Western MiG-29 drawings have shown the wing with flaps and possibly spoilers well inboard, without any of the slats or outboard ailerons of the 1980-82 drawings. The 1984 DoD drawing also omits the rear ventral fins present in earlier illustrations.

Left: MiG-29 "Fulcrum" all-weather, counter-air fighter-interceptor has an integrated look-down, shoot-down radar/AA-10 missile system.

Mitsubishi F-1

Origin: Japan.
Type: Single-seat close-support fighter-bomber.
Engines: Two Ishikawajima-Harima TF40-801A (licence-built Rolls-Royce/Turboméca Adour 102) augmented turbofans with max rating of 7,140lb (3,238kg).
Dimensions: Span 25·83ft (7·87m); length 58·5ft (17·86m); height 14·75ft; wing area 227·9sq ft (21·7m²).
Weights: 14,330lb (6,500kg); loaded (max) 30,200lb (13,700kg).
Performance: Max speed (clean) 1,056mph (1,700km/h, Mach 1·6); initial climb 19,680ft (6,000m)/min; service ceiling 50,025ft (15,250m); range (with eight 500lb bombs) 700 miles (1,126km).
Armament: One 20mm M61 multi-barrel gun under left side of cockpit floor; pylon hardpoints under centreline and inboard and outboard on wings, with light stores attachments at tips. Total weapon load normally 6,000lb (2,722kg) comprising 12 x 500lb bombs, eight 500lb plus two tanks

of 183gal, or two 1,300lb (590kg) ASM-1 anti-ship missiles, and four Sidewinders.

Development: Mitsubishi was strongly influenced by the Anglo-French Jaguar in designing the T-2 trainer, which has the same configuration and engines. Unusual in being a supersonic trainer, it has a simple Mitsubishi Electric search/ranging radar and a Thomson-CSF HUD. The JASDF bought a total of 28, plus 58 of the armed T-2A combat trainer version. All were delivered by 1984, one being grossly modified to serve as a CCV research aircraft.

The F-1, originally known as the FST-2, has the same airframe, engines and systems, but is a single seater, the rear cockpit being occupied by avionics. A Ferranti inertial nav/attack system is fitted, together with a different Mitsubishi radar with air/air and air/ground modes, a weapon-aiming computer, radar altimeter and RHAWS, the weapon system being modified from

1982 to handle the locally developed ASM-1 anti-ship missile.

Total procurement of the F-1 was 73, all of which have been delivered. The JASDF had planned to order a replacement (FS-X), but the daunting development cost has led to a decision to carry out a life-extension programme on the F-1 to keep them operational until at least 1993.

Below: A single-seat Mitsubishi F-1 close-support fighter procured by the JSDAR, derived from the T-2 trainer and showing strong Jaguar influence.

Myasishchyev M-4 ("Bison")

Origin: Soviet Union.
Type: Free-fall bomber and air-refuelling tanker.
Engines: Four 19,190lb (8,700kg) thrust Mikulin AM-3D turbojets.
Dimensions: Span 172·2ft (52·5m); length 162·0ft (49·38m), (with FR probe) about 170·0ft (51·8m); height 46·0ft (14·24m); wing area 3,400sq ft (320m²).
Weights: Empty (bomber or tanker) about 185,000lb (83,900kg), normal loaded 350,000lb (158,750kg).
Performance: Max speed at 36,100ft (11,000m), 625mph (1,005km/h, Mach 0·945); typical cruising speed (same height or above) 560mph (900km/h); service ceiling (normal weight) 52,000ft (15,850m); max operating radius (unrefuelled) 3,480 miles (5,600km); range with 12,000lb/5,450kg bombload 4,970 miles (8,000km); ultimate range with max overload fuel and no weapons, about 11,000 miles (18,000km).
Armament: (tanker) None, or twin NR-23 in tail turret; (bomber) six NR-23 in tail and forward dorsal and ventral barbettes, internal weapon load up to 33,070lb (15,000kg), believed all free-fall.

Development: Though it was never able to meet the ADD (Long-Range Aviation) requirement for mission radius, the big four-jet M-4 (also called Mya-4) was a tremendous achievement in aircraft engineering in its design period of 1951-53. Moreover, to a much greater extent than Western counterparts, it has proved capable of remaining in service into the 1980s, apparently without any significant structural fatigue problems.

The engine, almost the same as

that of the Tu-16, easily fitted inside the thick roots of the wing, which has a remarkably high aspect ratio for good subsonic range. The circular-section fuselage contains a pressurized capsule at the front for the flight crew (usually five), with a tail sighting station controlling three of the five twin-23mm gun turrets. Bogie main landing gears retract into fuselage bays ahead of and behind the three bomb bays amidships. The tip pods housing the retracted outrigger gears show marked nose-down incidence in flight, the wing twist (washout) being considerable. As in the B-52 the wing fuel bends the wings down to press these tip wheels on the ground on a full-load takeoff.

Production was on a smaller scale than expected, and curiously the much more powerful version with D-15 engines never went into service. There were several reconnaissance versions, but today the only M-4s still in service are believed to be 45 of the original bomber type,

with considerable additional avionics but often reduced defensive armament, and 30 of the tanker version with a hose-drum unit in the aft bomb bay (and occasionally with HDUs in pods under the outer wings as well).

Above: Small numbers of the M-4 "Bison-B" maritime reconnaissance variant were produced.

Below: Members of an M-4 bomber aircrew leave their aircraft after a training mission.

Nanzhang Q-5 "Fantan"

Origin: People's Republic of China.
Type: Close-support attack aircraft with air-combat capability.
Engines: Two 7,167lb (3,250kg) thrust Wopen-6 afterburning turbojets (Tumanskii R-9BF-811).
Dimensions: Span 31·83ft (9·7m); length (inc probe) 54·88ft (16·727m); height overall 14·8ft (4·51m); wing area 300·85sq ft (27·95m²).
Weights: Empty 14,317lb (6,494kg); loaded (clean) 21,010lb (9,530kg), (max) 26,455lb (12,000kg).
Performance: Max speed (hi, clean) 740mph (1,190km/h, Mach 1·27); (SL, clean) 752mph 61,210km/h, Mach 0·99); service ceiling 52,500ft (16,000m); TO run (max wt) 4,100ft (1,250m); combat radius (max bombs, no afterburner) (hi-lo-hi) 373 miles (600km), (lo-lo-lo) 248 miles (400km).
Armament: Two 23mm single-barrel guns, each with 100 rounds, in wing roots; internal bomb bay usually occupied by fuel tank, leaving four fuselage pylons each rated at 551lb (250kg) and four wing pylons, those inboard of landing gear being rated at 551lb (250kg) and those outboard being plumbed for 167gal (760lit) drop tanks. Max bomb or other stores load usually 4,410lb (2,000kg).

Development: The first major military aircraft to be designed in the People's Republic, the Q-5 was based on the J-6 but differs in almost every part. The chief change was to extend the forward fuselage,

Below: The Q-5 "Fantan" ground-attack derivative of the J-6 has a completely different apperance, with twin lateral intakes and elongated nose.

terminate the air inlet ducts in lateral inlets and add an internal weapons bay. The wings were extended at the roots, which also increased the track of the landing gear (which is strengthened to handle the much greater weights), and the vertical tail was made taller. The cockpit was redesigned, and enclosed by an upward-hinged canopy leading into a different fuselage spine. There were many systems changes, most important being an increase in internal fuel capacity of nearly 70 per cent.

Performance proved adequate with the original engines, though at high weights a good runway is needed, and a braking parachute is normally streamed (later aircraft have it housed in a pod below the rudder as in later J-6s).

At first it was thought the reason for using lateral inlets was to enable a radar to be installed, but this was mistaken. Q-5 development aircraft have flown with radar, and a small gunsight ranging set is fitted to some recent aircraft, but no variant has gone into production with a major radar. Avionics are described as fully adequate for visual attack missions.

The Pakistan AF, first customer for the A-5 export version, is delighted with the first batch of 42 received in 1983 and equipping Nos. 16 and 7 Sqns. Eventually the PAF expects to receive 140 A-5s, to arm eight squadrons and an OCU.

The photographs so far seen of the A-5 show it to be cleaner than regular Q-5s, though the avionic standards are said to be similar. Chinese Q-5s can carry nuclear bombs of from 5 to 20kT yield, and the usual method of delivery of these is a toss. Conventional weapons are usually aimed by the SH-1J optical sight in a dive attack.

Northrop F-5

Origin: USA.
Type: Light tactical fighter and attack/recon.
Engines: Two General Electric J85 afterburning turbojets, (A/B) 4,080lb (1,850kg) thrust J85-113 or -13A, (E/F) 5,000kg (2,270kg) thrust-21A.
Dimensions: Span (A/B) 25·25ft (7·7m) (A/B over tip tanks) 25·83ft (7·87m), (E/F) 26·66ft (8·13m), (E/F over AAMs) 27·92ft (8·53m); length (A) 47·16ft (14·38m), (B) 46·33ft (14·12m), (E) 48·16ft (14·68m), (F) 51·58ft (15·72m); height overall (E) 13·33ft (4·06m²); wing area (A/B) 170sq ft (15·79m²), (E/F) 186sq ft (17·3m²).
Weights: Empty (A) 8,085lb (3,667kg), (B) 8,361lb (3,792kg), (E) 9,683lb (4,392kg), (F) 10,567lb (4,793kg); max loaded (A) 20,576lb (9,333kg), (B) 20,116lb (9,124kg), (E) 24,722lb (11,214kg), (F) 25,225lb (11,442kg).
Performance: Max speed at 36,000ft (11,000m), (A) 925mph (1,489km/h, Mach 1·4), (B) 886mph (1,425km/h, Mach 1·34), (E) 1,077mph (1,734km/h, mach 1·63), (F) 1,011mph (1,628km/h, mach 1·53); typical cruising speed 562mph (904km/h, Mach 0·85); initial climb (A/B) 28,700ft (8,750m)/min, (E) 34,500ft (10,516m)/min, (F) 32,890ft (10,250m)/min; service ceiling (all) about 51,000ft (15,540m); combat radius with max weapon load and allowances (A, hi-lo-hi) 215 miles (346km), (E, lo-lo-lo) 138 miles (222km); range with max fuel (all hi, tanks dropped, with reserves) (A) 1,563 miles (2,518km), (E) 1,779 miles (2,863km).
Armament: (A/B) military load 6,200lb (2,812kg) including two 20mm M-39 guns each with 280 rounds, and wide variety of underwing stores, plus AIM-9

AAMs for air combat; (E/F) wide range of ordnance to total of 7,000lb (3,175kg) not including two (F-5F, one) M-39A2 guns each with 280 rounds and two AIM-9 missiles on tip rails.

Development: Cheap, simple, delightful to handle and supersonic, the small twin-jets from Northrop have found a ready market all over the world. Nobody is bothered at their obvious military shortcomings, and their customer list (currently 32 air forces) far exceeds that of any other non-Soviet fighter in the world, despite the rather major initial handicap that none of these fighters was ever adopted by its own country, except in small numbers for Aggressor roles and training foreign (customer) pilots.

The first generation F-5A Freedom Fighter and F-5B tandem dual two-seater each have two fuselage integral fuel cells with a capacity of 485·5gal (2,207lit), seven hardpoints for an external load of up to 6,000lb (2,722kg), the centreline being rated at 2,000lb (907kg) and landing gear for rough-field operations. Avionics are usually primitive.

Canadian CF-5s, now mostly sold to other countries such as Venezuela, have uprated (4,300lb/ 1,950kg) engines, FR probes, extensible nose legs and other changes, while Netherland NF-5s (now also being replaced) have manoeuvre flaps and larger drop tanks. Norway's aircraft, now flying in Turkey, have rocket ATO and arrestor hooks. Production totalled 879 by Northrop and 320 by the licensees.

Winner of an "F-5 successor" contest organized (with special eyes on SE Asia) by the DoD in 1969, the F-5E Tiger II has a wider fuselage increasing internal fuel to 563·7gal (2,563lit). Other changes include leading-edge root extensions, uprated engines, the previously mentioned extensible nose leg,

Above: One of 10 F-5Es used by the US Navy to simulate MiG-21s in "Aggressor"-type training.

Below: A Nellis-based F-5E of a USAF "Aggressor" squadron, seen from an F-15 Eagle.

manoeuvre flaps, gyro sight and arrestor hook, and — importantly — Emerson APQ-153 or 159 radar, a neat I/J-band set with fair capability against air or ship targets at ranges to 23 miles (37km). External load is raised to 7,000lb (3,175kg) on the five pylons, not including the added Sidewinder AAMs on tip rails.

Again individual customers have asked for extras, Saudi Arabia having a far more comprehensive avionic fit including Litton INS, comprehensive RWR and chaff/flare dispensers and Maverick ASMs. Switzerland specified antiskid brakes and a different ECM fit. Improvements that became standard during production included auto manoeuvre flaps on both leading and trailing edges, af flattened "shark nose" and larger leading-edge extensions.

The F-5F two-seater retains one gun and external weapons. The RF-5E Tigereye replaces the radar by a new, longer nose with a different profile in which can be installed a forward oblique KS-87D1 frame camera and any of a growing series

of pallets on which are mounted selected sensors. Pallet 1 mounts a KA-95B medium-altitude panoramic camera and RS-710E IR linescane. Pallet 2 combines a KA-56E with a KA-93B6 panoramic camera with 145° scan angle for use at heights from 10,000 to 50,000ft (3-15km). Pallet 3 is for Lorop (Long-Range Oblique Photo) missions and has a KS-174A Lorop camera of 66in (1·68m) focal length. Other pallets which might become available include one with a Zeiss mapping camera, two IR linescans and Elint receivers.

The pilot has advanced navigation and communications systems, including INS (updatable on overflying a recognized feature) and a TV display for visual correction of photo runs.

By 1984 Northrop and its licensees had delivered over 1,300 F-5Es and Fs, with production continuing at a reduced rate. This total includes many aircraft passed on to other users, such as Norway to Turkey and Iran/Jordan to Greece, but not a requested 12 for Honduras for

which Congressional approval had not been announced as this went to press.

In USAF service, about 60 F-5Es and a few Fs continue to help in the development of air combat techniques, in Aggressor roles, in the monitoring of fighter weapons meets and various hack duties. In addition, eight F-5Es are used by the Navy for Top Gun fighter pilot train-

Above: One of the Swiss Air Force's 98 F-5Es, and one of more than 1,300 E and F models delivered to 30 nations.

ing at the Naval Fighter Weapons School at NAS Miramar, California. Its small size and manoeuvrablity enables it to compete well even against opponents as advanced as the F-15.

Northrop F-20A Tigershark

Origin: USA.
Type: Multirole fighter.
Engine: One 17,000lb (7,711kg) thrust General Electric F404-100 augmented turbofan engine.
Dimensions: Span (excl AAMs) 26·66ft (8·13m); length 46·5ft (14·17m); height overall 13·84ft (4·22m); wing area 186sq ft (17·28m²).
Weights: Empty 11,220lb (5,089kg); loaded (clean) 18,005lb (8,167kg), (max) 27,500lb (12,474kg).
Performance: Max speed (hi, clean) about 1,320mph (2,125km/h, Mach 2); time to 40,000ft (12,200m) 2min 18sec; takeoff run (max weight) 4,200ft (1,280m); combat radius (seven Mk 82 bombs, two AAM, two tanks, hi-lo-hi plus large combat reserves) 345 miles (556km).
Armament: Two 20mm M39 guns each with 450 rounds; five weapon pylons (inner three plumbed for 22gal/100lit tanks), for total load of "over 8,300lb/3,765kg"; wingtip rails for Sidewinder AAMs. Loads can include nine Mk82 bombs, four Maverick ASMs or three GPU-5/A 30mm gun pods.

Development: For the third time Northrop has dramatically changed the same basic design without greatly changing its appearance. The Tigershark was originally styled F-5G, but its performance is so great that a new number (in the US DoD sequence, even though the US home market has shown no interest in this privately funded "export fighter") was assigned and it is being marketed as a new aircraft.

Features include a single engine giving almost double the thrust of the pair originally fitted, a GE APG-67 multimode lookup/lookdown radar with digital processing, a modern electronic-display cockpit, ring-laser INS, and the most modern

and comprehensive EW suite.

Certainly the F-20A's flight performance is most impressive, and it promises to be the fastest-climbing interceptor in the world. It is claimed that it can be airborne at Mach 0·9 more than 10nm (18·5km) from base in less than three minutes.

In most respects the F-20A overcomes the shortcomings of earlier Northrop fighters, and for air-combat performance (including surface attack) the cost/effectiveness of this already very reliable aircraft seems hard to beat.

The Tigershark first flew on 30 August 1982 but despite impressive air-show displays had not been ordered by mid-1985. The programme, which has cost Northrop at least $700 million has generated

much interest, but suffered a setback when the prototype crashed on 10 October 1984 during the final demonstration in a round-the-world tour. Two further prototypes were flying in 1984-85.

Below: An F-20 Tigershark shows both F-5 ancestry and new features — shark nose, splitter plates and redesigned rear fuselage.

Panavia Tornado ADV

Origin: International (Germany, Italy, UK).
Type: Long-range all-weather interceptor.
Engines: Two 16,000 + lb (7,258 + kg) thrust Turbo-Union RB199 Mk 104 augmented turbofans.
Dimensions: Span (25⁰) 45·6ft (13·9m), (67⁰) 28·21ft (8·6m); length 59·25ft (18·06m); height 18·7ft (5·7m); wing area not published.
Weights: Empty, equipped about 31,500lb (14,290kg); takeoff weight (clean, max internal fuel) 47,500lb (21,546kg); max not published.
Performance: Max speed (clean, at height) about 1,500mph (2,414km/h, Mach 2·27); combat mission with max AAM load, 2h 20min on station at distance of 375 miles (602km) from base with allowance for combat.
Armament: One 27mm Mauser cannon, four Sky Flash (later AMRAAM) recessed under fuselage and two AIM-9L Sidewinder AAMs (later ASRAAM).

Development: Though it is a purely British development, the ADV (air-defence variant) of Tornado is produced by the tri-national airframe and engine groupings as are the IDS series, and in due course it is likely to become the most widely used model, with greater export interest from more countries. It is unquestionably the most efficient long-range interceptor in the world, outperforming all known rivals in almost all respects (an exception is the sheer straight-line speed of the MiG-25 and 31), with engines of amazingly small size and low fuel burn.

Designated Tornado F.2 by the RAF, the interceptor has about 80 per cent commonality with the original IDS aircraft, and most of the airframe and aircraft systems are unchanged. The forward fuselage, made by BAe in any case, is completely new. This contains: a new pair of cockpits, with later electronic displays, different symbology, greater processing and storage capacity and a wet-film HDD recorder; Marconi/Ferranti AI.24 Foxhunter FMICW radar, with multimode lookup/lookdown TWS and missile guidance capability; deletion of one of the two guns; installation of a permanent FR probe fully retractable on the left side; installation of a ram-air turbine giving full hydraulic system power at high altitudes with main engines inoperative, down to below 230mph/ 370km/h; and addition of the Cossor 3500 series IFF, Singer ECM-resistant data link (with Nimrod AEW.3, for example, or ground stations), and a second INS.

Other airframe changes include forward extension of the fixed wing gloves, giving a major change in lift and agility, provision of 200gal (909lit) of extra fuel in the extended fuselage, and belly recesses for four Sky Flash (later AIM-120A) medium-range missiles, with twin-ram car-

tridge powered ejection giving clean launch at max negative g. The engines have various upratings which at high speeds and altitudes become large and significant, with extended jetpipes which improve afterbody shape and reduce drag; they also have digital control.

The only item not mentioned in connection with the production F.2 is the magnifying optical VAS (visual augmentation system) for positive identification of aircraft at great distances, which was regarded as a crucial item early in the programme.

By early 1985, British Aerospace had completed the basic flight development programme with the three prototypes, and had delivered a number of production aircraft, including two ATs (dual pilot ADV trainers), which after a spell at Boscombe Down cleared fully operational F.2s to form the OCU (Operational Conversion Unit) at Coningsby in September 1984.

The first production batch of 15 have also been completed, temporarily fitted with Mk 103 engines. The second and third RAF batches, numbering 52 and 92 aircraft, have the Mk 104 from the outset, as well as automatic schedule wing-sweep and manoeuvre-device (wing slats and flaps at 25⁰ to 45⁰ sweep) to give enhanced manoeuvrability with minimal pilot workload.

Updates planned for the future include still more powerful engines, even larger (495gal/2,250lit) drop

tanks, the AIM-120A (AMRAAM) and ASRAAM missiles and further improvements to the avionics. Allround performance has been demonstrated as equal to or better than prediction, with acceleration remaining "healthy" at Mach 2, and 800 knots (912mph/1,483km/h) indicated airspeed being registered at medium altitudes (true speed being much greater) and at heights down to 2,000ft (610m). Few if any other fighters can do this.

From early 1985 the Tornado F.2

Above: An early Tornado F.2 displays its mission load of four Skyflash and two Sidewinder missiles plus twin fuel tanks.

began to take its place in the RAF protecting the UK Air Defence Region. It will equip seven squadrons, the first five being at Leuchars (two) and Leeming (three).

Below: Tornado F.2 deploys flaps, spoilers and airbrakes; note the 27mm cannon muzzle.

Panavia Tornado IDS

Origin: International (Germany, Italy, UK).

Type: Two-seat multirole combat aircraft optimised for strike, (T) dual trainer.

Engines: Two Turbo-Union RB199 Mk 101 or 103 augmented turbofans each rated at 15,800lb (7,167kg) or over 16,000lb (7,258kg) thrust with full afterburner.

Dimensions: Span (25⁰) 46·6ft (13·90m), (67⁰) 28·21ft (8·60m); length 54·8ft (16·7m); height 18·7ft (5·7m); wing area not published.

Weights: Empty (equipped) 31,065lb (14,091kg); loaded (clean) about 45,000lb (20,411kg); max loaded, about 60,000lb (18,150kg).

Performance: Max speed (clean), at sea level, over 920mph (1,480km/h, Mach 1·2), at height, over 1,452mph (2,337km/h, Mach 2·2); service ceiling over 50,000ft (15,240cm); combat radius (8,000lb/3,629kg bombs, hi-lo-hi) 863 miles (1,390km).

Armament: Two 27mm Mauser cannon in lower forward fuselage; seven pylons, three tandem on body and four on the swinging wings, for external load up to 18,000lb (8,165kg).

Development: Most important military aircraft in Western Europe, the Tornado was the outcome of the first multinational collaborative programme to embrace design and development as well as manufacture, and to lead to a completely successful exercise in both management and hardware.

Though from the start a multirole aircraft, the Tornado IDS (interdiction strike) is optimized to the long-range all-weather blind first-pass attack mission against the most heavily defended surface targets, including ships. It is by far the most capable aircraft of its size ever built, and not least of its achievements is

Below: Wingtip Zeus ECM and Boz 100 chaff/flare pods plus 1,000lb (454kg) bombs and AIM-9s on a Luftwaffe Tornado IDS.

that on typical missions its fuel burn is roughly equal to that of an F-16, 60 per cent that of an F-4 and about 50 per cent that of an F-15 or Su-24, while carrying at least as heavy and varied a load of weapons as the best of these aircraft.

Features include a Texas Instruments multimode forward-looking radar with the option of various types of programmable software, a TFR (terrain-following radar), electrically signalled FBW (fly-by-wire) flight controls with artificial stability, fully variable supersonic inlets (which help make this the fastest aircraft in the world at low level, and one of the fastest at all heights, despite the extremely compact lightweight engines), advanced avionic systems to manage the array of stores which can be carried (which exceeds that of any other aircraft) and modern tandem cockpits with head-up and head-down displays in the front and three electronic displays in the back.

Among the stores which have been cleared are all tactical bombs of the four initial customers, nine rocket pods, Sidewinder AAMs and Sea Eagle, Kormoran, Maverick, Alarm, GBU-15, Paveway, AS.30 and AS.30L, Martel (seldom to be carried), Aspide, BL.755, JP.233

Above: Tornado GR.1s of the RAF's Honington-based 9 Sqn, each carrying four 1,000lb (454kg) bombs under the fuselage, plus drop tanks and Skyshadow ECM pods under the wings.

Below: Among the specialised weapons available for Tornados is the JP233 airfield attack system, seen here dispensing runway cratering (aft) and area denial submunitions.

and MW-1; Harpoon and possibly other cruise missiles may be carried later.

All aircraft have two guns, Martin-Baker Mk 10 automatic zero/zero seats, a gas-turbine APU which is self-cooling and can be left running on the ground, automatically scheduled lift-dumpers, pre-armed engine reversers, anti-skid brakes and (as a further option) a braking parachute. There is provision, so far exercised only by the RAF, to bolt on an FR probe package above the right side below the canopy. All sub-types have very comprehensive EW systems, with advanced RHAWS and either the Elettronica EL/73 deception jammer and ELT 553 ECM pod, or Marconi Avionics Sky Shadow.

Deliveries began in 1980 to the TTTE (Tornado Trinational Training Establishment), located at RAF Cottesmore in England. This has a strength of 50 aircraft, a high proportion being dual-pilot trainers for pilot conversion, the rest being for training navigators and complete crews. Aircraft from all three nations are used, the crews likewise being completely multinational until they actually pair up for final training as a team. By 1982 two further major training units were operational, that for the Luftwaffe (WaKo) being at Erding and for the RAF (TWCU) at Honington.

RAF squadrons began with Nos. 9 (Honington) and 617 (Marham), followed by eight in RAF Germany: 15 and 16 (ex-Buccaneer) at Laarbruch, all four squadrons in the former Jaguar wing at Brüggen, No. 9 (from Honington) and, in 1986, equipment of No. 2 (II) Sqn with aircraft specially configured for reconnaissance.

Marineflieger MFG 1 and 2 are both converted from the F-104G, while the Luftwaffe is converting four JaboG wings, Nos. 31 to 34. Italy's AMI is using 54 aircraft to replace the F/RF-104G in the 28º, 132º and 154º gruppi, with further aircraft equipping the 3º GEV (maintenance/training squadron) at Cameri.

Total national commitments, all delivered or in process of manufacture as this was written, comprise: RAF 219 with designation GR.1 (plus a pre-production machine brought up to GR.1 standard);

Above: The MW-1 system, seen here in operation aboard a Luftwaffe Tornado IDS, can dispense six different types of submunitions, including AP bomblets, anti-tank mines and anti-runway weapons.

Marineflieger, 112; Luftwaffe, 212; and AMI 99 plus one pre-production aircraft updated. This gives a total of 664, of which 420 had been delivered by 1985.

Rockwell OV-10

Origin: USA.
Type: (except B) Two-seat multirole counter-insurgency; (B) target tug.
Engines: (most) Two 715ehp Garrett T76-410/41 turboprops; (B(Z)) as other versions plus General Electric J85-4 turbojet of 2,950lb (1,338kg) thrust above fuselage; (D only) 1,040shp T76-420/421.
Dimensions: Span 40·0ft (12·19m); length (except D) 41·58ft (12·67m), (D) 44·0ft (13·4m); height 15·16ft (4·62m); wing area 291sq ft (27·03m²).
Weights: Empty (A) 6,969lb (3,161kg); max loaded (A) 14,466lb (6,563kg).

Performance: Max speed (A, sea level, clean) 281mph (452km/h); initial climb 2,300ft (700m)/min; (B(Z)) 6,800ft/min; service ceiling 30,000ft (9,150m); range with max weapon load, about 600 miles (960km); ferry range at 12,000lb (5,440kg) gross, 1,428 miles (2,300km).
Armament: Four 7·62mm M60C machine guns in sponsons; 1,200lb (544kg) hardpoint on centreline and four 600lb (272kg) points under sponsons; one Sidewinder missile rail under each wing; (OV-10D) as other versions plus M97 three-barrel 20mm cannon in remotely aimed ventral power turret.

Development: This twin-turboprop STOL aircraft was developed as a LARA (light armed reconnaissance aircraft) at a time when the USA was highly interested in so-called "brushfire wars" against primitive foes. It was, in fact, the first dedicated Co-In aircraft, intended to defeat insurgent forces whilst collaborating with friendly troops on the ground. Thus it offered use in the FAC role, with light weapons and plenty of markers and other pyrotechnics. At the rear of the central nacelle is a door for loading up to 3,200lb (1,452kg) of cargo, five paratroops (door left off in flight) or two stretchers and attendant.

Though obviously vulnerable to modern defences, the OV-10 combines good field length (around 1,200ft/366m), except at max

weight) with high agility and fair protection against small-arms fire.

A total of 271 were built for the USAF and Marines, followed by 18 for West Germany (12 being jet-boosted tugs), 40 OV-10Cs for Thailand, 16 OV-10Es for Venezuela and 16 OV-10Fs for Indonesia. Most complex are the 17 US Marine Corps' OV-10D night observation gunships, all converted OV-10As with uprated engines, FLIR, laser, ventral gun turret, APR-39 RHAWS, ALE-39 chaff/flare dispensers and other extras.

Below: OV-10 serving with the 19th TASS at Osan AB, South Korea, in low-visibility markings; PACAF OV-10s are now operated by the 22nd TASS at Wheeler AFB, Hawaii.

Saab 35 Draken

Origin: Sweden.
Type: (J35, F-35) single-seat all-weather fighter-bomber; (Sk35, TF-35) dual trainer; (S35) single-seat all-weather reconnaissance.
Engine: One Svenska Flygmotor RM6 (licence-built Rolls-Royce Avon with SFA afterburner) (D, E, F and export) 17,110lb (7,761kg) RM6C.
Dimensions: Span 30·83ft (9·4m); length 50·33ft (15·4m) (S35E, 52·0ft (15·8m); height 12·75ft (3·9m); wing area 529·6sq ft (49·2m²).
Weights: Empty (D) 16,017lb (7,265kg), (F) 18,180lb (8,250kg); max loaded (D) 22,663lb (10,280kg), (F) 27,050lb (12,270kg), (F-35) 35,275lb (16,000kg).
Performance: Max speed (D onwards, clean) 1,320mph (2,125km/h, Mach 2·0), (with two drop tanks and two 1,000lb bombs) 924mph (1,487km/h,

Mach 1·4); initial climb (D onwards, clean) 34,450ft (10,500m)/min; service ceiling (D onwards, clean) about 65,000ft (20,000m); range (internal fuel plus external weapons, typical) 800 miles (1,300km), (max fuel) 2,020 miles (3,250km).
Armament: (F) One 30mm Aden plus two RB27 Falcon (radar) and two RB28 Falcon (infra-red) missiles, plus two or four RB24; (F-35) two 30mm Aden plus nine stores pylons each rated at 1,000lb (454kg) all usable simultaneously, plus four RB24.

Development: Amazingly bold in its conception in 1949-50, the Saab 35 Draken (Dragon) matured as a supremely good and cost/effective supersonic interceptor which for all intents and purposes does the same job as the British Lightning on just half the number of the same type of engine. Indeed the final version —

the F-35 as bought by Denmark — does a great deal more, with the ability to put down nine 1,000lb (454kg) bombs with fair accuracy.

The main version still serving with the Swedish Flygvapen is the J35F (popularly Filip, and succeeding David, the J35D). The J35F was the most expensive of the six Swedish models, but also built in the largest numbers, and the new equipment including an outstanding pulse-doppler radar, Hughes-derived fire control system and armed with a gun and both IR and radar-guided AAMs. Saab went on to build 606 Drakens in all, the final batches being for Finland and Denmark. The Finns assembled their machines locally, by Valmet. In 1984 Finland funded "10 to 20" more Drakens for 1985.

Denmark found the Swedish machine not only extremely low-priced but also so good it is remaining in front-line use alongside the F-16 until at least 1987-88, while the F-100s and F-104s have been replaced by Fighting Falcons.

Three versions are in use, each having a counterpart in the Flygvapen with minor changes. The basic fighter-bomber is the F-35, serving with 725 Sqn in the Karup wing and backed by two-seat dual TF-35 conversion trainers which have limited weapons capability. The RF-35, whose Swedish partner is the S35E, is a reconnaissance fighter which normally carries the Red Baron night recon multisensor pod.

All Draken versions are cleared for off-airfield operation, and in Sweden it is routine to operate from country highways and even farm roads. All Drakens in service have been structurally audited to go on into the late 1980s, and avionics improvements have been incorporated including HUD sights. Danish Fs and RFs even have INS, laser ranger and many F-16 type subsystems.

Right: An F-35 fighter-bomber, one of 48 Drakens of various types in service with the Royal Danish Air Force.

Saab 37 Viggen

Origin: Sweden.
Type: (AJ) Single-seat all-weather attack; (JA) all-weather fighter; (SF) armed photo-reconnaissance; (SH) armed sea surveillance; (SK) dual trainer.
Engine: One Svenska Flygmotor RM8 (licence-built Pratt & Whitney JT8D turbofan redesigned in Sweden for Mach 2 with afterburner); (AJ, SF, SH and SK) 25,970lb (11,790kg) RM8A; (JA) 28,108lb (12,750kg) RM8B.
Dimensions: Span of main wing 34·77ft (10·6m); length (A) 53·47ft (16·3m); (JA with probe) 53·8ft (16·4m); height (most) 18·37ft (5·6m), (JA, SK) 19·37ft (5·9m); wing area 495·1sq ft (46·0m²).
Weights: Empty (all) about 27,000lb (12,250kg); loaded (clean JA) 33,070lb (15,000kg), (AJ normal armament) 35,275lb (16,000kg), (JA normal AAMs) 37,478lb (17,000kg), (AJ max weapons) 45,195lb (20,500kg).
Performance: Max speed (clean) about 1,320mph (2,135km/h, Mach 2), or Mach 1·1 at sea level; initial climb, about 40,000ft (12,200m)/min (time from start to take-off run to 32,800/10,000m = 100sec); service ceiling, over 60,000ft (18,300m); tactical radius with external stores (not drop tanks), hi-lo-hi profile, over 620 miles (1,000km).
Armament: Seven pylons (option: nine) for aggregate external load of 13,200lb (6,000kg), including RB04, 05 or 75 missiles for attack, and RB24 and 28 missiles for defence. In addition the JA37 has a 30mm Oerlikon KCA gun and carries two RB71 (Sky Flash) and four RB24 (Sidewinder) AAMs.

Development: Like all Swedish pro-grammes for combat aircraft the Type 37 Viggen (Thunderbolt) has been masterful and wholly suc-cessful, producing five sub-types of the same basic machine, each

tailored to a different primary role, within budget and on time. The first and still most numerous version is the AJ37, dedicated mainly to attack on surface targets and ships.

From the start the basic aircraft design was biased heavily in favour of STOL operations from rough strips, including straight stretches of country highways and dirt tracks. The big afterburning engine gives high thrust for quick getaway, though at the cost of temporary high fuel consumption. The large delta wing and canard foreplane form a powerful high-lift combination, and can also be used to pull very tight turns in close combat down to amazingly low airspeeds, though again the high drag means full after-burner must be used. On landing the Viggen can be brought in at only 137mph (220km/h), slammed on to the ground in a carrier-type no-flare impact, and at once given full reverse thrust (a feature shared only by Tornado) and maximum no-skid wheel braking.

The AJ37 carries up to 15,430lb (7,000kg) of various weapons on its seven pylons, including the RB04E cruise missile which homes on ships and packs a tremendous punch, the RB05A which can be used against ground, naval and certain airborne targets, and the RB75 Maverick TV-guided precision ASM. For air-to-air use gun pods, usually of 30mm Aden type, can be carried, as well as the RB24 Sidewinder and RB28 Falcon AAMs.

The AJ37 equips two squadrons of F6 wing at Karlsborg, two squadrons of F7 at Satenäs and one squadron of F15 at Söderhamm. Variants of the AJ37 are the SF37 and SH37 reconnaissance models.

The SH37, for maritime use, replaced the S32C Lansen in F13 Wing and in mixed SH/SF Wings F17 and F21. It is used primarily to survey, register and report all maritime activity near Sweden. It

has the basic airframe of the AJ37, with an LM Ericsson multimode radar, Marconi HUD and central digital computer, with an added camera for recording the radar displays. The three fuselage pylons carry a large tank on the centreline, a night reconnaissance pod with IR linescan and LLTV on the left and a Red Baron or long-range camera pod on the right. Inboard wing pylons can carry active or passive ECM jammer pods, and very com-plete Elint and EW recorders are car-ried, together with a tape recorder and a data camera which records film co-ordination figures, date, time, aircraft position, course, height, target location and other in-formation.

The SF37, which serves with the same three wings as the SH, has no nose radar, and its slim, pointed nose houses four vertical or oblique cameras for low-level use, two long-range vertical high-altitude cameras and VKA IR linescan. Also installed in the fuselage are the camera sight, an IR sensor and EW systems in-cluding and RWR and Elint recorders. The sensors give 180° horizon-to-horizon coverage and are specially arranged to work on wavelengths which reveal the presence of camouflaged targets.

External loads can include the cen-treline tanks and active or passive electronic counter measure pods on the inboard wing pylons.

Newest version, and the only one in production, the JA37 fighter is a much more extensive redesign than all previous variants, and its development cost more than the en-tire design and development of the original AJ model. Its performance is optimized to interception at a distance, using radar-guided medium-range Sky Flash AAMs, and to close combat at all altitudes. The engine has a different match of fans and compressors, and gives higher thrust.

The airframe has underwing fair-ings for four elevon power units on each wing instead of three, and the vertical tail (which folds flat for entry into low-ceilinged underground hangars) is the same taller swept-tip type as seen on the SK37 dual trainer version.

Though no longer new in con-cept, the avionics of the JA37 stand comparison with those of the F-15 or any other fighter in service today. The main radar is an Ericsson UAP-1023 pulse-doppler set operating in I/J-band and giving outstanding look-down performance against low-flying small targets in

adverse environments and in presence of intense ECM. The same can, of course, be said of the BAe Dynamics Sky Flash missile, used as the RB71, and the Swedish air force has placed a large follow-on order for this weapon. Other equipment includes an Advanced Marconi Avionics HUD, Singer-Kearfott main digital computer, Garrett digital air-data computer, advanced ILS and outstandingly comprehensive EW systems (which make those of many NATO aircraft look pathetic). The single gun, under the fuselage, is more powerful than any other 30mm gun in any Western fighter (except for the unique gun used in the anti-tank A-10A).

The JA37 normally flies with a large centreline drop tank, and it can carry various surface-attack weapons if necessary. Current production machines are painted off-white, and deliveries of half the 149 on order had been made as this was

written, to wings F13 at Norrköping, F17 at Ronneby and F21 at Lulea. By 1985 about 270 of the planned total of 329 Viggens had been

delivered, the aircraft still being built all being of the JA type. By 1987 a total of 17 squadrons will be fully equipped.

Above: A white JA37 of F13 shows the benefits of the canard layout at the high angles of attack essentail for air combat.

Below: JA37 Viggen fighter-bomber of the Brävalla-based F13 wing, RSAF. The second-generation Viggen is optimized for interception, as indicated by the armament of RB71 Sky Flash and RB24 Sidewinder AAMs, plus the 30mm KCA cannon.

Right: Pilot's-eye view as a Viggen prepares to take off. Most avionics are connected to a central computer, and the conventional instrumentation on the left is supplemented by two multi-mode head-down CRT displays plus a modern HUD.

Saab 39 Gripen

Origin: Sweden.
Type: Multirole fighter.
Engine: One 18,000lb (8,165kg) thrust Volvo Flygmotor RM12 augmented turbofan (licensed variant of GE F404).
Dimensions: Span 26·25ft (8·0m); length 45·91ft (14·0m); other figures not finalized.
Weights: The only figure is "normal max 17,645lb (8,000kg)".
Performance: Max speed (hi) about 1,320mph (2,124km/h, Mach 2), (SL) supersonic; required field length 3,280ft (1,000m).
Armament: One 27mm Mauser BK27 gun; four wing pylons for RB71 Sky Flash AAMs, RBS15F anti-ship missiles or various other attack loads; wingtip rails for RB24 Sidewinder AAMs; max weapon load not given.

Development: Fifth in the totally successful series of jet combat aircraft created by Saab, the JAS 39 actually has the maker's number 2110, the JAS 39 designation signifying Jakt (fighter), Attack, Spaning (reconnaissance). Partly because of economic inflation it is the smallest of the Saab fighters, but in capability it will be the greatest, with all-round capability exceeding even that of the Viggen, apart from the total weight of weapons carried.

Saab has had to tread a careful path between running too high a risk and failing to create an aircraft technically advanced enough to remain competitive into the 21st century. For example much of the airframe will be made of high-strength composites, and British Aerospace is the chief subcontractor for the wing, largely of carbon-fibre construction and provided with powered leading and trailing surfaces for maximum combat agility.

In conjunction with the fully powered swept canard foreplanes this wing should give the Gripen (Griffon) combat agility surpassing that of any aircraft flying in 1985, as well as the ability to operate safely, in the worst winter weather, from rough airstrips, highways and dirt roads. A further requirement for all Swedish military equipment is that front-line servicing is handled by short-service conscript personnel, so everything has to be reliable and foolproof.

The cockpit will of course be totally new, with a Martin-Baker S10LS seat, diffractive-optics HUD and three electronic displays, reprogrammable by the pilot to show just the items he needs at each point in the mission. Other features include fly-by-wire controls (tested in a Viggen), rear-fuselage airbrakes, plain lateral engine inlets, Ferranti/Ericsson multimode radar and pod-mounted FLIR (which together handle the reconnaissance mission, apart from pod-mounted optical cameras), and extremely comprehensive internal and external EW systems, with comprehensive jamming and dispensing capability.

Five prototypes are being built, and the inventory programme is at

Above: New-generation missiles are planned for JAS 39 Gripens — AIM-120 AMRAAMs (top) and anti-ship RBS15Fs; both aircraft are depicted with AIM-9s.

present 140 aircraft, of which about 25 are expected to be tandem dual two-seaters. All are to be delivered before the year 2000.

Saab 105

Origin: Sweden.
Type: Multi-role tactical aircraft.
Engines: (SK 60) two 1,640lb (743kg) thrust Turboméca Aubisque turbofans; (others) two 2,850lb (1,293kg) General Electric J85-17B turbojets.
Dimensions: Span 31·75ft (9·50m); length (SK 60) 34·4ft, (others) 35·44ft (10·80m); height 8·83ft (2·70m); wing area 175sq ft (16·3m²).
Weights: Empty (SK 60) 5,534lb (2,510kg), (G) 6,757lb (3,065kg); max (SK 60) 8,380lb (3,800kg) aerobatic or 8,930lb (4,050kg) other, (G) 14,330lb (6,500kg).
Performance: Max speed (clean, SL) (SK 60 at 8,820lb, 4,000kg) 447mph (720km/h), (G) 603mph (970km/h); initial climb (SK 60) 3,440ft (1,050m)/min, (G) 11,155ft (3,400m)/min; service ceiling (SK 60) 39,370ft (12,000m), (G) 42,650ft (13,000m); max range (hi, internal fuel) (SK 60) 1,106 miles (1,780km), (G) 1,230 miles (1,980km).
Armament: Six hardpoints carry total load of (SK 60) 1,543lb (700kg), (Ö) 4,409lb (2,000kg), (G) 5,180lb (2,250kg) including gun pods to 30mm calibre, 1,000lb bombs, missiles, tanks and target gear.

Development: This neat unswept machine began life as a trainer for the Swedish Flygvapen, seating pupil and instructor in side-by-side ejection seats in a pressurized cockpit with upward-hinged canopy. Eventually Saab delivered 150 in three models, the SK 60A trainer, the SK 60B with provision for attack missions, and the SK 60C with both attack and reconnaissance capability with a nose camera installation. Most important user is Austria, whose 40 Saab 105Ö — of which 36 remained in 1985 — comprise virtually the whole of that country's tactical airpower, equipping three squadrons. The Saab 105Ö carries a greater weapon load, yet the increased engine power permits single-engined takeoffs. The range of stores carried on six pylons is considerable, and the 105s, in both Sweden and Austria, are also used for liaison and target towing.

Above: The Saab 105G prototype attack/trainer was based on the Austrian Air Force 105Ö.

Below: The SK 60C variant of the Saab 105 has a reconnaissance camera installation in the nose.

SEPECAT Jaguar

Origin: France/UK.
Type: (GR.1 and A) single-seat all-weather attack.
Engines: Two Rolls-Royce/Turboméca Adour augmented turbofans: (A 7,305lb (3,313kg) Adour 102; (GR.1.) 8,040lb (3,647kg) Adour 104; (International) 9,270lb (4,205kg) Adour 811.
Dimensions: Span 28·5ft (8·69m); length 50·91ft (15·52m); height 16·04ft (4·89m); wing area 260·72sq ft (24·18m²).
Weights: Empty, about 15,432lb (7,000kg); "normal takeoff" (internal fuel and gun ammunition) 24,149lb (10,954kg); max 34,612lb (15,700kg).
Performance: Max speed (lo, some external stores) 840mph (1,350km/h, Mach 1·1), (hi, some external stores) 1,055mph (1,700km/h, Mach 1·6); attack radius, no external fuel, hi-lo-hi with bombs, 530 miles (852km); ferry range 2,614 miles (4,210km).
Armament: (A, Two 30mm (DEFA 553 each with 150 rounds; five pylons for total external load of 10,500lb (4,763kg); GR.1 as above but guns two 30mm Aden; (International) various, guns Aden

or DEFA, often overwing AIM-9P or Magic AAMs and various ASMs.

Development: Originally developed jointly in an Anglo-French project for a light attack aircraft and supersonic trainer, the Jaguar actually matured as one of the most effective attack aircraft of its day.

Features of the basic aircraft include a small but efficient high-mounted wing, with full-span double-slotted flaps, spoilers for roll control (supplemented by the tailer-ons) and powered slats which can be used in combat. The extremely small engines are fed by plain inlets, and instead of going for Mach 2 the designers aimed at good agility, comprehensive avionics, and the ability to operate from rough unsurfaced strips with heavy weapon loads. The main landing gears have long-stroke levered suspension, low-pressure twin tyres on each leg, and anti-skid brakes, the latter being backed up by a braking chute and arrester hook. The cockpit has an upward-hinged canopy (two in the tandem dual version) and a Martin-Baker Mk 9 zero-zero seat (except in French Jaguars which have the earlier Mk 4 seat). The two-seater has the same internal fuel capacity as other versions, the extra cockpit being added in the nose at the expense of some avionics and one gun.

Below: Launch of an R.550 Magic short-range AAM from the overwing pylon developed for the Jaguar International.

Above: Jaguar International S of the Oman Air Force.

The RAF Jaguar GR.1 has a full INS (recently updated together with a revised cockpit), HUD, projected-map display and laser in a "chisel nose", as well as engines uprated as a field modification. A total of 202 were delivered.

Two of the RAF squadrons, II and 41, can add a multisensor pod on the centreline for use in the armed recon role. A standard fitment has been the ARI.18223 passive RWR system, but there has been little ECM protection other than extremely low flying. In 1983 an interim fit of six jammer/chaff cartridges was added in a box under the parachute bay, and the RAF is installing a new RWR, jammer pod and Phimat chaff dispenser.

The Armée de l'Air Jaguar A has a simpler avionic system, with twin-gyro platform and doppler, but the final 30 of the 200 of both A and E (two-seat) Jaguars are equipped

with the Atlis II TV/laser pod used in conjunction with Matra smart bombs and the AS.30L laser-homing ASM. French Jaguars equip ECs (escadres de chasse) at St Dizier, Nancy, Toul-Rosières, Istres and Bordeaux.

There are numerous customer options in the export Jaguar International, most of which have uprated engines and avionics based on those of the GR.1. Those of Oman have a Marconi Avionics 920C computer and carry Sidewinder AAMs. The most advanced Jaguars so far exported are those of India, where 40 were delivered from BAe and 45 more are being assembled by HAL. Those tasked in the attack role have an RAF-type HUD or (on HAL-assembled aircraft) a new HUD and weapon-aiming system similar to that of the Sea Harrier, with the Magic AAM used for self-defence, Anti-ship Jaguars have the Agave nose radar (with optional chin laser), and AM39 Exocet missiles.

The latest order for new-build Jaguars was for 18 for Nigeria, with an option on a similar quantity. At the time of writing there had been no official confirmation of reports that Jaguars withdrawn from RAF Germany are to be purchased by Chile. One RAF GR.1 is serving as a research aircraft in a totally unstable CCV configuration in support of the BAe Advanced Combat Aircraft (ACA) programme.

Below: Jaguar GR.1 of 20 Sqn, RAF Germany, armed with laser-guided bombs, ALQ-101 ECM pod (port) and Phimat chaff dispenser.

Shenyang J-8 "Finback"

Origin: People's Republic of China.
Type: Multirole fighter.
Engine: (prototypes) Said to be one WP-7 (Tumanskii R-11) afterburning turbojet.

Development: In the early 1970s the People's Republic launched a programme to design its own supersonic fighter. The resulting J-8 has been inspected by Western visitors (including a US delegation in September 1980), but the information that has leaked out is sparse and contradictory. Described as a delta, larger than the J-7 (MiG-21), it is nevertheless said to be powered by the small engine fitted to that aircraft. Despite this, it is described as a Mach 2 aircraft, and to incorporate technology gleaned from inspecting MiG-23s supplied by Egypt.

According to a July 1981 report by the US Defense Intelligence Agency, the J-8 was not ready for production because the Chinese had not "produced adequate jet engines to power the aircraft". Yet more than a year earlier a Chinese-built Spey 202, built under Rolls-Royce licence, passed its full type test at Derby with flying colours, the rated thrust of this engine being 20,515lb (9,306kg). NATO has assigned the reporting name "Finback". Eventually reports of the design will begin to make sense. The Chinese aircraft industry has tended to produce adaptations of Soviet designs to avoid lengthy research and development programmes. Its capabilities, and potential for the future, should not, however, be underestimated.

Below: The J-8 "Finback" twin-engined delta-wing fighter is a Chinese development of the MiG-21/J-7; engine availability must be a major problem.

SIAI-Marchetti S.211

Origin: Italy.
Type: Light attack aircraft.
Engine: One 2,500lb (1,134kg) thrust P&WC JT15D-4 turbofan.
Dimensions: Span 27·66ft (8·43m); length 30·54ft (9·31m); height overall 12·46ft (3·8m); wing area 135·63sq ft (12·6m²).
Weights: Empty 3,560lb (1,615kg); loaded (clean) 5,511lb (2,500kg), (max) 6,834lb (3,110kg).
Performance: Max speed (clean) 460mph (740km/h); service ceiling 40,000ft (12,200m); takeoff to 50ft (15m) 1,640ft (500m); attack radius (hi-lo-hi, four rocket launchers, full mission reserves) 345 miles (556km); ferry range 1,543 miles (2,483km).
Armament: Total external load of 1,302lb (600kg) carried on four pylons rated at (inners) 660lb (300kg), (outers) 330lb (150kg); loads can include many bombs or rocket launcher types, recon or other pods, gun pods (20mm on inboard stations only) or 77gal (350lit) drop tanks (inboard only).

Development: Following the same configuration as the Alpha Jet, but with an unswept wing of supercritical profile, the S.211 is much smaller and has less than half the power. Despite this the two occupants sit in staggered Martin-Baker seats in a pressurized cockpit, and lack little in the way of sub-systems and avionics.

Customer options even include attack radar, HUD and a comprehensive EW/ECM suite. Features include electrically driven Fowler flaps, hydraulically powered ailerons, tailplane, airbrake and landing gear, and a sideways-hinged canopy through which the crew can eject.

Singapore, the fourth customer but the first to be named officially, is buying six and assembling four; the RSAF hopes to assemble a further 20, with locally manufactured components.

Below: One of three prototypes of the S.211 trainer and light attack aircraft, which features comprehensive equipment.

Soko Jastreb

Origin: Yugoslavia.
Type: Single-seat attack aircraft.
Engine: One Rolls-Royce Viper turbojet, 3,000lb (1,361kg) Mk 531.
Dimensions: Span (excl tanks) 34·35ft (10·47m); length 33·92ft (10·34m); height overall 10·75ft (3·28m); wing area 209·14sq ft (19·43m²).
Weights: Empty 6,217lb (2,820kg); max 11,243lb (5,000kg).
Performance: Max speed clean (19,680ft/6,000m) 510mph (820km/h); max range (tip tanks, no weapons) 945 miles (,520km).
Armament: Three 12·7mm guns with 135 rounds each, and eight wing stations, inners for loads up to 551lb (250kg) each, the rest for single 127mm rockets.

Development: First Yugoslavian jet to go into production, the G-2 Galeb (Seagull) is an all-metal tandem trainer, of which the Jastreb (Hawk) is the attack version. Galeb has Folland light ejection seats and modest weapon provisions. Flaps and gear are hydraulic, but flight controls manual, and each cockpit has its own side-hinged canopy. The G-2A-E of 1974, with updated avionics, serves in large numbers in Libya.

The single-seat Jastreb has a locally strengthened structure, and can have a drag chute and pressurization. Again Libya was a major customer, using J-1 Jastrebs in the Co-In role. There is also a two-seat TJ-1, which like the J-1 can carry recon cameras in the tank noses and tow air-firing targets.

Below: Soko G-2 Galeb two-seat armed trainer, of which the Jastreb is the attack version.

Soko Super Galeb

Origin: Yugoslavia.
Type: Trainer and light attack aircraft.
Engine: One 4,000lb (1,814kg) thrust Rolls-Royce Viper 632 turbojet licence-built by ORAO Yugoslavia.
Dimensions: Span 32ft 5in (9·88m); length 38ft 11in (11·86m); height overall 14ft 0·5in (4·28m); wing area 209·9sq ft (19·5m²).
Weights: Empty 7,165lb (3,250kg); loaded (training) 10,495lb (4,760kg), (combat) 13,470lb (6,110kg).
Performance: Max speed (19,680ft/6,000m) 565mph (910km/h); ceiling 49,200ft (15,000m); takeoff over 50ft (15m) 3,117ft (950m); combat radius (lo-lo-lo, gun pack and two rocket packs) 186 miles (300km).
Armament: Centreline gun pod

with 23mm GSh-23 gun with 200 rounds; four wing pylons rated at 7,721lb (350kg) inboard and 551lb (250kg) outboard for extremely wide range of weapons or other stores including photo/IR linescan recon pod or tanks.

Development: A totally new design, the G-4 Super Galeb is a thoroughly competitive tandem trainer which bears a remarkably close similarity to the BAe Hawk, though it has a turbojet instead of a turbofan, with rather less power. Flight controls (except for the rudder) are fully powered, the slab tail-planes having anhedral, and the stepped cockpits have Martin-Baker J10 rocket seats (which eject through the two side-hinged canopies) and are pressurised. The G-4 is a far more potent machine

than its predecessors, and is priced well below most rivals, but no foreign orders have yet been placed.

Above: Soko G-4 Super Galeb prototype shows its close resemblance to the BAe Hawk.

Soko Orao

Origin: Romania and Yugoslavia.
Type: Close-support, attack and reconnaissance aircraft.
Engines: Two Rolls-Royce Viper turbojets licence-built by Turbomecanica of Romania and ORAO of Yugoslavia, (development aircraft and 93A) 4,000lb (1,814kg) thrust Mk 632-41R, (production Orao and 93B) 5,000lb (2,268kg) Mk 633-47 with afterburners.
Dimensions: Span 31·56ft (9·62m); length (single-seat, inc probe) 48·88ft (14·9m), (two-seat, inc probe) 51·17ft (15·9m); height overall 14·6ft (4·45m); wing area 279·86sq ft (26·0m²).
Weights: Empty (single-seat 93A) 13,558lb (6,150kg), (93B target) 13,008lb (5,900kg); loaded (clean) (A) 19,458lb (8,826kg), (B target) 18,953lb (8,597kg); max (A) 22,765lb (10,326kg), (B target) 22,260lb (10,097kg).
Performance: Max speed at SL (A) 665mph (1,070km/h), (B target) 721mph (1,160km/h); service ceiling (A) 34,450ft (10,500m), (B target) 41,010ft (12,500m); takeoff/landing over 50ft (15m) (A) 5,250ft (1,600m); mission radius (A, max external weapons) (lo-lo-lo) 186 miles (300km), (hi-lo-lo) 224 miles (360km).
Armament: Two internal GSh-23 twin-barrel 23mm guns each with 200 rounds; five pylons, centreline rated 1,102lb (500kg), inners 992lb (450kg) and outers 551lb (250kg), for total load of 3,307lb (1,500kg).

Development: By far the biggest aircraft project ever undertaken in any Balkan country, the YuRom (so called from its participating countries) has been delayed by the decision to power it with a locally developed afterburning version of the Viper turbojet, and also by protracted arguments on precisely what the two air forces (one of them a member of the Warsaw Pact) really want. The design crystallized as a kind of lower powered Jaguar, without great pretentions as an air-

superiority fighter but capable of giving a good account of itself in front-line tactical missions in support of a land battle.

From the start the project has been split 50/50 between the two countries (though the fact that a prototype made its first flight in each country within 20 minutes of each other on the same day is said to have been coincidental). There is no duplication, CNIAR's plant at Craiova making the forward and centre fuselage and horizontal tail and SOKO's factory at Mostar the remainder. Thus the airframe is common to both nations, and so is the Messier-Hispano-Bugatti landing gear based on that of the Jaguar, as well as the Martin-Baker 10J seats and a few other items. Yet each country has elected to go ahead with quite different systems and equipment, so that what finally emerge are two quite different aircraft, without even considering avionics and weapons! To the Romanians it is the CNIAR 93 while the Yugoslavs call it the SOKO Orao (Eagle).

Basic features include large slotted flaps, outboard powered ailerons, powered full-span slats, and powered rudder slab tailplanes, the latter having anti-flutter tip masses on all aircraft seen by early 1984 but intended to be eliminated from production machines. Large perforated airbrakes are hinged ahead of the main gears beneath the bays oc-

Above: Yugoslav Air Force Orao 1 equipped with centreline reconnaissance pod spools up its engines on the runway.

cupied by the guns, which in turn are immediately below the engine inlet ducts which are very simple. A braking parachute is housed beneath the rudder. The production machine has a large Lerx (leading-edge root extension) as well as a narrow strake along each side of the nose to improve airflow in tight turns (7g is permissible).

The cockpit is pressurized and has an upward-opening canopy, and that of the Orao would be instantly familiar to an RAF Jaguar pilot; the IAR 93B has a more Soviet type avionic fit, including comprehensive EW systems. No radar has been requested but various weapon-aiming systems are being studied apart from the current Ferranti gyro sight.

An afterburning aircraft at last got into the air in the final days of 1983, and this enhances performance at

some cost in greater fuel burn. It is expected that a proportion of production machines will be two-seaters, which have dual pilot cockpits, the front cockpit moved slightly forward and the rear cockpit in place of a fuselage fuel cell giving reduced endurance. The production aircraft is also expected to have integral-tank wings, replacing small separate bladder cells in aircraft flying before 1984.

It is possible that after so much passage of time the original national requirements have changed, and the Yugoslavs are certainly keeping a very low profile on this programme, though Romania has ordered 201 IAR 93As and 165 IAR 93Bs for inventory service and these were expected to be delivered from late 1984. Totals are liable to vary, however.

Below: First prototype of the Orao, Yugoslav version of the aircraft developed jointly with Romania's IAR-93.

Sukhoi Su-7 "Fitter"

Origin: Soviet Union.
Type: Ground attack fighter.
Engine: One Lyul'ka afterburning turbojet, (7, 7B) 19,841lb (9,000kg) AL-7F (later variants) AL-7F-1 rated at 15,432/22,046lb (7,000/10,000kg).
Dimensions: Span 29·3ft (8·93m); length (incl probe) 57·0ft (17·37m), (7U) 58·71ft (17·7m); height 15·4ft (4·7m); wing area 297 sq ft (27·6m²).
Weights: (BMK, typical) empty 19,000lb (8,620kg); normal loaded 26,455lb (12,000kg); max 29,750lb (13,500kg).
Performance: Max speed (36,000ft/13,500m clean) 1,055mph (1,700km/h), Mach 1·6, (four loaded pylons) 788mph (1,270km/h, Mach 1·2); (SL) max afterburner 837mph (1,345km/h, Mach 1·1, (dry) 530mph (850km/h); max initial climb (afterburner, clean) 29,000ft (8,840m)/min; service ceiling 49,700ft (15,150m); combat radius (with tanks, hi-lo-hi) 200/300 miles (322/480km); range (two tanks) 900 miles (1,450km).
Armament: Two NR-30 guns each with 70 rounds; two side-by-side fuselage pylons and (7, B) two, (all

later variants) four, (many BMK) six, underwing pylons, normal load being two fuselage tanks (total 2,100lb/952kg) and up to 2,205lb (1t) ordnance on wings.

Development: This large single-engined machine was planned in 1953 as the fighter to counter the F-100 and other USAF Century-series supersonic aircraft. Much larger than same-shape prototypes from the MiG OKB, it featured a sharply swept wing with a plain leading edge and outboard ailerons, powered slab tailplanes half-way up the tubular rear fuselage, and a straight-through engine duct from the fixed double-shock inlet in the nose, bifurcated to pass each side of the cockpit and then rejoining to form a single tube passing through the centre of the fuselage tanks. The enormous engine promised abundant power, and after adding large fences near mid-span and at the tips, handling became superb. Production started in 1958, but by this time it was appreciated that Frontal Aviation's fighter missions could be flown by the much smaller MiG-21, and the Sukhoi Su-7 was thus deployed in the ground attack role.

From the start its worst failing was a combination of bombload and mission radius — poor by any standard, and ridiculous for a machine of such size and power. Fuel burn at sea level in afterburner is no less than 800lb (363kg) per minute, so the twin side-by-side belly drop tanks are needed on almost every mission. With these in place the maximum weapons load is only 2,205lb (1,000kg), though as this can comprise two SC500 bombs of 1,102lb (500kg) each the Sukhoi can pack a

Above: Indian Air Force Su-7s played a major role in the 1971 war with Pakistan, though several were lost through excessive fuel consumption.

fair punch. In addition, the 30mm guns also have devastating firepower, blasting 1,200 shells per minute each weighing almost 2·2lb (1kg) with high muzzle velocity, the destructive effect being more than double that of a 30mm DEFA or Aden and the AP rounds having

Sukhoi Su-9/-11 "Fishpot"

Origin: Soviet Union.
Type: All-weather interceptor.
Engine: (9) One 19,840lb (9,000kg) Lyul'ka AL-7F afterburning turbojet (11) one 22,046lb (10,000kg) AL-7F-1.
Dimensions: Span 27·66ft (8·43m); length (incl probe) (9) about 57·0ft (17·37m), (11) 60·0ft (18·29m); height 16·0ft (4·88m); wing area about 280sq ft (26m²).
Weights: Empty (9) about 19,000lb (8,620kg), (11) about 20,000lb (9,000kg); loaded (9) about 27,000lb (12,250kg), (11) 30,000lb (13,600kg).
Performance: Max speed (both clean 36,000ft/11km about 1,320mph (2,125km/h, Mach 2), (two tanks and AAMs), (Su-9) about 750mph (1,200km/h, Mach 1·14), (11) 840mph (1,350km/h, Mach 1·27); max initial climb (both) about 27,000ft (8·2km)/min; service ceiling, from about 55,000ft (17·76km) for Su-9 with AAMs to 62,000ft (18·9km) for Su-11 clean); range (both, high-altitude, two tanks and AAMs) about 7 miles (1,125km).
Armament: (9) Four K-5M (AA-1 Alkali) AAMs; (9U) usually not fitted; (11) two AA-3 Anab AAMs (normally one IR homing, one SAR homing).

Development: Having completed more than a quarter century of service with the Soviet Air Force, the Su-9/-11 series of interceptors is virtually obsolete. As a front-line fighter it has largely been replaced by the newer Su-15, which retains the several common parts. Many of the Su-9 single-seaters have been converted into drones and expended

in missile firings, but the Su-9U two-seat trainer was retained in service for conversion training of pilots assigned to the dwindling number of Su-11 units.

The Su-9 was developed from the mid-1950s T-3 tailed-delta prototype, a design evaluated against the swept-wing S-1 later adopted for service as the Su-7. The quickest way of getting a supersonic interceptor into service was to mate a developed T-3 airframe with the R1L Spin Scan radar and the AA-1 Alkali (K-5M) missile. By the late 1950s, the resulting Su-9 was in service, and the bureau was hard at work on the definitive Su-11.

This combined the proven wings, tail and centre/rear fuselage of the Su-9 with a revised nose section of

Right: The Su-9's ruggedness fits it for harsh operating conditions.

Below: Ten tonnes of thrust help an Su-11 get airborne.

considerable ability to pierce tank armour.

Best features of the Su-7 are its incredible all-round strength (which extends to the very long nose boom, from which any pilot can practise gymnastic pull-ups!) and its superb handling in all flight regimes. Even today large numbers of Su-7s, in both single- and two-seat versions, are still flying with Frontal Aviation simply as enjoyable hacks on which to get in flight time. This is despite the fact that the extremely reliable triplicated flight-control system still calls for large pilot input forces which have been likened to flying a Hunter in the manual mode. Another drawback is the extremely high takeoff and landing speed, which unless ATO rockets are used demands a long (10,000ft/3km) runway even at sea level, and the twin drag chutes of all current versions are mandatory.

The major users such as Egypt and India have decided it is not cost-effective to subject surviving Su-7s to a major avionic update, and in an era of high fuel price their numbers are diminishing fast. Some still remain, however, with the air forces of Afghanistan, Algeria, Iraq, North Korea, Syria and the Yemen People's Republic, as well as with most Warsaw Pact countries.

near-parallel form and a large radome sized to accommodate the much-improved Skip Spin (Uragan 5B) radar. The latter is a powerful 100kW set with a range of up to 25 miles (40km), and designed to illuminate targets for attack with the radar-guided version of the AA-3 Anab missile. Available in radar and infra-red guided forms, Anab is a massive weapon weighing around 600lb (275kg) and with a maximum range of at least 10 miles (16km), and perhaps as much as twice this figure. Each Su-11 carries two rounds — one under each wing. Normal Soviet Air Force practice is to carry one radar-guided round and one IR-homer. No cannon is fitted — a common but mistaken late-1950s design feature in several countries. (The later Su-15 entered service without a built-in gun but was later fitted with a clip-on gun pack.)

Like the Su-7, the Su-9/11 was an inelegant solution to the problem of creating a supersonic fighter, and one which substituted sheer brute force (in the form of a ten-tonne thrust engine) for aerodynamic refinement. A top speed of only Mach 1·27 was a disappointing performance for an aircraft in the weight class of the F-8 Crusader and with 10 per cent more thrust. Another negative feature was the need for long runways.

Probably as a result of these deficiencies, the type was never exported, but was retained only for home defence. When Soviet Air Force units were assigned to the defence of Egypt after the Six-Day War, the Su-15 was despatched rather than the Su-11. The latter could easily have been deployed with the Egyptian Air Force which had Su-7s.

Sukhoi Su-15 "Flagon"

Origin Soviet Union.
Type: All-weather interceptor.
Engines: Two Tumanskii afterburning turbojets, (A to D) 13,668lb (6,200kg) R-11F2-200, (E, F) 14,550lb (6,600kg) R-13F-300.
Dimensions: Span (A) about 30·0ft (9·14m), (others) 34·5ft (10·5m); length (incl probe) 68·0ft (20·5m); height 16·5ft (5·0m); wing area 385sq ft (35·7m²).
Weights: Empty (A) about 25,000lb (11,340kg), (F) about 27,000lb (12,250kg); loaded (A) about 35,275lb (16,000kg), (F) about 40,000lb (18,000kg), (F, max with external tanks) 44,900lb (20,000kg).
Performance: Max speed (clean, 36,000ft/11,000m), about 1,650mph (2,655km/h, Mach 2·5); (with AAMs) about 1,520mph (2,450km/h, Mach 2·3). initial climb about 45,000ft (13,700m)/min; service ceiling about 65,600ft (20,000m); combat radius (hi) about 450 miles (725km), (with tanks) 620 miles (1,000km); ferry range about 1,400 miles (2,250km).
Armament: Four wing pylons for AAMs, a typical load comprising two medium-range AAMs (AA-3 Anab or R-23R/AA-7 Apex, each in both IR and radar versions) on the outer pylons and two close-range AAMs (R-60/AA-8 Aphid) on the inners, the latter sometimes being paired to give a total of six missiles; two body pylons plumbed for drop tanks but also alternatively used to carry two GSh-23 gun pods, each with 200 rounds.

Development: A direct descendant of the original T-3 prototype of 1956, with a 57º pure-delta wing, this all-weather interceptor is almost certainly the last in an extremely long and generally very successful family. Via the single-engined Su-9 and -11, now no longer with PVO combat regiments, the Su-15 was developed with two smaller engines in order to provide twin-engine safety, greater thrust for even higher flight, make room for much more internal fuel, and also leave the nose free for a large radar (which Sukhoi's OKB had always found difficult to accommodate in a nose inlet system, especially one with variable-geometry features for a supersonic flight at high efficiency).

Like all PVO machines the Su-15 was designed to operate from long paved runways, so unlike FA aircraft it has an extremely high wing-loading. This goes well with the mission of long-range stand-off interception, but virtually eliminates good close dogfight performance and also results in takeoff/landing speeds close to 300mph (482km/h) at maximum weights, and anti-skid brakes with computer control are fitted together with a large cruciform brake chute.

All Su-15s in current service have a wing of increased span, with a midspan kink and reduced sweep outboard, which improves field length and low-speed handling. Until the mid-1970s the radar was the big J-band set called Skip Spin by NATO (previously used in the Yak-28P), housed in a giant nose radome of pure conical form. Today a much newer radar is used, in a curved ogival radome, as well as uprated engines fed by improved inlet ducts, and the missiles have been updated and gun pods added. All late versions have twin nose wheels.

About 700 have been in use for many years, but production ceased some time ago. The only major shortcoming of the Su-15 is that the mission radius, though good in a European context, is not long enough for this type to cover the whole of the Soviet Union's vast frontier.

It was an Su-15 from one of the incredibly numerous and powerful Voyska PVO regiments in the Far East that on 1 September 1983 shot down Korean Air Lines Flight 007, inward bound to Seoul, killing the 269 on board. Hours earlier a USAF RC-135 had been near the area, but there can have been no possible confusion over the identify of the 747, which obviously would have been tracked by surveillance radars over a great distance. An AAM fired at close range was used to destroy the airliner. In April 1984 Marshal of Aviation A. L. Koldunov, PVO Commander, wrote in *Pravda* that the achievement "was an historical example of PVO's high level of readiness to perform their military duty". The Soviets continued to display such deplorable callousness even during the search for wreckage.

Below: A Soviet Air Defence Force (PVO) pilot receives instruction in the cockpit of a "Flagon-D", first major operational version of the Su-15. Note the characteristic kinked wing leading edge and intakes for afterburner cooling.

Above: One of the early-model Su-15s shown at the Domodedovo flying display in July 1967.

Below: "Flagon-F", last known production version of the Su-15, with AA-3 and AA-8 missiles.

Sukhoi Su-17/-20/-22 "Fitter"

Origin: Soviet Union.
Type: Ground-attack fighter.
Engine: (most) One Lyul'ka AL-21F-3 afterburning turbojet with ratings of 17,200/24,700lb (7,800/11,200kg); (current variants) one Tumanskii R-29B afterburning turbojet with ratings of 18,960/27,500lb (8,600/12,500kg) (estimated).
Dimensions: Span (28°) 45·92ft (14·0m), (62°) 34·77ft (10·6m); length (basic -17 incl nose probes) 61·51ft (18·75m), (later variants) 62·75ft (19·13m), length of fuselage (inlet lip to nozzle) (-17) 50·53ft (15·4m), (later versions) 51·76ft (15·78m): wing area (28°) 431·6sq ft (40·1m²).
Weights: (estimated) Empty (Fitter-C) 22,050lb (10,000kg), (-H) 22,500lb (10,200kg); loaded (clean) (-C) 30,865lb (14,000kg), (-H) 34,170lb (15,500kg); max loaded (-C) 39,020lb (17,700kg), (-H) 42,330lb (19,200kg).
Performance: Max speed (clean, typical), (SL) 800mph (1,290km/h, Mach 1·05), (36,000ft/11,000m) 1,435mph (2,300km/h, Mach 2·17), (SL, with typical external stores) 650mph (1,050km/h); initial climb (clean) 45,275ft (13,800m)/min; service ceiling 59,050ft (18,000m); takeoff run at 17 weight, 2,035ft (620m); combat radius (-C, 2t bombload, hi-lo-hi) 391 miles (630km), (-H, 3t bombload, hi-lo-hi) 435 miles (700km); ferry range, four tanks (-C) 1,400 miles (2,250km), (-H) 1,700 miles (2,750km).
Armament: Two NR-30 guns each with 70 rounds and two K-13A (Atoll) or R-60 (Aphid) AAM; eight pylons (tandem pairs under fuselage, under LE of wing root and under wing-pivot fences) for total of 8,820lb (4,000kg) external ordnance/tanks.

Development: When the Su-22IG (IG meaning variable geometry in Russian) was demonstrated at an air show in 1967 few Western analysts took it seriously. It was basically an Su-7 with just the outer portions of its wings pivoted. Five years later it was suddenly discovered that large numbers of slightly improved models were in service, and from 1972 until 1977 many successively improved examples were put into use by Soviet Frontal Aviation, AVMF (Naval Aviation) and two WP air forces. Others have been exported, notably to Peru (52 remaining in use) and Libya (102).

This partial "swing-wing" configuration was developed in Moscow (TsAGI) in late 1963, and it was studied in connection with such machines as the M-4, Tu-22 and Tu-28/128 (and actually put into production with the Tu-22). Meanwhile, Sukhoi had already planned a series of improvements to the Su-7 involving later engines, dorsal spines of increasing size (as seen on the MiG-21), greater weapon loads and improved EW suites. The first stage was adoption of the big AL-21 engine which, despite its smoke at full power, burns less fuel than the AL-7 on most missions while giving much greater thrust.

Internal fuel was slightly increased, but the really big difference was that the swing wing and more thrust eliminated the previous range/payload shortcomings; in round figures this initial VG version, the Su-17 series, lifts twice the weapon load over mission radii increased by 30 per cent while eliminating ATO rockets and yet using airstrips half as long as previously! At the same time control input forces are reduced and inflight agility is improved, both turn radii and rate of roll being dramatically better.

Many of these early Su-17s are still in use, along with improved models with a longer nose housing a laser receiver in the conical inlet centre-body and a chin fairing which among other things accommodates a terrain avoidance radar. Almost all aircraft, including two-seaters (which have a down-tilted forward fuselage to improve view), have eight stores pylons. Some, including two-seaters, have only one gun, on the right.

The non-fighter designations Su-20 and -22 were applied to all subsequent variants, the former being the export Su-17 (including aircraft for WP countries), which all have various avionic deletions compared with those in Soviet service. Peru complained that its first batches, which have the designation Su-22 bestowed for export purposes though they are basically simplified -17s, were gravely lacking in navaids, had an almost useless Sirena 2 RWR and were fitted with IFF (not the usual SRO-2M) which was incompatible with SA-3 "Goa' SAMs supplied at almost the same time!

By this time Peru had signed for later variants, called "Fitter-F" and

Above: Two-seat Su-22U "Fitter-G" comes in to land with its variable-geometry wings in the fully-forward position.

Below: Poland was one of five countries to receive the Su-20 "Fitter-C" export version of the ground-attack Su-17.

"Fitter-J" by NATO, the latter being a member of the final sub-family with the smaller and much later-technology Tumanskii engine and a redesigned fuselage and tail. Features of these aircraft include a much deeper spine providing a major increase in internal fuel, a raised and redesigned cockpit giving a better pilot view and more room for additional avionics, two extra pylons, slightly bulging rear fuselage, and a vertical tail of increased height, with a dorsal fin, and added ventral fin.

Two-seaters of this family have a cockpit and canopy arrangement totally different from earlier versions, with a small metal rear canopy with a square window on each side. It is not known if a periscope is fitted; in earlier two-seaters (called "Moujik") this can be extended at below 373mph (600km/h).

All later variants have zero/zero seats, while earlier aircraft have rocket-assisted seats for zero height above 87mph (140km/h). A further important addition is air-conditioning.

Below: Su-22BKL "Fitter-J" of the Libyan Air Force armed with a pair of AA-2 Atoll (Soviet "Sidewinder") missiles.

Sukhoi Su-24 "Fencer"

Origin: Soviet Union.
Type: All-weather attack and reconnaissance.
Engines: Two afterburning engines (see below).
Dimensions: (estimated) Span (16⁰) 56·25ft (17·15m), (68⁰) 31·25ft (9·53m); length overall 69·83ft (21·29m); height 18·0ft (5·5m); wing area (16⁰) 50sq ft (46·4m²).
Weights: (estimated) Empty 39,700lb (18,000kg); loaded (clean) 64,000lb (29,000kg); max 87,080lb (39,500kg).
Performance: Max speed (clean, 36,000ft/11,000m) 1,590mph (2,560km/h, Mach 2·4), (clean, SL) about 870mph (1,400km/h, Mach 1·14); (max external load, hi) about 1,000mph (1,600km/h, Mach 1·5), (max external load, SL) about 620mph (1,000km/h); service ceiling (with maximum weapon load) 57,400ft (17,500m); combat radius (lo-lo-lo, 8t bombload) 200 miles (322km), (lo-lo-hi, 2·5t bombload) 590 miles (950km), (hi-lo-hi, 2,500kg bombload) 1,115 miles (1,800km); ferry range (six tanks) about 4,000 miles (6,440km).
Armament: Eight identical MERs (multiple ejector racks) each rated at 2,205lb (1,000kg), four under fuselage, two under fixed gloves and two pivoted to swing wings, for total load of 17,635lb (8,000kg); glove pylons plumbed for largest drop tanks seen on Soviet aircraft; two large blisters cover items installed in the underside of the fuselage, one of which (and possibly both) is a gun.

Development: Spurred by the USAF TFX (F-111) programme, this aircraft was planned at the same time and using the same excellent aerodynamics as the MiG-23, but using two engines instead of one. Despite the fact that it is a fundamentally very old and uncompetitive engine, the consensus of Western opinions is that the Su-24 is powered by the AL-21F-3, or a close relative. This is despite the fact that all the installational features familiar with this engine are absent. Thrust of each engine with max augmentation is in the 25,000lb (11,340kg) class.

From the start the Su-24 was a top-priority project, with nothing whatsoever compromised. Design engineers were made up into the biggest team ever seen, including many drawn from Poland's PZL, and in every detail the result is the very best that can be achieved. In many respects the Su-24 resembles the Tornado, though on a physically larger scale; it follows the F-111 in only one major feature: side-by-side seating for the pilot and weapon-systems officer. Unlike the American machine it has half its heavy weapon load under the fuselage, and the way the doors over the bays for the twin-wheel main gears fold down towards each other is particularly neat, the large panels immediately in front of them (covering the unidentified internal items) being the airbrakes.

There is no doubt the max wing loading of the Su-24 is greater than in any other combat aircraft at some 180lb/sq ft (878kg/m²), and combined with the max sweep of 68⁰ and near-absence of a fixed portion must result in outstandingly good buffet-free ride qualities in low-level missions at full power. In view of the totally uncompromised nature of this aircraft it is probable that the wing has a Mach/sweep programmer which continuously adjusts wing sweep, unlike the direct manual control needed on the Su-17/20/22.

Other airframe features include fully variable engine inlets, with auxiliary doors and ejectors; slender wings with full-span slats and double-slotted flaps; roll control by wing spoilers (at low speeds) and powered tailerons; a single vertical tail plus ventral fins at the chines of the wide flat-bottomed rear fuselage; and easily the best overall avionic fit seen on any tactical aircraft in service anywhere. Details are still the subject of speculation, and variations are already responsible for the so-far unexplained identification by NATO of three in-service variants called "Fencer-A, -B and -C". (All are believed to be side-by-side two-seaters.)

The main radar is unquestionably a pulse-doppler set of remarkable power and versatility. Terrain-following capability is provided by secondary TFR sets as in the F-111, and a separate doppler navigation radar is on the ventral centreline. The entire aircraft is covered with avionics, most of them flush or served by very small blisters. A laser ranger and marked-target seeker is in a small chin fairing, what are believed to be CW illuminators for radar missile guidance are ahead of the wing gloves on the top sides of the inlets, and the tail is a forest of flush aerials and small pods or blisters, varying from one aircraft to another.

It is probable that the internal EW/ECM suite is more comprehensive than in any other supersonic aircraft, though details are still being investigated by Western analysts. There is not the slightest doubt that the entire avionics installation was designed in parallel with the aircraft itself, and when any particular item was missing or unsuitable it was created from scratch.

Not least of the curious things about this ultra-important aircraft is that the dimensions given above were published in the West as estimates in 1974, when virtually nothing was known about what at that time was thought to be the "Su-19". They have never been altered, though their seeming exactness is questionable. NATO analysts have been franker when it comes to identifying the two ventral blisters, which cover permanently installed equipment. Both are probably guns, but of different types, in conformity with Soviet policy for use against different classes of target, but the official view is that only the left installation is a gun, the right-hand one being unidentified.

In 1984 the number of Su-24s in service was believed to be 800. They are serving in all peripheral Military Districts of the Soviet Union, the main concentration being in Europe but over 200 being around China and on the Pacific coastal areas. The two giant Su-24 forces are the 4th Air Army (Hungary and the Ukraine) and the 24th Air Army (Poland), each of which has five Su-24 polks (regiments) with 60 inventory aircraft apiece.

The three versions "Fencer-A, B and C" have not yet been described, but the most recent was only developed in 1981. According to the US DoD, some Su-24s are now being assigned to the ADD (long-range aviation) strategic force, "and the number assigned to this task is likely to increase by 50 per cent over the next few years".

A recent traveller to Riga Airport, Soviet Union, reported seeing rows of completely unpainted Su-24s obviously in combat service.

Below: Sukhoi Su-24 "Fencer" nuclear-capable, all-weather fighter-bomber of the Soviet Air Force. The very large fuel pods are attached to the fuselage (rather than to underwing pylons) adding to the already considerable range.

Sukhoi Su-25 "Frogfoot"

Origin: Soviet Union.
Type: Close-support attack and Co-In aircraft.
Engines: Two turbojets or turbofans (see below).
Dimensions: (estimated Span 50·83ft (15·5m); length 47·5ft (14·5m); height 16·75ft (5·1m); wing area 400sq ft (37m²).
Weights: (estimated) Empty 17,000lb (7,700kg), loaded (forward airstrip) 28,000lb (12,700kg); (max) 36,050lb (16,350kg).
Performance: (estimated) Max speed (SL, clean) 546mph (880km/h); field length (forward airstrip weight) 3,300ft (1,000m); combat radius (conditions not specified) 345 miles (556km).
Armament: At least one gun is mounted, including a large multibarrel cannon in a long chin fairing; other stores are carried on ten underwing pylons, total weight being put at 8,820lb (4,000kg).

Development: Many of the latest Soviet combat aircraft to enter service seem to have been inspired by US prototypes, and in this case the similarity to the Northrop A-9A (losing finalist in the AX competition) must be more than coincidental. Winner of the AX was the Fairchild A-10A, and by comparison with this

the Su-25 is significantly smaller and lighter, but it has roughly similar power and is faster. The main purpose is the same: attacks on ground targets in close support of friendly ground forces, with particular capability to take out heavy armour, fortifications and similar well protected targets. The design is thus biased in favour of short field length, independence of ground services, good low-level manoeuvrability and a high degree of immunity to ground fire up to about 23mm calibre.

Since 1982 small numbers of Su-25s — which curiously have a fighter (odd-number) designations, despite having little pretension to air-combat capability — have seen much action against the Mujaheddin in Afghanistan, who have commented on its long flight endurance at low level. It has operated with heavy bombs, very large numbers (a theoretical maximum of 320) of rockets and at least one high-velocity gun, and has often collaborated with Mi-24 "Hind-D" helicopters in making carefully combined attacks on the same target.

Though some dozens, if not hundreds of Su-25s are in FA service, a USAF official commented in late 1983 that the type had still not been committed to full production. Yet the 1984 edition of the DoD review,

Soviet Military Power, reported that Su-25s were coming off the production line at Tbilisi. Further, in June 1984 a Pentagon briefing, based on DIA assessments, reported that the Su-25 was one of the major current programmes, along with the MiG-29 and MiG-31.

In the 1983 edition of the same publication a detailed artist's impression of the Su-25 appeared, showing stumpy wings without tip pods, a humpbacked body with low cockpit flush with the upper line, and a very short nose with a large radar. As all these features are totally at variance with the actual aircraft it is difficult to comment on such things as avionic fit.

Available photographs confirm

Above: Su-25 "Frogfoot" in action in Afghanistan. Ten wing pylons are available for bombs or rocket pods, and a 30mm cannon is mounted internally. By early 1985 the Su-25 was also in Czechoslovakian service.

the long-span wing, tapered on the leading edge, for good STOL performance and long loiter endurance, high-mounted cockpit giving a good all-round view, dorsal spine leading to a delta-shaped fin, long downsloping nose (almost certainly without radar but with other surface-attack sensors) and wingtip pods which certainly are not mere antiflutter bodies. The engine bulges slope down from front to rear.

Sukhoi Su-27 "Flanker"

Origin: Soviet Union.
Type: Long-range multirole fighter.
Engines: Two augmented turbofans each in 28,000lb (12,700kg) thrust class.
Dimensions: (estimated) Span 45·92ft (14m); length (excl probe) 67·25ft (20·5m); height overall 19·66ft (6m); wing area 690sq ft (64m²).
Weights: (estimated) Empty 33,000lb (15,000kg); internal fuel 14,330lb (6,500kg); loaded (air-to-air mission) 49,600lb (22,500kg), (max, surface attack) 72,200lb (35,000kg).
Performance: (estimated) Max speed (hi, air-to-air mission) 1,350kt (DoD figure, converting to 1,555mph/2,500km/h, Mach 2·35); combat radius (air-to-air mission) 715 miles (1,150km).
Armament: Eight AAMs of various types including AA-X-10; probably at least one internal gun.

Development: Biggest and most powerful Soviet fighter apart from the MiG-25/31, the Su-27 is based on the same aerodynamics as the MiG-29, which in any case certainly owes much to the current crop of US fighters. Compared with the MiG-29 the Su-27 is almost exactly twice as big (in area terms, ie a 1·4 linear scale), twice as heavy and twice as powerful. Various analyses have been published since mid-1983, the figures above being based on those issued by the US DoD in March 1984. They show an aircraft significantly larger than November 1983 DoD estimates, but with slight-

ly less engine thrust and lower weights. Whether the 1983 estimates for turn rate — 17º/s at Mach 0·9 at 15,000ft/4,572m. sustained, and 23º/s peak instantaneous value — have since been subjected to revision has not been made public.

Despite its size, accepted in order to achieve long mission radius with many weapons, giving great persistence in air combat, the Su-27 is generally considered to be able to outfly the MiG-29, which itself was specifically designed to beat the F-14, F-15, F-16 and F-18 in close combat (and is generally accepted as doing so, the F-16 being the most difficult opponent in these circumstances). It is difficult to win by copying, and there is no question that the Soviet designers have carefully studied the US fighters before drawing the first line on paper, but with today's engines the US fighters are almost certainly unable to stay with the Su-27, which had the massive advantage of being started when the F-14 and F-15 were already flying. Obviously the Su-27 has a completely new pulse-doppler multimode radar with the greatest possible performance against low-flying targets.

What is much more serious than all the foregoing is the Su-27's armament. Not only is it now estimated — or known, because the information is published as fact — that this fighter carries eight AAMs, but they are partly or wholly of the AA-X-10 type. This is the first Soviet AAM which, in its radar-guided version,

has its own active seeker. Thus it can be fired against a distant hostile aircraft in the desired "fire and forget" manner, the Su-27 then either engaging other targets or turning away. There is no need to keep flying towards the enemy in order to illuminate the target with the fighter's own radar. The X-10 flies on strapdown inertial guidance until its own active radar switches on and locks-on to the target. This capability will not arrive in Western squadrons until the AIM-120A (AMRAAM) becomes operational in, it is hoped, 1986. The Soviet X-10 is part of the Su-27 weapon system which is already in preliminary service and was expected to be declared operational in spring 1984. It includes an IR search/track system and magnifying optics.

The aircraft may have a blended wing/body form, and as the inlets cannot be seen from above because

Above: The Su-27 "Flanker" has the power to outfly the latest American fighters, and the formidable AA-X-20 missile.

of the long leading-edge root extensions available information on them is largely speculative. The wing is well aft, the fins are vertical and overlap the wing and tail-planes, and the latter do not extend aft of the nozzles as in the MiG-29. How the eight AAMs, or DoD-announced alternative of 12 bombs, are carried is not yet clear; there is no room to get more than two bombs in tandem in tangential carriage on the fuselage, and the inevitable answer must include high-drag wing pylons.

There is probably a tandem two-seat Su-27, and probably a reconnaissance pod or pallet can be carried. This aircraft is reported by the DoD to be in production at Komsomolsk, in the Far East.

Tupolev Tu-16 "Badger" and H-16

Origin Soviet Union.
Type: Originally strategic bomber; today, various (see text).
Engines: Two 20,950lb (9,500kg) thrust Mikulin RD-3M turbojets.
Dimensions: Span (basic) 108·03ft (32·93m), (almost all current variants) 109·92ft (33·5m); length (basic) 114·17ft (34·8m), (D) 120·75ft (36·8m); height 35·42ft (10·8m); wing area (basic) 1,772sq ft (164·64m²), (current variants) 1,819sq ft (169m²).
Weights: Empty (basic -A) 81,570lb (37,000kg), (typical modern variant) 92,590lb (42,000kg) max (-A) 158,730lb (72,000kg), (all known current variants) 169,755lb (77,000kg).
Performance: Max speed (typical current variant, 30,000ft/9,000m and above) 587mph (945km/h); long-range cruise 485mph (780km/h); service ceiling (typical) 42,650ft (13,000m); range (typical, max fuel) 3,980 miles (6,400km), (with missile/bomb load) about 2,980 miles (4,800km).
Armament: (-A) Seven NR-23 cannon in three twin barbettes with one fixed firing ahead; bomb load up to 19,180lb (9,000kg); (modern versions) various, see text.

Development: Amazingly, this classic twin-jet has a design ancestry going back to the Soviet copy of the wartime B-29. The Tu-88 prototype flew in 1952, and 54 production bombers flew in the 1955 Aviation Day, yet hundreds are still in use, and a literal "Chinese copy" (H-6) is still in production!

The first version, called "Badger-A" by NATO, is a plain bomber carrying a bombload of up to 19,800lb (9,000kg) internally. The fuselage stemmed naturally from earlier Tupolev bombers, with the pressurized forward section housing the navigator/bomb aimer in the glazed nose, two pilots on the airline style flight deck and radio/gunner just to the rear with a dorsal sighting dome. At the tail is a second pressurized compartment for an observer (who also has gun aiming capabilities in this original variant) with large side blisters and a manned tail turret with radar gunlaying. The three twin-23mm turrets are lined electrically as in the original piston-engined bombers so that any of the three gunners can fire on any target he can see with whichever turrets he can command. A seventh gun is fixed on the right of the nose, aimed by the pilots.

The most striking feature when this bomber was new was its use of two giant engines recessed into each side of the fuselage inside the nacelles which curve inwards to improve airflow and also angle the jets away from the rear fuselage. Flight on one engine is no problem. Another notable feature was to provide bogie main gears retracting backwards into large pods projecting behind the wing; this became a Tupolev trademark on many subsequent aircraft.

From the start this large but easily flown bomber was very successful, and two plants delivered about 2,000. At the same time a similar number of B-47s were built, but these have not been seen for almost 20 years, whereas half the Tu-16s are still soldiering on, in many versions.

Some of those in Soviet service were converted as tankers, using a looped-hose hook-up between wingtips. Though no longer used, the left wingtip of almost all Soviet Tu-16s was modified with the receiver extension and socket. Other A-models still serve as tankers with a conventional HDU in the fuselage. Chinese H-6 bombers are painted white and have locally made WP-8 engines and symmetric wingtips. Several were used in testing seven free-fall nuclear weapons at Lop Nur prior to 1980, and the number in use today is put at about 140.

Other versions are known only from Western reporting names. "Badger-B" originally carried two "AS-1 Kennel" missiles on wing pylons, associated with Komet III

Above. The "Badger-C" version of the Tu-16 normally carries AS-6 "Kingfish" missiles under port or both wings.

radar working in I-band. Other additions include A322Z doppler and the RV-10 or -17 radar altimeter, also fitted to other models. The missiles have been removed and the same aircraft still fly with the DA (Long-Range Aviation) as free-fall bombers. Some, plus missiles, are stored in Indonesia.

"Badger-C" appeared in 1961 carrying the giant "AS-2 Kipper" anti-ship missile on the centreline, with a new nose filled by the giant "Puff Ball" radar and A-329Z guidance transmitter. Items eliminated were the weapon bay, nav/bomb compartment and nose gun. This continues in AVM service, but with "AS-6 Kingfish" missile(s) carried under the left wing, or under both wings. The same nose is used on "Badger-D", a major AVMF variant used for Elint missions with many special avionics including a row of three blister radomes along the belly.

"Badger-E" is a reconnaissance aircraft with a glazed nose and large camera pallets in the bomb bay and, in recent years, with other sensors in addition. The -F model is an Elint aircraft based on the -E with added receiver pods on deep wing pylons. "Badger-G" is an important anti-ship platform, originally armed with two "AS-5 Kelt" missiles on

modified AS-1 pylons and today with numerous (mainly internal) modifications and either one or two "Kingfish" missiles on the wing pylons (single missiles are carried either on the left or right), with a large belly radome and a large unidentified external device on the nose.

"Badger-H" is a strategic ECM/EW aircraft believed to be used mainly for dispensing large volumes of chaff cut to length on board according to signals received from hostile emitters via ventral aerials. The J variant (I was omitted) is a high-power jammer, with the main generation and transmitting system in a belly canoe radome. "Badger-K" is a special Elint model, again with front and rear ventral blisters.

The DA is at present flying some 300 with offensive capability, plus a few tankers and 90 ECM/EW models. The AVMF has 275 for maritime attack, 70 tankers and 40 recon and EW versions. The longevity of these aircraft is impressive, and the DoD recently commented that the "Kingfish" missile carries a nuclear or very large conventional warhead at highly supersonic speed for 135 miles (220km) at sea level, and much further at high altitudes, ending with pinpoint homing that is difficult to defeat.

Below: "Badger-H" is a chaff dispenser designed for stand-off or escort ECM protection.

Tupolev Tu-22 "Blinder"

Origin: Soviet Union.
Type: (A) Reconaissance bomber, (B) missile carrier, (C) maritime reconnaissance.
Engines: Two large afterburning turbojets (believed to be Koliesov VD-7 or -7F) with rating of at least 30,865lb (14,000kg) each.
Dimensions: (estimated) Span 91·83ft (28·0m); length 132·97ft (40·5m); height 34·0ft (10·4m); wing area 1,650sq ft (155m²).
Weights: (estimated) Empty 90,700lb (40,000kg); internal fuel 79,360lb (36,000kg); max 185,000lb (84,000kg).
Performance: Max speed (36,000ft/11,000m) 1,000mph (1,600km/h, Mach 1·5), (DoD estimate 800kt (920mph/1,480km/h); max speed at SL, 550mph (890km/h); typical hi-alt cruise 560mph (900km/h); service ceiling (afterburner) 60,000ft (18,000m), (dry power) 45,000ft (13,700m); unrefuelled combat radius (all-hi, US DoD estimate), 1,926 miles (3,100km); max range (all-hi, no supersonic dash) 4,040 miles (6,500km).
Armament: (A) one 23mm gun in tail; internal bay for up to about 17,600lb (8,000kg) of various bombs; (B) weapon bay configured for one "AS-4 Kitchen" ASM, and with larger radar; (C) usually none.

Development: Though by no means the first Tupolev supersonic bomber the Tu-105 (service designation Tu-22) was the first to go into production. Its most startling feature is the installation of the two large afterburning engines above the rear fuselage on each side of the fin. In most models the radar occupies the lower half of the pointed nose, above which is a removable flight-refuelling probe. The crew of three comprise navigator/bomb aimer in a glazed lower compartment aft of the radar, the pilot and the systems of-

ficer who also manages the single rear gun. The pilot seat ejects upwards and the others downwards. A total of almost 80,000 (36,000kg) of fuel fills most of the rest of the fuselage and integral wing tanks, though there is a long but shallow bay in the fuselage under the wing which in the basic free-fall bomber, called "Blinder-A" by NATO, is the bomb bay. It has left and right doors each split longitudinally to double-fold without significantly entering the slipstream. Flight controls are fully powered, and the main-gear pods also house strike cameras and chaff/flare dispensers. A few of these remain in service.

"Blinder-B" has the bomb bay converted to carry the "AS-4 Kitchen" long-range supersonic missile (see "Bear-G"). The doors are cut away to fit round the weapon and locked shut, and the nose is bulged to accommodate "Down Beat" radar. Surprisingly, as no missiles were supplied, this is the variant that was exported, presumably after conversion to carry free-fall bombs of various sizes. Libya received about 20 (though a recent assessment put the number still operating at only nine), while Iraq received one squadron of 12 (of which seven appear to be still flying). The Iraqi machines have seen considerable active service in the war against Iran, and one of the Libyan machines bombed Tanzanian targets in support of Uganda. About 125 remain in use in the Soviet

Union, plus about 12 converted for Elint missions.

"Blinder-C" is an AVMF reconnaissance version, with the entire lower fuselage from nose gear to tail bumper occupied by various cameras and other sensor installations. The simplest conversions have six large cameras mounted on the former weapon-bay doors, but there are several more complex arrangements which appear to include linescan and SLAR. About 40 of this version are in use, mainly in the European theatre. The Tu-22 was the starting point for the Tu-22M.

Below: "Blinder-D" training version of the Tu-22 shortly after takeoff, with afterburners on and undercarriage still only partially retracted.

Tupolev Tu-28P/128 "Fiddler"

Origin: Soviet Union.
Type: Long-range interceptor.
Engines: Two afterburner turbojets, almost certainly Lyul'ka AL-21F-3 with max rating of 24,250lb (11,000kg).
Dimensions: (estimated) Span 60·0ft (18·1m); length 89·25ft (27·2m); height 23·0ft (7m); wing area 860sq ft (80m²).
Weights: (estimated) Empty 54,000lb (24,500kg); internal fuel 30,000lb (13,000kg); loaded 88,000lb (40,00kg).
Performance: (estimated). Max speed (36,000ft/11,000m) (clean) 1,200mph (1,900km/h, Mach 1·8); (4 AAMs) 1,090mph (1,755km/h, Mach 1·65); initial climb 25,000ft (7,500m)/min; service ceiling 60,000ft (3,200km).
Armament: Four "AA-5 Ash" AAMs (see below).

Development: Despite its age this large and thus long-range intercep-

tor continues in service, though with numbers down to an estimated 120 (all, it is believed, in the Far East). Pilot and navigator sits in tandem, and the I-band "Big Nose" radar and four "AA-5 Ash" missiles have been standard for 20 years. It is possible that these worthy machines are being updated, for example with the long-range "AA-9" missiles carried by the MiG-31, but as the US DoD thinks the latter aircraft has an even greater combat radius than the old Tupolev this is unlikely. Their one advantage was their long range and endurance, and if the MiG-31 really has a radius of almost 940 miles (1,500km) the days of the Tu-128 are numbered.

Right: A row of Tu-28P (also known as Tu-128) "Fiddler" long-range interceptors, their intakes, canopies and engine bays protected by sheets and with caps over the engine nozzles.

Tupolev Tu-95 and Tu-142 "Bear"

Origin: Soviet Union.
Type: (Five variants) strike, (two) reconnaissance, (one) ASW.
Engines: Four 14,795ehp Kuznetsov NK-12MV turboprops.
Dimensions: Span 167·66ft (51·1m); length (most) 155·83ft (47·5m), ("Bear-F") 162·41ft (49·5m). height 39·75ft (12·12m). wing area 3,342sq ft (310·5m²).
Weights: (estimated) Empty (A) 165,400lb (75,000kg), (F) 178,600lb (81,000kg); max (A) 374,800lb (170,000kg), (F) 414,500lb (188,000kg).
Performance: Max speed (typical, half fuel, over 26,000ft/8,000m) 525mph (845km/h); hi-speed cruise 435mph (700km/h); service ceiling (typical) 44,300ft (13,500m); typical range/payload, 7,020 miles (11,300km) with military load exceeding 26,455lb (12,000kg); unrefuelled combat radius in recon/EW missions 5,150 miles (8,300km); endurance at econ cruise speed of 400mph (650km/h), 26 hours.
Armament: (A) six NR-23 guns in three remotely controlled barbettes; internal bay for ordnance load of up to 44,100lb (20,000kg), (later variants) see below.

Development: This aircraft used almost the same fuselage as the Tu-85 piston-engined bomber, itself derived from the B-29; but when it was mated with colossal swept wings carrying four of what even today are by far the biggest and most powerful turboprops in the world the result was truly dramatic. Instead of being a tricky machine riddled with problems — as one might have expected from a 15,000hp turboprop driving eight-blade contraprops of 18ft 4in (5·6m) diameter — the Tu-95 settled down as a multirole platform with capabilities which even today look impressive.

Year after year the DoD comments on the speed of 500kt and mission radius of 5,158 miles (8,300km) and notes that no other aircraft can do the same, not even the forthcoming even bigger machine called "Blackjack". It is

hardly surprising that a new ASW version went back into production in the 1970s, while today a completely new cruise-missile variant has also gone into production.

In general the circular-section fuselage resembles that of the Tu-16, though it is much longer and contains a large pressurized mid-section aft of the wing. No nose gun is fitted but the usual armament settled down as upper and lower rear turrets and manned tail station (different from Tu-16, and often with different radar), all with twin 23mm. Even the first version accommodated 16,053 gal (72,980lit) of fuel, housed in the structural box forming the basis of the wing, to which enormous track-mounted flaps are attached. Flight controls are powered, and ice protection is thermal.

In the original "Bear-A" free-fall bomber, a few of which remain operational with the DA, there are front and rear bomb bays each sized for a thermonuclear weapon, but able to carry 11,023lb (5,000kg) of conventional stores in each compartment. Most have an FR probe centred above the glazed nose. "Bear-B" was modified to carry and launch the "AS-3 Kangaroo", the largest ASM ever built, with range of 404 miles (650km); most of this variant have been converted to carry the smaller "AS-4 Kitchen" and several are on other duties including multi-sensor reconnaissance, air-cloud sampling and strategic meteorology. Total DA inventory of -A and -B is about 113.

"Bear-C" is a maritime reconnaissance aircraft with a blister fairing on both sides of the rear fuselage (often seen on the right side only on the -B variant) and with the FR probe invariably installed. "Bear-D" covers a wealth of variations on an important AVMF type carrying a very large electronic payload but no weapons. The weapon bay is occupied by "Big Bulge", the largest radar of its type in history (exceeded only by AWACS radars); there are many other aerials, blisters or dielectric fairings and in some aircraft the

Above: "Bee Hind" tail warning radome and twin 23mm cannon in the tail of a Tu-142 "Bear-D".

tail turret is replaced by a long fairing for further electronics. About 50 are used for reconnaissance and providing mid-course guidance to cruise missiles launched from distant warships. "Bear-E" is a rebuild for strategic reconnaissance, usually with the rear part of the bomb bay occupied by six/seven large cameras and with various other sensors.

There is some confusion over "Bear" type numbers. The original Tupolev OKB number was 95 (the Tu-16 having been the Tu-88), and when "Bear-A" entered service it received the VVS designation Tu-20. In recent years this has fallen into disuse, the original Tu-95 being more common despite the fact odd numbers are reserved for fighters in the VVS system. To confuse matters further it is now claimed that another OKB number, Tu-142, is used by the AVMF for all its naval versions. One must conclude that the military services no longer assign their own designations, but simply use those of the original design OKB.

Originally it was thought the Tu-142 designation applied only to the totally new "Bear-F", first seen in 1973. To the surprise of Western observers these were not rebuilds but new aircraft, and tailored entirely for the open-ocean ASW mission. Fuel capacity is increased, and the fuselage completely redesigned with a longer forward section, a tactical compartment, crew rest bunks and a rear section containing two stores bays with various sonobuoy launchers. The fuselage is cleaner than on any other variant, and the

dorsal and ventral turrets are absent.

There are two main radars, one with a ventral blister smaller than that of "Bear-D" and further forward, and the other with a flush installation; in some aircraft there is a third radar in the cabin position but again unlike that of "Bear-D". Yet further puzzles are the bulged nosewheel doors, the fact that early F aircraft had extended main-gear fairings, and the replacement in a few machines of the tailplane tip pods by a single MAD stinger projecting aft of the tip of the vertical tail (the tailplane pods had nothing to do with MAD). Small numbers appear each year, the 1984 total being about 48.

"Bear-G" is simply the "B" after conversion to carry the "AS-4 Kitchen" missile, which compared with "AS-6 Kingfish" carried by other aircraft is rather bigger, heavier, slower (Mach 2) and longer-ranged, with a low-level mission up to 186 miles (300km). Probably far more important is the completely new AS-X-15 cruise missile, a 24ft (7·3m) weapon which has close relatives for use from ships and mobile land tubes and delivering a large conventional or thermonuclear warhead with great accuracy over a distance of 1,860 miles (3,000km). The only existing carrier for this formidable weapon is the equally new "Bear-H". This has an airframe very close to that of the "Bear-F", though obviously its fuselage is completely rearranged. The new H model is in full production, and will hold the fort, probably alongside converted older versions, until "Blackjack" arrives in 1987.

Below: Tu-142 "Bear-F" anti-submarine aircraft with MAD sting on the tailfin tip.

Vought A-7 Corsair II

Origin: USA.
Type: Attack aircraft.
Engine: (D, H) one 14,250lb (6,465kg) thrust Allison TF41-1 turbofan; (E) one 15,000lb (6,804kg) TF41-2; (P) one 12,200lb (5,534kg) Pratt & Whitney TF30-408 turbofan.
Dimensions: Span 38·75ft (11·8m); length (D) 46·12ft (14·06m), (K) 48·96ft (14·92m); height overall 16·07ft (4·9m); wing area 375sq ft (34·83m²).
Weights: Empty (D) 19,781lb (8,972kg), max (D) 42,000lb (19,050kg).
Performance: Max speed (D, clean, SL), 690mph (1,110km/h) (5,000ft/1,525m, with 12 Mk 82 bombs) 646mph (1,040km/h); tactical radius (with unspecified weapon load at unspecified height), 715 miles (1,151km); ferry range (internal fuel) 2,281 miles (3,671km), (max with external tanks) 2,861 miles (4,604km).
Armament: One 20mm M61A-1 gun with 1,000 rounds, and up to 15,000lb (6,804kg) of tactical weapons on eight hardpoints (two on fuselage each rated 500lb/227kg, two inboard wing pylons each 2,500lb/1,134kg, four outboard wing pylons each 3,500lb/1,587kg).

Development: One of the most cost/effective dedicated attack aircraft ever built, the A-7 demonstrated the ability to carry such heavy loads and deliver them so accurately that in 1966 it was selected as a major USAF type, and 457 were delivered of the uprated A-7D version with the TF41 engine (derived from the Spey) and a new avionic suite providing for continuous solution of nav/attack problems for precision weapon delivery in all weather. This version also introduced the M61 gun and a flight-refuelling boom receptacle, preceding Navy models (now retired from front-line units) having probes.

Like earlier models the D had landing gears retracting into the fuselage, a large door-type airbrake, folding wings and triplex power for the flight controls. The pylons on the cliff-like sides of the fuselage are usually used by Sidewinder self-defence AAMs. Most A-7 pilots have a high opinion of their air-combat capability, and agility (at least rate of roll) is good even with a heavy load of bombs.

In turn the USAF A-7D, 375 of which now equip many attack units of the Air National Guard, was the basis of the current A-7E which is still a major type in the US Navy. Powered by a slightly uprated TF41, this has the usual Navy folding FR probe and if anything even more comprehensive all-weather and EW avionics than the D. The A-7 is being replaced by the F-18, but at the time of writing still served with both Atlantic and Pacific carrier air groups and in a total of 23 operational plus two shore training squadrons.

A total of 596 of this type were delivered, of which 222 are being equiped to carry the Texas Instruments FLIR pod on the inboard pylon on the right side, linked to a new Marconi raster-type HUD for improved night attack capability. Budget limitations have held actual supply of these pods to 110.

Newest of all the US variants is the two-seat A-7K, 42 of which have been distributed in pairs to 11 of the 13 ANG combat-ready A-7D units plus a further 16 to the 162nd Tac Fighter Training Group at Tucson. A direct-view tube provides for Walleye and similar TV ASM guidance, and Pave Penny pods are carried for laser-guided stores, but even with "iron bombs" accuracy is under 10ft (3m).

Among European customers, Greece purchased 60 A-7H new from Vought, and these are virtually A-7Es with the more powerful model of TF41, and the pilots trained with the US Navy and not the USAF. The H equips three mira (squadrons), all in the maritime support and anti-ship role: 340 and 345 at Souda Bay, Crete, and 347 at Larissa in the north. Greece also bought six two-seat TA-7H similar to the A-7K.

Portugal was eager to obtain more effective combat aircraft but had little money, and eventually selected 20 A-7As well-used by the US Navy and before delivery completely refurbished by Vought. The engine remains the TF30 but improved to P-408 standard, and the avionics have been brought up virtually to A-7E standard. These aircraft equip Esc 302 at Monte Real, replacing the F-86 but tasked primarily in the strike role; the FAP still needs an air-combat fighter. Vought expects to find other second-hand customers, and is trying to set up a re-engining programme, for example with the F110 engine.

Above: A USAF A-7D lets go a stick of Mk 82 Snakeye retarded bombs in a shallow dive.

Below: Although replaced in USAF front-line service the A-7D is still in service with the ANG.

Right: With its full-span leading edge flaps extended, a US Navy F-4C of CVW-8 approaches a deck landing.

Below: Latest Corsair in US service is the two-seat A-7K for the ANG; this converted A-7D acted as the prototype.

Yakovlev Yak-28 "Firebar"

Origin: Soviet Union.
Type: Interceptor, bomber and electronic warfare aircraft (see below).
Engines: (most) Two 10,410/13,670lb (4.6/6.2t) thrust Tumanskii R-11 afterburning turbojets.
Dimensions: Span (most) 41·0ft (12·5m); length overall from 71·0ft (21·6m) to (28P) 75·0ft (22·9m); height overall 12·96ft (3·95m); wing area 405sq ft (37·6m²).
Weights: (estimated) Empty (28P) 29,000lb (13,150kg); max loaded (28P) 44,000lb (20,000kg).
Peformance: (estimated) Max speed (hi, typical, clean) 1,240mph (2,000km/h Mach 1·88), (SL) 646mph (1,040km/h, Mach 0·85); service ceiling (afterburner) 55,000ft (16,750m); combat radius (all-hi) 560 miles (900km); ferry range 1,550 miles (2,500km).

Armament: (most) None, (28P) two "AA-3 Anab" and two "AA-2-2 Advanced Atoll" AAMs.

Development: Russians have a reputaton for never discarding anything that can be used, and this seemingly obsolescent supersonic twin-jet has been described as "being withdrawn from service" since 1968! Yet a recent outline of Far East PVO (air defence forces) bases showed two complete regiments of Yak-28 ("Firebar") all-weather interceptors at Anadyr and Providenya and another at Leonydovo on Sakhalin island, and in the 1984 edition of the US DoD's *Soviet Mi itary Power*, the 28P is very much in evidence, together with its lately assessed speed of "1,080 knots" (compared with the figure of 650 accepted during the previous 20 years!).

This rakish tandem-seat version has a very long and pointed radome and, like other variants, tandem bicycle main gears, with small outriggers near the wingtips. The code-name "Brewer" identifies former bomber and reconnaissance

Above: Yak-28 "Brewer-C" EW/attack aircraft with nose compartment roof hatches open for the navigator/bombadier.

versions now used mainly for EW, Elint and active ECM jamming.

Yakovlev Yak-36MP "Forger"

Origin: Soviet Union.
Type: Shipboard strike fighter.
Engines: One vectored-thrust turbojet (believed to be an AL-21F) rated at about 17,500lb (7,940kg) and two lift jets in the forward fuselage (believed to be of Koliesov design) each rated in the 8,000lb (3,630kg) thrust class.
Dimensions: (estimated) Span 24·0ft (7·32m); length (A) 50·0ft (15·25m), (B) 58·0ft (17·68m); height overall 14·33ft (4·37m); wing area 170sq ft (15·8m²).
Weights: (estimated) Empty (A) 16,200lb (7,350kg), (B) 18,000lb (8,165kg); max (both) 25,500lb (11,565kg).
Performance: (estimated) Max speed (hi, clean) 627mph (1,009km/h, Mach 0·95), (SL, clean) 608mph (978km/h, Mach 0·8); initial climb 14,750ft (4,500)/min; service ceiling 39,370ft (12,000m); combat radius (max weapons) (lo-lo-lo) 150 miles (240km), (hi-lo-hi) 230 miles (370km); ferry range (four tanks) 1,600 miles (2,575km).
Armament: All carried on four pylons on fixed inner wing, including GSh-23 gun pods, "AA-8 Aphid" (R-60) close-range AAMs, rocket launchers or bombs to total of 3,000lb (1,361kg), or "AS-7 Kerry" ASMs.

Development: During the past year the performance of this VTOL shipboard aircraft has been revised downwards, it no longer being credited with level supersonic speed. The version called "Forger-B" differs from the "A" model in being a much longer tandem seater, thought to be used for pilot training but possibly serving in EW and other operational roles.

Takeoff and landing are usually vertical, but rolling launches have recently been flown. The main pair of nozzles are vectored forwards up to 10º to balance the rearward thrust of the slightly inclined tandem lift jets behind the cockpit, between

the inlet ducts to the main engine. Landings are clearly electronically guided from the ship, such is their precision. Avionics are fairly limited, the two-seater not even having the small gunsight ranging radar in the nose of "Forger-A".

Primary missions are believed to be destruction of ocean patrol and ASW aircraft, anti-ship attack and reconnaissance. Up to 12 of these machines have been observed aboard the large VTOL carriers *Kiev*, *Minsk* and *Novorossisk*.

Right: Single-seat "Forger-A" V/STOL attack fighters aboard the carrier *Minsk*; note the absence of tie-down points.

Below: Operational Yak-36 units aboard the Kiev class carriers include 12 single-seaters and one two seat "Forger-B".

COMBAT HELICOPTERS

Aérospatiale Dauphin

Origin: France.
Type: Multirole helicopter with special-purpose variants.
Engines: Two 710hp Turboméca Arriel IC turboshafts; (366G) two 680hp Avco Lycoming LTS101-750A-1 turboshafts.
Dimensions: Diameter of four-blade main rotor 44ft 2in (13·46m); length of fuselage (basic 365N) 36ft 6·39in (11·44m), (F) 39ft 8·77in (12·11m); height 13ft 2in (4·01m). main rotor disc area 1,532sq ft (142·29m²).
Weights: Empty (basic N) 4,447lb (2,016kg), (F) 4,720lb (2,141kg), (G) 5,992lb (2,718kg); max (N, external load) 8,818lb (4,000kg), (F) 8,598lb (3,900kg), (G) 8,928lb (4,050kg).
Performance: Max speed at SL (N) 174mph (280km/h), (F) 156mph (252km/h), (G) 160mph (257km/h); range with max normal fuel (N) 546 miles (880km), (F) 558 miles (898km), (G) 471 miles (760km).

Development: This huge programme is underpinned by large civil sales and licence deals which easily exceed 500 aircraft not including licence production in China. The basic twin-engined SA 365N is an attractive machine seating a pilot and up to 13 passengers but normally used with role equipment which can include various casevec models usually with four stretcher casualties. Tricycle landing gear retracts fully, the tail rotor is multi-bladed and rotates inside a duct form in the fin (the so-called Fenestron arrangement) and Aérospatiale have continually updated the basic airframe and rotors with composite construction using glassfibre/Nomex honeycomb, glassfibre/Kevlar, glass-fibre/Rohacell, light alloy/Nomex (35 per cent of the fuselage). and carbonfibre (main rotor blades). Equipment in the basic 365N can include a duplex autopilot with a navigation coupler, a 3,527lb (1600kg) cargo sling and a 605lb (275kg) winch.

The SA365F was developed under contract to Saudi Arabia. The first four of this model are SAR helicopters with an OMERA ORB 32 radar, auto hover coupler, winch, searchlight and other special role equipment. The next 20, all for Saudi Arabia, are anti-ship missile carriers, equipped with pylons for four AS.15TT missiles guided by the nose-mounted Agrion 15 radar developed from the Iguane radar of the Atlantic ATL2. Aérospatiale claim the range of the small AS. 15TT to exceed 9.3 miles (15km). The radar can be used for various surveillance tasks, as well as for over-the-horizon target designation for missiles launched from shore or surface ships.

Aérospatiale are also developing an ASW version of the 365F, initially with MAD and sonobuoys, plus a homing torpedo, but eventually with dipping sonar. The Irish 365F has a Bendix radar and is used for offshore surveillance and SAR from ship and shore platforms. The USCG's SA 366G, known as the HH-65A Dolphin, is expected to be a 90-helicopter programme, though delivery has been severely delayed. This is mainly because the tremendous burden of mission equipment made it impossible to carry enough fuel and payload, but by various modifications (for example by fitting a large Fenestron in a new tail-end made entirely of carbon fibre and Nomex or Kevlar) the empty weight has been made more acceptable and deliveries are expected to continue through 1986.

Above: Launch of an AS.15 TT radar-guided anti-ship missile from an SA 365F Dauphin 2, developed from the SA 365N initially for Saudi Arabia.

Aérospatiale Ecureuil

Origin: France (UH-12, Brazil).
Type: Multirole light helicopter.
Engines: (350B) One 641hp Turboméca Arriel 1 turboshaft, (350D) one 615hp Avco Lycoming LTS101-600A-2 turboshaft, (355) two 425hp Allison 250-C20F turboshafts.
Dimensions: Diameter of three-blade main rotor 35ft 0·86in (10·69m); length of fuselage 35ft 9·53in (10·91m); height 10ft 4in. (3·15m); main rotor disc area 966sq ft (97·524m²).
Weights: Empty (B) 2,348lb (1,065kg), (D) 2,359lb (1,070kg), (355) 2,811lb (1,275kg); max (with external load) 4,630lb (2,100kg), (355) 5,511lb (2,500kg).
Performance: Max cruising speed (B) 144mph (232km/h), (D) 143mph (230km/h), (355) 149mph (240km/h); range with max fuel at SL, no reserve, (B) 435 miles (700km), (D) 472 miles (760km), (355) 528 miles (850km).
Armament: None, except French Armée de l'Air 355 version, see below.

Development: Designed as a successor to the Alouette, this family of attractive small helicopters will certainly outsell even its predecessor, and the 2,000th sale may be announced before this book appears. Most are civilian, but many air forces, navies and other military users have picked the Ecureuil (Squirrel) for various duties. The basic machine is well streamlined and has the patented Starflex rotor hub made of glassfibre, with each blade held in a steel/rubber balljoint requiring no attention or maintenance. Normal landing gear is two skids, with emergency flotation gear optional. A typical seating plan has two bucket seats in front and two two-place bench seats behind, but a two-stretcher plus attendant casevac model is available and a slung load of 1,650lb (750kg) can be carried (355 carries up to 2,300lb, 1045kg).

One of the early military customers was the government of Australia, which bought 12 AS350Bs for pilot training with the RAAF and six for survey/utility with the RAN, followed later by another six for RAAF SAR and liaison. In Brazil Helibras produces the 350B under licence as the HB 350B Esquilo (Squirrel). Deliveries began in 1979, Brazilian customers including the Brazilian Navy and two other Latin American air forces.

The first military customer for the twin-engined 355 is the Armée de l'Air, which in 1983 ordered 50 of a specially equipped version to be delivered in 1983-85. Early batches are being used for the surveillance of strategic military bases and for various support missions, but later batches are to be equipped to fire the SATCP anti-aircraft missile. Originally for use by infantry and from land vehicles, the SATCP has been specially adapted for use from the AS 355 and has multispectral IR/optical homing guidance and full IFF. Targets would mainly be hostile helicopters.

Below: First AS 355M Ecureuil, later models of which will have HATCP air-to-air missiles.

Aérospatiale Gazelle

Origin: France.
Type: Multiple and utility helicopter.
Engine: One Turboméca Astazou turboshaft; (most 341s) 590shp Astazou IIIA, (342) 858shp Astazou XIVH or (342M) XIVM.
Dimensions: Diameter of three-blade rotor 34ft 5·5in (10·5m); length of fuselage 31ft 3·19in (9·35m); height overall 10ft 5·2in (3·18m); main rotor disc area 932·05sq ft (86·59m^2).
Weights: Empty (341G) 2,022lb (917kg), (342L) 2,150lb (975kg); max (341G) 3,970lb (1,800kg), (342L) 4,188lb (1,900kg).
Performance: Max cruising speed (all) 164mph (264km/h); hovering ceiling (341) 6,561ft (2,000m), (342) 9,430ft (2,875m); range at SL max fuel, (341) 416 miles (670km), (342) 469 miles (755km).
Armament: (342) Can include four or six HOT missiles, or two AS.12 missiles, or two rocket pods (various types), a 20mm GIAT

cannon, two 7·62mm machine guns firing ahead, a side-firing GE Minigun or an Emerson MiniTAT or similar chin turret.

Development: A natural refined successor to the pioneer Alouette, the Gazelle was one of three types named in a major Anglo-French agreement of February 1967 under the terms of which Westland helicopters shared in production and assembled Gazelles for British customers. The Gazelle has proved most sucessful, and was still selling in small numbers in 1985, with more than 1,100 of many versions already delivered.

The French assembly line at Marignane has been supplemented by licence production by SOKO of Yugoslavia, which is continuing to build the SA 342L after delivering a substantial number of earlier versions (not included in the previous total for deliveries). Gazelle 342s are also produced in Egypt.

The chief military variants are: SA 341B, British Army Gazelle AH.1, Astazou IIIN; 341C, Royal Navy trainer HT.2; 341D, RAF trainer HT.3; 341F, French ALAT (army light aviation) model Astazou IIIC; 341H, military export version, Astazou IIIB; 342K, second-generation export military variant; 342L, improved Fenestron tail rotor and Astazou XIVH; 342M, French ALAT armed model.

The 342M is certainly the most advanced variant, with extremely comprehensive avionics, sight systems and weapons, but the Gazelle AH.1s of Britain's Army 656 Sqn and the Royal Marines No 3 Commando Brigade were hurriedly upgraded for Falklands War duty in April 1982 with armour (as used on Gazelles in Northern Ireland), SNEB rocket pods, IFF, radar altimeter and a foldable main rotor for shipboard use. At Wideawake (Ascension) further modifications included manually aimed GPMGs firing from the left door and four pylons for Matra rocket pods and night landing flares.

Above: SA 342L Gazelle armed with 68mm rocket boxes but without roof-mounted sight.

Aérospatiale Puma/Super Puma

Origin: France.
Type: Multirole tactical transport helicopter.
Engines: (Puma) Two Turboméca Turmo turboshaft engines, of IIIC4 type rated at 1,328hp, IVB rated at 1,400hp or IVC rated at 1,575hp; (Super) two Turboméca Makila IA each rated at 1,780hp.
Dimensions: Diameter of four-bladed main rotor (P) 49ft 2·5in (15·0m), (S) 51ft 2·17in (51·6m); length of fuselage (P) 46ft 1·5in (14·06m), (S) 48ft 5in (14·76m), (M) 50ft 11in (15·52m); height overall (S) 16ft 1·7in (4·92m); main rotor disc area (P) 1,905sq ft (177m^2).
Weights: Empty (P) 7,795lb (3,536kg), (S) 9,260lb (4,200kg); max (P) 16,315lb (7,400kg), (S) 20,615lb (9,350kg).
Performance: Max cruising speed (P) 168mph (271km/h), (S) 173mph (280km/h); hovering ceiling (P at max wt) 5,575ft (1,700m), (S at 18,410lb, 8,350kg), 6,890ft (2,100m); range (P max fuel, no reserves) 355 miles (572km), (S standard fuel, no reserves) 394 miles (635km).
Armament: (P) Usually none, but provision for hand-aimed 20mm GIAT cannon or various other weapons, (S) can include one 20mm GIAT, two 7·62mm pivoted or firing ahead, two AM39 Exocet, six AS15TT or one AS39 and three AS15TT, two torpedoes and sonar, or two launchers for either 19 rockets of 70mm or 36 of 68mm.

Development: The Puma was designed to meet a French Army (ALAT) need for a capable tactical airlift helicopter able to operate in all climates and all weathers. It also met a 1967 need of the RAF, and accordingly was a mainstay of the Anglo-French helicopter co-production agreement of that year. Portions were assigned to Fairey Aviation

(Westland) at Hayes.

Features of the basic machine include compact size (appreciably smaller than the Sikorsky S-61 family), a traditional articulated hub and conventional (as opposed to Aérospatiale's patented Fenestron) tail rotor. The twin-wheel retractable tricycle landing gears have even been tested with a clutch-in power drive to facilitate taxiing under camouflaged hides in battle conditions.

The original Puma became operational with the French ALAT at Mulhouse in June 1970, and with RAF No 33 Sqn a year later. Some export customers specified such extras as weather radar, special electronic navigation systems and emergency flotation bags for overwater use. Later models, such as the SA 330L, have composite main-rotor blades which enable greater loads to be carried.

Accommodation is provided for 16 troops (20 in high density) or six stretchers and six seated casualties, while up to 6,614lb (3,000kg) can be slung externally. The Puma was one of the first helicopters to be available with total protection against icing.

In 1974 Aérospatiale began development of the AS 332 Super Puma. This is a carefully refined machine with new engines driving more efficient rotors, a slightly longer fuselage and new high-energy landing gears with increased track and wheelbase, and with single-wheel main units able to kneel for loading heavy stores or to reduce height aboard ships.

Accommodation is provided for 21 troops, but the AS 332M has a longer cabin seating 25, or accommodating nine stretchers with three seats. The external load is increased to an impressive 9,921lb (4,500kg) and there has been a great improvement in combat equipment, crash-

worthiness and protection against hostile fire.

One of the major users is Indonesia, where several versions are licence-built by Nurtanio. Local assembly has been offered to Argentina, which since 1983 has received 24 army models from France.

Above: Nap-of-the-earth (NOE) flight demonstration by a 21-seat AS 332B Super Puma.

Below: AS 332F naval Super Puma armed with two AM39 Exocet missiles for use in the anti-ship role.

Agusta A109A

Origin: Italy.
Type: Multirole light helicopter.
Engines: Two 420hp Allison 250-C20B turboshaft engines.
Dimensions: Diameter of four-blade main rotor 36ft 1in (11·0m); length of fuselage (without nose sensors or other adds-on) 35ft 1·5in (10·706m); height overall 10ft 10in (3·3m); main rotor disc area 1,022·9sq ft (95·03m²).
Weights: Empty (basic) 3,232lb (1,466kg), (special, eg EW version) 4,199lb (1,905kg); max (all) 5,732lb (2,600kg).
Performance: Max cruising speed (max wt) 169mph (272km/h); economical cruise (max wt) 143mph (230km/h); hovering ceiling (max wt) 4,900ft (1,495m); range with max fuel 345 miles (556km).
Armament: See text.

Development: One of the most attractive of all helicopters, the A109A has a beautifully streamlined form, with fully retractable landing gear and the engines inside the upper rear fuselage. Over 300 have been ordered, with manufacture of the MkII (which incorporates numerous, mostly minor, improvements) shared in part with the Hellenic Aerospace Industries, which in 1984-87 is making major fuselage parts for 77 of the helicopters.

Almost all by 1984 had been of various basic civil versions, despite the fact that Agusta offer an extraordinary variety of military versions, which include: utility, for seven troops, or 2,000lb (907kg) cargo, or two stretchers and two seated passengers, with hoist; command/control, with target designation sensors and lasers, and various attack weapons and special communications; light attack, with 12 different arrangements of machine guns as well as a range of rocket pods and other stores; anti-tank, demonstrated with a nose TOW sight and eight TOW missiles; aerial scout with special observation and radio systems and machine guns and rockets; ESM/ECM, available in several forms with passive receiver and analyser, and optional active jammers, dispensers and weapons; a very wide range of naval and coastal patrol versions which can carry search radar, or AM 10 missiles, or equipment for guiding ship-launched Otomat cruise missiles; a Mirach RPV platform carrying two drones for surveillance, ECM or other missions, retrieving them after use; and numerous police and SAR models with every kind of emergency equipment. In July 1984 the heavier 109K was announced, with 723hp Turboméca Arriels.

Below: This Italian Army Light Aviation A109A Mk II has a nose-mounted sight for use with the pairs of TOW anti-tank missile launchers carried each side of the fuselage.

Agusta A129 Mangusta

Origin: Italy.
Type: Anti-armour and scout helicopter.
Engines: Two 1,035hp Rolls-Royce Gem 2Mk 1004D turboshaft engines.
Dimensions: Diameter of four-blade main rotor 39ft 0·5in (11·9m); length of fuselage 40ft 3·27in)12·275m); height overall 10ft 11·9in (3·35m); main rotor disc area 1,197sq ft (111·2m²).
Weights: Empty (equipped) 5,575lb (2,529kg); max 8,157lb (3,700kg).
Performance: Max speed (SL at 8,080lb, 3,600kg) 168mph (270km/h); cruising speed 155mph (250km/h); hovering ceiling 7,850ft (2,390m); endurance (anti-tank, no reserve) 2h 30min.
Armament: Four weapon attachments, inners 661lb (300kg) rating, outers 441lb (200kg), all able to tilt +3°/−12° from horizontal. Outers will normally carry two quad launchers of TOW missiles, with the option of HOT or Hellfire as alternative; various gun pods can be carried, or up to 52 rockets of 70mm size in four launchers.

Development: The first dedicated battlefield anti-armour helicopter to be produced in Western Europe, the Mangusta (Mongoose) grew in power and capability during development, the 450hp Allison engines originally planned being replaced by ones of twice the power. The machine now flying has roughly the power of HueyCobras, and should in many ways be superior, but it remains to be seen whether it will find wide acceptance in the face of this and other rivals, notably the far more expensive and much larger AH-64A Apache. Indeed, Agusta has even studied the latter machine's 1,700hp T700 engines as a possible power unit for a further-uprated export version.

Today's A129 is a refined machine which from the start has naturally been planned for maximum effectiveness and lethality on the battlefield, with the greatest possible protection and resistance to hostile gunfire up to 23mm. Almost the only surprising feature is location of the TOW telescopic sight, with laser and FLIR, very low in the nose where the entire machine must be exposed in order to see the enemy. Agusta has worked with Martin Marietta on an MMS (mast-mounted sight).

No requirement exists for a chin turret or other trainable guns, but the Italian army has called for a comprehensive EW installation including radar and laser warning receivers, IR jammer and chaff/flare dispenser. Equipment includes a PNVS as on the Apache, integrated helmet sight system, and an IMS (integrated multiplex system) to centralize management of all subsystems.

Right: The Italian Army will have two or three 30-aircraft squadrons of the A129 Mangusta; a mast-mounted sight may be fitted.

Bell Huey and Grifon

Origin: USA.
Type: Multirole utility helicopter.
Engines: (204) One Avco Lycoming T53 turboshaft rated at 770, 825 or 930shp, or 1,250shp Rolls-Royce Gnome; (205) usually one T53-13B flat rated at 1,100shp; (212) Pratt & Whitney Canada PT6T-3 twin turboshaft flat-rated at 1,250shp but with each engine section able in emergency to deliver 900shp; (412) PT6T-3B-1 flat rated at 1,308shp and with each engine able to supply 1,025shp in emergency.
Dimensions: Diameter of main rotor (two blades except 412) (204, UH-1B, -1C) 44ft 0in (13·41m), (205, 212) 48ft 0in (14·63m), (212 tracking tips) 48ft 2·3in (14·69m), (412, four blades) 46ft 0in (14·02m); overall length (rotors turning) (early) 53ft 0in (16·15m) (virtually all modern versions) 57ft 3·2in (17·46m), height overall (modern, typical) 14ft 4·8in (4·39m).
Weights: Empty (XH-40) about 4,000lb (1,814kg), (typical 205) 4,667lb (2,116kg), (typical 212) 5,549lb (2,517kg); maximum loaded (XH-40) 5,800lb (2,631kg), (typical 205) 9,500lb (4,309kg), (212/UH-1N) 10,500lb (4,762kg), (412) 11,500lb (5,121kg).
Performance: Maximum speed (all) typically 127mph (204km/h); econ cruise speed, usually same; max range with useful payload, typically 248 miles (400km).
Armament; See below.

Development: When Bell flew the prototype XH-40 in 1956 it was just a small machine for the US Army. Nobody could have foreseen that derived versions would be used by more air forces, and built in greater numbers, than any other military aircraft since World War II. Over 20 years the gross weight has been almost multiplied by three, though the size has changed only slightly.

Early Model 204s, called Iroquois by the US Army, seated eight to ten, carried the occasional machine-gun, and included the TH-1L Seawolf trainer for the US Navy and the Italian-developed Agusta-Bell 204AS with radar and ASW sensors and torpedoes. The Model 205 (UH-1D, -1H etc) have more power and carry up to 15 passengers. Dornier built 352 for the W German Army, and similar versions are still in production at Agusta, Fuji and AIDC. Canada sponsored the twin-engined 212 (UH-1N, Canada CH-135), which again is made in Italy in an ASW version, with a new radar, AQS-13B variable-depth sonar and two torpedoes.

Many Hueys (called thus from the original "HU" designation, later

Above: The AB412 Grifon shows its capabilities in the assault troop transport role.

changed to UH) carry guns, anti-tank missiles and night-fighting gear. The 412, with four-blade rotor, is licensed to Agusta and China. The improved AB412 Grifon is a multirole model; those for the Turkish navy will carry BAe Sea Skua anti-ship missiles.

Bell OH-58 Kiowa

Origin: USA.
Type: Light multirole helicopter.
Engine: One 317shp Allison T63-700 or 250-C18 turboshaft; (206B models) 420shp Allison 250-C20B or 400shp C20.
Dimensions: Diameter of two-blade main rotor 35ft 4in (10·77m), (206B) 33ft 4in (10·16m), (206L) 37ft 0in (11·28m); length overall (rotors turning) 40ft 11³/₄in (12·49m), (206B) 38ft 9¹/₂in (11·82m); height 9ft 6¹/₂in (2·91m).
Weights: Empty 1,464lb (664kg), (206B slightly less), (206L) 1,962lb (809kg); maximum loaded 3,000lb (1,361kg), (206B) 3,200lb (1,451kg), (206L) 4,000lb (1,814kg).
Performance: Economical cruise (Kiowa S/L) 117mph (188km/h), (206B 5,000ft, 1,525m) 138mph (222km/h). max range S/L no reserve with max useful load, 305 miles (490km), (206B and L) 345 miles (555km).
Armament: See text.

Development: In 1962 Bell's OH-4A was a loser in the US Army's LOH (light observation helicopter) competition. Bell accordingly marketed the same basic design as the Model 206 Jet-Ranger, this family growing to encompass the more powerful 206B and more capacious 206L LongRanger. In 1968 the US Army re-opened the LOH competition, naming Bell now winner and buying 2,200 OH-58A Kiowas similar to the 206A but with larger main rotor. US navy trainers are TH-57A Sea Rangers, Canadian designation is CH-136, and Australian-assembled models for Army use are 206B standards. Agusta builds AB206B JetRanger IIs, many for military use (Sweden uses the HKP 6 with torpedoes) and the big-rotor AB206A-1 and B-1.

Sales of all versions exceed 5,500, most being five-seaters (206L, seven) and US Army Kiowas having the XM27 kit with 7·62mm Minigun and various other weapons. Bell has rebuilt 272 US Army OH-58As to OH-58C standard with many changes including an angular canopy with flat glass panels, the T63-720 (C20B) engine with IR suppression, new avionics and instruments and a day optical system.

Bell has produced a military version of the stretched 206L Long-Ranger known as TexasRanger. Able to seat seven, it can fly many missions but is marketed mainly in the attack role with uprated C30P engines, four TOW missiles, roof sight, FLIR (forward-looking infrared) and laser range-finder/designator.

In 1981 Bell's Model 406 was named winner of the US Army AHIP (Army Helicopter Improvement Program), for a near-term scout, with designation OH-58D. New features include the much more powerful engine, an MMS (mast-mounted sight) and completely new avionics and cockpit displays, most of the latter by Sperry. The ball-type MMS contains a TV and a FLIR, and equipment includes inertial navigation, night vision goggles and an airborne-target handoff system.

Two FIM-92 Stinger close-range AAMs will be packaged on the right side, and some $290 million is likely to be spent on updating the existing Kiowas with Stingers, and the basic plan is that 578 should be brought

Above: The proposed Model 406C Combat Scout incorporates many of the features of the AHIP 406.

up to OH-58D standard by 1991. Of the five machines in the prototype programme, No 3 (flown 6 October 1983) was the first to have an operative MMS. Bell already has long-lead contracts for the first 12 OH-58Ds.

Below: Third and fourth prototypes of the US Army's new OH-58D scout helicopter.

Bell AH-1 HueyCobra

Origin: USA.
Type: Close-supporter and attack helicopter.
Engines: (AH-1G) one 1,400shp Lycoming T53-13 derated to 1,100shp for continuous operation, (-1J) 1,800shp P&W Canada T400 Twin Pac with transmission flat-rated at 1,100shp, (-1R, -1S) 1,800shp T53-703, (-1T) 2,050shp T400-WV-402; (-1T$) two 1,625shp General Electric T700-401 turboshafts.
Dimensions: Diameter of two-blade main rotor 44ft 0in (13.41m), (-1T) 48ft 0in (14.63m), (Model 249) four blades, diam as original; overall length (rotors running) (G, Q, R, S) 52ft 11¹/2 (16.14m), (J) 53ft 4in (16.26m), (T) 58ft 0in (17.68m); length of fuselage/fin (most) 44ft 7in (13.59), (T) 48ft 2in (14.68m); height (typical) 13ft 6¹/4in (4.12).
Weights: Empty (G) 6,073lb (2,754kg), (J) 7,261lb (3,294kg), (S) 6,479lb (2,939kg), (T) 8,608lb

(3,904kg); maximum (G, Q, R) 9,500lb (4,309kg). (J, S) 10,000lb (4,535kg), (T) 14,000lb (6,350kg).
Performance: Maximum speed (G, Q) 172mph (277km/h), (J) 207mph (333km/h), (S, with TOW) 141mph (227km/h); max rate of climb, varies from 1,090ft (332m)/min for J to 1,620ft (494m)/min for S; range with max fuel, typically 357 miles (574km).
Armament: Typically one 7·62mm multibarrel Minigun, one 40mm grenade launcher, both in remote control turrets, or 20mm six-barrel or 30mm three-barrel cannon, plus four stores pylons for 76 rockets of 2·75in calibre or Minigun pods or 20mm gun pod, or (TOWCorba) eight TOW missiles in tandem tube launchers on two outer pylons, inners being available for other stores.

Development: Bell was a pioneer of the concept of the armed battlefield helicopter, initially using the small

Model 47 as the basis for the Sioux Scout. In 1965, after only six months of development, the HueyCobra emerged as a combat development of the UH-1 Iroquois family. It combines the dynamic parts — engine, transmission and rotor system — of the original Huey with a new streamlined fuselage providing for a gunner in the front and pilot above and behind him, and for a wide range of fixed and power-aimed armament systems.

The first version was the US Army AH-1G, with 1,100hp T53 engine, of which 1,124 were delivered, including eight to the Spanish Navy for anti ship strike and 38 as trainers to the US Marine Corps. The AH-1Q is an anti-armour version often called TOWCobra because it carries eight TOW missile pods as well as the appropriate sighting system. The AH-1J SeaCobra of the Marine Corps and Iranian Army has twin engines, the 1,800hp Twin Pac having two T400 power sections driving one shaft. Latest versions are the -1Q, -1R, -1S and -1T, with more power and new equipment. All

Cobras can have a great variety of armament.

The first Fuji-assembled AH-1S for the JGSDF flew in Japan in June 1979. Dornier of W Germany is rebuilding US Army AH-1Gs to -1S standard. The company-financed Model 249 is an advanced Cobra with the four-blade Model 412 rotor, reduced in diameter. Another new development is the US Marine Corps' AH-1T +, sometimes written -1T Plus, by far the most powerful of all Cobras with the two GE T700 engines. First flown on 16 November 1983, this combines enhanced weapon load with sparkling performance, even with one engine out. The Marines are buying two batches of 22 each.

Right: One of the 44 AH-1T+ SuperCobras which the USMC will use for attack, escort and other firepower missions.

Far right: TOW-armed Modernised AH-1S HueyCobra anti-armour helicopters of the Japan Ground Self-Defense Force.

Hughes AH-64A Apache

Origin: USA.
Type: Anti-armour and attack helicopter.
Engines: Two 1,696shp General Electric T700-701 turboshafts.
Dimensions: Diameter of four-blade main rotor 48ft 0in (14.63m); length (rotors turning) 58ft 3·1in (17.76m); length of fuselage 48ft 2in (14.68m); height overall 13ft 11·7in (4.26m); wing span 17ft 2in (5.23m); main-rotor disc area 1,809·5sq ft (168·11m²).
Weights: Empty 11,015lb (4,996kg); max 17,650lb (8,006kg).
Performance: Max speed 186mph (300km/h); max cruising speed 182mph (293km/h); hovering ceiling 10,200ft (3,110m); range (internal fuel) 428 miles (689km); endurance in anti-tank mission 1h 50min.
Armament: One 30mm Hughes M230A1 Chain Gun with 1,200 rounds in remotely aimed ventral mounting; four wing pylons for 16 Hellfire missiles or four 19-tube FFAR launchers or any combination.

Development: Probably the most expensive helicopter ever put into production, the AH-4 is the ultimate expression of the US Army's need for an all-can-do attack helicopter. A generation later than the Lockheed AH-56 Cheyenne, which was cast in the same mould but cancelled (on grounds of complexity and cost), it has considerably greater power than HueyCobras now in use, but Bell is fighting back with a Cobra powered by the same engines as the Apache. This contest will thus be interesting because the uprated Cobra would probably have a lower price tag and better flight performance and manoeuvrability.

At present the T700 engines give the AH-64A a decided edge in payload capability, which is

translated not only in weapons but also in all-weather and night sensors and weapon aiming systems.

The main sensors are PNVS (pilot's night vision system) and TADS (target acquisition and designation sight) jointly developed by Martin Marietta and Northrop. Both crew members have the Honeywell IHADSS (integrated helmet and display sight system) and each can in emergency fly the helicopter and control its weapons. The nose sight incorporates day/night FLIR (forward-looking infra-red), laser ranger/designator and laser tracker.

Avionics and other equipment are more comprehensive than in any other helicopter, in the Western world at least. Navigation systems include a strapdown inertial system, doppler and ADF, which with a digital autostabilization system enables the pilot — who sits above and behind the copilot/gunner — to fly NOE (nap-of-Earth) missions in visual conditions without bothering about location. Protection is enhanced by careful structural design to withstand gunfire up to (it is intended) 23mm calibre, together with Black Hole IR suppression, an

advanced radar warning receiver (RWR), and a laser detector, radar jammer, IR jammer and chaff/flare dispenser.

The chief mission sensors are the TADS (target acquisition and designation sight) and PNVS (pilot's night vision sensor), which surprisingly are in the nose, normally exposing the entire machine when in use. TADS enables any surface target to be tracked manually or automatically during attack by the gun, rocket or the Hellfires.

The US Army plans to buy 515 of

Above: "Black Hole" infra-red suppressors and armoured nacelles protect the Apache's engines.

these machines. Initial operational capability was due in 1985, and delivery is building up to a peak rate of 12 per month.

In 1985 there were proposals for a marinized version of Apache for a variety of attack roles.

Below: Principal anti-tank armament of the AH-64 is the Hellfire laser-homing missile.

Hughes OH-6 and Defender

Origin: USA.
Type: (OH-6A) Observation (500) light multirole helicopter, (Defender) variants for close support reconnaissance, ASW or dedicated anti-armour warfare.
Engine: One Allison turboshaft; (OH-6A) T63-5A flat rated at 252·5shp, (500M) 250-C18A flat-rated at 278shp, (Defender) 420shp 250-C20B.
Dimensions: Diameter of four-blade main rotor 26ft 4in (8·03m); length overall (rotors turning) 30ft 3·8in (9·24m); height overall 8ft 1·5in (2·48m).
Weights: Empty (OH) 1,229lb (557kg), (500M) 1,130lb (512kg); max loaded (OH) 2,700lb (1,225kg), (500M) 3,000lb (1,361kg).
Performance: Max cruise at S/L 150mph (241km/h); typical range on normal fuel 380 miles (611km).
Armament: See blow.

Development: Outstanding because of its compact design and agility, the tadpole-like OH-6A Cayuse remains a major utility, training and liaison helicopter in the US Army, which originally bought 1,434 to serve in the observation mission. From it Hughes has developed the

world's most successful family of light military helicopters, all sharing a common airframe, engine and systems but differing in equipment and weapons.

There are seven current production models: Model 500M, the basic variant chosen by most customers (those of the Spanish Navy are equipped for ASW duties with a MAD "bird" and two torpedoes); 500MD Defender, with more powerful engine (in many machines flat-rated at 375shp), particle inlet separator, IR suppression, self-sealing tanks and optional armour (available with seven seats, or two stretchers, or a wide range of weapons including TOW missiles in two twin pods); 500MD Scout Defender, with various machine guns, 14 FFAR rockets or a 40mm grenade launcher; 500MD /TOW Defender, with four TOW missiles and the stabilized TOW sight installation in the nose; MMS-TOW, the option of having an MMS (mast-mounted sight); the 500 MD/ASW, with full integrated ASW systems and two torpedoes, equipment normally including rapid-inflating emergency floats, deck-hauldown gear and a nose radar as well as MAD and special radio; and the Defender II

uprated multi-mission version with quiet rotors, APR-39 RWR, laser ranger/designator and video link to cockpit TV displays.

In 1983 the National Guard Association of the United States urged Congress to provide funds beginning FY1985 to upgrade the Army National Guard OH-56As over a five-year period which would keep the helicopter operational beyond the year 2000. Recommended modifications would affect the airframe, alighting gear, electrical systems and avionics, and would include lightweight armour, Black Hole IR suppression system and provision of

Above: The multi-mission 530MG Defender features TOWs, mast-mounted sight and very advanced avionics and cockpit.

Stinger missiles with visual sighting.

In 1985 Hughes, now taken over by McDonnell Douglas, was studying the market for military versions of the stretched Model 500E and uprated 530F, the latter having the 650hp Allison C30 engine.

Below: TOW missile firing by an anti-tank 500MD/MMS-TOW Defender, another of the wide range of Model 500/530 helicopters.

Kaman SH-2 Seasprite

Origin: USA.
Type: Ship-based multirole helicopter (ASW, anti-missile defence, observation, search/rescue and utility).
Engines: Original versions, one 1,050 or 1,250hp General Electric T58 turboshaft, all current versions, two 1,350hp T58-8F.
Dimensions: Main rotor diameter 44ft 0in (13·41m); overall length (blades turning) 52ft 7in (16m); fuselage length 40ft 6in (12·3m); height 13ft 7in (4·14m).
Weights: Empty 6,953lb (3,153kg); max 13,300lb (6,033kg).
Performance: Max speed 168mph (270km/h); max rate of climb (not vertical) 2,440ft (744m)/min; service ceiling 22,500ft (6,858m); range 422 miles (679m).
Armament: See below.

Development: For well over 20 years the neat Seasprite has been standard equipment aboard US Navy frigates, in various versions including the HH-2C rescue/utility with armour and various armament including chin Minigun turret and waist-mounted machine guns or cannon; others are unarmed HH-2D. One has been used in missile firing (Sparrow III and Sidewinder) trials in the missile-defence role. All Seasprites have since 1970 been drastically converted to serve in the LAMPS (light airborne multipurpose system) for anti-submarine and anti-missile defence.

The SH-2D has more than two tons of special equipment including powerful chin radar, sonobuoys, MAD gear, ECM, new navigation and communications systems and Mk 44 and/or Mk 46 torpedoes. All are being brought up to SH-2F standard with improved rotor, higher gross weight and improved sensors and weapons.

Though only the interim LAMPS platform, the SH-2 is a substantial programme. The first of 88 new SH-2F Seasprites became operational in 1973, and by 1982 Kaman had delivered 88, plus 16 rebuilt SH-2Ds. Moreover, despite the existence of the bigger and much more costly SH-60B LAMPS III, 18 new-production SH-2Fs were ordered in 1981 and these were delivered during 1984-5. The final nine have increased max weight (this may be later cleared on the whole fleet after small changes). A second batch of 18 new Seasprites was part-funded at the time of writing.

Above: Originally a UH-2A, this Seasprite is seen here acting as an interim-LAMPS test aircraft with two radars before being rebuilt to SH-2F standard.

Below: A sonobuoy is dropped from an SH-2F Seasprite, originally a UH-2B rebuilt to serve in the interim-LAMPS ASW and missile-defence role.

Kamov Ka-25 "Hormone"

Origin: Soviet Union.
Type: Ship-based ASW, search/rescue and utility helicopter.
Engines: Two 990shp Glushenkov GTD-3BM turboshafts.
Dimensions: Main rotor diameter (both) 51ft 8in (15·75m); fuselage length, about 34ft (10·36m); height 17ft 8in (5·4m).
Weights: Empty, about 11,023lb (5,000kg); max 16,535lb (7,500kg).
Performance: Max speed 130mph (209km/h); service ceiling, about 11,000ft (3,350m); range (standard) 250 miles (400km), (external tanks) about 400 miles (650km).
Armament: One or two 400mm AS torpedoes, nuclear or conventional depth charges or other stores, carried in internal weapon bay.

Development: Since 1965 about 500 of these well-proven machines have been built, their unusual coaxial-rotor design being adopted partly because of Kamov's long experience with this arrangement and also because it enables overall dimensions to be kept down, facilitating shipboard operation, and it eliminates casualties caused by tail rotors.

The basic machine began life with 900hp engines, but it has been continually updated and fitted with special role equipment for three main missions. Designations are not yet known, but NATO calls the three variants by the names given below.

"Hormone-A" is an ASW helicopter, operated from many Soviet surface ships as well as from shore bases. Over the years the all-weather and night capability has been augmented, and auto-hovering and anti-icing are now installed together with very comprehensive communications. Mission equipment includes a chin radar, MAD "bird" (not always carried), a dunking sonar, box of sonobuoys and dye markers. It has been said that lack of night and all-weather dipping (dunking) sonar is a major shortcoming, but it would be "un-Russian" to build a night/all-weather helicopter and then omit such a basic item of ASW gear. Various EW and IFF fits have been seen, as well as an EO sensor in a turret on the tail boom, a

Below: "Hormone-A", with sonobuoy canister on fuselage side, operating from Moskva.

Kamov Ka-27 "Helix"

Origin: Soviet Union.
Type: Shipboard helicopter for (A) ASW or (B) missile guidance.
Engines: Two 2,225hp Isotov TV3-117V turboshafts.
Dimensions: (estimated) Diameter of each three-blade main rotor 54ft 11·4in (16·75m); length of fuselage 36ft 1in (11·0m); height overall 18ft 0in (5·5m).
Weights: (estimated) Empty 12,125lb (5,500kg); max 25,353lb (11,500kg).
Performance: Estimated (by US DoD), cruising speed 161mph (260km/h); operating radius 186 miles (300km).
Armament: Assumed to include ASW weapons in internal underfloor bay in "Helix-A" version.

Development: A natural second-generation development of the Ka-25, the helicopter called "Helix" by NATO is known to be the Ka-32 in its civil form, this being used not only as a 25-seater but also as a cargo and flying crane machine for construction work, able to fly 115 miles (185km) with a slung load of 11,020lb (5,000kg). A Ka-32 surveillance model with all-weather avionics and anti-icing, is used aboard various ships including ice-breakers.

The military models are designated Ka-27 by the US Department of Defense. First seen aboard the then-new Udaloy missile ship in 1981, the ASW version ("Helix-A") is only slightly larger in overall dimensions than the Ka-25, but the cabin has almost twice the volume and the engines are more than twice as powerful. There are only two fins, though these have slats, and there are numerous other minor differences. Like the Ka-25 the three-blade rotors have powered folding, and there is a typically comprehensive array of avionics and sensors. The latter have not been positively

MBB BO105

Origin: West Germany.
Type: (C) Multirole all-weather helicopter, (M) liaison and light observation, (P) dedicated anti-armour.
Engines: Two 430shp Allison 250-C20B turboshafts.
Dimensions: Diameter of four-blade main rotor 32·3ft (9·84m); length overall (rotors turning) 38·91ft (11·86m); height overall 9·84ft (3·0m); main rotor disc area 820sq ft (76·2m²).
Weights: Empty (C) 2,622lb (1,189kg) (P) 3,817lb (1,731kg); max loaded 5,291lb (2,400kg).
Performance: Max speed 167mph (270km/h); cruising speed 144mph (232km/h); max rate of climb 1,773ft (540m)/min; range with max payload (no reserves) 357 miles (575km) (408 miles 565km, at height of 5,000ft, 1,525m).
Armament: Various options including six HOT or TOW missiles and their associated stabilized sight system; Spanish and Iraqi BO 105s have cannon, in some cases in addition to HOT missiles.

Development: Though expensive, this quite small helicopter has always had many major advantages including twin turbine engines, an advanced rotor with a forged titanium hub and hingeless blades of flexible GRP (glass reinforced plastics) construction, and comprehensive avionics, so that the basic helicopter is cleared for one-pilot operation in IFR conditions in airways or other controlled airspace.

Most-comprehensively equipped model is the 105P anti-armour version, 212 of which were supplied to the West German army. This seats a pilot and weapons operator side-by-side, and has six Euromissile HOT missile tubes, each round being guided by a stabilized sight mounted in the cabin roof. A primary navaid is Singer ASN-128 doppler. Another anti-tank variant carries eight Hughes TOW missiles. Normal seating is for five, the triple rear set being replaced if necessary by light cargo or two stretchers.

The burden of extra equipment carried by the P-series machines is evident from the empty weight figure. Another heavy model is the Mexican Navy variant used from corvettes in various patrol duties, with nose radar, folding blades, flotation gear, deck fixtures and various other systems.

Many of the Indonesian and Philippine BO 105s are military, sold to third parties, while CASA is also assembling machines for other customers, notably including Iraq. The latter's machines, in use against Iran, have a different ammunition chute for the ventral cannon compared with the Spanish army model, the latter being subdivided into 28 anti-tank, 14 observation and 18 armed reconnaissance. The BK 117, which was developed and produced jointly by MBB and Kawasaki of Japan, is a larger and more powerful machine first flown in June 1979, and is likely to evolve in military forms in due course.

Above. TOW-armed BO 105CB, standard production version of this useful lightweight.

Below: BO 105P, designated PAH-1 by the German Army, with HOT anti-tank missiles.

---

sensor pod on the base of the ventral centre fin and various other unidentified blister fairings. Weapons are normally carried in a long internal bay under the cabin floor, which in some aircraft is enlarged (it is said, to house wire-guided torpedoes).

"Hormone-B" is a specialized model whose main purpose is to acquire over-the-horizon targets for warships and then provide guidance for the ship-fired missiles (of such types as SSN-3, "Shaddock", SSN-12 "Sandbox" and SSN-19). It is characterized by absence of ASW gear and weapon bay but addition of a much larger and more bulbous chin radar and a second scanner, in a ventral "dust-bin" at the rear of the cabin. Special data links are pro-

Above: Ka-25s are standard equipment aboard *Kara, Kirov, Kiev,* "Kresta II", and Moskva class warships; this "Hormone-A" is on the deck of *Moskva.*

vided to pass target co-ordinates to the ship missiles.

"Hormone-C" machines, often painted bright red and white, are simple utility transports used for plane-guard, SAR, vertrep, VIP and COD missions. The cabin seats up to 12, and a winch is fitted.

The Indian Navy operates five ex-Soviet Ka-25s aboard its three Krivak class destroyers. The Syrian Arab Air Force operates nine Ka-25s on coastal ASW duties, and an unknown number are in use by Yugoslavia.

identified in the West, but clearly are likely to reflect 20 years of experience with the Ka-25.

The "Helix-B" model, replacing "Hormone-B" in the over-the-horizon targeting and missile guidance role, is known to exist but no illustrations had been released as this book was published.

Below: One of the pair of Ka-27 "Helix-A" helicopters carried by the ASW guided missile destroyer *Udaloy* is wheeled into the ship's port hangar with its rotor folded. The aircraft carrier *Novorossiysk* carried at least 16 Ka-27s on her 1983 maiden deployment.

Mil Mi-8 and Mi-17 "Hip"

Origin: Soviet Union.
Type: Armed attack helicopter; other variants for general utility, assault transport, and electronic warfare.
Engines: (-8) Two 1,700shp Isotov TV2-117A turboshafts; (-17) two 1,900shp TV3-117MT.
Dimensions: Main rotor diameter 69·85ft (21·29m); overall length, rotors turning, 82·8ft (25·24m); fuselage length 60·08ft (18·31m); height 18·53ft (5·65m).
Weights: Empty (-8T) 15,026lb (6,816kg), (-17) 15,653lb (7,100kg); max (-8) 26,460lb (12,000kg), (-17) 28,660lb (13,000kg).
Performance: (-8T) Max speed 161mph (260km/h); service ceiling 14,760ft (4,500m); range (-8T, full payloads, 5 per cent reserve at 3,280ft) 298 miles (480km).
Armament: Optional fitting for external pylons for up to 12 stores carried outboard of fuel tanks (always fitted); typical loads eight pods of 57mm rockets, or mix of gun pods and anti-tank missiles (Mi-8 not normally used in anti-tank role), with max load four triple 16 x 57 launchers for 192 rockets, plus four AT-2 Swatter or AT-3 Sagger missiles; 12·7mm gun.

Development: In 1985 the Mi-8 and the more powerful Mi-17 were still in full production at Kazan and Ulan-Ude, with total deliveries in excess of 10,000 to some 39 air forces around the world. This is all the more impressive when it is remembered that this family of helicopters

are more powerful than a Sea King or Black Hawk, and the cabin can easily seat 32 passengers, or in tactical roles carry vehicles within the normal payload limit of 8,820lb (4,000kg) internally.

The most common model is the Mi-8T utility transport, which has circular windows, full provisions for loading and positioning heavy cargo, 28 tip-up seats along the sides on quickly removable mounts, and attachments for 12 stretcher casualties.

This model has the NATO name Hip-C. Hip-D is an EW (Comint/Elint) platform equipped with extra aerials (antennae) and large boxes on the outboard pylons. Hip-E is the extra-heavily armed assault model, which is widely used by Frontal Aviation. Hip-F is the corresponding export model, usually with the AT missiles changed to AT-3 Saggers. Hip-G is a communications relay platform with prominent aerials. Hip-H is the uprated Mi-17 (visually identified by having the anti-torque rotor on the left of the tailfin). Hip-J and -K are both Mi-8 EW platforms with various special electronics. All Mi-8 and -17 helicopters have full provision for flight at night or in bad weather, including icing conditions. They have not been seen with inlet filters, though plain momentum-type particle separators are common on the Mi-17.

Below: Mi-8 "Hip-E", standard Soviet armed assault transport, with nose gun, rocket pods and rails for anti-tank missiles.

Above: Egyptian Mi-8 with pairs of rocket pods on rails. Not all of the 39 countries to receive Mi-8s are happy with them.

Below: Another "Hip-C" of the Egyptian Air Force, which still has some 50 in service despite heavy losses in the 1973 war.

Mil Mi-24 "Hind"

Origin: Soviet Union.
Type: Armed assault helicopter (variants, see below).
Engines: Two 2,200shp Isotov TV-117A turboshafts (early series, 1,700shp TV-117A).
Dimensions: Diameter of five-blade main rotor, about 55·75ft (17m); length overall (ignoring rotors) 55·75ft (17m); height (to tip of fin) 14·0ft (4·25m); main rotor disc area 2,443sq ft (226·9m²).
Weights: (estimated) Empty 14,300lb (6,500kg); normal loaded 22,046lb (10,000kg); max 25,400lb (11,500kg).
Performance: Max speed 199mph (320km/h) (A-10 helicopters built at Arsenyev have set many records including speeds up to 228·9mph/368·4km/h, and a women's time-to-height record of 2min 33sec to 9,842ft/3,000m); range with max weapons given as only 99 miles (160km).
Armament: (A) usually one 12·7mm DShK gun aimed from nose, two stub wings providing rails for four wire-guided anti-tank missiles and four other stores (bombs, missiles, rocket or gun pods); (B) two stub wings of different type without anhedral and only four pylons; (C) similar to (A) without nose gun and usually without wingtip missile rails; (D) four-barrel 12·7mm gun in remotely directed chin turret, wings equipped with racks for four AT-6 Spiral anti-tank missiles and either four 32-round 2·24in (57mm) rocket pods, four 55lb (250kg) bombs, gun pods or other stores;

(E) four-barrel turret and six AT-6 missiles, and (sub-variant) two-barrel 23mm cannon firing ahead.

Development: This outstanding helicopter was originally designed as a derived Mi-17 able to carry more sensors and weapons at the expense of having a smaller cabin. All versions so far seen have essentially the same cabin, able to carry a squad of eight troops and their equipment. This leaves abundant spare payload which can be made up with weapons hung on various arrangements of wings and pylons.

The result is rather large for an armed helicopter, and using an articulated rotor, even of reduced size with five blades, limits in-flight agility to manoeuvres more akin to the Sea King than the Lynx. At the same time the versatility and firepower of these machines, especially in their later versions with a redesigned forward section, puts them in a class of their own.

The original TV2-powered production version, called "Hind-B", had the tail rotor on the right of the fin, and in front of the main cabin was a crew compartment with windows all round and seats for a pilot, copilot, nav/gunner and forward observer (who might be the troop squad commander). Access was gained by a sliding bulged window/door at the rear on the left, and an upward-hinged window by the gunner's seat. Ahead of the gunner was a bullet-proof windscreen with a wiper, and there were plenty of areas of armour. Access for the

eight troops was through large doors on each side, hinged into upper and lower sections, the lower portions incorporating steps for the fastest possible boarding. Weapons were carried on four pylons on small horizontal wings just behind the cabin. The tricycle landing gear was fully retractable.

Few of this type were built, and the first major model, "Hind-A", was fitted with much larger weapon "wings", sloping sharply down and carrying not only the four normal pylons (for 32 x 57 rocket pods or other stores) but also pairs of launch rails for AT-2 Swatter anti-tank missiles.

Some hundreds of this type were built, some of them being exported. Production was expanded to include both Arsenyev and Rostov-on-Don, and in about 1976 numerous improvements were introduced, including the TV3 engine, the tail rotor

Above: Mil Mi-24 "Hind-A" was the first production version, with a conventional cabin and right-mounted tail rotor.

on the left side, improved glassfibre main-rotor blades and widespread substitution of steel and titanium for parts previously of aluminium alloy, in the interest of greater resistance to hostile gunfire.

Though a third model in this original series was identified — "Hind-C", with no nose gun or chin blister, and without the AT-2 missile rails — the main change was to switch (almost completely) to a series of "gunship" models with a redesigned cockpit. These are lavishly equipped with avionics and

Below: An Mi-24 "Hind-D" gunship is refuelled. The type has demonstrated its impressive firepower in Afghanistan.

sensors, and are in most respects the most formidable battlefield helicopters in the world.

First of this series, "Hind-D" introduced the new forward fuselage with even greater skin thickness and armour protection, and just two crew seats, the weapon operator in the nose, under a canopy hinged to the right, and the pilot higher to the rear with a door on the right. Under the nose is a powered turret with a new four-barrel gun with a wide range of fire and very effective against other helicopters or other aerial targets as well as those on the surface.

Full analysis of the sensors has yet to be published, but the big turret on the right under the nose includes a FLIR and an LLTV; inside the cockpit is an optical sight system with haze filter and x10 magnifying option; and on the left is a ventral pod containing electronics concerned with anti-tank missile guidance, as related later. At the tip of the left weapon "wing" is a laser designator which is also often seen on "Hind-A" but at the top of the left inboard pylon.

The primary anti-armour weapons, also used effectively against troops hiding in mountainous regions in Afghanistan, are the two types of missile, AT-2C Swatter and AT-6 Spiral. Though hundreds of the former type have been captured in recent years, no account of its guidance has been published (it is widely believed to home on IR radiation). Despite this self-sufficiency, "Hind-D" has a small pod, as already mentioned, associated with missile guidance. When AT-6 is fitted the helicopter designation changes to "Hind-E". This much later weapon is tube launched and has laser semi-active homing guidance, using the wing-mounted designator. Yet again, there is an associated pod, mounted like the other pattern on a swivelling pivot on the lower left side of the nose.

Other equipment of "Hind-D" and "-E" includes full thermal anti-icing, a low-airspeed sensor boom giving precise pitch/yaw angles for accurate firing of the unguided rockets, and extremely comprehensive IFF, ILS and other navaids.

Since 1981 a proportion of Mi-24s, still called "Hind-E", have appeared with a twin-barrel 23mm gun fixed on the left side firing ahead, and with a rounded nose faired into the sensor turrets replacing the original gun turret.

Surprisingly, under 2,000 Mi-24s were thought to have been built by late 1984, with production continu-

Above: "Hind" is now believed to carry a flight engineer as well as a pilot and air gunner.

ing. Western analysts continue to debate the vulnerability and effectiveness of these powerful machines, and it is is significant that a dedicated gunship (Mil Mi-28 "Havoc") has now appeared.

Mil Mi-28 "Havoc"

Origin: Soviet Union.
Type: Believed to be air-combat helicopter.
Engines: Believed to be two 2,200shp Isotov TV3 turboshaft engines.
Dimensions: Diameter of main rotor believed to be same as Mi-24 (estimated 55ft 9in, 17·0m); length of fuselage (estimate 57ft 1in 17·4m).
Weights: (Estimate) Empty 17,640lb (8,000kg); maximum 24,250lb (11,000kg).
Performance: (Estimate) Maximum speed 220mph (354km/h); combat radius 149 miles (240km).

Development: The 1984 edition of

the US Department of Defense booklet "Soviet Military Power" contained a crude drawing (traced over an AH-64 Apache, so far as one could tell) labelled "Mi-28 Havoc". The existence of a slim-bodied Mil gunship had been suspected for a long time, and prototypes may have been under test for years. Now it appears that the helicopter given the NATO name "Havoc" is nearing operational service. If the number is 28 it suggest a primary role other than air-combat, though possibly the rule that fighters always receive odd numbers does not apply to helicopters.

Little is known about "Havoc", though Jane's All the World's Aircraft suggests that it actually more

closely resembles the Lockheed AH-56A Cheyenne than an Apache. One would expect to find a tandem stepped cockpit followed by a slender fuselage and, probably, engines, rotors and tail similar to the Mi-24. This would be a wholly logical, if not obvious, development. Such a helicopter would have outstanding flight performance (the large-cabin Mi-24 already holds various speed records at speeds up to 228.9mph, 368.4km/h) and also outstanding manoeuvrability, even when heavily loaded with weapons. A "Havoc" could fly all the missions flown by "Hind" versions except those requiring transport of a squad of troops. It would combine tremendous weapons capability in air/air and air/ground roles with improved survivability and probably the highest speed of any combat helicopter.

In April 1985, the US Department of Defense issued its annual review of Soviet Military Power, but the artwork is still highly speculative. The large single-barrel gun is nothing like any Soviet weapon previously known and appears to be of some 30mm calibre, but perhaps the oddest thing of all is that nothing except the main rotor seems to bear the slightest commonality with the Mi-24. Why almost everything proven and reliable should be totally altered is not explained.

Below: The Pentagon's latest artist's impression of new Mil Mi-28 "Havoc", which is expected to carry modified AT-6 missiles with millimetre-wave seekers, giving an effective range of up to 5 miles (8km) against tanks.

Sikorsky S-70 (H-60 Hawk family)

Origin: USA.
Type: (UH) combat assault transport, (EH) electronic warfare and target acquisition, (HH) combat rescue, (SH) ASW and anti-ship helicopter.
Engines: (UH, EH) two 1,560shp General Electric T700-700 turboshafts, (SH, HH) two 1,690shp T700-401 turboshafts.
Dimensions: Diameter of four-blade rotor 53·66ft (16·36m); length overall (rotors turning) 64·83ft (19·76m); length (rotors/tail folded) (UH) 41·33ft (12·6m), (SH) 41·04ft (12·5m); height overall (UH) 16·8ft (5·13m), (SH) 17·16ft (5·23m).
Weights: Empty (UH) 10,624lb (4,819kg), (SH) 13,648lb (6,191kg); max loaded (UH) 20,205lb (9,185kg) normal mission weight 16,260lb, 7,375kg)m (HH) 22,000lb (9,979kg), (SH) 21,884lb (9,926kg).
Performance: Max speed (184mph (296km/h); cruising speed (UH) 167mph (269km/h), (SH) 155mph (249km/h); range at max wt. 30min reserves, (UH) 373 miles (600km), (SH) about 500 miles (805km).
Armament: (UH) See below, (EH) electronic only, (SH) two Mk 46 torpedoes and alternative dropped stores, plus offensive avionics.

Development: This helicopter family owes little to any previous Sikorsky type, and when the first version (the UH-60A Black Hawk) was designed in the late 1970s it was packed with new technology. The original requirement came from the US Army, which needed a UTTAS (utility tactical-transport aircraft system) for general battlefield supply duties.

The basic role called for a crew of three and a cabin for 11 troops with full equipment, but the UH-60A can readily be fitted with 14 troop seats, or alternatively with six stretchers. The cargo hook can take a load of 8,000lb (3,629kg), and a typical front-line load can be a 105mm gun, 50 rounds of packaged ammunition and the gun crew of five. The UH-60A can also be used for command and control or for reconnaissance missions. The design was made especially compact so that it would fit into a C-130 Hercules; a C-5 Galaxy can carry six.

The UH-60A proved to be a particularly successful machine, with surprising agility and good all-weather avionics. As with all later versions most of its critical parts are designed to withstand 23mm gunfire, and in a gruelling kind of combat life it has proved one of the most "survivable" of all helicopters.

Originally there was no call for armament, apart from provision for the troops to fire a 7·62mm M60 LMG from a pintle mount on each side of the cabin (and standard kit includes chaff/flare dispensers for self defence). In the 1980s, however, the Army began re-equipping its Black Hawk fleet with the ESSS (external stores support system) which adds large "wings" with four pylons, on which can be hung anything from motorcycles to 16 Hellfire precision missiles. Gun pods are another alternative, as is the M56 mine-dispensing system. For self-ferry duties four external tanks can be attached, giving a range of 1,323 miles (2,130km).

As expected the UH-60A is a big programme, with deliveries about 610 at the time of writing, with a predicted eventual total for the Army of 1,107. Another 11 have gone to the USAF, nine being used for recovery and rescue missions at Eglin AFB. The EH-60A Black Hawk is a specialized ECM version packed with 1,800lb (816kg) of equipment designed to detect, monitor and jam enemy battlefield communications. The installation is called Quick Fix II, the main item being the ALQ-151 ECM kit.

The Army expects to buy 77 of this version as its SEMA (special electronics mission aircraft). It also hopes to fund 78 of another version, the EH-60B, fitted with the SOTAS (stand-off target acquisition system), main element of which is a large target-indicating radar whose signals are relayed to a ground station. This programme has run into cost and schedule problems, and a cheaper radar may be fitted.

The main USAF model is the HH-60D Night Hawk, which uses almost the same airframe but has uprated engines and transmission more akin to that of the SH-60B Sea Hawk. The Night Hawk is planned as the next-generation all-weather combat rescue and special missions vehicle. Features include external tanks and additional internal fuel to achieve a full-load combat radius of 288 miles (463km), flight-refuelling probe, various new weapons provisions in the fuselage, added protection systems, and a mass of avionics including all-weather terrain following with a FLIR and the LANTIRN, as well as special communications and a pilot's helmet display like that of the AH-64A Apache. The USAF hopes to afford 89, which have Europe 1 camouflage, as well as 66 HH-60Es which will save money by having no FLIR, radar-map display or helmet display.

The major armed version is the US Navy's SH-60B Seahawk. This, the Sikorsky S-70L, wea built to meet the neet for a LAMPS III (light airborne multi-purpose system), packaged into a shipboard helicopter. In fact the SH-60B is much larger and more costly than most shipboard machines, because of the strenuous demands for equipment and weapons to handle ASW and ASST (anti-ship surveillance and targeting) in all weather.

Apart from all the obvious changes needed for operation from destroyers and frigates, the Seahawk has uprated engines and transmission and a totally new fuselage with APS-124 search radar under the cockpit and forward part of the cabin, side chin-mounted ESM pods, a MAD station on the right, a battery of 25 sonobuoy launchers on the left with four more reloads inside the fuselage (125 buoys in all), and an impressive array of sensors, navaids, communications, processing systems and self-defence measures. Normal ASW armament comprises two torpedoes. There is no FR probe, but the Seahawk can be coupled to a refuelling hose from a ship and refuel in the hover or while maintaining station at sea.

The Navy achieved operational capability in 1984, and expects to receive 203 of these big machines for use from 106 surface warships. They can also fly Vertrep (vertical replenishment), medevac, fleet support, SAR and communications

Above: EH-60A, a specialised ECM version, carries 1,800lb (816kg) of electronic gear.

relay missions. The Navy expects to buy a further 175 of the SH-60F type, for its CV-helo missions. These will operate from carriers in the direct protection of the battle fleet, using AQS-13 dipping sonar instead of buoys, and also flying plane-guard SAR missions. Japan has naturally chosen the SH-60B as its next naval helicopter, and will probably arrange licence production later in the 1980s. In autumn 1984 it was also chosen by Australia.

Above: UH-60As provide much-needed mobility for the infantry, and are a great improvement on the UH-1 series.

Below: SH-60B of the US Navy, currently the major armed version of this adaptable and effective machine.

Sikorsky AUH-76

Origin: USA.
Type: Armed utility helicopter.
Engines: Two 682shp Allison 250-C30 turboshafts.
Dimensions: Diameter of four-blade main rotor 44·0ft (13·41m); length of fuselage 43·37ft (13·22m); height overall 14·82ft (4·52m); main-rotor disc area 1,257sq ft (116·77m²).
Weights: Empty 5,600lb (2,540kg); max 10,300lb (4,672kg).
Performance: Max cruising speed about 167mph (269km/h); speed for best range 144mph (232km/h); range highly variable according to load and mission but typically 450 miles (724km).
Armament: MPPS (multi-purpose pylon system) can be installed for a very wide range of gun and/or rocket pods, 16 TOW missiles (with mast-mounted sight), Hellfire missiles, four Sea Skua anti-ship missiles, Stinger ADSM multirole missiles or Mk46 torpedoes. At all times two machine-guns can be pintle-mounted in each doorway.

Development: The excellent global sales of the civil S-76 high-speed passenger machine prompted development of the multirole armed AUH version (the designation is the company's, and no US sale has yet been announced).

The main cabin can seat ten armed troops, 12 passengers or 16 people sitting on the floor, and among a wide range of options are a rescue hoist and three stretchers plus two attendant seats. An MMS can be installed as an alternative to a sight on the cockpit roof, and many weapons can be hung on the MPPS which in 30 minutes can be installed on the main floor projecting on each side with four pylons. There are many provisions for self-defence, including a roof cable cutter, chaff/flare dispensers, self-sealing tanks and run-dry bearings.

To increase the potential of the military S-76, the TOW missile was tested in a firing programme during 1984. With eight TOWs, the AUH-76 has a 114 mile (184km) radius with an hour loiter time.

Above: TOW missile firing test at Fort Rucker, Ala, for the new Sikorsky H-76 helicopter.

A naval version is also being developed, aimed at maritime surveillance, ASW, SAR and utility missions. It will operate from frigate-sized ships and will have optional radar, ECM, sonar, etc.

The first sale is to the Philippine AF, which will receive 12 armed multirole machines, two SAR-equipped and three VIP models. Several air forces have bought the civil S-76.

Westland Lynx

Origin: France (30 per cent) and UK.
Type: (AH.1 and army version) tactical helicopter for transport, utility, electronic warfare, anti-armour attack, search/rescue, multi-sensor reconnaissance, armed escort, casevac and command missions; (HAS.2 and naval versions) multi-role shipboard anti-submarine and anti-ship search, classification and strike, vertrep, troop transport, fire support, reconnaissance, liaison and other duties. (Lynx 3, army version anti-armour).
Engines: (early variants) Two 900shp Rolls-Royce Gem 2 turboshafts, (current) two 1,120shp Gem 41-1, (L3) two 1,346shp Gem 60.
Dimensions: Diameter of four-blade main rotor 42·0ft (12·80m); length overall (rotors turning) 49·75ft (15·16m), (L3) 50·75ft (15·47m); height overall (rotors turning) 12·0ft (3·66m).
Weights: Empty (army, typical, bare) 5,683lb (2,578kg), (naval) 6,040lb (2,740kg); max loaded (army) 10,000lb (4,535kg), (navy) 10,500lb (4,763kg), (L3) 12,000lb (5,443kg).
Performance: Max speed 200mph (322km/h); cruising speed (army) 161mph (259km/h), (naval) 144mph (232km/h); max climb at sea level 2,480ft (756m)/min; typical range with full payload and reserves 336 miles (540km); ferry range (army) 834 miles (1,342km).
Armament: (army) eight HOT, Hellfire or TOW or six AS.11 missiles and associated stabilized roof sight, Emerson Mini TAT ventral turret with 3,000 rounds, 20mm gun in cabin or two 20mm externally, pintle-mounted 7·62mm Minigun in cabin or wide range of rocket pods; (naval) four Sea Skua or AS.12 missiles, two Sting Ray or Mk44 or 46 torpedoes, two Mk 11 depth charges and various other options; (L3) see text.

Development: One of the most important of all modern helicopters, the Lynx has been developed into a large and prolific family which includes three main groups: army Lynx for battlefield missions; navy Lynx for shipboard duties; and the Westland 30. The latter, with a greatly enlarged cabin, is likely eventually to be ordered in various military forms (usually with the more powerful T700 engine) but is at present used only as a civil passenger transport.

All versions of Lynx are characterized by extreme combat agility conferred by the advanced semi-rigid main rotor with blades of stainless steel and plastics fixed to a one-piece hub of forged titanium. Thus despite its size the Lynx has been looped repeatedly, rolled at 100°/sec and flown sideways and backwards at up to 80mph (129km/h). The wide cockpit is representative of the very latest practice, and all versions have most comprehensive avionics for safe one-pilot operation in the worst weather (as repeatedly proved during Falklands blizzards). The basic cabin is normally able to seat 10 armed troops, with a large sliding door on each side; in the casevac role there is normally a crew of two plus three stretchers and an attendant, and in the transport role the slung load limit is a very useful 3,000lb (1,361kg).

The British Army uses the AH.1, with landing skids and missions as outlined above under "Type". Equipment can include mine dispensing chutes, ALE-39 chaff/flare dispensers, Magic or Stinger AAMs and a range of rocket launchers, and a team of nine carrying Milan missiles can be accommodated in the cabin. Westland delivered 114, 60 of which have TOW missiles, and is now supplying the AH.5 with Gem 41-1 engines. The Mk 28, with 41-1 engines, is used by Qatar.

By 1984 about 330 Lynx had been ordered, and the next major development stage is the Lynx 3, called L3 in the data. This is being produced in both army and navy versions, both with 1,346shp (contingency rating) engines and cleared to operate at 12,000lb (5,443kg). The rotor has new aerofoil tips to increase efficiency, while the tail rotor is derived from that of the Westland 30 and is quieter than that of today's Lynx. Avionics are based on a digital databus and completely revised and

Below: Pre-production Lynx-3 armed with Hellfire missiles and large-calibre cannon. Production examples will have a new main rotor, mast-mounted sight and air-to-air missiles.

Westland Scout and Wasp

Origin: UK.
Type: (S) Multirole tactical
helicopter, (W) general utility and
ASW helicopter for use from small
surface vessels.
Engine: (S) One 685shp Rolls-
Royce Nimbus 102 turboshaft; (W)
710shp Nimbus 503 (both flat-rated
from 968shp).
Dimensions: Diameter of four-
blade main rotor 32·25ft (9·83m);
length overall (rotors turning)
40·33ft (12·29m); length of fuselage
30·33ft (9·24m); height (rotors
turning) 11·66ft (3·56m).
Weights: Empty (S) 3,232lb
(1,465kg); (W) 3,452lb (1,566kg);
max (S) 5,300b (2,405kg); (W)
5,500lb (2,495kg).
Performance: Max speed (S)
131mph (211km/h); (W) 120mph
(193km/h); rate of climb (S)
1,670ft (510m)/min; (W) 1,440ft
(439m)/min; range with four
passengers and reserves (S) 315
miles (510km); (W) 270 miles
(435km).
Armament: (S) Various options
including manually aimed guns of
up to 20mm calibre, fixed GPMG
installations, rocket pods or guided
missiles such as SS.11; (W)
normally, two Mk44 torpedoes.

Development: Originally designed
by Cierva (Saro), these neat
machines were among the first turbine helicopters in production, yet
many are still active in both versions.
The Scout has two front seats and a
three-seat rear bench; in the casevac
role it carries two stretchers internally and two externally in panniers.
Skid landing gear is standard.

British Army Scouts have a
stabilized magnifying sight in the
cabin roof, and this was put to good
use in the battle for the Falklands
where SS.11 missiles were guided
against point targets. One Scout was
shot down by a Pucara.

The navalized Wasp has four long
legs with castoring lockable wheels
and deck hauldown gear. They are
used for every kind of duty from
small vessels. In 1983 refurbished
machines were supplied to the
RNZN.

Above; British Army Scout fires SS.11 ATGW from a very low hover, during a test on the Salisbury Plain training area.

Below: Naval Wasp HAS.1 has castoring wheel undercarriage in place of the land versions' skids. This is an RN machine.

updated with the most comprehensive navaids, mission equipment and
EW/ECM systems.

The army Lynx 3 is a dedicated
anti-armour machine with a mast-
mounted sight, probably the
TADS/PNVS and helmet sight as in
the AH-64A, and with eight or more
Hellfire missiles plus Stinger or other
AAMs. The projection above the
windshields is a cable cutter. The
corresponding navy Lynx 3 is configured for small-ship operation
carrying a much greater weight of
active and passive sonobuoys,
dunking sonar, 360°radar, MAD and
weapons (eg. one Sting Ray and
two Sea Skuas) than current Lynx,
while offering greater range and
time on station.

Other Lynx in service stem from
the basic navy HAS.2, which has
tricycle landing gear and special
equipment for deck operation. Missions are again outlined above, and
the HAS.2 has Seaspray nose radar
for maritime search and tracking,
with Orange Crop ESM in a box just
above. All versions can carry four
Sea Skua anti-ship missiles (which
worked flawlessly in the Falklands)
and many have been fitted with
MAD and/or dunking sonar and AS
weapons. The MK 2(FN) is the
French Aéronavale version with
French avionics including ORB-31W
radar and Alcatel dunking sonar and
torpedoes. The HAS.3 is the RN
model now in production with 41-1
engines, the corresponding French
model being the Mk 4.

Other marks are: 21 (Brazil), 23
(Argentina), 25 (Netherlands UH-
14A for SAR), 27 (Netherlands SH-
14B for ASW with Alcatel sonar),
80 (Danish ASW and maritime
patrol), 81 (Netherlands SH-14C for
ASW with MAD), 86 (Norwegian AF
for SAR), 87 (Argentina, 41-2
engines), 88 (West German ASW
model with Bendix AQS-18 sonar)
and 89 (Nigerian navy for ASW and
SAR with 43-1 engines, RCA radar
and other updates).

Above: Royal Navy Lynx HAS.2 armed with four Sea Skua anti-ship missiles. Other wire-guided missiles (eg, AS.12) can be carried; torpedoes or depth charges form the ASW armament.

Below: British Army Lynx AH.1 with roof-mounted sight for its TOW ATGW. The British Army has 60 TOW-armed Lynx; all are stationed in Germany as part of the NATO anti-tank force.

ELECTRONIC WARFARE AIRCRAFT

Boeing E-3 Sentry

Origin: USA.
Type: Airborne Warning and Control System (AWACS) platform.
Engines: (except Saudi) Four 21,000lb (9,526kg) thrust Pratt & Whitney TF33-100/100A turbofans; (Saudi) four 22,000lb (9,979kg) CFM56-2-A2 turbofans.
Dimensions: Span 145ft 9in (44·42m); length 152ft 11in (46·61m); height 41ft 4in (12·6m) (over fin); wing area 3,050sq ft (283·4m²).
Weights: Empty, not disclosed but about 162,000lb (73,480kg), loaded 325,000lb (147,400kg).
Performance: Maximum speed 530mph (853km/h); normal operating speed, about 350mph (563km/h); service ceiling, over 29,000ft (8·85km); endurance on station 1,000 miles (1,609km) from base, 6h.
Armament: None.

Development: Back in the early 1950s the USAF pioneered the concept of the overland radar surveillance platform, mainly using EC-121 Warning Stars (based on the Super Constellation, and continuing in unpublicised service until almost 1980). During ·the 1960s radar technology had reached the point at which, with greater power and rapid digital processing, an OTH (over the horizon) capability could be achieved, plus clear vision looking almost straight down to detect and follow high-speed aircraft flying only just above the Earth's surface.

One vital ingredient was the pulse-doppler kind of radar, in which the "doppler shift" in received frequency caused by relative motion between the target and the radar can be used to separate out all reflections except those from genuine moving targets. Very clever signal processing is needed to eliminate returns from such false "moving targets" as leaves violently distributed by wind, and the most difficult of all is the motion of the sea surface and blown

spray in an ocean gale. For this reason even more clever radars are needed for the overwater mission, and the USAF did not attempt to accomplish it until well into the 1980s.

While Hughes and Westinghouse fought to develop the new ODR (overland downlock radar), Boeing was awarded a prime contract on 8 July 1970 for the AWACS (Airborne Warning And Control System). Their proposal was based on the commercial 707-320; to give enhanced on-station endurance it was to be powered by eight TF34 engines, but to cut costs this was abandoned and the original engines retained though driving high-power electric generators.

The aerial for the main radar, back-to-back with an IFF (identification friend or foe) aerial and communications aerials, is mounted on a pylon above the rear fuselage and streamlined by adding two D-shaped radomes of glassfibre sandwich which turn the girder-like aerial array into a deep circular rotor-dome of 30ft (9·14m) diameter. This turns very slowly to keep the bearings lubricated; when on-station it rotates are 6rpm (once every ten seconds) and the searchlight-like beam is electronically scanned under computer control to sweep from the ground up to the sky and space, picking out every kind of moving target and processing the resulting signals at the rate of 710,000 complete "words" per second. The rival radars were flown in two EC-137D aircraft built from existing 707s, and the winning Westinghouse APY-1 radar was built into the first E-3A in 1975.

The first E-3A force was built up in TAC, to support quick-reaction deployment and tactical operation by all TAC units. The 552nd AWAC Wing received its first E-3A at Tinkor AFB, Oklahoma, on 24 March 1977, and went on operational duty a year later. Subsequently the 552nd have operated in many parts of the world. It was augmented from 1979 by NORAD (North American Air Def-

Above: One of the 18 E-3A Sentry Airborne Warning and Control System aircraft produced for service with NATO in Europe.

Below: Main Situation Display Console aboard an E-3; the original E-3A had nine, and another five have been added.

ense) personnel whose mission is the surveillance of all North American airspace and the control of NORAD forces over the Continental USA.

Production Sentries have been delivered to five standards. The so-called Core E-3A identifies the first 24 aircraft delivered to the USAF, with limited capability except against aerial targets. They had nine SDCs (situation display consoles), two ADUs (auxiliary display units) and 13 available communications channels.

The E-3B designation applies to the E-3As after complete updating with JTIDS (joint tactical information distribution system), the CC-2 computer, more radios and Have Quick anti-jamming, five more SDCs, a tele-typewriter and an "austere" (limited) maritime surveillance capability.

The US/NATO E-3A applies to 18 aircraft paid for by NATO nations (Belgium, Canada, Denmark, W Germany, Greece, Italy, Netherlands, Norway, Portugal, Turkey and the USA) and operated in NATO markings with crews of 15 men drawn in

rotation from the 11 nations participating. They are based at Geilenkirchen, with FOBs (forward operating bases) at Preveza (Greece), Trapani (Italy) and Konya (Turkey), plus an FOL (location) at Oerland (Norway). This was also the original standard of the USAF aircraft Nos 25-34, which in fact are being updated to E-3C standard with five more SDCs, five more UHF radios and Have Quick anti-jamming. The modifications to both Core and USAF Standard aircraft are being made with Boeing kits being delivered by April 1987.

Five E-3A/Saudis are being delivered as the major part of a $4,660 million package in 1985, which have the quiet fuel-efficient CFM56 engine.

Below: Tanker's eye view of a USAF E-3A, all 24 of which are being updated to E-3B standard: even without refuelling the Sentry can travel 1,000 miles (1,610km), spend 6hr on station and still return to its base.

Grumman E-2C Hawkeye

Origin: USA.
Type: Carrier- and land-based airborne early warning and control (AWACS) platform.
Engines: Two 4,910ehp Allison T56-A425 turboprops.
Dimensions: Span 80ft 7in (24·56m); length 57ft 6·75in (17·54m); height 18ft 3·8in (5·58m); wing area 700sq ft (65·03m²).
Weights: Empty 37,945lb (17,211kg). max take-off 51,817lb (23,503kg).
Performance: Max speed 374mph (602km/h); cruising speed 310mph (500km/h); service ceiling 30,800ft (9,390m); endurance (max fuel) 6·1h.
Armament: None.

Development: Unique in offering a valuable and comprehensive AEW package in an aircraft of compact dimensions and moderate operating cost, the Hawkeye has managed to make the transition from being a highly specialized aircraft for US Navy carrier air wings to become a major contender for sales in the world market for operation from land airfields.

Rivals would point out, truthfully, that it is today possible to build a land-based AEW platform markedly better than the Hawkeye, partly by using nose/tail radomes instead of a rotodome, and partly by using fan engines in an airframe not compromised by the need for compatibility with carriers. But no such aircraft exists, and it would cost so much to create it, and take so long, that the Grumman product has the market to itself, apart from the handful of customers able to afford an E-3.

Together with some EC-121 variants, the original E-2A of 1961 was the first aircraft to have the new style of rotodome in which the aerial itself is given a streamlined fairing instead of being housed inside a radome. The size of airframe needed to house the APS-96 radar was such that considerable ingenuity was needed; for example, four fins and rudders are used, all mounted at 90° to the dihedral tailplane and all well below the wake of the rotodome. The rotodome itself is set at a positive incidence to lift at least its own weight, while on a carrier it is retracted a short distance to enable it to clear the roof of the hangar.

The dome rotates once every 10 seconds when in operation, and the radar gives surveillance from a height of 30,000ft (9,144m) within a radius of 300 miles (480km). Ten years into the programme, the APS-96 was replaced by the APS-125 with an Advanced Radar Processing System (ARPS) which gives a much improved discrimination and detection capability over both land and water. With this new radar the aircraft designation changed to E-2C; it entered service as such in 1973 and remains the standard "eyes" of the US Navy at sea, with numerous updates in the subsequent years.

The two pilots occupy a wide flight deck. Behind them, amidst a mass of radar raking and the high-capacity

vapour-cycle cooling system (the radiator for which is housed in a large duct above the fuselage), is the pressurized Airborne Tactical Data System (ATDS) compartment. This is the nerve-centre of the aircraft, manned by the combat information officer, air control officer and radar operator. They are presented with displays and outputs not only from the main radar but also from some 30 other electronic devices, including passive detectors and communications systems. These combine to give a picture of targets, tracks and trajectories, and signal emissions, all duly processed and, where appropriate, with IFF interrogation replies. Passive Detection System (PDS) receivers are located on the nose and tail, and on the tips of the tailplane for lateral coverage.

Production of the E-2C is slow but steady, with the 95th airframe due to be delivered in 1987. Export customers have been happy to take the aircraft as it is; in fact, the Israelis have turned the wing-folding capability to advantage as it has enabled them to park their E-2Cs in hard shelters.

An E-2C mission can last six hours, and at a radius of 200 miles (322km) the time on station at 30,000ft (9,144m) can be almost 4 hours. This is appreciably shorter than the 10 hours at this radius of the E-3A and Nimrod, but Grumman claim a 2:1 price differential in acquisition and operating costs (the E-2C costs £39 million about $50 million — but Singapore is paying $601 million for four, with support and training).

The E-2C radar can detect airborne targets anywhere in a 3,000,000 cubic mile surveillance envelope, and it is claimed that a target as small as a cruise missile can be detected at ranges over 115 miles (185km), fighters at ranges up to 230 miles (370km), and larger aircraft at 289 miles (465km). All friendly and enemy maritime movements can also be monitored. The AN/ALR-59 PDS can detect the presence of electronic emitters at ranges up to twice that of the radar system. High speed data processing enables the E-2C automatically to track more than 250 targets at the same time, and to control more than 30 airborne in-

Below: With landing gear, flaps and arrester hook deployed, an E-2C prepares to touch down; approach speed is 103kt (191km/h).

Above: The four E-2Cs bought by Israel in 1981 have proved invaluable, seeing constant use in the air battles over Lebanon.

tercepts. A new Total Radiation Aperture Control antenna (TRAC-A) is now under development, and this will enable the range to be increased, reduce the sidelobes and enhance the ECCM capability.

Grumman has for many years had a team working on improved Hawkeyes, some with turbofan propulsion, and even on a completely new replacement (E-7).

Above: Despite severe compromises to fit aircraft carriers, the E-2C's capabilities and ready availability have brought sales to several land-based air forces.

Below: Turboprops were chosen to give the necessary flight performance combined with long endurance on limited fuel, but some radar interference resulted.

Grumman EF-111A Raven

Origin: USA.
Type: EW aircraft.
Engines: Two 18,500lb (8,390kg) thrust Pratt & Whitney TF30-3 afterburning turbofans.
Dimensions: Span (fully spread) 63ft (19·2m), (fully swept) 31ft 11·6in (9·74m); length 77ft 1·5in (25·51m); height overall 20ft 0in (6·1m); wing area (gross, 16°) 525sq ft (48·77m²).
Weights: Empty 53,418lb (24,230kg); loaded 87,478lb (39,680kg).
Performance: Max speed 1,160mph (1,865km/h, Mach 1·75) at 36,000ft (11,000m); cruising speed (penetration) 561mph (919km/h); initial climb 3,592ft (1,095m)/min; service ceiling (combat weight, max afterburner) 54,700ft (16,670m); range (max internal fuel) 2,484 miles (3,998km); take-off distance 3,250ft (991m).
Armament: None.

Development: Though the USAF was a pioneer of fighters converted to act as "Wild Weasel" defence suppression aircraft, with comprehensive radar receivers and anti-radar missiles, it has been clear since 1945 that any serious air force must also deploy dedicated EW aircraft able to jam hostile emissions across the board. The US Navy were the first with such aircraft (the EA-6B), and after prolonged study the USAF decided to use the same tactical jamming system but to repackage it into surplus F-111A aircraft.

It was a matter of chance that it was found possible to convert the F-111 into an EW platform, though it was said at the time that this would be not only the most cost-effective but also the lowest-risk solution. Grumman, the Navy's prime contractor for the EA-6B, was at first by no means certain that the conversion was possible, a particular problem being the need for two extra seats with neither unacceptable aerodynamic penalties nor severe effects on range.

Considerable improvements in the electronic suite, however, not only made the ALQ-99 able to handle hostile threats more quickly but also, by means of increased automation, enabled the operating crew to be reduced to one man. This system, the ALQ-99E, features inflight-adaptable aerials, digital jamming, and the complete isolation of active and passive systems: the ALQ-99E jamming subsystem detects, identifies, locates, records and, where desired, jams every kind of hostile emitter using computer control over direction and time. It is generally considered to be the best tactical electronic warfare system in the world at present.

Turning an F-111A into an EF-111A is a major rebuild operation. The main changes are the fin receivers and the installation of the jamming equipment pallet. Canadair supplies the fin, which is reinforced to carry 370lb (168kg) of pod structure loaded with 583lb (264kg) of electronic equipment. Grumman assembles the main jammer installation which, mounted on its pallet, weights 4,274lb (1,939kg), while the canoe radome and door add a further 464lb (210kg). The EF-111A, therefore, flies like an F-111A with a 6,000lb (2,700kg) bomb load, and on Red Flag and other exercises the aircraft has repeatedly demonstrated its ability to fly in formation with F-111As on high-speed attack runs.

Forty-two EF-111As have been ordered; of these, 24 are assisgned to the 388th Electronic Combat Squadron at Mountain Home in the USA and 12 to a second ECS at RAF Upper Heyford in England. A further six aircraft will be held against training and attrition requirements. Intitial operational capability was achieved in November 1983.

In the future it is likely that the EF-111A will be fitted with JTIDS, which will enable it to establish realtime, secure links with ground stations or with AWACS aircraft. UK-based EF-111As represent the only combat EW force in Western Europe.

Right: Turning F-111As into EFs involved rather more than simply festooning the airframe with antennas. In particular, the 370lb (168kg) fin-tip pod and its 583lb (264kg) of electronic receivers posed structural as well as aerodynamic problems.

Below: NATO Raven One, the first EF-111A assigned to Europe, at Pease AFB, New Hampshire, after flying in from Grumman's Bethpage, Long Island, plant. The next stage of its journey took it to RAF Upper Heyford, England, where the 42nd Electronic Combat Squadron operates 12 "Electronic Foxes".

Ilyushin Il-18 "Coot"

Origin: Soviet Union.
Type: (Il-18) Passenger transport, many converted as freighters, (Il-20 "Coot-A") Elint platform.
Engines: Four 4,250ehp Ivchyenko Al-20M turboprops.
Dimensions: Span 122ft 8·5in (37·4m); length 117ft 9in (35·9m); height 33ft 4in (10·17m); wing area 1,507sq ft (140m²).
Weights: (18D) Empty 77,160lb (35,000kg); max 141,100lb (64,000kg).
Performance: Max cruising speed (18D) 419mph (675km/h), (Il-20) 325mph (523km/h); range (max fuel, both) 4,040 miles (6,500km).
Armament: None.

Development: Quite large numbers of Il-18 long-range turboprop transports, mainly in the 18D version, still fly with many air forces. Some are VIP machines for heads of state; others retain up to 122 seats, while others are cargo or trials aircraft.

The Il-20, known to NATO as "Coot-A", is one of the Soviet Union's many dedicated EW aircraft. Upwards of 20 are in service, apparently all with the naval AVMF. One makes a weekly run from the Kola region down the Norwegian coast; others cover the Baltic, and several have visited Britain.

The biggest feature is a 33ft 8in (10·25m) canoe radar, almost certainly a SLAR. Other similar installations occupy two 14ft 5in (4·4m) pods on each side of the forward fuselage, and no fewer than 32 other visible avionics-related items, most of them blade aerials, bulges or flush dielectric panels. Most Il-20s have two prominent domes projecting under the rear fuselage, but one photographed in May 1981 by the US Navy had what looked like search-lights (they are probably giant passive spiral-conical receivers) facing at different angles diagonally down to the right.

The purpose of the several flush, light-coloured rectangular panels visible on the fuselage is not clear. They are probably made from glass-fibre, and could act as radio transparencies for micro-wave antennas designed to sample Western radar transmissions and data links.

Il-20s are estimated to have a flight crew of five and a team of at least ten to man the many sensing and recording systems. Some are natural shiny metal, and others are painted grey.

The modifications to the basic Il-18 have resulted in a multi-sensor reconnaissance and stand-off jammer platform of great power and endurance, but not able to accompany attacking aircraft in hostile airspace.

Further dedicated military versions of "Coot" can be expected as Aeroflot phases-out its Il-18 transports.

Above: Detail of the cylindrical housings, probably for spiral-conical receiving antennas, replacing the large domes normally carried by "Coot-A".

Below: "Coot-A" carries a large canoe radome under its fuselage, probably housing a side-looking airborne radar, plus large fairings on the forward fuselage.

Tupolev Tu-126 "Moss"

Origin: Soviet Union.
Type: AWACS-type surveillance and control.
Engines: Four 15,000ehp Kuznetsov NK-12MV turboprops.
Dimensions: Span 168·0ft (51·2m); length overall 181·08ft (55·2m). height 52·66ft (16·05m). wing area 3,350sq ft (311 m²).
Weights: (estimated) Empty 200,000lb (90,000kg), loaded 375,000lb (175,000kg).
Performance: max speed (36,000ft/11,000m and above) about 460mph (740km/h); patrol speed 320mph (515km/h); service ceiling 39,400ft (12,000m); unrefuelled max range 7,800 miles (12,550km); max endurance 24 hours.
Armament: None.

Development: When Aeroflot, the Soviet civil-aviation organization, replaced its fleet of giant Tu-114 long-range passenger liners with the Il-62 the airframes were not scrapped but converted into these AWACS-type aircraft. Bearing a close kinship with the Tu-95 "Bear", apart from having a much larger fully pressurized fuselage, they have served since about 1967 as the first aircraft of their type in Soviet service.

Features include a surveillance radar and IFF, with data link served by a rotating "rotodome" aerial roughly the same diameter as that of the E-3A but deeper, giving aerial faces of greater area. There are numerous other avionics installations, as well as a flight-refuelling probe. The reskinned fuselage almost certainly accommodates a crew larger than the 12 suggested in Western accounts.

This aircraft was never more than a temporary surveillance and control platform, with the disadvantages associated with having a pylon-mounted rotodome which suffers not only from all the usual shortcomings of such an arrangement but also from the reflections from the 32 blades of the large propellers. Moreover the radar is not of the pulse-doppler type, and was described by US DoD analysts as having "limited effectiveness over water and none over land". This assessment is certainly unrealistic; Russian hardware is never anything but tough and useful. The single Tu-126 detached with its crew to serve with the Indian AF during the 1971 war with Pakistan proved of tremendous value and operated round the clock in both defensive and attacking roles.

Above: The dark area of the Tu-126's rotodome is the dielectric panel for the antenna; effectiveness is presumed to be degraded by propeller interference.

About 10 of the original 12 or 13 aircraft are still in use, but are expected to be replaced by a version of Il-76 or -86.

MARITIME PATROL AND ASW AIRCRAFT

BAe Nimrod

Origin: UK.
Type: (MR) Maritime reconnaissance; (R) electronic intelligence, (AEW) airborne early warning.
Engines: Four 12,140lb (5,507kg) thrust Rolls-Royce Spey 250 turbofans.
Dimensions: Span 114ft 10in (35·0m), (with tip ESM pods) 115ft 2in (35·09m); length (MR) 126ft 9in (38·63m), (MR with probe) 129ft 1in (39·3m), (R) 117ft (35·6m), (R with probe, one aircraft only) 120ft 2in (36·6m), (AEW) 137ft 8·6in (41·97m); height (MR, R) 29ft 9·48in (9·08m), (AEW) 35ft 0in (10·67m); wing area 2,121sq ft (197m²).
Weights: Empty (MR, typical) 86,000lb (39,101km); max 192,000lb (87,090kg).
Performance: Max speed (MR, R) 575mph (926km/h); patrol speed (two engines) 230mph (370km/h); normal operating ceiling 42,000ft (12,800m); typical takeoff/landing distance 4,800ft (1,463m); endurance with max mission payload 12h; max range 5,755 miles (9,265km).
Armament: (MR) Up to 13,500lb (6,120kg) of many types of store carried in internal weapons bay in six lateral rows, including nine torpedoes plus bombs (mines and depth charges not carried by RAF); external pylons for two pairs of self-defence Sidewinders, or two Harpoon missiles, various pods including cannon pods or other stores, (R, AEW) none.

Development: Despite the NATO design of the Breguet Atlantic, the RAF eventually accepted a British proposal for a replacement for the piston-engined Shackleton MR.3. Hawker Siddeley hastily prepared the HS.801 design to combine the wings and fuselage of the Comet airliner with the Rolls-Royce Spey 250 powerplant. An unpressurized ventral section was added to the fuselage in order to create the internal volume for a large weapons bay.

A total of 46 production Nimrod maritime reconnaissance aircraft were ordered by the RAF, deliveries beginning in 1969. Forty-three were delivered to MR.1 standard, the other three being diverted as development aircraft for the AEW.3.

In service with the RAF the MR.1 proved to be a splendid and very popular machine, with abundant performance, the ability to overshoot on one engine, supreme reliability and excellent handling under all conditions. Despite intensive flying in often outrageous weather only one aircraft was lost, as a result of a multiple birdstrike immediately after takeoff. Thanks to the RAF's diminishing oceanic requirements surplus MR.1s are being rebuilt as AEW.3s, while 32 are being completely refurbished as MR.2s

with EMI Searchwater radar, vastly augmented computer memory and data-processing, Ferranti inertial navigation, enhanced communications, new displays and an on-board crew training system. One of the biggest advances is the use of three separate processors for radar, acoustics and tactical navigation.

Redelivery of MR2s began in August 1979, and for the Falklands war many aircraft were urgently fitted with probes, self-defence AAMs and augmented attack capabilities with bombs, Stingray torpedoes and Harpoon missiles. Several missions lasted up to 19h, and today these extra capabilities are standard on the MR.2. The only feature delayed is the addition of Loral EWSM wingtip pods, extensively tested on XV241, the ESM development aircraft.

The R.1 Elint version has no tail MAD boom but instead is fitted with three large passive receivers, one on the nose of each wing tank and the third facing aft at the tailcone.

By 1977 the RAF was becoming desperate for a replacement for the Shackleton AEW Mk 2 and there was increasing disenchantment with the protracted delays affecting NATO's plans for an E-3A force, and so the British Government took the decision to "go it alone" with an early warning version of the Nimrod, a decision which was very welcome for the British avionics industry since the resulting programme includes the most complex avionics equipment ever fitted to a front-line RAF aircraft.

A total of eleven AEW Mk 3 aircraft are being built, using Nimrod airframes originally manufactured for maritime patrol use. Twin antenna assemblies for the Marconi Avionics search radar and Cossor Jubilee-Guardsman IFF are mounted in the nose and tail radomes, each covering 180° in azimuth. Each covers 180° sequentially with the other, thus giving 360° coverage and avoiding the aerodynamic and radar obscuration problems inherent

Above: Nimrod MR.2, featuring greatly enhanced sensor and computing capacity compared with the original MR.1 version.

Below: Other MR.1s are being rebuilt as AEW.3s, with nose and tail radomes each providing 180° of coverage in azimuth.

in rotodomes. The structural changes affect aircraft performance only slightly, and directional stability is maintained by a 3ft (0·91m) increase in the height of the fin. Despite its size, the nose radome does not interfere with the forward view of the pilot during take-off and landing. A Loral ESM system is mounted in the wingtip pods, which detects and analyses radio and radar transmissions.

The programme certainly has its problems, and its critics. Early 1985

reports suggest that the 11 aircraft will have cost over £1,300 million in total by the time the project is finished some two-and-a-half years later than expected. Development problems appear to be centred on radar reliability and excess heat generation, and overall weight of the fully manned and fuelled aircraft.

Below: The complexity of the AEW.3's new systems has caused in-service date to slip from 1982 to an estimated 1987-88.

Dassault-Breguet Atlantique

Origin: France.
Type: Maritime patrol and ASW aircraft.
Engines: Two 5,665shp Rolls-Royce Tyne 21 turboprops.
Dimensions: Span (over pods) 122ft 9·23in (37·42m). length 110ft 4in (33·63m). height overall 35ft 8·74in (10·89m). wing area 1,295·3sq ft (120·34m²).
Weights: Empty 56,218lb (25,500kg); max 101,850lb (46,200kg).
Performance: Max speed (medium height) 400mph (644km/h); cruising speed 345mph (555km/h); patrol speed 195mph (315km/h); typical ASW mission 8h patrol at 690 miles (1,110km) from base or 5h at 1,150 miles (1,850km) for mission time 12h 31min; ferry range 5,635 miles (9,075km).
Armament: Internal bay for wide range of stores including eight AS torpedoes or three torpedoes and one AS39 Exocet, plus up to 7,716lb (3,500kg) on four wing pylons.
Development: Though Dassault-Breguet has emphasized that the

original Br. 1150 Atlantic was the only aircraft in the world designed as a long-range maritime patrol aircraft, instead of merely being modified for the job from a passenger airliner, the Atlantic has had only limited success and some of the original users are replacing them with P-3 Orions. A case in point is the Netherlands, whose Atlantics, offered at only $1·7 million each, have yet to find a customer. The Atlantic continues in service in France, West Germany, Italy and Pakistan.

The updated aircraft was originally called ANG (Atlantic Nouvelle Génération), then Atlantic ATL2 and now simply the Atlantique, Dassault Breguet having for the time being accepted that it is only going to be ordered by France. Again the manufacturer emphasizes that today's Atlantique is not a revamped Atlantic but a "totally new aircraft", though this is misleading. What *has* been completely transformed are the essential mission avionics, equipment including Thomson-CSF Iguane search radar, a new Crouzet MAD installation in the tailboom, totally new sonics installations and

vastly more capable data processing and communications. Sensors include over 100 sonobuoys, and a FLIR turret under the nose.

Without change the Atlantique can lay mines or transport 20 passengers or cargo, and it is clearly an excellent basis for an AEW platform or tanker. At pressnt, however, the French Aéronavale's rquirement for 42 is having to carry the whole programme.

Production of 16 aircraft began in 1984, to meet the 1988 in-service data. The SECBAT consortium

Above: Originally known as Atlantic Nouvelle Generation, the Atlantique ATL2 features chin FLIR turret, retractable radome and tail MAD boom; the four missiles shown on the wing hardpoints would be in addition to weapon bay stores.

comprises Dassault-Breguet (France), Aérospatiale (France), Dornier and MBB (West Germany), SABCA (Belgium) and Aeritalia (Italy); Fokker (Netherlands) is no longer involved.

Fokker F27 Maritime

Origin: Netherlands; formerly licensed to USA.
Type:. Maritime, oversea patrol.
Engines: Two Rolls-Royce Dart turboprops, (current) 2,140shp Mk 536-7R, (from late 1984) 2,200shp Mk 551.
Dimensions: Span 95ft 2in (29·0m); length (most) 77ft 3·5in (23·56m), (500) 82ft 2·6in (25·06m); height overall (most, standard landing gear) 27ft 11in (8·5m); wing area 753·5sq ft (70·0m²).
Weights: Empty 27,600lb (12,519kg); max 45,900lb (20,820kg) (Maritime cleared to "operational necessity" weight of 47,500lb/21,320kg).
Performance: Cruising speed

287mph (463km/h); low-level patrol speed 172mph (277km/h) takeoff run at sea level 3,200ft (975m); endurance 10-12 hours; range 3,107 miles (5,000km).
Armament: Options include fuselage pylons for Exocet, Sea Eagle or Harpoon missiles, and additional pylons inboard and outboard of the existing tank pylons which most customers would use for torpedoes (inboard) and rockets or ESM pods (outboard).
Development: Apart from the Soviet An-24/26 family the F27 is the most successful turboprop transport in history, despite the handicap of a very small home market.

The basic aircraft was designed for high cruising efficiency rather than short field-length, and features include a level floor at truck-bed height, fully pressurized interior and pneumatic-boot de-icing of all leading edges.

The usual military version is the Mk 400M, with accommodation for up to 13,283lb (6,025kg) of cargo loaded through a large side door, or 46 paratroops or 24 stretchers and nine attendants. A survey (cartographic) version is available with inertial navigation. The Mk 500 is stretched to seat up to 60, some having large doors on both sides.

The Maritime was originally provided only with surveillance equipment for offshore patrol and SAR, but later customers have asked for weapons, and Exocet is among the ordnance carried by some

customers (reportedly including Pakistan). All Maritimes have extremely comprehensive avionics for all-weather navigation and communications, and from 1982 a Sperry auto flight-control system has been standard. Customer options have always included Litton APS-504 surveillance radar, and a new radar is soon to be offered. Other 1985 options include a MAD tail stinger and a fin-cap ESM pod. A Royal Australian Navy F27 is using lidar (laser radar) to map depth of ocean water, and surviving Iranian 400Ms have been modified for target towing.

Below Peruvian F27 Maritime on test over the North Sea. Five other countries have ordered the type, and the armed Maritime Enforcer is also available.

Ilyushin Il-38 "May"

Origin: Soviet Union.
Type: Long-range shore-based maritime patrol and ASW aircraft.
Engines: Four 4,250ehp Ivchyenko A1-20M turboprops.
Dimensions: Span 122ft 8·5in (37·4m); length 129ft 10in (39·6m); height overall 33ft 4in (10·17m); wing area, 1,507sq ft (140m²).
Weights: (estimated) Empty 80,470lb (36,500kg); max 143,300lb (65,000kg).
Performance: (estimated) Max cruising speed 400mph (644km/h) at about 27,000ft (8,230m); patrol speed 200mph (322km/h) at 1,000ft (300m); max range 4,500 miles (7,240km); loiter endurance 12h.
Armament: Main weapon bay immediaely ahead of wing box in underfloor area of fuselage, with twin outward-opening doors; possible weapon stowage in rear internal bay or on underwing pylons, but no firm evidence.
Development: Still the standard long-range ocean patrol and ASW aircraft of the AVMF (Soviet naval aviation), the Il-38 has in recent years been backed up by the ultra-long-range Tu-142 ("Bear-F"), though the latter is even older in concept. A new aircraft, perhaps based on the Il-76, is expected. Even the existing Il-38s may be rebuilt Il-18 civil transports, the timing being correct, but so far there is no hard evidence of this.

Compared with the Il-18, the fuselage is unchanged in cross-section, and no deep lobe has been added to house weapons and sensors. Instead weapons – anti-submarine torpedoes, mines, depth bombs and normal anti-ship bombs – are accommodated in the shallow space under the floor ahead of the wing.

There has been speculation that there are four hardpoints on the wing for stores pylons, which would enable rockets to be fired and air/surface missiles to be launched, but such pylons have not been seen. An alternative explanation is that the internal bay extends the full depth of the fuselage, with walkways past it on each side. This would enable the weapon load to reach the kind of level (6,600 to 13,200lb/3,000 to 6,000kg) expected for an aircraft of this size and power, and would also help explain the extraordinary for-

ward shift of the wing, relative to the Il-18, which could not possibly be accounted for by the mass of the extra radar under the forward fuselage. At the same time, a large weapon load ahead of the wing would result in a gross change in centre of gravity position when the stores were dropped.

The fuselage is almost certainly pressurized, and has few windows, though it is reasonable to suppose there is a large tactical compartment amidships (which in this aircraft means behind the wing) with navigation and attack displays and readouts from the various types of sensor. Ducts on each side of the forward fuselage probably draw in and expel air to "sniff" for traces of diesel smoke.

The rear fuselage certainly houses a store of sonobuoys, dropped through a tube and fired from a retro-launcher (which cancels aircraft speed so that the buy drops straight down), but most drops observed by RAF and US aircraft have been from the main weapon bay whose large doors have to be opened for the purpose. In the extreme tail is a MAD installation which accounts for most of the extra length relative to the Il-18. The surveillance radar looks like that carried by the "Hormone-B" version of Ka-25 helicopter, but is probably a different and more powerful installation.

Only about 60 of these machines are thought to be currently operating with the AVMF, though they have been detached to bases far from the Soviet Union including the Yemen People's Republic and Asmara, Ethiopia. Almost standard Il-38s equip No. INAS 135 based at Dabolim, Goa, India. They were ex-AVMF.

In 1984 a new version was seen with a second large ventral blister fairing under the weapon bay behind the radar; this May-B is in AVMF service.

Below: Like the Il-20 "Coot", the Il-38 "May" is derived from the Il-18 airliner but, unlike its ELINT counterpart, "May" is of remarkably clean appearance, with surveillance radar antenna housed in the dome under the forward fuselage and MAD equipment in the tail boom.

Lockheed P-3 Orion

Origin: USA.
Type: Maritime patrol and ASW aircraft, (EP) EW platform; data are for P-3C Update.
Engines: Four 4,910ehp (4,510shp) Allison T56-14 turboprops.
Dimensions: Span 99ft 8in (30·37m); length 116ft 10in (35·61m); height overall 33ft 8·5in (10·27m); wing area 1,300sq ft (120·77m²).
Weights: Empty 61,491lb (27,890 kg); normal loaded 135,000lb (61,235kg); max 142,000lb (64,410kg).
Performance: Max speed (15,000ft/4,570m at 105,000lb/47,625kg) 473mph (761km/h); patrol speed 237mph (381km/h); takeoff over 50ft (15m) 5,490ft (1,673m); mission radius (3h on station at low level) 1,550 miles (2,494km), (no time on station) 2,383 miles (3,825km).
Armament: Internal bay can accommodate eight AS torpedoes, or two Mk 101 nuclear depth bombs plus four torpedoes or a variety of mines and other stores; ten underwing pylons carry mines, depth bombs, torpedoes, Harpoon anti-ship missiles or other stores. Total expendable load 20,000lb (9,072kg).

Development: Derived from the L-188 Electra passenger airliner, the P-3 Orion was specifically ordered as an off-the-shelf type, for operation in the ASW role from shore bases. Nobody expected it to have a production run of over 30 years; moreover, it has become virtually the standard aircraft in its class and is now replacing the Atlantic which was a "clean sheet of paper" design.

The P-3 inherited from the L-188 good short-field performance, outstanding handling even with three of the broad-bladed propellers feath-

ered, thermal de-icing (bleed air on the wings, electric on the tail) and a pressurised fuselage with a circular section giving a max cabin width of 10ft 10in (3·3m).

Instead of adding a giant weapon bay as was done with the Nimrod, Lockheed merely put in a shallow bay under the floor and made up the required payload with external pylons. Typically the P-3 is flown by a flight crew of five, with the centre fuselage occupied by the tactical crew, also numbering five. There is the usual dinette and two folding bunks at the extreme rear.

Early P-3A and B versions are still serving with some export customers, though all have had some updating and a few in the USA have been converted for other roles. US Navy squadrons VQ-1 and -2 replaced their EC-121 Warning Stars with the EP-3E, gross rebuilds of early P-3s to serve in the Elint role. Distinguished by their giant "doghouse" radomes and other aerials above and below the fuselage, the EP-3Es are equipped for detecting, fixing and recording emissions from unfriendly ships, their main sensors being passive receivers, direction finders and signal analysers.

The US Navy and Lear Siegler developed a modification kit to update early P-3s, especially the P-3B, with new navaids and sensors; the RNZAF aircraft have been thus modified. The only current P-3A operator is Spain, whose aircraft are ex-USN and replaced leased examples.

Standard model today is the P-3C Update III. This has APS-115 radar, ASA-64 MAD in the tail boom, a battery of sonobuoys and other sonics installations in the rear fuselage, and launchers for A- and B-sized buoys immediately aft of the wing.

In many Orions, including P-3Cs of early vintage, the chin position is oc-

Lockheed S-3 Viking

Origin: USA.
Type: Carrier-based ASW aircraft.
Engines: Two 9,275lb (4,207kg) thrust General Electric TF34-400A turbofans.
Dimensions: Span (over pods) 68ft 8in (20·93m); length 53ft 4in (16·26m); height overall 22ft 9in (6·93m); folded dimensions, span 29ft 6in (8·99m), length 49ft 5in (15·06m), height 15ft 3in (4·65m); wing area 598sq ft (55·56m²).
Weights: Empty 26,650lb (12,088kg); normal loaded 42,500lb (19,277kg); max 52,539lb (23,831kg).
Performance: Max speed (SL, clean) 518mph (834km/h); max cruising speed 426mph (686km/h); loiter speed 184mph (206km/h); mission radius (no time on station, high-level transit) 1,300 miles (2,092km); ferry (two 250 gal/1,136lit drop tanks) 3,600 miles (5,794km).
Armament: Split internal weapon bays normally accommodate four of various stores (AS torpedoes, destructors, Mk 82 bombs, mines or depth bombs); wing pylons carry

Harpoon anti-ship missiles or (with triple ejector racks) total of six rocket pods, cluster bombs, mines, Mk 82 bombs, Mk 36 destructors or other stores, including tanks.

Development: Perhaps the world's best aviation example of squeezing a quart into a pint pot, the S-3 seats four men in McDonnell Douglas Escapac seats in a pressurised nose cabin, with APS-116 radar ahead, a retractable FR probe above, a retractable FLIR turret below, batteries of sonobuoy launchers to the rear and a MAD boom so long that when it is retracted the front almost reaches the rear of the retracted probe. On the tips of the folding wings are ESM pods housing IBM ALR-47 receivers which give direction of source and IFM (instantaneous frequency measurement). The entire amidships section of fuselage is packed with avionics (fuel being in the integral-tank wing boxes), the chief item being the Univac 1832 GP computer.

Lockheed delivered 187 S-3A aircraft, which have since been the

cupied by a glass-paned gondola housing a KA-74A gimbal-mounted surveillance camera; today, this is replaced by a retractable FLIR. Another variable feature is the choice of pods on the two inboard pylons under the wing roots. It is common to find the ALQ-78 ESM (passive receiver, with what look like anhedralled delta wings) on the left pylon and the AXR-13 LLLTV on the right. Today the latter is usually replaced by the nose FLIR, and the ESM will eventually be replaced by a completely new AIL system in wingtip pods which will also provide targeting data for the Harpoon missiles, provision for which was incorporated at the Update II stage.

The three Update stages have dramitcally multipled processor speed and memory, added navaids such as VLF/Omega to the original INS and mix of doppler and Loran, and completely new subsystems for managing and processing the sonics, including a new receiver and the IBM Proteus acoustic processor.

RAAF P-3Cs are equipped to launch the Barra buoy family. The Canadian Armed Forces had various special mission requirements and selected the CP-104 Aurora, which packaged into the P-3 airframe a completely different set of ASW and other avionics, most of which were based on those of the S-3 Viking (including the APS-116 radar and Univac AYK-10 nav/tac computer), the central computer of the P-3C being the ASQ-114. As an offset Canadair was given the job of making major airframe items, and this continues with the P-3C programme. Canada wanted improved passenger seating, SAR capability and equipment for ice recon, fisheries surveillance and, with a weapon-bay sensor pallet, resource survey and pollution control.

The CAF received 18 CP-104s, and in 1984 was reported negotiating for a further ten, of which perhaps half

might be of a long-planned AEW version. Marconi has proposed a P-3 with front and rear radomes, but Lockheed had not yet moved away from the older rotodome technology in early 1984, and has proposed such aircraft to various customers, notably including the RAAF.

The AEW Orion uses almost the whole APS-125/ARPS installation of the E-2C, and is said to have a similar price of some $40m. In view of the complexity of AEW systems this is the only way an Orion could be made operational by 1986, and despite Lockheed's own rapid funding, and use of an ex-RAAF P-3B as prototype, this timescale is still thought to be optimistic.

Iran's six P-3Fs suffer from shortages of spares but two are currently operating on offshore surveillance in their camouflage of three shades of blue. Argentina's Comando de Aviacion Naval has seven former civil L-188 Electras. Two are being converted with APS-20 type AEW radar (and, it is said, air-surface missiles) for stand-off surveillance and guidance of friendly attack aircraft; two

more are being comprehensively rigged out as Elint platforms, but also fitted with chaff/flare dispensers, jammers and search radar; and the final three will be simpler machines optimized for the surveillance and transport roles.

Above: Aerodynamic prototype of the Lockheed P-3 airborne early warning and control Orion.

Below: A P-3A of Air Test and Evaluation Squadron VX-1 launches a Harpoon missile.

---

standard ASW aircraft with all Carrier Air Wings of the Atlantic and Pacific fleets. This was a smaller programme than expected, and evaluations of S-3As converted into examples of the US-3A COD transport and KS-3 tanker did not lead to continued procurement. In 1982, however, three Vikings were rebuilt as US-3As, and these are now in service alongside the C-2A (which will remain the standard COD transport). The US-3A carries a crew of two and six passengers in an unchanged-size fuselage, as well as 4,600lb (2,087kg) cargo, of which about 3,000lb (1,361kg) can be accommodated in large cargo pods on the wing pylons; all-cargo payload is 7,500lb (3,402kg).

From 1985 Lockheed expects to redeliver 160 aircraft updated as S-3Bs, with provisions for the Harpoon missile on the wing pylons, expanded ESM capability, a new sonobuoy receiver, better radar processing and greatly enhanced sonics processing capacity with the IBM AYS-1 Proteus system. Should a 15th Carrier Air Wing be needed, Lockheed proposes a follow-on run of either 82 or 103 new S-3Bs.

Right: S-3As of ASW squadron VS-22 on patrol; the profusion of blade antennas on the upper and lower fuselage is indicative of the extensive electronics carried in their compact airframes.

Below: An S-3A of VS-41 Shamrocks ASW training squadron at its NAS North Island, San Diego, base with MAD boom extended; for carrier stowage the tail and wings are folded.

Mil Mi-14 "Haze"

Origin: Soviet Union.
Type: Shore-based ASW helicopter.
Engines: Two 2,200hp Isotov TV3-117MT turboshafts.
Dimensions: Assumed to be the same as the Mi-8 (rotor has five blades with diameter 69·85ft (21·29), and fuselage length 59·54ft (18·15m); height about 21·5ft (6·55m).
Weights: Empty, probably about 17,650lb (8,000kg); maximum loaded probably 26,460lb (12,000kg).
Performance: Max speed probably similar to Mi-8 (161mph, 260km/h); cruising speed at max weight probably about 120mph (193km/h); range with full combat gear probably about 311 miles (500km).
Armament: Includes homing torpedoes and an option of mines, depth charges or, possibly, anti-ship missiles.

History: First flight (V-14) not later than 1973; service delivery (Mi-14) prior to 1977.

Development: Though it uses a powerplant group and rotor system essentially similar to that of the Mi-17, the Mi-14 has a modified air-frame, and is equipped solely for ASW missions. It forms a direct parallel with the HSS-2 (S-61), in having a boat-type hull. The landing gear is of the tricycle type, however, so the main gears retract into sponsons located at the rear of the cabin section. All landing gears have twin wheels, the steerable nose gear folding into a watertight box in the hull immediately behind the surveillance radar. The cockpit is almost the same as that of the Mi-17, but the cabin is arranged as a tactical compartment with seats for three sensor operators and the tactical commander.

Two main sub-types have been

identified, in ASW model ("Haze-A") and an MCM (mine countermeasures) version ("Haze-B"). Considering the vast extent of the offshore regions where hostile submarines might lurk around the Soviet Union, the number of Mi-14s deployed (a little over 100) is trivial. It may be that the Ka-27 will prove superior and later be deployed from shore bases as well as from ships (a role in which the Mi-14 is too large and unwieldy).

Above: The Mi-14 "Haze" carries a MAD "bird" stowed against the rear of the fuselage pod.

The Mi-14's amphibious capability is apparently not normally used, and it is probable that alighting on water is for emergencies only. Certainly this helicopter has not yet been seen in an SAR role. The Mi-14 is in service with the Soviet Navy, and has been exported to Bulgaria and Libya.

Shin Meiwa SS-2

Origin: Japan.
Type: (PS-1) ASW flying boat.
Engines: Four 3,060ehp T64-IHI-10 turboprops plus one 1,250shp T58-IHI-10-M1 turboshaft for boundary-layer control.
Dimensions: Span 108·75ft (33·15m); length 109·77ft (33·46m); height overall (land) 32·64ft (9·95m); wing ara 1,462sq ft (135·82m²).
Weights: Empty 51,367lb (23,300kg); max (water) 94,797lb (43,000kg); (land) 99,206lb (45,000kg).
Performance: Max speed 299mph (481km/h); takeoff to 50ft (15m), (land) 2,035ft (620m), (water) 1,970ft (600m); max range 2,614 miles (4,207km).
Armament: Weapons compartment on upper deck normally housing four 331lb (150kg) AS bombs or other stores;

two wing pylons each for two AS torpedoes; wingtip rails for triple 5in (127mm) rockets.

Development: After a prolonged research programme Shin Meiwa has delivered a small number of two versions of these technically unique aircraft. All are characterized by an ambitious BLC (boundary-layer control) scheme energized by a T58 engine in the hull which feeds compressed air to blow over the powerful flaps, rudder and elevators. The wing also has slats. As a result the takeoff is short and slow-flying ability extraordinary, the original objective being to fly not much faster than a submerged nuclear submarine and search by dipping a powerful sonar into the sea, even in bad weather.

Shin Meiwa delivered 23 of the SS-2 version (JMSDF designation PS-1), an ASW flying boat with a crew of ten in service with the 31st

Air Group. Later the company delivered six US-1 search and rescue amphibians, and three US-1As with more powerful engines.

All these have fully retractable landing gear (the PS-1 merely having retractable tricycle beaching chassis which cannot be used for landing on a runway). The US machines have a crew of nine and 12 stretchers or 20 seated survivors. They serve with the JMSDF 71st Squadron.

Above: The PS-1 flying boat has exceptional slow-flying qualities, needed while tracking SSNs and during the deployment of dunking sonar equipment.

Below: The US-1 amphibian was derived from the PS-1 as a search and rescue aircraft; using boundary layer control, it can take off from water in a distance of 1,970ft (600m).

Sikorsky S-61

Origin: USA.
Type: General purpose helicopter with various specialist missions including ASW, Elint, transport/gunship, SAR, anti-ship (see below).
Engines: Two General Electric T58 turboshafts; (SH-3A and derivatives) 1,250shp T58-8B; (SH-3D and derivatives) 1,400shp T58-10; (S-6IR versions) 1,500 T58-5.
Dimensions: Diameter of main rotor 62·0ft (18·9m); length overall 72·66ft (22·15m); (61R) 73·0ft; height overall 16·83ft (5·13m); main-rotor disc area 3,019sq ft (280·5m²).
Weights: Empty (simple transport versions, typical) 9,763lb (4,428kg); (ASW, typical) 11,865lb (5,382kg); (armed CH-3E) 13,255lb (6,010kg); max (ASW) about 18,626lb (8,449kg); (transport) usually 21,500lb (9,750kg); (CH-3E) 22,050lb (10,000kg).
Performance: Max speed (typical, max weight) 166mph (276km/h); initial climb (max) 2,200-1,310ft (670-400m)/min, depending on weight; service ceiling, typically 14,700ft (4,480m); range with max fuel, typically 625 miles (1,005km).
Armament: Very variable.
Development; When it first flew in 1959 the S-61 brought a dramatic increase in capability over its piston-engined predecessors — though of course it was a midget compared with the Mi-6 which had flown two years earlier. The S-61 featured an amphibious hull, twin turbine engines above the hull and an advanced flight-control system. First versions carried anti-submarine warfare (ASW) sensors and weapons, and were developed for the US Navy, entering service in 1961-62 as the SH-3 series, with the name Sea King. By the 1960s variants were equipped for transport duties, minesweeping, drone or spacecraft recovery (eg, lifting astronauts from the sea), electronic surveillance and (S-61R series) transport/gunship and other combat roles.

The S-61R family has a tricycle landing gear, the main wheels retracting forwards into sponsons and the cabin having a full-section rear ramp/door and a 2,000lb (907kg) roof-rail winch. The USAF model in this family was the CH-3E, 50 of which were rebuilt for combat operations with armour, self-sealing tanks, various weapons, rescue hoist and retractable flight-refuelling probe, and designated HH-3E Jolly Green Giant. The Coast Guard name for the HH-3F sea search version is Pelican.

It could be said that the HH-3E "came of age" during the Vietnam War when these reliable machines flew countless missions into the North to rescue shot-down aircrew. Operating out of Udorn or Da Nang, the Jolly Green Giants could fly to any point in North Vietnam and return home, some even demonstrating an eight-hour mission with the HC-130Ps.

Some customers for Agusta-built versions have specified the Italian Marte anti-ship weapon system with Sistel radar and Sea Killer Mk 2 missiles. Especially since Sikorsky itself ceased manufacturing the S-61, the Italian licensee has cleaned up a growing market, five of the more recent orders being repeat orders for South American countries which originally came within Sikorsky's marketing area. Almost all Agusta's recent export sales have been of the SH-3D model, though some (for the Italian Marinavia, for example) are to the latest SH-3H standard fully equipped for ASW, anti-ship and SAR missions, while others, such as the latest batch for the FAM (Brazilian Navy), are austerely equipped and used as utility transports. Agusta is still producing the S-61R (called the HH-3F) for export, but has not named the customers.

Mitsubishi production, which still adheres to the obsolete HSS-2 series of designations, has been mainly for the JMSDF, total deliveries to this service for ASW

Above: US Navy SH-3D Sea King ASW helicopter with MAD "bird", normally stowed behind the starboard sponson, deployed.

Below: The USAF's HH-3E, developed for combat rescue over Vietnam, now serves with AFRES and ANG ARRS units.

and SAR amounting to 118, with T58-IHI-10 licence-built engines. Production by this licensee is now giving way to licensed manufacture of the even more expensive Sikorsky SH-60B. Total production of the S-61, including a relatively small number of civil examples, reached 770 by Sikorsky and over 400 by licensees.

The Westland Sea King is derived from the Sikorsky S-61 design, covered in a separate entry under that company.

Below: An Agusta-built SH-3D of the Italian Navy launches a Sea Killer Mk 2 anti-ship missile, part of the Marte system developed for the type.

Westland Sea King

Origin: UK.
Type: (Sea King) anti-submarine helicopter.
Engines: Two Rolls-Royce Gnome turboshafts; past production, mostly 1,500shp Gnome H.1400; current, 1,600shp H.1400-1.
Dimensions: Diameter of five-blade main rotor 62·0ft (18·9m); length overall (rotors turning) 72·66ft (22·15m); length of fuselage 55·83ft (17·02m); height (rotors turning) 16·83ft (5·13m).
Weights: Empty (Sea King ASW) 15,474lb (7,019kg); (Commando) 12,222lb (5,543kg); max (H.1400-1 engines) 21,000lb (9,525kg).
Performance: Max speed 143mph (230km/h); typical cruising speed 131mph (211km/h); max (not vertical) rate of climb (ASW) 1,770ft (540m)/min; (Commando) 2,020ft (616m)/min; approved ceiling 10,000ft (3,048m); range (max load) about 350 miles (563km), (max fuel) 937 miles (1,507km).
Armament: See below.

Development: Based on the S-61, these helicopters have been developed by Westland to meet many customer requirements, the initial order being 56 Sea King HAS.1 for the Royal Navy. Unlike the US counterpart, they were provided with a tactical compartment for autonomous operation independent of friendly ships, using radar, dunking sonar and other equipment. All were converted to HAS.2 standard with more fuel, greater power, six-blade tail rotor and higher gross weight limit. These are in turn being converted to HAS.5 standard, supplementing 25 built new to this standard, distinguished by a long flat-topped radome for Sea Searcher radar and packed with additional or upgraded equipment including Dec-

ca 71 doppler and TANS navaids and passive sonar with Marconi Lapads (lightweight acoustic processing and display system).

HAS.5s can detect and pinpoint hostile submarines at far greater ranges, and also handle information from buoys dropped by other aircraft such as Nimrods. The lengthened cabin houses a crew of four plus Lapads operator, while weapons include four Sting Ray (previously, Mk 46) AS torpedoes or Mk 11 depth charges.

Other British marks include the HAR.3 and AEW. The former is an uprated SAR version, 19 of which are used in bright yellow livery by the RAF. Crewed by two pilots, electronics/winch operator and loadmaster/winchman, they carry 19 rescuees, or two stretchers and 11 seated or six stretchers, and have very comprehensive all-weather navaids for use in appalling conditions offshore or among mountains.

The AEW (for which no mark number has yet been assigned) was suggested in 1970 but ignored until desperate need over the Falklands for airborne radar resulted in a "crash programme". In 11 weeks two HAS.2s were fitted with Searchwater surveillance radar, the aerial being in a kettledrum radome stabilized by internal pressure, swung down from the right side of the helicopter to have a clear view below in the operating position. It is linked with Cossor IFF. Six more AEWs are to be produced by modification.

Standard SAR helicopter of the German Marineflieger, the Sea King 41 is being updated in a major programme, managed by MBB. In May

1984 MBB ordered the new Sea-spray Mk 3 radar with multiple-target track-while-scan, 360º surveillance and advanced displays for these helicopters.

The Mk42 is the Indian Navy version, the 42A having hauldown gear for small ships and the 42B being equipped to fire Sea Eagle anti-ship missiles. The 20 Mk 42Bs will be

delivered in 1986 with other updates including the Super Searcher command/control radar. The Mk 43 serves the Norwegian AF in the SAR role, the 45 the Pakistan Navy (ASW), Mk 47 Egypt (in 1984 anti-ship missiles were to be added to these or to existing Commandos), Mk 50 the RAN (Royal Australian Navy) tasked in multiple roles.

Above: Originally an HAS.1, then an HAS.2, this Royal Navy Sea King was one of the first Searchwater AEW conversions.

Below: An AM39 Exocet anti-ship missile is launched from an ASH-3H operated by the French government for trials purposes.

Below: ASW equipment of the Egyptian Navy's six Sea King Mk 47s includes dipping sonar, plus a search radar in dorsal radome.

AIR-LAUNCHED MISSILES

Air-to-Air Missiles

AA-2 Atoll

Origin: Soviet Union.
Type: Close-range AAM.
Propulsion: Solid motor resembling those used in Sidewinder, nozzle diameter 3·15mm (80mm).
Dimensions: Length (IR) about 110in (2·8m), (radar) about 114in (2·9m); body diameter 4·72in (120mm); span (early AA-2 canard) 17·7in (450mm), (AA-2-2 canard and all tails) 20·9in (530mm).
Weight: At launch (typical) about 154lb (70kg).
Performance: Speed about Mach 2·5; range about 4 miles (6·5km).
Warhead: Blast-fragmentation, 13·2lb (6kg).

Development: Unlike most Russian weapons this AAM is beyond doubt a copy of a Western original, the early AIM-9B Sidewinder. When first seen on 9 June 1961, carried by various fighters in an air display, it was almost identical to the US weapon. Since then it has followed its own path of development, and like Sidewinder has diversified into IR and SARH versions. Body diameter is even less than that of Sidewinder, and so far as is known all models have the nose-to-tail sequence of AIM-9B. The 13·2lb (6·0kg) warhead is a BF type with smooth exterior.

Believed to be designated K-13A or SB-06 in the Soviet Union, several early versions have been built in very large numbers as standard AAM for most models of MiG-21, which carry two on large adapter shoes (which house the seeker cooling system in later models) on the underwing pylons. Licence production by the MiG complex of Hindustan Aeronautics has been in progress since the

early 1970s, and it is believed there is also a Chinese version.

Since 1967 there have been later sub-types called AA-2-2 or Advanced Atoll by NATO. Some reports ascribe these designations to the SARH versions, but the consensus of opinion is that there are IR and radar versions of the first-generation missiles, in various sub-types, and IR and radar versions of the Advanced model. Several photographs indicate that later models have quite different control fins. These fins are driven in opposite pairs through 30°, and the later fin is unswept, has a cropped tip and greater area and is fitted after loading on the launcher. Like AIM-9 versions, IR missiles have hemispherical noses transparent to heat, and radar versions slightly tapered noses that appear opaque.

Current carriers include all later fighter MiG-21s, with four missile shoes instead of two, and the MiG-23S swing-wing fighter which also carries later AAMs.

AA-3 Anab

Origin: Soviet Union.
Type: Medium/long-range AAM.
Propulsion: Probably solid motor.
Dimensions: Length (IR) about 161in (4·1m), (radar) 157·5in (4·0m); body diameter 11in (280mm); span 51in (1·3m).
Weight: At launch about 600lb (275kg).
Performance: Speed about Mach 2·5; range (IR) about 12 miles (19km), (radar) at least 15 miles (24km).
Warhead: Unknown but probably large.

Development: This second-generation AAM was the first long-range all-weather missile to reach the PVO, which it did at about the time dummy examples were displayed carried by an early Yak-28P interceptor at Tushino at the 1961 Soviet Aviation Day display. At that time it was at first thought by the West to be an ASM, but gradually it was identified as a straightforward AAM carried in both IR and SARH versions, usually one of each. The carriers are the Yak-28P in all versions except trainers, Su-11 and Su-15. All these aircraft have the radar called Skip Spin by NATO, a much more capable installation than those associated with the earlier AAMs and probably derived from the Scan Three fitted to the Yak-25. Believed to be designated RP-11, it operates in I-band between 8690+8995 MHz at peak power of 200 kW, with a PRF of 2700+3000pps and pulse-width of about 0·5 microsec. It is assumed that CW illumination is provided for missile homing.

AA-3 has large rear wings indexed in line with cruciform canard controls, and solid propulsion is assumed. Aerodynamics may be derived from AA-1, though as there appear to be no wing control surfaces it is probable that the canards can be driven as four independent units for roll control. There is no information on either type of homing head; the motor has a single central nozzle, and may have boost/sustainer portions, and the warhead is amidships, with a proximity fuze. An AA-3-2 Advanced Anab has been identified since 1972, but how it is "advanced" is not known.

The Yak-28P is believed to have been withdrawn to reserve and training units, and the Su-11 is also gradually being phased out of first-line PVO service, but this missile remains the primary armament of the Su-15 deployed in large numbers (1982 estimate, 700 in IA-PVO regiments excluding spares and reserves for attrition).

AA-5 Ash

Origin: Soviet Union.
Type: Medium- + long-range AAM.
Propulsion: Probably solid motor.
Dimensions: Length (IR) 18ft (5·5m), (radar) 17ft (5·2m); body diameter 12in (305mm); span 51in (1·3m).
Weight: At launch about 860lb (390kg).
Performance: Speed about Mach 3; range (IR) about 13 miles (21km), (radar) about 35 miles (55km).
Warhead: Unknown but probably large, perhaps 100lb (45kg).

Development: This large AAM was developed in 1954-59 specifically to arm the Tu-28P long-range all-weather interceptor, and genuine missiles were seen carried by a development aircraft of this family at the 1961 Aviation Day display at Tushino. (This aircraft had a very large ventral bathtub believed to house side-looking or early-warning radar, not seen subsequently). Early versions of Tu-28P at first mistakenly reported as Blinder but corrected to Fiddler, carried two of these missiles on underwing pylons. So far as one can tell, they were SARH guided, associated with the Big Nose radar of the carrier aircraft, a very large and powerful I-band radar which had no counterpart operational in the West until the AWG-9 of 1974. The missile is matched to the radar in scale, being large than any Western AAM. For many years Western estimates of AA 5 range were ludicrously low, but they are creeping up and may now be about half the true value for the radar version.

Early Tu-28Ps are thought to have entered PVO·service soon after 1961, filling in gaps around the Soviet Union's immense frontier. By 1965 the Tu-128 was being armed with the newly introduced IR version of this missile. This aircraft has four under-wing pylons carrying the IR version, with Cassegrain optics

Left: An Su-22 of the Libyan AF with AA-2-2 Advanced Atoll AAMS.

Below left: Radar (left) and IR versions of the AA-5 Ash under the wing of a Tu-28.

Below: A pair of AA-3 Anabs carried by an Su-15 Flagon. The IR-guided version (nearest camera) is slightly longer, despite the radar version's distinctive pointed nose.

behind a small nose window, on the inners and the SARH model, with opaque (usually red-painted) conical nose on the outers. Early versions of MiG-25 Foxbat interceptor were also armed with this missile, usually one of each type also had Bis Nose or an early model of Fox Fire radar.

This large but obsolescent weapon remains in the IA-PVO inventory as the standard armament of the Tu-128 Fiddler, the production aircraft, which despite (despite repeated rumours of an interceptor version of the Tu-22 Blinder) has no known replacement offering the same combat radius.

AA-6 Acrid

Origin: Soviet Union.
Type: Medium/long-range AAM.
Propulsion: Unknown but probably solid motor with very long-burn sustainer.
Dimensions: Length (IR) 248in (6·3m), (radar) 232in (5·9m); body diameter 15·7in (400mm); span 88·6in (2·25m).
Weight: At launch (both) about 1,765lb (800kg).
Performance: Speed about Mach 4; range (IR) about 15·5 miles (25km), (radar) about 50 miles (80km).
Warhead: Unknown but US estimate 132-200lb (60-90kg), blast/fragmentation.

Development: Largest AAM in the world, this awesome weapon family was designed around 1959-61 originally to kill the B-70 Valkyrie (which instead was killed by the US Congress) and entered PVO service as definitive armament of the Mach 3·2 MiG-25 "Foxbat A" interceptor. With four missiles, two IR-homers on the inner pylons and two SARH on the outers, this aircraft is limited to Mach 2·8. Like the Tu-28, the MiG-25 was intended to detect targets at long range, using the Markham ground-air data link to give a cockpit display based on ground surveillance radars, switching to its own Fox Fire radar at about 100 miles (160km) range. This equipment, likened to an F-4 AWG-10 in character but greater in power. inc-ludes CW aerials in slim wingtip pods to illuminate the target for the SARH missiles, which could probably lock-on and be fired at ranges exceeding 62 miles (100km); both peak-pulse/CW power and receiver-aerial size are considerably greater than for any Sparrow and closely similar to AWG-9/Phoenix. The IR version has much shorter range, though there is no reason to doubt that current Soviet technology is increasing IR fidelity as is being done elsewhere.

Acrid has a large long-burning motor, giving a speed generally put at Mach 4 (the figure of 2·2in one report is nonsense) and manoeuvres by canard controls, with supplementary ailerons (possibly elevons) on all four wings. The latter have the great area needed for extreme-altitude interception, for the B-70 cruised at well over 70,000ft (21km); but early Acrid missiles did not have look-down capability. Soviet films suggest that, when the range is close enough, it is usual to follow national standard practice and ripple missiles in pairs, IR closely followed by SARH. The two homing heads are different in shape, the IR head being a curved ogive and the radar type being a pointed cone.

Despite reports that it is carried by the later versions of Su-15, no missile of this family has been seen on any interceptor other than the PVO MiG-25. Later versions of this aircraft are believed to carry the AA-X-9.

AA-7 Apex

Origin: Soviet Union.
Type: Medium/long-range AAM.
Propulsion: Advanced boost/sustain solid motor.
Dimensions: Length (radar, only known version) 181in (4·6m); body diameter (front) 7·7in (195mm), (main section) 8·8in (223mm); span 39in (1m).
Weight: At launch, about 705lb (320kg).
Performance: Speed about Mach 3; range (radar) estimated from 12·4 miles (20km) to at least 25 miles (40km).
Warhead: Unknown but one US report states 88lb (40kg).

Development: Standard medium-range AAM carried on the glove pylons of the MiG-23 in all its interceptor versions, this missile is aerodynamically similar to the AA-5 Ash. The large cylindrical body rides on four large delta wings and four rear control fins indexed in line around the nozzle of the motor. The newer weapon is, however, somewhat smaller, and has a totally different front end. Curiously, the body diameter is reduced over almost the first metre from the tip of the nose, so the guidance section has to fit in a constricted portion.

As in other Soviet AAMs there are believed to be IR and SARH versions, and while a smaller diameter may not reduce IR seeker performance it obviously must restrict the diameter of the receiver aerial in the SARH model. In fact, there may not be any SARH dish, because surrounding the guidance section are four projections which were at first wrongly identified by Western observers as extra control surfaces but which in fact are almost certainly SARH receiver aerials working on an interferometer principle (as do the four aerials spaced around the nose of the Sea Dart missile). It should be possible to home on the signals received by these four shallow fin-like projections, and dispense with the need for an internal radar dish.

At the same time, the reason for the reduced diameter of the forebody of this missile is obscure, unless it is to provide an area-ruled ogival platform to carry the four blade aerials. Immediately to the rear, at the upstream end of the full-diameter section, a dark ring probably locates the windows of the proximity fuzing system.

The warhead is estimated at no less than 88lb (40kg). The tail controls are cropped at the tips at an angle appropriate to about Mach 3, and carry near the tips forward-facing bullet fairings similar to those on the wings and control fins of certain other Soviet anti-aircraft missiles.

The only real puzzling thing about AA-7 is its very poor Western estimated performance, which may simply be yet another case of rather childish wishful thinking. There is no evident reason why this missile should not have performance appreciably greater (in range, ceiling and manoeuvrability) than most versions of Sparrow, its nearest Western counter-part. The High Lark J-band radar of most MiG-23 interceptors is a set in the 150-kW class with a plate aerial of about 30in (0·76m) diameter. If the low estimates of AA-7 range are accurate, one explanation might be devotion of much of the missile body to duplicated guidance methods in each round, with extremely sophisticated electronic counter-counter measures (ECCM) circuits to ensure that the homing lock is never broken.

The only alternative is a sharp up-

Left: MiG-25 Foxbat interceptors armed with early versions of the AA-6 Acrid long-range missiles.

Below left: Probably destined to replace the AA-2 Atoll on Soviet fighters, the AA-8 Aphid shows its novel configuration under the wings of a MiG-21bis.

Below: AA-6 Acrids have been supplied to export users of the MiG-25, including the Libyan AF.

Right: One of the initial test batch of 94 AIM-120 AMRAAMs installed on the glove pylon of an F-14 Tomcat for trials.

Far right: MiG-23MF interceptor with wing-mounted AA-7s and AA-8s carried under the fuselage.

Below right: First customer for the AMRAAM will be the F-16, which unlike other current USAF fighters has no BVR missile.

ward revision of the estimated flight performance. AA-7 was probably developed in 1971-74.

AA-8 Aphid

Origin: Soviet Union.
Type: Close-range AAM.
Propulsion: Advanced boost/sustain motor, probably solid.
Dimensions: Length (IR, only known version) 84·6in (2·15m); body diameter 4·72in (120mm); span 15·75in (400mm).
Weight: At launch about 121lb (55kg).
Performance: Speed about Mach 3; range estimated 0·3-3·5 miles (500m-5·5km).
Warhead: Estimated at 13·2lb (6kg).

Development: Supplementing and eventually probably to replace the vast stock of AA-2 Atoll close-range missiles, this interesting weapon is one of the smallest guided AAMs ever built, and is being produced in extremely large quantities as the air-combat missile of the PVO, FA and possibly AV-MF for the 1980s.

Very similar in shape and size to the USAF Hughes Falcon of 30 years ago, AA-8 is a technically novel canard with delta wings right at the tail, canard delta control fins and, at the extreme nose and immediately ahead of the control fins, four rectangular blades of very low aspect ratio and with span considerably less than that of the controls. These are believed to be fixed aerodynamic surfaces to enhance combat manoeuvrability, but they are by no means obviously linked with this purpose as are the two sets of surfaces in the French Matra Magic missile, and it is possible that the fixed blades serve a different

function concerned with guidance or ECM. The only known AA-8 has a hemispherical glass nose for an IR seeker, though some Western reports state that there is also an SARH model.

Almost certainly AA-8 propulsion is of the boost/coast type, and it may have better manoeuvrability than any other AAM over ranges up to about 5 miles (8km). A black stripe between the guidance section and motor almost certainly locates the proximity fuze for the warhead, which has been estimated at barely half the size of the warhead carried by all current models of Sidewinder.

AA-8 was first seen in the West in 1976 but no clear photograph appeared until 1981. It is carried by the MiG-21, MiG-23 (on body pylons), Yak-36MP and probably others.

AA-9

Not yet publicly associated with a NATO code-name, this large and advanced AAM is the primary weapon carried by Foxhound, the new-generation two-seat derivative of the MiG-25. No picture has yet been published of this missile, but it was said by Washington sources to have done well in prolonged look-down shoot-down tests in 1978 at Vladimirovka. Impressive simulated kills were scored against target RPVs simulating US cruise missiles flying at heights within 200ft (91m) of the ground, after acquisition by the Foxhound at heights from 20,000ft (6,100m) upwards. Maximum range is expected to be in the neighbourhood of 80 miles (130km).

AA-X-10

This Western designation has been published for a further new AAM

which is probably to be carried by the MiG-29 Fulcrum or other advanced fighter. No details are available.

AA-XP-1

American designation for another alleged new Soviet AAM; described by *Aviation Week* as "All-aspect, look down, shoot down, IOC (initial operational capability) 1984". Range said to be 11 to 19 nautical miles (12·7-21·9 miles, 20·4-35·2km).

AA-XP-2

American designation for another alleged Soviet AAM; described by *Aviation Week* in same terms as XP-1 but with range estimated at 21·5 to 38 nautical miles (24·9-43·75 miles, 40·3-70·4km).

AMRAAM

Origin: United States.
Type: Medium-range AAM.
Propulsion: Advanced internal rocket motor.
Dimensions: Length 145·7in (3·7m); body diameter 7·0in (178mm).
Weight: At launch 326lb (148kg).
Performance: Speed probably about Mach 4; maximum range in excess of 30 miles (48km).
Warhead: Expected to be lighter than 50lb (22kg).

Development: Also called BVR (Beyond Visual Range) missile, the Advanced Medium-Range AAM is the highest-priority AAM programme in the United States, because AIM-7F is becoming long in the tooth and is judged urgently in need of replacement in the 1980s by a completely new missile. AMRAAM is a joint USAF/USN programme aimed at

producing a missile having higher performance and lethality than any conceivable advanced version of Sparrow, within a package that is smaller, lighter, more reliable and cheaper.

AMRAAM will obviously be matched with later versions of F-14, -15, -16 and -18 equipped with programmable signal processors for doppler beam-sharpening and with advanced IR sensors able to acquire individual targets at extreme range. The missile will then be launched automatically on inertial mid-course guidance, without the need for the fighter to illuminate the target, the final terminal homing being by a small active seeker.

The task clearly needs a very broad programme to investigate not only traditional sensing and guidance methods but also new ones such as target aerodynamic noise, engine harmonics and laser scanning to verify the external shape and thus confirm aircraft type. Multiple-target and TWS (track-while-scan) are needed, and AMRAAM has a high-impulse motor giving rapid acceleration to a Mach number higher than 4, with subsequent manoeuvre by TVC and/or tail controls combined with body lift, wings not being needed. Mid-course guidance is Nortronics inertial, and the small Hughes active terminal radar now uses a TWT (travelling-wave tube) transmitter.

The original list of five competing groups was narrowed to two in February 1979, and at the end of 1981 Hughes was picked over Raytheon to build 94 test missiles, with options on 924 for inventory plus follow-on production (which, because the US buy alone is expected to exceed 13,000 for the USAF and 7,000 for the Navy/Marines, is expected to be split between two contractors,

Raytheon probably becoming second-source).

In 1980 West Germany and the UK signed a memorandum of understanding assigning AMRAAM to the USA and ASRAAM to the two European nations. Since then work has gone ahead on integrating the US missile into the RAF Tornado F.2, replacing Sky Flash, and the Luftwaffe F-4F (the latter possibly being refitted with APG-65 or improved APG-66 radar under the Peace Rhine programme). The Tornado Foxhunter radar may need a small L-band transmitter to provide mid-course updating.

Testing of full-scale development rounds started in 1984, and the production run begins with 174 in 1985 and 1,042 in 1986.

Aspide

Origin: Italy.
Type: Medium-range AAM.
Propulsion: Solid motor developed from Rocketdyne Mk 38 by Difesa e Spazio.
Dimensions: Length 12ft 1in (3·7m); body diameter 8in (203mm); wing span 39·37in (1·00m).
Weight: At launch 485lb (220kg).
Performance: Speed Mach 4; range 31-62 miles (50-100km).
Warhead: Difesa e Spazio 72·75lb (33kg) fragmentation.

Development: Though a wholly Italian development, and the largest single missile programme in the country, this impressive weapon was designed to be compatible with systems using Sparrow. This extends to AAM applications, for which the immediate prospect is the Italian Air Force F-104S Starfighter originally tailored to AIM-7E, and several surface-launched applications. The Italian SAM system using this missile is named Spada in its mobile land form; a different ship-to-air system is named Albatros.

Similar to Sparrow in basic configuration, Aspide is powered by an advanced single-stage motor by SNIA-Viscosa Difesa e Spazio (which made the motors for Italian Sparrows) giving higher thrust and a speed of Mach 4 at burnout. The all-round performance is claimed to exceed that of even AIM-7F, and the guidance is likewise claimed to have significant advantages over that of the American missile. Matched with an I-band monopulse fighter radar, it is said to have greater ECCM capability, to offer increased snapdown performance and to be markedly superior at very low atitudes. The seeker aerial system is driven hydraulically.

The radome and forebody are described as redesigned for more efficient operation at hypersonic speeds, and in the AAM role the moving wings are said to have extended tips with greater span. The fragmentation warhead is positioned ahead of the wings.

Following carry-trials in 1974 and prolonged static testing of the seeker, firing trials at Salto di Quirra, Sardinia, began in May 1975. By 1977 fully representative Aspide missiles, including the AAM version, had completed qualification firings and production began in 1978. Final verification trials took place in 1979-80. The

AAM is replacing Sparrow in the Italian Air Force and is also being deployed on Italian Air Force Tornados.

ASRAAM

Origin: UK/Germany.
Type: Close-range AAM.
Propulsion: Advanced solid motor, not yet defined.
Dimensions: Not defined but probably smaller than Sidewinder.
Weight: Probably lighter than typical Sidewinder.
Performance: Range limits probably to be about 0·6-9·3 miles (1 to 15km); speed over Mach 3.
Warhead: Probably small, see below.

Development: The obvious need for a completely new close-range AAM was made more acute by the progressive obsolescence of the AIM-9 family and cancellation of the German Viper and British SRAAM. After years of talking, the decision was taken at government level to develop an Asraam (Advanced Short-Range AAM) in Europe for use by NATO. The MoU (memo of understanding) was signed by the USA, France (which has merely a watching brief), the UK and Federal Germany. Following a "pre-feasibility" stage which lasted to the end of 1981 the three actively participating governments authorized BAe Dynamics and BGT, the team leaders, to proceed to the feasibility study stage (up to 1983), with engineering development lasting from 1985 to 1990 and production deliveries beginning "in the early 1990s". Clearly, the official view is that time and the effects of sustained inflation do not matter, and that Sidewinder can meet all challenges for another decade!

Though US industry has no share in programme management it can bid at all stages, and is almost certain to play a major role in the eventual weapon. An excellent move was the invitation, accepted in September 1982, for Hughes to join the team to smooth compliance with US requirements.

A severe, and seemingly unnecessary, handicap is that Asraam must fit launchers already in use for Sidewinder and Magic, even though the latter have large external controls and tailfins, while Asraam relies on body lift. Midcourse guidance could be by a simple strapdown inertial unit, giving greatly extended range especially in adverse weather. Terminal homing is expected to be by a staring focal-plane array IR seeker, all processing being digital by microminiature electronics. Nothing has yet leaked out regarding possible use of active radar homing, which could certainly be accommodated within a body diameter slimmer than Sidewinder. It might even be possible for each round to use both IR and active radar. The West has in the past lacked the Soviet Union's ability to select either type of guidance to suit prevailing weather.

Objectives include minimum cost, zero maintenance over a long shelf life, all-aspect engagement and high kill probability with accuracy sufficient for direct-hitting to be

guaranteed, eliminating need for a proximity fuze or large warhead. A possible configuration was shown in triple mock-ups on a single Sidewinder launcher at the 1982 SBAC show at Farnborough. One of the encouraging aspects of the programme is that, should development appear to lag, the USA could withdraw and produce its own Asraam; this should be a powerful incentive.

Kukri

Origin: Republic of South Africa.
Type: Close-range AAM.
Propulsion: Double-base solid-propellant rocket motor.
Dimensions: Length 9·66ft (294·4cm); body diameter 5in (127mm); tail span 20·86in (530mm); canard span 16·5in (420mm).
Weight: 161·8lb (73·4kg).
Performance: Maximum speed, Mach 1·84. Maximum range 1·24 miles (2km) at sea level, 2·5 miles (4km) at high altitudes; minimum range 328 yards (300m).
Warhead: High explosive (HE) fragmentation.

Development: The V3B IR-guided dogfight missile has been in service for some years with the South African Air force; the Kukri is the export version, available (as the publicity coyly puts it) to "selected" customers. Both missiles are highly manoeuvrable, but their most unusual characteristic is that they are coupled to the pilot's helmet sight. All that is required is for the pilot to look at the target and wait for acquisition

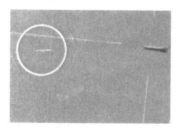

to be signalled by an audio tone in his earphones. He then designates the target and the missile commences tracking; thereafter he can fire at will. This feature enables targets to be acquired off the axis of the aircraft's datum.

An interesting aerodynamic feature is the use of an asymmetric double canard configuration, required to confer the high manoeuvrability required for dogfighting. The triangular canards give roll control, while the double-delta canards control pitch acceleration.

The missile is operational on SAFF Dessault-Breguet Mirage III and Mirage F1 fighters. The manufacturers state that the installation can be modified quickly and simply for many other aircraft types.

Magic

Origin: France.
Type: Close-range AAM.
Propulsion: SNPE Romeo (Magic I) or Richard (Magic II) Butalane high-impulse solid motor.
Dimensions: Length 109in (2770mm); body diameter 6·2in (157mm); span 26·3in (668mm).
Weight: At launch 198lb (89·8kg).
Performance: Speed about Mach 3; range 0·2-6·2 miles (0·32-10km).
Warhead: Conventional rod/fragmentation, 27·6lb (12·5kg) with all-sector proximity fuze or impact-loop detonation.

Development: Alone among European companies Matra took on the Sidewinder in head-on competition and has not merely achieved

technical success but has also established 14 export customers and an output rate exceeding that of any other AAM ever produced in Western Europe. Wisely the weapon was made installationally interchangeable with Sidewinder, but the design requirements were greater than those of presently available versions of the US missile, including launch anywhere within a 140° forward hemisphere at all heights up to 59,000ft (18,000m) and with limitations at higher altitudes; ability to engage from almost any target aspect (head-on will shortly be achieved); ability to snap-fire at ranges down to 984ft (300m); ability to fire from a launch platform flying at any speed (no minimum) up to over 808mph (1300km/h) whilst pulling up to 6g; and ability to pull 3·5g and cross in front of the launch aircraft only 164ft (50m) ahead.

The IR guidance uses the SAT type AD.3601, the PbS seeker being cooled prior to launch by a liquid-nitrogen bottle in the launch rail. Its output drives the electric control section with four canard fins (almost the reverse shape of those of Super 530) stationed immediately downstream of four fixed fins with the same span as the tips of the controls.

The tail fins are free to rotate around the nozzle. Propulsion is by an SNPE Romeo single-stage composite-DB motor which gives high acceleration for 1·9sec. The warhead weights 27·6lb (12·5kg) of which half is the explosive charge detonated by IR proximity and DA fuzes.

Matra began development as a

company venture in 1968, receiving an Air Ministry contract in 1969. After various simpler air trials a missile with guidance was fired from a Meteor of the CEL against a CT-20 target in a tight turn on 11 January 1972. On 30 November 1973 a Magic was fired from a Mirage III in an extreme test of manoeuvrability. IOC was reached in 1975, since when production at Salbris has built up to the rate of 100 per month. Unit price is in the order of $15,000, and more than 6,000 rounds have been delivered.

Phoenix, AIM-54

Origin: United States.
Type: Long-range AAM.
Propulsion: Aerojet (ATSC) Mk 60 or Rocketdyne Flexadyne Mk 47 long-burn solid motors.
Dimensions: Length 157·8in (4·01m); body diameter 15in (380mm); span 36·4in (925mm).
Weight: 985lb (447kg).
Performance: Speed Mach 5-plus; range over 124 miles (200km).
Warhead: Continuous-rod (132lb, 60kg) with proximity and impact fuzes.

Development: By far the most sophisticated and costly AAM in the world this missile provides air defence over an area exceeding 12,000 square miles (31,000km²) from near sea level to the limits of altitude attained by aircraft or tactical missiles. But it can be fired only from the F-14 Tomcat in association with the highly sophisticated AWG-9 radar tracking and fire-control system, and costs nearly half a million dollars.

Following the classic aerodynamics of the Falcon family, Phoenix was originally AAM-N-11 and Hughes aircraft began development in 1960 to replace the AIM-47A and Eagle as partner to the AWG-9 for the F-111B. This advanced fire-control system was the most capable ever attempted, and includes a very advanced radar (derived from the ASG-18 carried in the YF-12A) of high-power PD type with the largest circular aerial (of planar type) ever carried by a fighter. It has look-down capability out to ranges exceeding 150 miles (241km), and is backed up by an IR tracker to assist positive target identification and discrimination.

AWG-9 has TWS capability, and, had it been fitted, an F-111B with the maximum load of six Phoenix missiles would have been able to engage and attack six aircraft at maximum range simultaneously, weather and conditions and target aspect being of little consequence; indeed the basic interception mode assumed is head-on, which is one of the most difficult at extreme range.

Propulsion is by a long-burning Rocketdyne (Flexadyne) Mk 47 or Aerojet Mk60 motor, giving a speed to burnout of Mach 3·8. Combined with low induced drag and the power of the large hydraulically driven tail controls this gives sustained manoeuvrability over a range not even approached by any other AAM, despite the large load of electrical battery, electrical conversion unit, autopilot, electronics unit, transmitter/receiver and planar-array seeker head (all part of the DSQ-26 on-board guidance) as well as the 132lb (60kg)

annular blast fragmentation warhead with Downey Mk 334 proximity fuze, Bendix IR fuze and DA fuze.

Hughes began flight test at PMIC in 1965, using a DA-3B Skywarrior, achieving an interception in September 1966. In March 1969 an F-111B successfully engaged two drones, and subsequently Phoenix broke virtually all AAM records including four kills in one pass (out of a six-on-six test, there being one no-test and one miss), a kill on a BQM-34A simulating a cruise missile at 50ft (15m), and a kill on a BQM-34E flying at Mach 1·5 tracked from 153 miles (246km), the Phoenix launched at 127 miles (204km) and impacting 83·5 miles (134km) from the launch point. The first AWG-9 system for the F-14A Tomcat, which replaced the F-111B, was delivered in February 1970. Production of Phoenix AIM-54A at Tucson began in 1973, since when output averaged about 40 per month. By the third quarter of 1978 output had passed 2,500; it then slowed sharply and production ended in 1980.

Since late 1977 production missiles were of the AIM-54B type with sheet-metal wings and fins instead of honeycomb structure, non-liquid hydraulic and thermal-conditioning systems, and simplified engineering. In 1977 Hughes began a major effort to produce an updated Phoenix to meet the needs of the 1990s. This missile, AIM-54C, has totally new all-digital electronics, more reliable and flexible in use than the analog unit, with a solid-state radar replacing the previous klystron tube model. Accuracy is improved by a new

Far left top: Sequence showing an Aspide actually striking an Aerospatiale CT.20 target drone on trials in Sardinia.

Far left bottom: Mock-ups of ASRAAM, the parallel European development to the AMRAAM.

Left: Magic AAM mounted on the wingtip launcher of a Mirage F1.C. A direct competitor of the AIM-9 Sidewinder, Magic has achieved considerable success.

Sequence below: A Phoenix launch by an F-14 from USS *Constellation*. Left, the missile is ejected through the F-14's airflow by explosive charges and the motor ignites; right, the missile accelerates towards its target. On May 5, 1975, a Phoenix-armed F-14 intercepted a Mach 2.8 Bomarc target at 72,000ft (21,950m); the AIM-54 was launched some 51nm (95km) from the target and was totally successful.

strapdown inertial reference unit from Nortronics, and ECCM capability is greatly enhanced. Another improvement is the new proximity fuze developed by the Naval Weapons Center. Following the test and delivery of 30 pilot-production rounds in the second half of 1981, full production started in 1982 and annual deliveries were: 1983-108, 1984-265, 1985-400. A total of 567 of this version are scheduled to be delivered in 1986.

Python 3

Origin: Israel.
Type: Close-range AAM.
Propulsion: Rafael solid motor.
Dimensions: Length 9·84ft (3m) approx; body diameter 5·9in (15cm) approx.
Weight: 264·5lb (120kg).
Performance: Claimed to be superior in speed and turning capability to Sidewinder AIM-9L; maximum effective range 9·3 miles (15km), minimum effective range 547 yards (500m).
Warhead: HE 24·25lb (11kg).

Development: First exhibited at the 1981 Paris air show, the Python 3 has been developed as a successor to the Isaeli Shafrir II and is claimed to surpass its predecessor in all respects. The Rafael Armament Development Authority is responsible for most parts of Python 3, including the new IR seeker cell which is claimed to have exceptional sensitivity and a wider lock angle than that of most other IR-homing missiles.

The delta canard controls are large, and are probably pneumatically actuated, as with Shafrir. The fixed tailfins are sharply swept on both leading and trailing edges, and carry the roll-control aerodynamic surfaces and slipstream-driven rollerons.

The Python 3 is now fully operational with the Israeli Air Force. It has seen service in the many engagements over Lebanon in the past three years.

Sidewinder, AIM-9

Origin: USA.
Type: Close-/medium-range AAM.
Propulsion: Solid motor (various, by Rockwell, Aerojet or Thiokol, with Aerojet Mk 17 qualified on 9B/E/J/N/P and Thiokol Mk 36 or reduced-smoke TX-683 qualified on 9L/M).
Dimensions: Types vary between 111·4in and 120·9in (2,830 and 3,070mm long).
Weight: Between 155 and 195lb (70·4 and 88·5kg) for various types at launch.
Performance: Range between 2 and 11 miles (3·2 and 17·7km); mission time, between 20 and 60 seconds.
Warhead: (B/E/J/N/P) 10lb (4·5kg) blast/fragmentation with passive IR proximity fuze (from 1982 being refitted with Hughes DSU-21/B active laser fuze), (D/G/H) 22·4lb (10·2kg) continuous rod with IR or HF proximity fuze, (L/M) 25lb (11·4kg) advanced annular blast/fragmentation with active laser IR proximity fuze.

Development: One of the most influential missiles in history, this slim AAM was almost un-American in development for it was created out of nothing by a very small team at NOTS China Lake, operating on the proverbial shoe-string budget. Led by Doctor McLean, this team was the first in the world to attack the problem of passive IR homing guidance, in 1949, and the often intractable difficulties were compounded by the choice of an airframe of only 5in (127mm) diameter, which in the days of vacuum-tube electronics was a major challenge. In 1951 Philco was awarded a contract for a homing head based on the NOTS research and today, 28 years later, the guidance team at Newport Beach, now called Ford Aerospace and Communications, is still in production with homing heads for later Sidewinders. The first XAAM-N-7 guided round was successfully fired on 11 September 1953. The first production missiles, called N-7 by the Navy, GAR-8 by the USAF and SW-1 by the development team, reached IOC in May 1956.

These early Sidewinders were made of sections of aluminium tube, with the seeker head and control fins at the front and four fixed tail fins containing patented rollerons at the back. The rolleron is similar to an air-driven gyro wheel, and one is mounted in the tip of each fin so that it is spun at high speed by the slipstream. The original solid motor was made by Hunter-Douglas, Hercules and Norris-Thermador, to Naval Propellant Plant design, and it accelerated the missile to Mach 2·5 in 2·2 sec.

The beauty of this missile was its simplicity, which meant low cost, easy compatibility with many aircraft and, in theory, high reliability in harsh environments. It was said to have "less than 24 moving parts" and "fewer electronic components than the average radio". At the same time, though the guidance method meant that Sidewinder could be carried by any fighter, with or without radar, it was erratic in use and restricted to close stern engagements at high altitude in good visibility. The uncooled PbS seeker gave an SSKP of about 70 per cent in ideal conditions, but extremely poor results in bad visibility, cloud or rain, or at low levels, and showed a tendency to lock-on to the Sun, or bright sky, or reflections from lakes or rivers.

The pilot energised his missile homing head and listened for its signals in his headset. It would give a growl when it acquired a target, and if it was nicely positioned astern of a hot jetpipe the growl would become a fierce strident singing that would rise in intensity until the pilot let the missile go. There were plenty of QF-80, Firebee and other targets that had early Sidewinders up their jetpipe in in the 1950s, but unfortunately real-life engagements tended to have the wrong target, or the wrong aspect, or the wrong IR-emitting background. In October 1958, however, large numbers of Sidewinders were fired by Nationalist Chinese F-86s against Chinese MiG-17s and 14 of the latter were claimed in one day. This was the first wartime use of AAMs.

The staggering total of nearly 81,000 of the original missile were built in three almost identical versions which in the new 1962 scheme were designated AIM-9, 9A and 9B. Nearly all were of the 9B form, roughly half by Raytheon. A further 15,000 were delivered by a European consortium headed by BGT, which in the late 1960s gave each European missile a new seeker head of BGT design known as FGW Mod 2. This has a nose dome of silicon instead of glass, a cooled seeker and semi-conductor electronics, and transformed the missile's reliability and ability to lock-on in adverse conditions.

By 1962 SW-1C was in use in two versions, AIM-9C by Motorola and -9D by Ford. This series introduced the Rocketdyne Mk 36 solid motor giving much greater range, a new airframe with tapered nose, long-chord controls and more swept leading edges on the tail fins, and completely new guidance. Motorola produced the 9C for the F-8 Crusader, giving it SARH guidance matched to the Magnavox APQ-94 radar, but for various reasons this odd man out was unreliable in performance and was withdrawn. In contrast, 9D was so successful it formed the basis of many subsequent versions, as well as MIM-72C Chaparral. The new guidance section introduced a dome of magnesium fluoride, a nitrogen-cooled seeker, smaller field of view, and increased reticle speed and tracking speed. The control section introduced larger fins, which were detachable, and high-power ac-

tuators fed by a longer-burning gas generator. The old 10lb (4·54kg) warhead with passive-IR fuze was replaced by a 22·4lb (10·2kg) annular blast fragmentation head of the continuous-rod type, fired by either an IR or HF proximity fuze.

AIM-9E was fitted with a greatly improved Ford seeker head with Peltier (thermoelectric) cooling, further-increased tracking speed and new electronics and wiring harnesses, giving increased engagement boundaries especially at low level. AIM-9G has so-called SEAM (Sidewinder Expanded Acquisition Mode), an improved 9D seeker head, but was overtaken by 9H. The latter introduced solid-state electronics, even faster tracking speed, and double-delta controls with increased actuator power, giving greater manoeuvrability than any previous Sidewinder as well as limited all-weather capability. AIM-9J is a rebuilt 9B or 9E with part-solid-state electronics, detachable double-delta controls with greater power, and long-burning gas generator. Range is sacrificed for high acceleration to catch fast targets.

There are J-1 and J-3 improved or "all-new" variants. A major advance came with Sidewinder 9L, with which NWC (as NOTS now is) at last responded to the prolonged demands of customers and the proven accomplishments of BGT. The latter's outstanding seeker head developed for Viper was first fitted to AIM-9L to give Alasca (All-Aspect Capability), a great missile that was merely used by

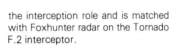

Above left: AIM-9N Sidewinder being loaded onto USAF F-16A Fighting Falcon, using old-fashioned musclepower.

Left: F-16A carrying two AIM-9J and two AIM-9L Sidewinders.

Top: Lengthening the Tornado ADV enabled two pairs of Sky Flash AAMs to be carried.

Above: Israeli Python 3 AAM has seen much operational service.

Above right: Viggen shown with two Sky Flash, and two Sidewinder AAMs probably of three carried.

Right: Test-firing of AIM-7F Sparrow from F/A-18 Hornet, which carries two such AAMs.

Germany as a possible fall-back in case 9L failed to mature. AIM-9L itself, in full production from 1977, has long-span pointed delta fins, a totally new guidance system and an annular blast fragmentation of preformed rods, triggered by a new proximity fuze in which a ring of eight GaAs laser diodes emit and a ring of silicon photodiodes receive.

About 16,000 of the 9L series were expected to be made by 1983, and at least a further 9,000 are likely to be made by a new BGT-led European consortium which this time includes BAe Dynamics and companies in Norway and Italy. Pilot production deliveries began in 1981, and BAe received its first production contract (for £40 million) in February 1982. No European missiles had reached British squadrons in April 1982 and 100 AIM-9L were supplied for use by Harriers and Sea Harriers in the South Atlantic from US stocks, gaining 25 known victories against Argentinian aircraft.

AIM-9M is a revised L. 9N is the new designation of J-1 (all are 9B or 9E rebuilds). 9P are rebuilds of 9B/E/J. and additional 9P missiles are being made from new.

Sky Flash

Origin: United Kingdom.
Type: Long-range AAM.
Propulsion: Aerojet or Rockwell Mk 52 PB/AP solid motor.
Dimensions: Length 145in (3680mm); body diameter 8in (203mm); span 40in (1020mm).

Weight: At launch 425lb (193kg).
Performance: Speed Mach 4, range 31 miles (50km).
Warhead: Sparrow-7E type 66lb (30kg) continuous-rod pattern, with proximity and DA fuzes.

Development: While the US industry developed its own monopulse seeker for Sparrow, the UK industry began such work in 1969, leading to a brilliant series of test firings in November 1975 and production delivery to the RAF by BAe Dynamics in 1978. Originally XJ.521, and later named Sky Flash, this missile is a 7E2 with a completely new MSDS homing head operating in I-band with inverse processing by all-solid-state microelectronics. The warm-up time has been reduced from about 15 sec to less than 2 sec. The short range of the basic 7E2 is considered acceptable for European conditions, though the 7F motor could be fitted if needed.

The trials programme from Point Mugu is judged the most successful of any AAM in history; more than half actually struck the target, often in extremely difficult conditions of glint or evasive manoeuvres, while the miss-distance of the remainder averaged "about one-tenth that of most radar-guided AAMs". Moreover, the warhead is triggered by a deadly EMI active-radar fuze placed behind the seeker, the warhead being behind the wings.

Sweden has adopted Sky Flash as RB 71 for the JA 37 Viggen. Sky Flash is carried by RAF Phantoms in the interception role and is matched with Foxhunter radar on the Tornado F.2 interceptor.

However, Sky Flash Mk2 was unfortunately abandoned by the British government at an advanced stage in early 1981, neatly destroying work which had put the BAe/MSDS team ahead of the world. Instead the American AMRAAM will be purchased, with first deliveries being in 1985.

Sparrow, AIM-7

Origin: USA.
Type: Medium-/long-range AAM.
Propulsion: (7E) Aerojet or Rockwell Mk 52 Mod 2 PB/AP solid motor, (7F, M) Hercules or Aerojet Mk 58 high-impulse solid motor.
Dimensions: Length (E, F) 144in (3660mm), (M) 145in (3680mm); body diameter 8in (203mm); span 40in (1020mm).
Weight: At launch (E) 452lb (205kg), (F, M) 503lb (228kg).
Performance: Speed (both) about Mach 4; range (E) 28 miles (44km). (F, M) 62 miles (100km).
Warhead: (E) 66lb (30kg) continuous-rod warhead, (F, M) 88lb (40kg) Mk 71 advanced continuous-rod warhead, in each case with proximity and DA fuzes.

Development: Considerably larger than other contemporary American AAMs, this missile not only progressed through three fundamentally different families, each with a different prime contractor, but late in life mushroomed into totally new versions for quite new missions as an ASM (Shrike) and a SAM (two types of Sea Sparrow).

Sperry Gyroscope began the programme as Project Hot Shot in 1946, under the US Navy BuAer contract. By 1951 Sperry had a contract for full engineering development of XAAM-N-2 Sparrow I, and the suffix I was added because by that time there was already a Sparrow II. The first representative guided flight tests took place in 1953. This missile was a beam rider, with flush dipole aerials around the body which picked up the signals from the fighter radar beam (assumed to be locked-on to the target) and drove the cruciform delta wings to keep the missile aligned in the centre of the beam. At the tail were four fixed fins, indexed in line with the wings. Propulsion was by an Aerojet solid motor, and missile assembly took place at the Sperry-Farragut Division which operated a Naval Industrial Reserve plant at Bristol, Tennessee.

IOC was reached in July 1956, and Sparrow I was soon serving in the Atlantic and Pacific Fleets, and with the Marine Corps.

In 1955 Douglas obtained limited funding for Sparrow II, as main armament for the proposed F5D-1 Skylancer. Amazingly, however, the company did not switch to SARH guidance but to fully active radar, and this was tough in a missile of 8in (203mm) diameter, a figure common to all Sparrows. In mid-1956 the Navy decided to terminate Sparrow II, but it was snapped up by the Royal Canadian Air Force as armament for the Arrow supersonic interceptor. After severe difficulties Premier Diefenbaker cancelled Sparrow II on 23 September 1958, and the Arrow itself the following February.

Three years previously Raytheon had begun to work on Sparrow III, taking over the Bristol plant in 1956. Sparrow III uses almost the same airframe as Sparrow II but with SARH guidance. By the mid-1950s Raytheon had become one of the most capable missile companies, possibly because its background was electronics rather than airframes. It built up a missile engineering centre at Bedford, Massachusetts, with a test base at Oxnard (not far from Point Mugu), California; production of Sparrows was finally shared between Bristol and a plant at South Lowell, near Bedford.

Most of the airframe is precision-cast light alloy. Early Sparrow III missiles had an Aerojet solid motor, not cast integral with the case, and introduced CW guidance. AIM-7C, as it became, reached IOC in 1958 with Demons of the Atlantic and Pacific fleets. AIM-7D introduced the Thiokol (previously Reaction Motors) prepacked liquid motor, and was also adopted by the Air Force in 1960 as AIM-101 to arm the F-110 (later F-4C) Phantom. All fighter Phantoms can carry four Sparrows recessed into the underside of the fuselage, with target illumination by the APQ-72, APQ-100, APQ-109, APQ-120, or APG-59 (part of AWG-10 or -11) radar. In the Italian F-104S Starfighter the radar is the Rockwell R-21G/H, and in the F-14 Tomcat the powerful Hughes AWG-9. The AIM-7D was also the basis for PDMS Sea Sparrow.

AIM-7E, the next version (also used in the NATO Sea Sparrow system), uses the Rocketdyne free-standing solid motor with Flexadyne propellant (Mk 38), which gives a slightly increased burnout speed of Mach 3·7. The warhead is of the continuous-rod type, the explosive charge being wrapped in a tight drum made from a continuous rod of stainless steel which shatters into about 2,600 lethal fragments. DA and proximity fuzes are fitted. Many thousands of 7E missiles were used in Vietnam by F-4s, but, owing to the political constraints imposed on the American fighters, were seldom able to be fired. Accordingly AIM-7E2 was developed with shorter minimum range, increased power of manoeuvre and plug-in aerodynamic surfaces requiring no tools. The AIM-7C, D and E accounted for over 34,000 missiles.

Introduced in 1977, AIM-7F has all-solid-state guidance, making room for a more powerful motor, the Hercules Mk 58, giving further-enhanced flight speed and range, as well as a larger (88lb, 40kg) warhead. Claimed to lock-on reasonably well against clutter up to 10 db, -7F is compatible with CW PD radars (and thus with the F-15 and F-18), and has a conical-scan seeker head. In 1977 GD Pomona was brought in as second source supplier and with Raytheon is expected to deliver about 19,000 missiles by 1985, split roughly between the Navy and Air Force, plus

hoped-for exports.

In 1982 both contractors switched to AIM-7M, developed by Raytheon. This has an inverse-pocessed digital monopulse seeker generally similar to Sky Flash in giving greatly improved results in adverse conditions. GD's first contract was for 690, following 3,000 of the 7F type.

SRAAM

Origin: United Kingdom.
Type: Close-range AAM.
Propulsion: Advanced solid motor by IMI Summerfield with control actuation (see text) by Sperry Gyroscope (now part of BAe).
Dimensions: 107·25in (2724mm); body diameter 6·5in (165mm).
Weight: Not published.
Performance: Typical of IR-homing dogfight missiles.
Warhead: Not published.

Development: Experience in Vietnam rammed home the urgent need for close-range air-combat weapons. HSD (now BAe Dynamics Group) put company money into a study of close-range AAMs. By 1970 the missile, named Taildog, had been completely designed other than details of the solid motor and IR seeker. Later in that year a small MoD contract was received and development proceeded under the name SRAAM (short-range AAM). The contract was terminated in 1974, and replaced by a low-key technology-demonstration programme to be undertaken without urgency and involving merely eight firings, from ground launchers and a Hunter fighter.

The first shot with guidance in April 1977 passed within lethal range of a difficult target, triggering the novel BAe fuze. Subsequent firings proved SRAAM's unparalleled manoeuvrability, which can include a 90° turn immediately on leaving the launcher.

The objective of the designers, abundantly achieved, was to produce a simple AAM system, of low cost, that could be attached to any aircraft without needing modification of either the aircraft or launcher; to give the pilot unprecedented SSKP in a dogfight, while greatly reducing his workload; and to offer high snapshot lethality against targets in previously impossible situations such as crossing at minimum range.

SRAAM is instantly available and fired automatically as soon as the seeker acquires a target coming into view ahead. It is wingless, and carried in a lightweight twin-tube launcher whose adapter shoe houses the fire-control system. The chosen missile tube flicks open its nose doors, fires the round, and closes the doors to reduce drag. The passive IR seeker commands the missile by motor TVC. In August 1977, when AIM-9L was chosen for British use, it was stated that SRAAM would be "kept alive" to provide a coherent design base. An outstanding successful interception took place on 18 August 1980, using swivel-nozzle TVC and with four motor-bleed tangential jets for roll control. Much SRAAM knowledge is expected to go into Asraam.

Super 530

Origin: France.
Type: Long-range AAM.
Propulsion: SNPE Angèle Butalane high-impulse (composite CTPB) solid motor, 2-sec boost and 4-sec sustain.
Dimensions: Length 139·4in (3·54m); body diameter 10·35in (263mm); span (wing) 25·2in (640mm); (tail) 35·43in (900mm).
Weight: At launch 551lb (250kg); early production 529lb (240kg).
Performance: Speed Mach 4·6; range (early production) 22 miles (35km), (530F) "several dozen km".
Warhead: Thomson-Brandt fragmentation, over 66lb (30kg).

Development: By January 1971, when development of this missile started, Matra was a mature AAM producer able to take a studied look at the requirements and secure in the knowledge that R.530 would probably remain in production almost a further decade. Though to a slight degree based on the R.530, as reflected in the designation, this is in fact a to·lly new missile marking very large advances in flight performance and offering doubled acquisition distance and effective range and also introducing snap-up capability of 25,000ft (7;600m), since increased to 29,500ft (9,000m), believed to exceed that of any other AAM other than Phoenix.

From the start only one method of guidance has been associated with Super 530, SARH. This uses the EMD Super AD-26, matched with the Cyrano IV radar of the Mirage F1. Electric power comes from a silver/zinc battery with 60-sec operation. Thomson-Brandt developed the Angèle propulsion motor, with Butalane composite propellant of much higher specific impulse than that of earlier French motors. This can accelerate the missile rapidly to Mach 4·6, thereafter sustaining approximately this speed to sustainer burnout. Wings are not necessary at this speed, but Super 530 does have four wings of very low aspect ratio, manoeuvring by the cruciform of tail fins which have an unusual shape. It can pull 20g up to 56,000ft (17km) and 6g at 82,000ft (25km).

The homing head was test-flown in September 1972, and an inert missile airframe was air-launched in July 1973. Firing trials from a Canberra of the CEV began in 1974, progressing to trials with guidance in 1975. Firing trials from a Mirage F1.C began at Cazaux in 1976, and evaluation firing at CEAM has been in progress since 1975.

Super 530F, the version for the Mirage F1, entered service in December 1979, and more than 2,000 rounds had been delivered by mid-1982. In 1984 production began on the Super 530D (D for doppler) matched to the RDI radar of the Mirage 2000. This version has good snap-down performance.

Air-to-Surface Missiles

ALARM

Origin: UK.
Type: Air-launched anti-radar missile.
Propulsion: Two-stage solid propellant motor.
Dimensions: None released.
Weight: Approximately 390lb (177kg).
Performance: Not released.
Warhead: Not released.

Development: The BAe-developed Air-Launched Anti-Radar Missile (ALARM) has been selected to meet the Royal Air Force's Air Staff Target (AST) 1228; the only challenge came from the American HARM missile. BAe's Sky Flash missile proved to be slightly too small to act as the basis for ALARM and a slightly larger missile has been designed, but still very much lighter than the US HARM. It is thus capable of being carried by tactical aircraft in addition to their normal interdiction payload; for example, Tornado can carry up to four ALARMs as well as its normal attack weapon load, which typically would be two JP.233 airfield attack weapons, two long-range fuel tanks, an ECM pod and flare dispensers.

Configuration resembles that of Sky Flash, but with considerably smaller wings, since violent manoeuvres are not required. The weapon is launched as the aircraft approaches defended hostile airspace, zooms up on the power of its two-stage motor to about 40,000ft (12,190m) and then pitches over to point, nose-down, the whole time under simple strapdown inertial guidance. It then falls slowly under a drogue parachute, while searching for hostile emitters. The long search time ensures that every defence system in the area is detected. The advanced microprocessor analyses the received signals, selects the most important target, and the parachute is then jettisoned, the wings and tail flick open and the missile homes at high speed on to the selected radar emitter.

It is thought that if an enemy SAM radar, for instance, "switches off", ALARM slowly descends then homes on its target immediately if it is switched back "on".

There are several interesting features of the ALARM system. The first is that it is a software-controlled weapon system; this not only confers considerable growth potential and operational flexibility, but also means that changes to the system during its life can be achieved by software (rather than hardware) updating and modification. ALARM can either be autonomous, or can have an "intelligent interface" with its carrying aircraft; this enables the missile system to receive inputs from the aircraft's data bus until the actual moment of launch.

The first flight of ALARM on a Tornado GR 1 took place on 13 February 1985, with the maximum load of nine mounted on the underwing pylons. First air firings were expected to take place in late 1985. The UK requirement is for about 2000 missiles, and ALARM should enter service in 1987/88.

AS-4 Kitchen

Origin: Soviet Union.
Type: Air-to-surface missile.
Propulsion: Rocket, believed to be liquid propellant.
Dimensions: Length, about 37ft (11·3m); body diameter 35·4in (0·9m); span about 9ft 10in (3m).
Weight: At launch, about 13,000lb (5,900kg).
Performance: Speed up to Mach 3·5 at high altitude; range variable up to 286 miles (460km) on all high-altitude profile.
Warhead: Nuclear 350 kiloton or large conventional, 2,200lb (1000kg).

Development: Yet another disclosure at the 1961 Soviet Aviation Day fly-past was this much more advanced and highly supersonic ASM, carried recessed under the fuselage of one of the ten Tu-22 "Blinder" supersonic bomber/reconnaissance aircraft that took part. This aircraft, dubbed "Blinder B" by NATO, had a larger nose radome, and other changes, as have several other Tu-22s seen in released photographs. Most aircraft of this sub-type have the outline of the AS-4 missile visible on their multi-folding weapon-bay doors, but the missile appears seldom to be carried today and in any case most remaining Tu-22s are of other versions, serving with the ADD and AV-MF.

The missile itself has slender-delta wings, a cruciform tail and, almost certainly, a liquid-propellant rocket. Prolonged discussion in the West has failed to arrive at any degree of certainty concerning the guidance, though the general consensus is that it must be inertial, possibly with mid-course updating by a Tu-95 or other platform. A homing system is obviously needed for moving targets such as ships.

Both versions of the swing-wing Tu-22M "Backfire" multi-role platform are believed to have carried this missile, probably in AV-MF service. Surprisingly, AS-4 has been seen on these new bombers frequently, whereas the later AS-6 has been seen more often on aged Tu-16s.

AS-5 Kelt

Origin: Soviet Union.
Type: Air-to-surface missile.
Propulsion: Single-stage liquid-propellant rocket with pump feed.
Dimensions: Length about 28ft 3in (8·6m); body diameter 35·4in (0·9m); span 15ft (4·57m).

Left: Matra Super 530 AAM being fired from a Mirage F1.C. The Super 530 is currently the standard AAM of the French Armée de l'Air.

Below left: The BAe SRAAM (here seen from the launch aircraft) has unrivalled manoeuvrability and is the basis for ASRAAM.

Right: Tu-16 "Badger-G" of the Egyptian Air Force with two AS-5 Kelt ASMs. Used in the Yom Kippur War with mixed success, AS-5 is now dated.

Below: A bevy of missiles in under-wing mounts. From left to right: Martel, ALARM, Sea Eagle, Sky Flash and SRAAM.

Below right: Despite its age the AS-4 Kitchen (here seen recessed into the weapons bay of a Soviet "Backfire") is still in service.

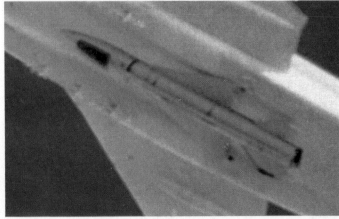

Weight: At launch, about 6,600lb (3000kg).
Performance: Speed Mach 1·2 at high altitude (subsonic at low level); range up to 143 miles (230km) at high altitude (112 miles, 180km, at low altitude).
Warhead: Conventional, 2,200lb (1,000kg).

Development: First seen in a released photograph of September 1968, showing one of these missiles under the wing of a Tu-16, AS-5 is based on the airframe of AS-1 and some may even be rebuilds. In place of the turbojet and nose-to-tail duct there is a rocket with extensive liquid-propellant tankage. In the nose is a large radome. Superficially the nose and underbody fairing appear to be identical to those of the ship-launched SS-N-2 Styx and AS-5 thus is credited with the same choice of active radar or passive IR homing, having cruised to the vicinity of the target on autopilot, with initial radio-command corrections.

By the early 1970s deliveries are thought to have exceeded 1,000, all of them carried by the so-called "Badger G". This launch platform has the same pylons as the "Badger B", and a nose navigator compartment.

In the early 1970s about 35 of these aircraft, plus missiles, were supplied to the Egyptian Air Force, possibly with Soviet aircrew and specialist tradesmen. In the Yom Kippur War in October 1973 about 25 missiles were launched against Israeli targets. According to the Israelis 20 were shot down en route, at least one by an F-4; five penetrated the defences. A supply dump was hit by one missile, but at least two homed automatically onto the emissions from Israeli radar sta-

tions. All the missiles were released at a height of some 29,500ft (9,000m), reaching a speed of about Mach 0·95, although at lower altitudes speed fell off to about Mach 0·85.

AS-6 Kingfish

Origin: Soviet Union.
Type: Anti-ship missile.
Propulsion: Rocket motor, said by US DoD to be solid-propellant.
Dimensions: Length about 33ft (10m); body diameter 35·4in (0·4m); span about 98in (2·5m).
Weight: At launch, about 11,000lb (5000kg).
Performance: Speed (DoD estimate), Mach 3 at high altitude; range (DoD) 155-135 miles (250-560km).
Warhead: Nuclear 350 kilotons or 2,205lb (1000kg) conventional.

Development: At first thought to be a development of AS-4, this completely new missile gradually was reassessed as the first Soviet ASM publicly known that offers precision guidance over long ranges. It is still largely an enigma in the West, but has a very large fuselage with pointed nose, low-aspect-ratio delta wings and quite small aircraft-type tail controls. The fin is above the body, whereas in AS-4 it is on the underside. Propulsion is by an advanced rocket, and key features of AS-6 are much higher flight performance and dramatically better accuracy than any previous Soviet ASM.

It clearly reflects vast advances in inertial guidance and nuclear-warhead design, and it is generally believed to possess terminal-homing. According to the DoD one version has an active radar, while another

homes on enemy radar signals. Area-correlation has been suggested as a third (unconfirmed) possibiity.

Development appears to have been protracted, and though reported prior to 1972 AS-6 was still not on wide service in 1975. By 1977 it was carried under the wings of both the Tu-16 "Badger-G" and Tu-22M "Backfire". User services certainly include the AV-MF and possibly the ADD.

Launched at about 36,000ft (10,973m) the missile climbs rapidly to about 59,000ft (17,983m) for cruise at about Mach 3. It finally dives on its target, or it can approach just above the sea or surface of the land.

AS-7 Kerry

Origin: Soviet Union.
Type: Air-to-surface missile.
Propulsion: Rocket, said by DoD to be single-stage solid-propellant.
Dimensions: No estimates yet published.
Weight: At launch (DoD estimate) 2,645lb (1,200kg).
Performance: Speed, Mach 1 range up to 6·8 miles (11km).
Warhead: (DoD) conventional, 220lb (100kg).

Development: Though the Soviet Union has clearly been testing tactical ASMs for at least 20 years, not one is known to have entered service until the late 1970s, a very strange fact. For over a decade AS-7 has been reported to be carried by the Su-24 "Fencer", and it is probably part of the armament of the MiG-27 and several other FA (Frontal Aviation) types.

Guidance was originally thought to be radio command, an outdated method which normally requires the directing aircraft to loiter in the vicini-

ty of the target. The DoD opinion is that AS-7 is a beam rider, though whether the beam is radar or laser has not been divulged. The traditional form of beam-riding guidance is a most odd choice for a missile intended to attack battlefield targets. It appears safer to regard the question of guidance as unknown.

According to some reports it is carried by the Yak-36MP "Forger", in this case presumably against ship targets.

AS-8

Origin: Soviet Union.
Type: Air-to-surface missile.
Data: Not available.

Development: As yet not publicly associated with a NATO reporting name, this is said to be a "fire and forget" missile to be carried by all Soviet attack helicopters, such as the so-called "Hind-D" version of Mi-24 and the Sukhoi Su-25. Described as similar to the American Hellfire, it is reported to have a solid rocket motor, passive radiation seeker (Hellfire has a seeker that homes on laser radiation) and range of 5-6¼ miles (8-10km) at a flight Mach number of 0·5-0·8. IOC was apparently achieved in 1977 when AS-8 missiles began to appear on Mi-24 units in East Germany.

AS-9

Origin: Soviet Union.
Type: Anti-radiation missile.
Data: Not available.

Development: Originally reported as an ARM (anti-radar missile) with a rocket motor for use from the Su-24 and similar FA tactical aircraft, AS-9

Left: AS.15TT mock-ups on an AS 365F Dauphin helicopter. Agrion radar is under the nose.

Bottom left: The highly successful AS.12, here mounted on an Alouette III helicopter.

Below: Far East-based Soviet "Badger-G" carrying an AS-6 Kingfish ASM.

Right: This large missile on the glove pylon of a Soviet Su-17 has not been officially identified as an AS-7, but probably is.

Right below: An AS.30L, with sustainer motor burning, just before hitting the target in a French trial.

is now known to be a large cruise-type weapon carried by the Tu-22 "Blinder" and Tu-16 "Badger", and almost certainly by "Backfire".

Recent DoD assessments describe it as a winged weapon with air-breathing (probably turbojet) propulsion, with a range of 62 miles (100km) at Mach 0·8 and carrying a 330lb (150kg) warhead. Other reports give it an even greater range of 124 miles (200km), which probably implies autopilot or strapdown inertial mid-course guidance before the seeker head locks on to a suitable emitting target.

AS-X-10

Origin: Soviet Union.
Type Air-to-surface missile.
Data: Not available.

Development: Another of the virtually unknown ASMs which have proliferated in US reports in the past few years, this is said to be an EO-homing (semi-active laser) precision missile, with a length of 10ft (3m), and a range of 6·2 miles (10km) at Mach 0·8 on the thrust of a solid rocket motor. Such a low speed seems odd for a weapon with such guidance intended to penetrate targets which are likely to be defended by modern weapons, and like the rest of the "data" should be viewed with suspicion. Carrier aircraft are said to include the MiG-27, Su-17 and Su-24.

AS.12

Origin: France.
Type: General purpose air-to-surface guided missile.
Propulsion: SNPE boost/sustain solid motor.

Dimensions: Length 73·9in (1870mm); body diameter (max at warhead) 8·25in (210mm); span 25·6in (650mm).
Weight: At launch 168lb (76kg).
Performance: Speed 210mph (338km/h); range, max (measured relative to Earth) 5 miles (8km).
Warhead: Usually OP, 3C, explodes 63lb (28·4kg) charge after penetrating 1·57in (40mm) armour; alternative hollow-charge AP or fragmentatgion anti-personnel types.

Development: Developed in 1955-7 by Nord-Aviation, this missile was a natural extrapolation of the original SS.10 and 11 system to a bigger weapon, with a warhead weighing roughly four times as much and suitable for use against fortifications or ships. Trials began in 1958, and production of surface-launched SS.12 started in late 1959, with AS.12 following in 1960, the original planned carrier aircraft being the French Navy Etendard and Super Frelon.

AS.12 can be use with the APX260 (Bezu) or SFIM 334 gyrostabilized sight and with IR night vision equipment, but the wire-transmitted guidance system is the basic command-to-line-of-sight (CLOS) type with optical (flare) tracking; the TCA semi-automatic system is not available with AS.12. Maximum airspeed at launch is 230mph (370km/h).

Over 10,000 missiles have been produced, and the type remains in wide-scale service. AS.12 has been carried by the Alizé, P-2 Neptune, Atlantique, and Nimrod aircraft, as well as Alouette, Wasp, Wessex, Gazelle and Lynx helicopters. Several were fired by both sides in the South Atlantic War, one crippling the

Argentinian submarine *Sante Fe*, which was running on the surface on the way into Grytviken harbour.

AS.15TT

Origin: France.
Type: Air-launched anti-ship missile.
Propulsion: SNPE Anubis, Nitramite solid motor, smokeless, 45.2sec burn time.
Dimensions: Length 85·04in (2·6m); body diameter 7·28in (185mm); span 22·22in (564mm).
Weight: At launch 220lb (100kg).
Performance: Speed over 628mph (1,010km/h); range over 9·3 miles (15km).
Warhead: Derived from AS.12, conventional 66lb (30kg).

Development: Though Lasso (AM.10) meets the requirements of the French Navy, it is right on the limits of what can be accomplished with wire guidance and could lose export sales to the British Sea Skua. To rival the British missile Aérospatiale is developing AS.15, in at least two versions, using the same warhead as AM.10 but with radio command guidance.

The basic AS.15 has much in common with AM.10 but the body is slimmer and there are flip-out rear fins. It can be launched from existing AS.12 installations provided they have been updated to AM.10 standard with a stablized sight and, preferably, Flir or imaging IR. Like other Aérospatiale tactical missiles of this series the basic AS.15 has to be steered all the way to the target by the operator. AS.15TT (Tous Temps, all-weather), on the other hand, is a substantially different missile, though again carrying the standard warhead. It is not roll-stabilized and is guided semi-automatically.

The basic system depends on Thomson-CSF Agrion 15 radar (derived from the Iguane developed as a retrofit to the Alizé aircraft), with pulse-compression and frequency agility to improve behaviour in the presence of ECM. This radar continuously compares the sightlines to the target and missile, and a digital radio link drives the difference to zero. After a programmed descent to sea-skimming height on the radio altimeter the missile runs to within 1,000ft (300m) of the target and is then commanded to sink to immediately above the sea surface to be sure of hitting the target.

AS.15TT has been integrated with two Aérospatiale helicopters, Dauphin 365N and the Super Puma. Both thus acquire a long-range surveillance capability with auto-digital link to the missiles or surface vessels. The first complete air-firing test took in place in October 1982.

To date only Saudi Arabia has ordered AS.15TT.

AS.30/30L

Origin: France.
Type: Air-to-surface missile.
Propulsion: SNPE solid with composite boost and CDB sustainer (max time 21sec).
Dimensions: Length (X12 warhead) 151in (3839mm), (X35) 153in (3,885mm), (AS.30L) 143in

(3,650mm); body diameter 13·5in (340mm); span 39·4in (1m).
Weight: At launch 1,146lb (520kg).
Performance: Speed Mach 1·5; range up to 7 miles (11·25km).
Warhead: Conventional, 529lb (240kg), with optional impact or delay fuzes.

Development: A logical scale-up of AS.20, this hard-hitting missile has a higher wing loading yet can be launched at Mach numbers down to 0·45 compared with the lower limit of 0·7 for the earlier missile. Originally the Nord 5401, it was developed in 1958 and disclosed on the Mirage III and Northrop N-156F in 1960. AS.30 was produced to meet a French DTE requirement for an ASM with range of at least 6·2 miles (10km) without the launch aircraft having to come within 1·86 miles (3km) of the target (today unacceptably close). CEP was to be 33ft (10m) or less, and all these demands were exceeded. Early AS.30 missiles, tested from Canberras and Vautours at Colomb-Béchar and Cazaux, were aerodynamically similar to AS.20.

The missile is not roll-stabilized and the sustainer motor is equipped with two nozzles, one on each side. The operator keeps tracking flares on the missile aligned with the target by a radio link which sends signals to bias two vibrating spoilers that intermittently interrupt the jets from the nozzles. The autopilot interprets the command to interrupt the correct jet to steer left/right or up/down.

In 1964 an improved AS.30 was produced with four flip-out tail fins indexed in line with the wings, and without spoilers on the sustainer nozzles. At the same time the TCA semi-automatic guidance system was introduced, with an SAT tracker in the aircraft continuously monitoring an IR flare on the missile and the pilot keeping the target centred in his attack sight, an onboard computer zeroing any difference between the two sightlines without the need to work a pitch/yaw joystick control. About 3,870 AS.30 missiles were delivered, most of them exported; only the Armée de l'Air used the TCA guidance.

As a company venture, Thomson-CSF and Aérospatiale began to work on a laser-guided AS.30 in 1974 (Ferranti in Britain proposed this with company hardware almost a decade earlier). Using Martin-Marietta licensed technology Thomson-CSF developed the Atlis (automatic tracking laser illumination system) target-designation pod and a complementary Ariel seeker head able to fit any missile of 3·94in (100mm) or greater diameter.

Aérospatiale produced the AS.30L (AS.30 Laser) to make use of this more modern guidance system. In late 1977 an Armée de l'Air Jaguar A tested an Atlis 1 pod at Cazaux, in the course of which unguided AS.30L prototype missiles were fired. These had roll-stabilization and were programmed to fly on a gyro reference in a pre-guidance phase, prior to picking up the radiation from the target.

In 1980 trials began using pre-production missiles homing on radiation from targets illuminated by the Cilas ITAY-71 laser in the Atlis 2 pod,

which also includes a TV target tracker to assist accurate designation. In 1981 Aérospatiale claimed the system, linked with the Jaguar, was the only one in the world to allow autonomous firing with laser guidance from single-seat aircraft. Deliveries began in 1983, initially on 300 missiles to arm the last 30 Armée de l'Air Jaguars.

ASM-1

Origin: Japan.
Type: Anti-ship missile.
Propulsion: Nissan Motors single-stage solid rocket.
Dimensions: Length 157·5in (4·0m); body diameter 13·75in (350mm); span 47·25in (1·2m).
Weight: At launch 1,345lb (610kg).
Performance: Speed Mach 0·9; range, maximum 56 miles (90km).
Warhead: Conventional anti-ship, 440lb (200kg).

Development: Mitsubishi was selected as prime contractor for this large anti-ship missile in 1973, and after successful development the basic air-launched version was accepted by the Defence Agency in December 1980. Work has also begun on other versions, including a surface-launched variant with a tandem boost motor and an extended-range model with turbojet propulsion.

The basic ASM-1 has mid-course guidance provided by a Japan Aviation Electronics strapdown inertial system, with a TRT radio altimeter which holds altitude just above the tops of the largest waves. Near the target the Mitsubishi Electronics active radar seeker is switched on to home on to the largest reflective target. Guidance tests were flown with a C-1 transport in 1977, and in December that year unguided rounds were fired from an F-1 over Waseka Bay. Guided flight trials began in July

1978, at which time a US report gave the estimated unit price as $384,000. Production was initiated in late 1979, and from 1982 about 30 rounds per year were being delivered to F-1 squadrons. Other possible carrier aircraft include the P-2J and P-3C patrol aircraft.

CWS

Origin: West Germany.
Type: Container Weapon System.
Propulsion: (when fitted) one solid rocket motor.
Dimensions: Length 154in (3910mm), (with rocket) 163·4in (4150mm), body width 28·7in (728mm); body depth 16·5in (420mm); span (when fitted) 78·74in (2m).
Weight: At launch, (captive dispenser) 2,205lb (1,000kg), (jettison) 2,425lb (1,100kg), (stand-off) 2,646lb (1,200kg), (standoff,

powered) 3,307lb (1,500kg).
Performance: Speed, high-subsonic range (stand-off) up to 12·4 miles (20km).
Warhead: Dispenses submunitions.

Development: The Container Weapon System is purpose-designed for dispensing bomblets for use against armour or other surface targets. Minimum cost and maximum versatility are assured by the modular concept, the one unvarying central feature being the "warhead" comprising a container with 42 submunition ejection tubes fired by an inter-valometer.

The submunitions are of the same types as previously produced for MBB's MW-1 simple dispenser used by German Tornados. To the central box may be added an upper hardback adapter fitting the NATO 30in (762mm) store station, or the same adapter fitted with wings for stand-

Far left top: West German MBB Apache Container Weapon System (CWS) is intended to launch bomblets against tanks and other surface targets.

Far left centre: F-4E of the Israeli Air Force carrying out a test launch of a Gabriel III anti-ship missile, which has been developed from the ship-borne missile system.

Far left bottom: Durandal pavement penetrator missiles being launched during a high-speed test flight over an airfield runway.

Left top: French Aéronavale Super Frelon helicopter launches an AM-39 Exocet ASM.

Left centre: Guided-Bomb Unit (GBU-15) mounted on an F-4 Phantom, with electro-optical sensor above the pylon.

Below: AGM-88 HARM missile mounted on an underwing hardpoint of an F-4G Wild Weasel aircraft of the USAF. Although very effective, HARM procurement has been reduced due to the high costs of the system.

off use. On the nose goes a fairing, or a fairing containing an autopilot and height sensor (set to 164 or 328ft, 50 or 100m), or containing these plus a 3/6 mile, (5/10km) inertial platform, or these three units plus a computer. On the back goes a fairing which can have fins, or two powered control fins, or four (two-axis) control fins, or two-axis control plus the rocket.

No orders have yet been placed, but the CWS is being used as the basis for the Franco-German Armet project.

Durandal

Origin: France.
Type: Runway penetration bomb.
Propulsion: SNPE Hector solid (Epictète-filled) smokeless motor.
Dimensions: Length 106·3in (2,700mm); body diameter 8·78in (223mm); span 16·9in (430mm).
Weight: At launch 430lb (195kg).

Performance: Normally released at down to 185ft (56m) at up to 685mph (1,100km/h).
Warhead: Conventional, 220lb (100kg), 1-sec delay fuze of 33lb (15kg) TNT charge.

Development: Originally gaining notoriety in 1969 as the "concrete dibber" used by Israel, Durandal was developed in collaboration with SAMP as a weapon tailored to the task of causing maximum damage to concrete pavements and other hard targets such as concrete aircraft shelters.

It is a simple tube which tactical aircraft can carry in multiple (16 on the Mirage 2000). After release it is braked by Matra's standard parachute braking kit (45,000 delivered) and tilted nose-down before the motor fires. Acceleration is extremely high, ground impact taking it through reinforced concrete 15·75in (400mm) thick before detonation of the warhead. Standard runway damage area is 2,153ft² (200m²).

A single run by two aircraft is claimed to neutralize runway, taxiway, manoeuvring aprons and numerous shelters. Delayed-action fuzing, designed to prevent or hamper runway repair work.

By late 1982 well over 6,000 rounds had been sold, and Durandal was the winner of a long competitive evaluation by the USAF.

Exocet AM.39

Origin: France.
Type: Air-launched anti-ship missile.
Propulsion: Condor 2-sec boost motor and SNPE Hélios solid (Nitramite-filled) smokeless 150-sec burn sustainer.
Dimensions: Length 15ft 4½in (4690mm); body diameter 13·75in (350mm); span 43·3in (1·1m).
Weight: At launch, 1,444lb (655kg).
Performance: Speed, high subsonic range 31-43·5 miles (50-70km) depending on launch altitude.
Warhead: Serat hexolite/steel block, 364lb (165kg), penetrates armour at contact angles to 70°, proximity and delay fuzes.

Development: Exocet was designed as a ship-launched sea-skimming missile, fed with target data before launch and provided with inertial midcourse guidance, flying at Mach 0·93 at a height of about 8ft (2·5m), and finally switching on the EMD Adac X-band monopulse active radar seeker to home on the ship target. Exocet was obviously a potential ASM and inert rounds were dropped by an Aéronavale Super Frelon in April 1973, followed by cutgrain powered launches in June of that year.

In May 1974 the decision was taken to put the air-launched Exocet into production for the Aéronavale, and since the Aérospatiale has sold this missile to an increasing list of export customers.

Originally almost identical to MM.38, and designated AM.38, the ASM developed into AM.39 with a new propulsion system (see data) and reduced overall missile length and weight giving increased performance. The wings and fins are reprofiled to facilitate carriage at supersonic speeds, and because of the greater range and flight-time the Adac seeker radar operates over a greater angular scan. AM.39 entered Aéronavale service in July 1977 carried aboard the Super Frelon (two missiles), followed by Pakistani Sea Kings.

The Super Etendard followed in mid-1978 with either one or two on underwing pylons, and AM.39s fired from such aircraft of the Argentine navy gained world-wide notoriety in the 1982 Falklands conflict (the missile which hit HMS *Sheffield* failed to explode, the ship being lost due to a fire started by the still-burning sustainer). The successes have boosted an already good order-book, and new versions are under development.

Aérospatiale claims AM.39 to be the only long-range missile in the West that can be launched from helicopters (the latest platform is the Super Puma). A one-second delay allows the missile to drop clear before boost-motor ignition.

Gabriel III

Origin: Israel.
Type: Air-launched anti-ship missile.
Propulsion: Boost/sustain solid rocket motor.
Dimensions: Length 151in (3·84m); body diameter 13in (330mm); wing span 43in (1·1m).
Weight: c 1,322lb (600kg).
Performance: Speed, transonic; range over 37 miles (60km).
Warhead Blast/frag 331lb (150kg), delay action fuze.

Development: The original family of Gabriel ship-launched missiles stemmed from various earlier weapons which included an air-launched version, and the wheel has now turned full circle with the perfection of Gabriel III A/S (air/surface).

Autonomous after launch, it can be carried by F-4, F-16 or A-4 aircraft, and launched in either a fire-and-forget or a fire-and-update mode. In the latter the pre-programmed point in the sea-skimming run at which the inertial guidance is replaced by active radar homing is delayed to a point nearer the target, so the active search covers a smaller geographical sector and operates for a shorter time.

IAI, the manufacturers, emphasize the excellent ECCM capability of Gabriel III and its exceptionally low cruise height which is set pre-launch according to sea state.

GBU-15 CWW

Origin: USA.
Type: Modular glide (smart) bomb.
Propulsion: None.
Dimensions: Length 154in (3·91m); body diameter 18in (457mm); span 59in (1499mm).
Weight: At launch 2,450lb (1,111kg).
Performance: Speed, subsonic: range, typically 5 miles (8km) but highly variable with launch height and speed.
Warhead: Mk 84 bomb, 2,000lb (907kg).

Development: The CWW (cruciform-wing weapon) is the modern successor to the Vietnam-era Pave Strike Hobos (homing-bomb system), of which GBU-8 (guided bomb unit) was the chief production example. Like GBU-8, GBU-15 is a modular system comprising standard GP (general purpose) bombs to which a target-detecting device and trajectory-control fins are added. The full designation of the basic production missile is GBU-15(V)/B, and it is also called a modular guided glide bomb (MGGB) or modular guided weapon system.

Though the payload and structural basis may be the CBU-75 cluster munition the normal basis is the Mk 84 2,000lb (907kg) bomb. To the front are added an FMU-124 fuze, a tubular adapter and either of two target-detecting devices, TV or IRR (imaging infra-red). At the rear are added an autopilot, displacement gyro, primary battery, control module and data-link module, and the weapon is completed by attaching four canard fins and four large rear wings with powered control surfaces on the trailing edges. (An alternative PWW, planar-wing weapon, by Hughes, is no longer active).

GBU-15 is launched at medium to extremely low altitudes. In the former case it is guided over a direct line of sight to the target. In the latter it is launched in the direction of the target, while the carrier aircraft gets away at very low level. It is steered by a data-link by the operator in the aircraft who has a display showing the scene in the seeker in the nose of the missile (TV is the usual method). The missile climbs until it can acquire the target, and then pushes over into a dive. The operator has the choice of steering the missile all the way to the target or locking-on the homing head. Extensive trials from F-4, F-111 and B-52 aircraft are complete and substantial deliveries have been made to USAF.

HARM, AGM-88A

Origin: USA.
Type: Anti-radiation missile.
Propulsion: Thiokol single-grain (280lb, 127kg, filling of non-aluminized HTPB) reduced-smoke boost/sustain motor.
Dimensions: Length 13ft 8½in (4·17m); body diameter 10in (254mm); span 44in (1,118mm).
Weight: At launch 796lb (361kg).
Performance: Speed over Mach 2; range/height variable with aircraft to about 11·5 miles (18·5km).
Warhead: Fragmentation with proximity fuze system.

Development: Neither Shrike nor Standard ARM is an ideal air-launched ARM and in 1972 the Naval Weapons Center began R&D and also funded industry studies for a High-speed Anti-Radiation Missile (HARM). Among the objectives were much higher flight speed, to lock-on and hit targets before they could be switched off or take other action, and to combine the low cost and versatility of Shrike, the sensitivity and large launch envelope of Standard ARM, and completely new passive homing using the latest microelectronic digital techniques and interfacing with new aircraft systems. In 1974 TI was selected as system integration contractor, assisted by Hughes, Dalmo-Victor, Itek and SRI (Stanford Research Institute).

The slim AGM-88A missile has double-delta moving wings and a small fixed tail. The TI seeker has a simple fixed aerial (antenna) yet gives broadband coverage, a low-cost autopilot is fitted, and Motorola supply an optical target detector forming part of the fuzing for the large advanced-design warhead. Carrier aircraft include the Navy/Marines A-6E, A-7E and F/A-18, and the Air Force APR-38 Wild Weasal F-4G and EF-111A, with Itek's ALR-45 radar warning receiver and Dalmo-Victor's DSA-20N signal analyser both interfaced. Proposed carriers include the B-52, F-16 and Tornado. HARM can be used in three modes. The basic use is Self-protect, the ALR-45 detecting threats, the launch computer sorting the data to give priorities and pass to the missile a complete set of digital instructions in milliseconds, whereupon the missile can be fired. In the Target of Opportunity mode the very sensitive seeker locks-on to "certain parameters of operation and also transmissions associated with other parts of a radar installation" which could not be detected by Shrike or Standard ARM. When configured in the pre-briefed mode a HARM can be fired "blind" in the general direction of known enemy radiation emitters; if the latter are "silent" the missile will self-destruct; however, if one of them should radiate the HARM will immediately home on to it.

Test flights began in 1976; redesign followed and following prolonged further tests delivery to user units began in early 1983.

Harpoon, AGM-84A

Origin: USA.
Type: Anti-ship missile.
Propulsion: One Teledyne CAE J402-400 turbojet, sea-level thrust 661lb (300kg).
Dimensions: Length 12ft 7in (3·84m); body diameter 13·5in (343mm); span 30in (762mm).
Weight: At launch 1,160lb (526kg).
Performance: Speed Mach 0·75, range over 57 miles (92km).
Warhead: NWC 500lb (227kg) penetration/blast with impact/delay and proximity fuzing.

Development: This potentially important weapon system began as an ASM in 1968, but three years later was combined with a proposal for a ship- and submarine-launched missile system. McDonnell Douglas Astronautics (MDAC) was selected as prime contractor in June 1971. The main development contract followed in July 1973, and of 40 prototype weapon systems 34 were launched in 1974-5, 15 being the RGM-84A fired from ships (including the PHM *High Point* whilst foilborne) and three from submarines, the other 16 being air-launched. At first almost wholly trouble-free, testing suffered random failures from late 1975, and the clearance for full-scale production was delayed temporarily. Production of all versions amounted to 315 in 1976, and about 2,100 by early 1983. Target data, which can be OTH if

supplied from a suitable platform, are fed before launch to the Lear-Siegler or Northrop strapdown inertial platform which can steer the missile even if launched at up to 90° off the desired heading. Flight control is by cruciform rear fins. A radar altimeter holds the desired sea-skimming height, and no link with the aircraft is required. Nearing the target the Texas Instruments PR-53/DSQ-58 active radar seeker searches, locks-on and finally commands a sudden pull-up and swoop on the target from above.

The Naval Weapons Center and MDAC are also studying possible versions with supersonic speed, torpedo-carrying payload, imaging IR homing, passive radiation homing, nuclear warhead, vertical launch, midcourse guidance updating and other features.

MDAC expects to make at least 5,000 systems by 1988 despite the delayed start. Of these well over 2,000 will be for the US Navy, for surface ships, submarines, and P-3C, A-6E, S-3B, A-7E and F/A-18A aircraft. The S-3 carries two missiles and the other types four. Production is at the rate of 40 missiles per month. Among the aircraft systems is a missile firing simulator.

JP-233

Origin: UK.
Type: Submunition dispenser.
Propulsion: None.
Dimensions: Not disclosed.
Weight: (Largest size) in the region of 5 long tons (11,200lb, 5,080kg).
Performance: Free fall.
Warhead: Payload of various bomblets.

Development: Originally known as the LAAAS (Low-Altitude Airfield Attack System). JP.233 is a series of submunition dispensers for parachute-retarded payloads which include pavement-cratering bomblets and anti-personnel mines, with or without delay-action fuzing.

Aircraft with wing pylons can carry either short-finned containers for bomblets or medium-length finned containers for mines. The F-111 can carry a pair of each type, as can the F-16, but the Jaguar and Harrier can carry only one pair which must be of the same type. The most important carrier aircraft will be the Tornado IDS, which carries both types of payload in a single giant tandem pod on the centreline (which has significantly lower drag than the German MW-1 dispenser).

Development began in November 1977 as a 50/50 programme with the USA, but the latter pulled out in 1982. Engineering development was completed in 1984.

Kormoran

Origin: West Germany.
Type: Air-to-surface anti-ship missile.
Propulsion: SNPE double-base solid motors, twin Prades boost motors and central Eole IV sustainer.
Dimensions: Length 173·2in (4·4m); body diameter 13·39in (340mm); span 39·37in (1m).
Weight: At launch 1,323lb (600kg).

Performance: Speed Mach 0·95; range up to 23 miles (37km).
Warhead: Advanced MBB type, 352lb (160kg) with 16 radially mounted projectile charges and fuze delayed for passage through 3·5in (90mm) of steel plate.

Development: The first major post-war missile programme in West Germany, this began life in 1964 to meet a Marineflieger (Navy Air) requirement for a large anti-ship missile. Based on a Nord project, the AS.34, using the Sfena inertial guidance planned for the stillborn AS.33, it became a major programme in the new consortium MBB, with Aérospatiale participation.

The basic weapon exactly follows Nord/Aérospatiale principles, but incorporates more advanced guidance. After release from the F-104G or Tornado carrier aircraft the boost motors give 6,063lb (2750kg) thrust each for almost 1 sec, when the sustainer takes over and gives 628lb (285kg) for 100 sec. Sfena/Bodenseewerk inertial mid-course guidance is used with a TRT radio altimeter to hold less than 98ft (30m) altitude. The missile then descends as it nears the pre-inserted target position, finally descending to wavetop height as the Thomson-CSF two-axis seeker (operating as either an active radar or passive receiver) searches and locks-on. Impact should be just above the waterline, and the warhead projects liner fragments with sufficient velocity to penetrate up to seven bulkheads.

Flight trials from F-104Gs began on 19 March 1970. The first of an initial 350 Kormoran production missiles was delivered in December 1977, and by mid-1978 the Marineflieger MFG 2 at Eggbeck was fully equipped. The first Tornado-Kormoran unit was MFG 1 at Schleswig-Jagel in 1982. It is also carried by Italian Tornados.

Kormoran 2 is under development, and this will include greater hit probability, improved ECM protection, increased range and simpler operation. If R&D is successful production should start about 1990.

Martel

Origin: France/UK.
Type: Air-to-surface missile.
Propulsion: Solid motor; (AS.37) SNPE Basile boost (2·4s burn) and Cassandre sustain (22·2s), both composite, (AJ.168) SNPE composite boost and cast double-base sustainer.
Dimensions: Length (AS.37) 162·2in (4·12m), (AJ.168) 152·4in (3·87m); body diameter 15·75in (400mm); span 47·25in (1·2m).
Weight: At launch (AS.37) 1,213lb (550kg), (AJ.168) 1,168lb (530kg).
Performance: Speed, see text; range (treetop-height launch) 18·6 miles (30km), (hi-alt launch) 37·2 miles (60km).
Warhead: Conventional 331lb (150kg) with DA or (AS.37) proximity fuze.

Development: This excellent weapon grew from studies by HSD in Britain and Nord-Aviation and Matra in France in 1960-3. In September 1964 the British and French governments

Above: A-10 with two AGM-65 Maverick ASM (and Pave Penny and ALQ-119) under each wing.

Far right top: AGM-65B Scene-Magnification Maverick has improved optical sensors.

Far right upper: Tornado IDS JP.233 pod can dispense two types of submunition.

Far right lower: AGM-45A Harpoon mounted under the wing of a P-3B Orion 1972.

Far right bottom: F/A-18A with four Harpoons on wing pylons.

Right: AJ.168 "British Martel" under the wing of a Buccaneer of RAF Strike Command.

agreed to develop the weapon system jointly, in one of the first examples of European weapon collaboration. In the event it was Engins Matra that became the French partner, responsible for the AS.37 anti-radar Martel. HSD developed the AJ.168 version with TV guidance. The name stems from Missile Anti-Radar TELevision.

Having a configuration similar to the AS.30, Martel has French propulsion. Flight mach number is typically about 0·9, though this depends on angle of dive. Several sources state that Martel is supersonic.

The operator of AJ.168 studies the target area as seen on his control screen in the cockpit of the launch aircraft, fed by the MSDS vidicon camera in the nose of the missile. When he acquires a target he manually drives a small graticule box over it to lock-on the TV seeker

before launch. The weapon is then fired, holding height constant by a barometric lock, and steered by the operator's control stick via a streamlined underwing pod which also receives the video signals from the missile. Special features assist the operator to steer the missile accurately to the target.

AS.37 has an EMD A.37 passive radiation seeker, with steerable inverse-Cassegrain aerial. If the rough location of a hostile emitter is known, but not its operating frequency, the seeker searches up and down a preset band of frequencies; when it detects the enemy radiation the aerial sweeps through 90° in azimuth to pinpoint the location. When it has locked on the missile is launched, thereafter homing automatically. Alternatively, if the hostile radiation is known before takeoff the seeker can be fitted with a matched aerial and

receiver to pinpoint the source. AS.37 continues to home no matter how the hostile radiation may change frequency so long as it remains within the preset band.

Both versions of Martel have the same warhead, AS.37 having a Thomson-CSF proximity fuze. AS.37 is carried by the Mirage III, Jaguar, Buccaneer and Atlantic; AJ.168 is used only by the RAF Buccaneers, but could be made compatible with the Phantom, Tornado and two-seat Jaguar or Harrier, and has been mentioned in a weapon list for Nimrod. Production terminated in the late 1970s.

Maverick, AGM-65

Origin: USA.
Type: Air-to-surface missile.
Propulsion: Thiokol boost/sustain solid motor, from 1972 TX-481 and

from 1981 TX-633 with reduced smoke.
Dimensions: Length 98in (2,490mm); body diameter 12in (305mm); span 28·3in (720mm).
Weight: At launch (AGM-65A, shaped-charge) 436lb (210kg), (65A, blast/frag) 635lb (288kg).
Performance: Speed classified but supersonic; range 0·6-10 miles (1-16km) at sea level, up to 25 miles (40km) after Mach 1·2 release at altitude.
Warhead: Choice of Chamberlain shaped charge (83lb, 37·6kg, charge) or Avco steel-case penetrator blast/frag.

Development: Smallest of the fully guided or self-homing ASMs for US use, AGM-65 Maverick was approved in 1965 and, following competition with Rockwell, Hughes won the programme in June 1968. An initial

17,000-missile package was fulfilled in 1975, and production has continued at reduced rate on later versions. The basic missile, usually carried in triple clusters under the wings of the F-4, F-15, F-16, A-7, A-10 and Swedish AJ37A Viggin, and singly by the F-5 and the BGM-34 RPV, has four delta wings of very low aspect ration, four tail controls immediately behind the wings, and a dual-thrust solid motor.

In mid-1978 Hughes completed production of 26,000 AGM-65A Mavericks and for three years had no production line (though other versions followed, see below). Unguided flights began in September 1969, and the missile has been launched at all heights down to treetop level.

The pilot selects a missile, causing its gyro to run up to speed and light a cockpit indicator. The pilot then visually acquires the target, depresses his uncage switch to remove the pro-

123

tective cover from the missile nose, and activates the video circuitry. The TV picture at once appears on a bright display in the cockpit, and the pilot then either slews the video seeker in the missile or else lines up the target in his own gunsight. He depresses the track switch, waits until the cross-hairs on the TV display are aligned on the target, releases the switch and fires the round. Homing is automatic, and the launch aircraft at once escapes from the area.

In the 1973 Yom Kippur war AGM-65A was used operationally, in favourable conditions. It requires good visibility, and the occasional $48,000 A-model breaks its TV lock and misses its target — for example, because of overwater glint.

AGM-65B, Scene-Magnification Maverick, has new optics, a stronger gimbal mount and revised electronics. The pilot need not see the target, but instead can search with the seeker and cockpit display which presents an enlarged and clearer picture. Thus he can identify the target, lock-on and fire much quicker and from a greater slant range. AGM-65B was in production (at up to 200 per month) from May 1980 to May 1983.

AGM-65C Laser Maverick was for close-air support against laser-designated targets, the lasers being the infantry ILS-NT200 or the airborne Pave Knife, Pave Penny, Pave Spike, Pave Tack or non-US systems. Flight testing began in January 1977, using the Rockwell tri-Service seeker. Troop training has established the method of frequency and pulse coding to tie each missile to only one air or ground designator, so that many Mavericks can simultaneously be homed on many dif-

ferent sources of laser radiation.

AGM-65C was replaced by AGM-65E with "tri-Service" laser tracker and digital processing which in 1982 was entering production for the US Marine Corps with heavy blast/frag warhead. Westinghouse tested Pave Spike with the Minneapolis-Honeywell helmet sight for single-seat aircraft.

In May 1977 engineering development began on AGM-65D IR-Maverick, with Hughes IIR tri-Service seeker. Considerably more expensive than other versions, the IIR seeker — especially when slaved to an aircraft-mounted sensor such as FLIR, a laser pod or the APR-38 radar warning system — enables the Maverick to lock-on at least twice the range otherwise possible in northwest Europe in mist, rain or at night. Of course, it also distinguishes between "live targets" and "hulks". Using the centroid seeker in place of the original edgelock optics, AGM-65D was tested from an F-4 in Germany in poor weather in January-March 1978.

While Hughes continues to produce the common centre and aft missile sections, delay with the laser-seeker E-version means that AGM-65D got into pilot production first.

All AGM-65A Mavericks have the same 130lb (59kg) conical shaped-charge warhead, but different warheads are in prospect. The Mk 19 250lb (113kg) blast/fragmentation head is preferred by the Navy and Marines, giving capability against small ships as well as hard land targets, and may be fitted to C and D versions with new fuzing/arming and a 4 in (102mm) increase in length. Another warhead weighs 300lb (136kg), while in December 1976 the

Air Force expressed a need for a nuclear warhead.

Hughes' Tucson, Arizona, plant is likely to be hard-pressed to handle TOW, Phoenix and residual Roland work on top of enormously expanded Maverick production. By far the largest numbers are expected to be of the IIR Maverick, AGM-65D, of which well over 30,000 rounds are expected to be produced, at a rate of 500 per month.

Prolonged tests have confirmed the long range, which at last matches the flight limitations of the missile itself, and AGM-65D is the standard missile for use with the Lantirn night and bad-weather sensor system now being fitted to F-16s and A-10s.

The Navy is expected to procure AGM-65F, which is almost the same missile but fitted with the heavy penetrator warhead of AGM-65E, and with modified guidance software exactly matched to give optimum hits on surface warships. With this missile family Hughes has achieved a unique capability with various guidance systems and warheads, resulting in impressively large production and interchangeability.

MRASM, AGM-109

Origin: USA.
Type: Air-to-surface missile.
Propulsion: Modified Teledyne CAE J402-400 turbojet (660lb, 300kg, sea-level thrust).
Dimensions: Length (H,K) 234in (5·94m), (I) 192in (4·88m); body diameter 21in (533mm); span (wings extended) 103in (2·616m).
Weight: At launch (H) 2,900lb (1,315kg), (I) 2,225lb (1,009kg), (K) 2,630lb (1,193kg).

Performance: Speed 550mph (885km/h); range (sea level, Mach 0·6) (H) 293 miles (472km), (I)350 miles (564km), (K) 316 miles (509km).
Warhead: (H) 58 TAAM bomblet/mine payloads, 1,060lb (481kg); (I) WDU-7B or -18B unitary warhead, 650lb (295kg); (K)WDU-25A/B unitary warhead, 937lb (425kg).

Development: The General Dynamics AGM-109 was one of the chief versions of the Tomahawk strategic nuclear cruise missile, first tested in air drops from P-3 Orions and A-6 Intruders in 1974. It differed from the ship/submarine/GLCM versions in having no rocket boost motor or launch capsule/box. Main propulsion, at first a J402, switched like other versions to a Williams F107 turbofan in competition with the Boeing AGM-86B as the ALCM for SAC. When the Boeing missile was chosen, Tomahawk was recast in different roles, and eventually in 1981 the naval versions were all terminated, chiefly on cost grounds. GLCM continued as a tactical weapon of the Air Force, and a completely new version, MRASM (Medium-Range ASM) was launched in 1981 as a non-nuclear cruise missile for wide use by the Air Force arming many types of aircraft beginning with the B-52 and F-16.

MRASM has been taken to a high pitch of development — interestingly enough with the original pure-jet engine, but in a much modified form able to fly 8-hour missions burning the new JP-10 fuel and with a positive oil storage, retapered turbine, oxygen start system and

Left: RAF Jaguar GR 1 carrying 1,000lb "iron bomb" fitted with Paveway II Laser-Guided Bomb (LGB) kit. Guidance comes from nose-mounted Ferranti laser.

Above: Originally designated Saab 305E, the RB 05A is used only by the Swedish Air Force. This example is about to be loaded onto an AJ37A Viggen.

Right: In production for 28 years, the RB 04E was originally made for the A32A Lansen, but is still in service and now equips the AJ37 Viggen.

zirconium-coated combustor. The basic missile has been developed in three forms, differing in payload and guidance. AGM-109H is the baseline airfield attack missile, with DSMAC II (digital scene-matching area-correlation) guidance and carrying a heavy payload of 58 TAAM (tactical airfield attack missile) bomblets or delayed-action mines, discharged from upward-facing tubes along the fuselage.

This version is in competition with short-range or free-fall anti-airfield weapons, and justifies its high cost by the fact it is a launch-and-leave missile which eliminates the need for the carrier aircraft to come within 300 miles (483km) of the target.

AGM-109I is a dual-role weapon proposed to the Navy for use by A-6E squadrons. It has a large unitary warhead and both DSMAC II and IIR (imaging IR) guidance for either anti-ship or land attack missions. AGM-109K is a pure sea-control missile with only IIR guidance; the scene-matching and large fuel-cell power plant are replaced by an enlarged warhead. GD states that all versions could have IOC in 1986.

Paveway LGBs

Origin: USA.
Type: Laser-guided unpowered bombs.
Propulsion: As for original bombs plus from 6 to 20in (152-500mm) length and with folding tailfins.
Weight: As for original bombs plus about 30lb (13.6kg).
Performance: Speed, free-fall; range, free-fall so varies with release height, speed.
Warhead: As in original bombs.

Description: This code-name identifies the most diverse programmes in history aimed at increasing the accuracy of tactical air-to-surface weapons. This USAF effort linked more than 30 separately named systems for airborne navigation, target identification and marking, all-weather/night vision, weapon guidance and many other functions, originally for the war in SE Asia. In the course of this work the "smart bombs" with laser guidance managed by the Armament Development and Test Center at Elgin AFB, from 1965, were developed in partnership with TI, using the latter's laser guidance kit, to form an integrated family of simple precision weapons. The first TI-guided LGB was dropped in April 1965.

All these bombs are extremely simple to carry, requiring no aircraft modification or electrical connection; they are treated as a round of ordnance and loaded like a free-fall bomb. Carrier aircraft have included the A-1, A-4, A-6, A-7, A-10, A-37, F-4, F-5, F-15, F-16, F/A-18, F-100, F-105, F-111, AV-8A, B-52 and B-57. Targets can be marked by an airborne laser, in the launch aircraft or another aircraft, or by forward troops. Like almost all Western military lasers the matched wavelength is 1.064 microns, the usual lasers (in Pave Knife, Pave Tack or various other airborne pods) being of the Nd/YAG type. More recently target illumination has been provided by the Atlis II, LTDS, TRAM, GLLD, MULE, LTM, Lantirn and TI's own FLIR/laser designator.

In all cases the guidance unit is the same, the differences being confined to attachments and the various enlarged tail fins. The silicon detector array is divided into four quadrants and is mounted on the nose of a free universal-jointed housing with an annular ring tail. As the bomb falls this aligns itself with the airstream, in other words the direction of the bomb's motion. The guidance computer receives signals from the quadrants and drives four control fins to equalize the four outputs. Thus, the sensor unit is kept pointing at the source of laser light, so that the bomb will impact at the same point. Electric power is provided by a thermal battery, energised at the moment of release, and power to drive the fins comes from a hot-gas generator.

Users include the RAF for use on Mk13/18 1,000lb (454kg) bombs carried by Buccaneers, Tornados and Jaguars. Total production of Paveway guidance units has been very large; in the early 1970s output was at roughly 20,000 per year, at a unit price of some $2,500.

Since 1980 the Paveway II weapons have been in production including a simpler and cheaper seeker section, and a folding-wing aerofoil group. Portsmouth Aviation has integrated the system with RAF bombs used from Harriers over the Falklands. The Paveway III version, which is now in full-scale production, has flick-out lifting wings and a microprocessor and can be dropped at treetop height.

RB 04

Origin: Sweden.
Type: Air-to-surface missile.
Propulsion: Solid rocket motor with boost/sustain charges fired consecutively.

Dimensions: Length 14ft 7in (4.45m); body diameter 19.7in (500mm); span (RB 04C, D) 80.3in (2.04m), (E) 77.5in (1.97m).
Weight: At launch (C, D) 1,323lb (600kg), (E) 1,358lb (616kg).
Performance: Speed high-subsonic; range variable with launch height up to 20 miles (32km).
Warhead: Unitary 661lb (300kg) conventional, DA and proximity fuzes.

Development: This hard-hitting ASM has enjoyed one of the longest active programmes of any guided missile, for the requirement was finalised in 1949, and missile hardware was being manufactured for 28 years (1950-78). Planned as a primary weapon to be carried by the Saab A32A Lansen, this missile was originally designed and developed by the Robotavdelningen (guided-weapons directorate) of the national defence ministry, whose first missile, RB302, was flight-tested in 1948 from a T18B bomber.

The original RB (Robotbyran) 04 was made large enough to carry an active radar seeker, giving all-weather homing guidance earlier than for any other ASM apart from Bat. The configuration is of aeroplane type, with a rear delta wing with end fins and four control fins around the forebody. The two-stage solid cast-DB motor is by IMI Summerfield Research Station in Britain. The radar is by PEAB (Swedish Philips) and the autopilot, originally the XA82, is a Saab design with pneumatically driven gyros and surface servos.

The first launch took place from a Saab J29 fighter on 11 February 1955, and following very successful development the first production version, RB 04C, entered service with the Swedish Air Force in 1958, equipping all A32A aircraft of attack wings F-6, -7, -14 and -17. In the early 1960s the Robotavdelningen developed a version with improved motor and guidance, RB 04D, which was in production in the second half of that decade.

On 1 July 1968 the bureau became part of the Air Material department of the Forsvarets Materielverk (Armed Forces Material Admin, FMV), and the ultimate development of this missile, RB 04E, was assigned to Saab (now Saab Bofors). Produced mainly to arm the AJ37 Viggen, which carries up to three, RB 04E has a reduced span, modernised structure and more advanced guidance. All versions have the same very large fragmentation warhead.

RB 05A

Origin: Sweden.
Type: Air-to-surface missile.
Propulsion: Volvo Flygmotor VR-35 liquid rocket motor, 5,620lb (2,550kg) boost, 1,124lb (510kg) sustain.
Dimensions: Length 11ft 10in (3.6m); body diameter 11.8in (300mm); span 31.5in (0.8m).
Weight: At launch 672lb (305kg).
Performance: Speed supersonic; range up to 5.6 miles (9km).
Warhead: Conventional warhead by Forenade Fabriksverken, proximity fuzed.

Development: When the decision was taken to restrict what had been the Robotavdelningen (national missile directorate) to R&D only, Saab was the natural choice for this missile, prime responsibility for which was placed with the company in 1960.

Originally known as Saab 305A, RB 05A is a simple command-guidance weapon readily adaptable to many types of launch aircraft. One unusual feature is supersonic flight performance, conferred by advanced aerodynamics and a pre-packaged liquid motor fed with Hidyne and RFNA, pumped by a gas-pressurized piston and collapsible aluminium bladder to burn rapidly in the boost phase and slower in the sustainer mode. Motor performance is independent of missile attitude or acceleration, and there is no visible smoke.

The missile automatically centres itself dead ahead of the launch aircraft and is then steered by a microwave link from the pilot's miniature joystick. The guidance is claimed to be highly resistant to jamming, to be usable at low altitudes over all types of terrain, and to be able to attack targets at large offset angles.

RB 05A is carried by the AJ37A Viggen and by various versions of the Swedish Air Force's Saab 105 trainer.

RBS 15F

Origin: Sweden.
Type: Air-to-surface anti-ship missile.
Propulsion: Microturbo TRI 60-2 Model 077 turbojet, sea-level thrust 831lb (377kg).
Dimensions: Length 171·25in (4·35m); body diameter 19·7in (500mm); span 55in (1·4m).
Weight: At launch 1,318lb (598kg).
Performance: Speed high-subsonic; range, classified but several tens of kilometres.
Warhead: Large FFV blast/fragmentation with DA and proximity fuzes.

Development: RBS 15 was designed as a ship-to-ship weapon for use aboard the *Spica* II FPBs. In August 1982 Saab Bofors announced a Swedish defence materiel administration (FMV) order, worth some SK500 million, for the RBS 15F air-launched version, for use by Viggens and the forthcoming JAS.

The 15F has no launch canister or boost motors, but is carried on external pylons and launched when over the horizon from the target. A fire-and-forget weapon, it has pre-programmed mid-course guidance and an advanced radar seeker by PEAB with digital processing and frequency agility, selectable search patterns and modes, target-choice logic (to pick the most important of a group of hostile ships) and variable ECCM facilities. Saab announced RBS 15 will have quick reaction time, high kill probability and high efficiency against all naval targets in any weather. IOC will probably be in 1986.

First export order for RBS 15 (SSM version) came from Finland in 1983, for use on Helsinki missile craft.

Sea Eagle

Origin: UK.
Type: Air-to-surface anti-ship missile.
Propulsion: Microturbo TRI 60-1 Model 067 turbojet, sea-level thrust 787lb (357kg), mainly made in UK.
Dimensions: Length 13ft 5½in (4·1m); body diameter 15·75in (400mm); span 47·25in (1·2m).
Weight: Classified but clearly in the 1;200lb (550kg) class.
Performance: Speed, probably about Mach 0·9; range, classified but several tens of miles.
Warhead: ROF product, suitable to disable the largest surface warships.

Development: Originally designated P3T, Sea Eagle is an over-the-horizon fire-and-forget missile developed from Martel by switching to air-breathing propulsion and adding active radar and sea-skimming capability. Launched to meet Air Staff Requirement 1226, it progressed swiftly to the project definition phase in 1977, and development has been remarkably trouble-free.

The airframe is basically that of Martel, with an underbelly air inlet. Guidance is initially on autopilot, with the on-board microprocessor storing the target's last known position and velocity, with height maintained by a Plessey radar altimeter just above the waves (or following a programmed profile). The target is then acquired by the very advanced MSDS active radar seeker.

Electric power comes from Europe's first production lithium batteries. Features include full night and all-weather capability against the most powerful and electronically sophisticated targets, low life-cycle costs and "round of ammunition" storage and maintenance.

Launch trials began in November 1980, and a full-range sea-skimming flight was made in April 1981. In early 1982 a £200 million production contract was announced. Sea Eagle will enter service in the mid-1980s, initially on RAF Buccaneers (four missiles), RN Sea Harriers (two), and Indian Navy mark 42B Sea King helicopters. It can be carried by all other tactical aircraft and is almost certain to be issued to RAF Tornado GR.1 squadrons. It is also seen as the basis for a long-range missile for use against high-value land targets. P5T is a proposed ship-mounted SSM version.

Sea Skua

Origin: UK.
Type: Air-to-surface anti-ship missile.
Propulsion: BAJ Vickers solid boost/sustainer motors.
Dimensions: Length 98·5in (2·5m); body diameter 9·75in (250mm); span 28·5in (720mm).
Weight: At launch 320lb (145kg).
Performance: Speed high-subsonic; range over 9·3 miles (15km).
Warhead: Blast/frag, 44lb (20kg), exploded within ship hull.

Development: Originally known by its MoD project number of CL-834, this missile is significantly more ad-vanced than the comparable French missiles being dedveloped as successors to AS.12. Instead of having wire or radio command guidance Sea Skua is based on semi-active radar homing, and in its first application aboard the Lynx helicopter the target is illuminated by the Ferranti Seaspray over-water radar.

The weapon system is intended to confer upon helicopter-armed frigates and similar surface ships the capability to destroy missile-carrying FPBs, ACVs, PHMs and similar agile small craft at ranges greater than that at which they can launch their own missiles. This range is also said to be great enough to provide the launch helicopter with considerable stand-off protection from SAMs.

The Decca Tans navigation system of the Lynx can combine with ESM cross-bearings to identify and fix the target, backing up the position on the Seaspray display. The Sea Skua, treated as a round of ammunition needing only quick GO/NO GO checks, is then launched and swoops down to one of four preselected sea-skimming heights, depending on wave state, using a BAe-manufactured TRT radio altimeter. Near the target a pre-programmed or command instruction lifts the missile to target-acquisition height for the MSDS homing head to lock-on. Trials began in 1978.

BAe has privately financed studies matching Sea Skua to other aircraft (including fixed wing) and in coastal defence installations. The company expects that this missile roughly one-tenth as heavy as Exocet but able to destroy ships of 1,000 tons with one shot, will find very wide use. In the South Atlantic War two Lynx helicopters, operating in appalling weather and heavy seas, each ripple-fired a pair of Sea Skuas, sinking an 800-ton warship and crippling another.

Shrike, AGM-45

Origin: USA.
Type: Anti-radar missile.
Propulsion: Rockwell (Rocketdyne) Mk 39 or Aerojet (ATSC) Mk 53 (polybutadiene) or improved Mk 78 (polyurethane, dual-thrust) solid motor.
Dimensions: Length 120in (3·05m); body diameter 8in (203mm); span 36in (914mm).
Weight: At launch (approximately depending on sub-type) 390lb (177kg).
Performance: Speed Mach 2; range 18-25 miles (29-40km).
Warhead: Blast/frag, 145lb (66kg), proximity fuze.

Development: Based in part on the Sparrow AAM, this was the first anti-radar missile (ARM) in the US since World War II. Originally called ARM and designated ASM-N-10, it was begun as a project at NOTS (later Naval Warfare Center) in 1961, and in 1962 became AGM-45A. Production by a consortium headed by Texas Instruments (TI) and Sperry Rand/Univac began in 1963 and Shrike was in use in SE Asia three years later with Wild Weasel F-105Gs and EA-6As. Early experience was disappointing and there have since

Right: Sea Eagle ASMs mounted on a Buccaneer strike aircraft. The missiles' engine inlets are faired over to reduce drag.

Far right: AGM-78 Standard ARM mounted on a Wild Weasel Republic F-105G Thunderchief. Production ended in 1976 but AGM-78 is still in service.

Right below: AGM-62 Walleye glide bomb mounted on a USN A-7. Although unpowered the Walleye has excellent guidance.

Below: Sea Skua missiles mounted on a Royal Navy Lynx, an airframe/missile combination that proved very successful in the South Atlantic War in 1982.

been numerous models, identified by suffix numbers, to rectify faults or tailor the passive homing head to a new frequency band identified in the potential hostile inventory.

Carried by the US Navy/Marines A-4, A-6, A-7 and F-4, the Air Force F-4, F-105 and EF-111 and the Israeli F-4 and Kfir, Shrike is switched on while flying towards the target and fired as soon as the TI radiation seeker has locked-on. After motor cutoff Shrike flies a ballistic path until control-system activation.

The seeker has a monopulse crystal video receiver and continually updates the guidance by determining the direction of arrival of the hostile radiation, homing the missile into the enemy radar with its cruciform centre-body wings driven in "bang/bang" fashion by a hot-gas system.

There were at least 18 sub-types in the AGM-45-1 to -10 families, with over 13 different tailored seeker heads, of which the USAF bought 12,863 by 1978 and the Navy a further 6,200: In the Yom Kippur war Israel used Shrike tuned to 2965/2990 MHz and 3025/3050 MHz to defeat SA-2 and SA-3 but was helpless against SA-6. In 1978-81 additional procurement centred on the -9 and -10 for the USAF to be carried by F-4G and EF-111A platforms, together with modification kits to equip existing rounds to home on to later SAM and other radars.

Standard Arm, AGM-78

Origin: USA.
Type: Anti-radar air-to-surface missile.
Propulsion: Aerojet (ATSC) Mk 27 Mod 4 boost/sustain solid motor.
Dimensions: Length 180in (4·57m) body diameter 13·5in (343m); span (rear fins, greater than strake wings) 43in (1·09m).
Weight: At launch, typically 1,400lb (635kg).
Performance: Speed Mach 2·5; range (depending on launch height) up to 35 miles (56km).
Warhead: Conventional blast/fragmentation, direct-action and proximity fuzes.

Development: In September 1966 the Naval Air Systems Command contracted with Pomona Division of General Dynamics for an ARM having higher performance, longer range and larger warhead than Shrike, which at that time was giving indifferent results. Unlike Shrike the whole programme was developed in industry, the basis being the Standard RIM-66A ship-to-air missile. Flight testing took place in 1967-8; production of AGM-78 Mod 0 began in late 1968.

AGM-78 Mod 0 was carried by the Air Force Wild Weasel F-105F and G and the Navy A-6B and E. The missile flies on a dual-thrust motor, steering with tail controls and very low aspect ratio fixed wings. The Mod 0 AGM-78A of 1968 was fitted with the TI seeker used in Shrike. This was soon replaced by the Maxson broad-band seeker of the main (Mod 1) production version, AGM-78B. This has capability against search, GCI, SAM and other radar systems, and is intended to give the launch platform freedom to attack from any direction and turn away "outside the lethal radius of enemy SAMs". Carrier platforms preferably have a TIAS (Target Identification and Acquisition System) able to measure "specific target parameters" and supply these to the seeker head before launch.

The Mod 1 missile is compatible with the APR-38 system carried by the USAF F-4G Wild Weasel which supplies this need. AGM-78C, D and D-2 have further-increased capability and reduced unit cost, but in 1978 production was not funded, (deliveries then about 700) and effort has since been devoted to improving missiles with field mod kits. Navy and Marine A-6E squadrons carry this missile, as would the Wild Weasel F-16.

Walleye, AGM-62

Origin: USA.
Type: Guided unpowered bomb.
Propulsion: None.
Dimensions: Length (I) 135in (3·44mm), (II) 159in (4·04m); body diameter (I) 12·5in (317mm); (II) 18in (457mm); span (I) 45·5in (1·16m), (II) 51in (1·3m).
Launch weight: (I) 1,100lb (499kg), (II) 2,400lb (1,089kg).
Performance: Speed subsonic; range (I) 16 miles (26km), (II) 35 miles (56km).
Warhead: (I) 825lb (374kg), (II) based on Mk 84 bomb.

Development: An unpowered glide bomb with TV guidance, AGM-62 Walleye was intended to overcome the aircraft-vulnerability hazard of visual radio-command ASMs, Walleye quickly proved successful, and in January 1966 Martin was awarded the first production contract. This was later multiplied and in November 1967 the need for Walleye in SE Asia resulted in Hughes Aircraft being brought in as second-source. In 1969 the Navy described this missile as "The most accurate and effective air-to-surface conventional weapon ever developed anywhere".

Walleye I has a cruciform of long-chord delta wings with elevons, a gyro stabilized TV vidicon camera in the nose, and ram-air windmill at the tail to drive the alternator and hydraulic pump. The pilot or operating crew-member identifies the target, if necessary using aircraft radar, aims the missile camera at it, focusses it and locks it to the target using a monitor screen in the cockpit. The aircraft can then release the missile and turn away from the target, though it must keep the radio link with the missile. In theory the missile should glide straight to the target, but the launch operator has the ability to break into the control loop and, watching his monitor screen, guide it manually into the target.

In 1968 the Navy funded several developments — Update Walleye, Walleye II, Fat Albert and Large-Scale Walleye among them — which led to the enlarged Walleye II (Mk 5 Mod 4) for use against larger targets. In production by 1974, Walleye II was deleted from the budget the following year and replaced by the first procurement of ER/DL (Extended Range/Data-Link) Walleye II (Mk 13 Mod 0). The ER/DL system was originally planned in 1969 to allow a launch-and-leave technique at greater distance from the target, the missile having larger wings to improve the glide ratio, and the radio data-link allowing the operator to release the missile towards the target and when the missile was much closer acquire the target on his monitor screen, focus the camera and lock it on.

Operations in SE Asia showed that it would be preferable to use two aircraft, the first to release the Walleye (if possible already locked on the approximate target position) and then escape and the second, possibly miles to one side, to update the lock-on point and monitor the approach to the target.

Land Weapons

The major trend in ground forces weapons is towards ever-increasing mechanization, with tracked vehicles gaining at the expense of wheels. The main battle tank (MBT) remains the symbol of army power, with all major armies now adopting guns of at least 120mm calibre, while weights (despite great efforts to reduce) have crept up to an average around the 50 ton mark. Some revolutionary designs are now starting to appear, but none has yet been committed to production.

Artillery is also now almost entirely mounted on tracked chassis, with towed guns relegated to "out-of-area" operations and Third World armies. Most armies now consider 155mm to be the minimum effective calibre for modern warfare, while mobile ballistic missile systems continue to proliferate. Great attention is also being paid to air defence, with tracked missile and gun systems entering service in some numbers.

Anti-tank defences are another major capability area, with the tank gun and missile (ATGW) getting equal attention. The new armour (eg, British "Chobham" armour) has significantly increased tank protection against the hollow-charge warhead used on every ATGW to date, and a new look at this area is being taken by the major armies.

A factor of major concern to field commanders is the mechanism for command, control and communications (the so-called C^3). As armies become more mobile, more complicated and armed with ever more sophisticated weapons and sensors, the whole question of C^3 becomes more difficult and yet more important. This is being offset by computer-based systems of great complexity and considerable expense. A further complication is that as electronic devices become so widespread and so critical to tactical success, so, too, is the importance of Electronic Warfare (EW) enhanced. Indeed, the great battlefield of the future may not be on the ground but in the invisible reaches of the electromagnetic spectrum.

Above: The machine gun remains a basic focus of infantry firepower.

MAIN BATTLE TANKS

AMX-30/-32/-40

Origin: France.
Type: MBT.
Crew: 4.
Armament: One 105mm gun; one 20mm cannon or one 12·7mm machine-gun coaxial with main armament; 1 7·62mm machine-gun on commander's cupola; 2 smoke dischargers on each side of turret.
Dimensions: Length (including main armament) 31·1ft (9·48m); length (hull) 21·62ft (6·59m); width 10·17ft (3·1m); height (including searchlight) 9·35ft (2·85m).
Combat weight: 81,569lb (37,000kg).
Engine: HS-110 12-cylinder water-cooled multi-fuel engine developing 720hp at 2,000rpm.
Performance: Speed 40mph (65km/h); range 280 miles (450km); vertical obstacle 3·05ft (0·93m); trench 9·5ft (2·9m); gradient 60 per cent.

Development: After the end of World War II France quickly developed three vehicles, the AMX-13 light tank, the Panhard EBR 8 x 8 heavy armoured car and the AMX-50 heavy tank. The last was a very interesting vehicle with a hull and suspension very similar to the German PzKpfw V Panther tank used in some numbers by the French Army in the immediate postwar

period. The AMX-50 had an oscillating turret, a feature that was also adopted for the AMX-13 tank. The first AMX-50s had a 90mm gun, this being followed by a 100mm and finally a 120mm weapon.

At one time it was intended to place the AMX-50 in production, but as large numbers of American M47s were available under the US Military Aid Program (MAP) the whole programme was cancelled. In 1956 France, Germany and Italy drew up their requirements for a new MBT for the 1960s. The basic idea was good: the French and Germans were each to design a tank to the same general specifications; these would then be evaluated together; and the best tank would then enter production in both countries, for use in all three. But like many international tank programmes which were to follow, this came to nothing: France placed her AMX-30 in production and Germany placed her Leopard 1 in production.

The AMX-30 is built at the *Atelier de Construction* at Roanne, which is a government establishment and the only major tank plant in France. The first production AMX-30s were completed in 1966 and entered service with the French Army the following year. The type has now replaced the American M47 in the French Army

and has also been exported to a number of countries. The hull of the AMX-30 is of cast and welded construction, whilst the turret is cast in one piece. The driver is seated at the front of the hull on the left, with the other 3 crew members in the turret. The commander and gunner are on the right of the turret with the loader on the left. The engine and transmission are at the rear of the hull, and can be removed as a complete unit in under one hour. Suspension is of the torsion-bar type and consists of 5 road wheels, with the drive sprocket at the rear and the idler at the front, and there are 5 track-return rollers. These support the inner part of the track.

The main armament of the AMX-30 is a 105mm gun of French design and manufacture, with an

Above: AMX-30 with its turret traversed. Note the main infra-red searchlight on the left of the turret front and another on the commander's cupola.

elevation of 20° and a depression of -8°, and a traverse of 360°, both elevation and traverse being powered.

A 12·7mm machine-gun or a 20mm cannon is mounted to the left of the main armament. This installation is unusual in that it can be elevated independently of the main armament to a maximum of 40°, enabling it to be used against slow flying aircraft and helicopters. There is a 7·62mm machine-gun mounted on the commander's cupola and this can be aimed and fired from within the turret. Two smoke dischargers

Challenger

Origin: UK.
Type: MBT.
Crew: 4.
Armament: One 120mm gun; one 7·62mm coaxial machine-gun; one 7·62mm AAMG; 8 smoke dischargers.
Dimensions: Length (with gun forwards) 37·89ft (11·55m); length (hull) 27·52ft (8·39m); width 11·55ft (3·52m); height 9·48ft (2·89m).
Combat weight: 132,275lb (60,000kg).
Engine: Rolls-Royce Condor 12V-1200 diesel; 12,000bhp at 2,300rpm.
Performance: Road speed 34·8mph (56km/h); vertical obstacle 2·95ft (0·9m); trench 9·84ft (3m); gradient 60 per cent.

Development: With the Chieftain in service the British Army started work on a successor in the mid-1960s. This turned into a bilateral programme with Germany from 1970 to 1977, but when this (like so many international efforts) failed, work continued on a purely national MBT-80 project. Meanwhile the Chieftain itself had been developed for export to Iran, emerging as the very advanced Shir 2. The British, therefore, faced with many other armies having Chobham-armour tanks in service before the country that invented it, decided to develop Shir 2 into an MBT suitable for British Army service, and use it to replace

approximately half the Chieftain fleet. The result — the Challenger — is a very fine tank, which proved itself during Exercise Lionheart in mid-1984.

Challenger is armed with the standard LIIA5 120mm rifled gun, but this will be replaced in the late 1980s by a new high-pressure 120mm rifled gun. Both guns fire the full range of current ammunition plus the new APFSDS round, of which 52 are carried. An MSDS Improved Fire Control System (IFCS) is fitted, which, in conjunction with the laser sights, gives a very high probability of a first round hit. Currently under development is the Thermal Observation and Gunnery System (TOGS), which will offer yet further improvements.

The power unit is the Rolls-Royce Condor 12V-1200 diesel, fitted with Garret-AiResearch turbo-superchargers. This was selected after very thorough consideration of the US AVCO-Lycoming AGT-1500 gas turbine, which is fitted to the M1 tank. The gas turbine is lighter than a diesel of comparative power, but its much greater fuel requirements more than offset this, and its demands on the logistics system are correspondingly greater.

The British Army now finds itself with an interesting dilemma. The aim with Challenger is to re-equip about half the existing Chieftain fleet, leaving the remainder to be

Above: The uncompromising appearance of Challenger is a consequence of using the highly effective Chobham armour, which has resulted in a slab-sided turret.

Right: 12th Hussars Challenger on exercise in Germany in September 1984, when it proved an impressive fighting vehicle.

replaced by a completely new tank sometime in the mid-1990s. But, will the pressure to replace the balance of the Chieftains, plus the need to keep the Royal Ordnance Factories' production lines busy, allow this, or will the Challenger end up as the complete replacement for Chieftain?

are mounted each side of the turret. Forty seven rounds of 105mm, 500 rounds of 20mm and 2,050 rounds of 7·62mm ammunition are carried. There are 5 types of ammunition available for the 105mm gun: HEAT, HE, Smoke, Illuminating and Practice. The HEAT round is the only anti-tank round carried. This weights 48·5lbs (22kg) complete, has a muzzle velocity of 3,281ft/s (1,000m/s) and will penetrate 14·17in (360mm) of armour at an angle of 0°.

Most other tanks carry at least 2, and often 3, different types of anti-tank ammunition, for example HESH, APDS and HEAT. The French HEAT round is of a different design to other HEAT rounds and the French claim that it is sufficient to deal with any type of tank it is likely to encounter on the battlefield. Other HEAT projectiles spin rapidly in flight as they are fired from a rifled tank gun, but the French HEAT round has its shaped charge mounted in ball bearings, so as the outer body of the projectile spins rapidly, the charge itself rotates much more slowly. In 1980 an APFSDS projectile entered production, and this can penetrate 1·96in (50mm) of armour at an incidence of 60° and a range of 5,470yds (5,000m).

The AMX-30 MBT is in service with France, Greece, Iraq, Lebanon, Qatar, Saudi Arabia, Spain, United Arab Emirates, Venezuela and Chile. Many of these armies also operate specialized versions of the basic vehicle (for example, ARV, AVLB,

SP howitzer, missile carriers).

Latest French Army version is AMX-30B2, which has a number of improvements, including an integrated fire-control system, laser rangefinder, LLTV system, and a new gearbox. 271 new production AMX-30B2s are now being delivered, and 730 of the 1,173 strong AMX-30 fleet will be brought up to the new standard.

France has also developed the AMX-32, with 120mm gun and laminated armour. This is intended for export but no orders have been forthcoming. Another new tank — the AMX-40 — was revealed in 1983. A totally new design, as opposed to a rehash of the AMX-30/32 series, it is also intended for export, but no orders have yet been announced. Meanwhile, the French Army is looking forward to the *Engin Principa de Combat* (EPC), a 50-tonne replacement for the AMX-30 and due to enter service in 1989.

Above left: the latest French MBT, the AMX-40, shows off its 120mm smooth-bore gun. Fire control system is the COTAC as used by the AMX-32, but otherwise this is a completely new design.

Left: LIke the AMX-40, the AMX-32 was designed specifically for export; based on the AMX-30B2, it features laser rangefinder, LLLTV system and a stabilised periscope sight for the commander.

Chieftain/Khalid

Origin: UK.
Type: MBT.
Crew: 4.
Armament: One 120mm L11 series gun; one 7·62mm machine-gun coaxial with main armament, one 7·62mm machine-gun in commander's cupola; one ·5in ranging machine-gun; 6 smoke dischargers on each side of turret.
Dimensions: Length (gun forward) 35·42ft (10·795m); length (hull) 24·66ft (7·518m); width overall (including searchlight) 11·99ft (3·657m); height overall 9·5ft (2·895m).
Weight: 121,250lb (55,000kg).
Engine: Leyland L.60 No 4 Mk 8A

12-cylinder multi-fuel engine developing 750bhp at 2,100rpm.
Performance: Road speed 30mph (48km/h); road range 280 miles (450km); vertical obstacle 3ft (0·914m); trench 10·33ft (3·149m); gradient 60 per cent.

Development: In the 1950s the British Army issued a requirement for a new tank to replace the Centurion tank then in service. The army

Below: British Army Chieftains wade ashore from a tank landing craft. This 1950s design is now being at least partially replaced by the Challenger.

required a tank with improved firepower, armour and mobility. The Chieftain was designed by the Fighting Vehicles Research and Development Establishment (now the Military Vehicles and Engineering Establishment), and the first prototype was completed in 1959.

The Chieftain (FV4201) was preceded by a tank known as the FV4202, however. This was designed by Leyland; and two of them were built and used to test a number of features later adopted for the Chieftain. The FV4202 used some Centurion automotive components. The Chieftain prototype was followed by a further six prototypes in 1961-2, and after more development work the Chieftain was accepted for army use in 1963. The Chieftain finally entered service with the British army only in 1967 as there were problems with the engine, transmission and suspension.

Total production for the British Army amounted to some 900 tanks. In 1971 the Iranian Army placed an order for 700 Chieftains, this order being followed by a further order for a new model called the Shir Iran. In 1976 Kuwait placed an order for about 130 Chieftains.

The Chieftain has a hull front of cast construction, with the rest of the hull of welded construction, and the turret is of all cast construction. The driver is seated in the front of the hull in the semi-reclined position, a feature which has enabled the overall height of the hull to be kept to a minimum. The commander and gunner are on the right of the turret, with the loader on the left. The commander's cupola can be traversed independently of the main turret by hand. The engine and transmission are at the rear of the hull. Suspension is of the Horstmann type, and consists of six road wheels, with the idler at the front and the drive sprocket at the rear, and there are three track-return rollers.

The main armament consists of a 120mm gun with an elevation of 20° and a depression of -10°, traverse being 360°. A GEC-Marconi stabilization system is fitted, enabling the gun to be fired while the tank is moving across country with a good chance of a first-round hit. A 7·62mm machine-gun is mounted coaxially with the main armament and there is a similar weapon in the commander's cupola, aimed and fired from within the cupola. When originally introduced, the gunner aimed the 120mm gun using a ·5in ranging machine gun, but this has now been removed from British Chieftains and the gunner now uses the Barr and Stroud laser rangefinder to obtain correct range to the target. A 6-barrelled smoke discharger is mounted on each side of the turret. Some 64 rounds of 120mm and 6,000 rounds of 7·62mm ammunition are carried (Chieftain Mk5 only).

The 120mm gun fires a variety of ammunition, of the separate-loading type, including High-Explosive Squash Head (HESH), Armour-Piercing Discarding Sabot (APDS), Smoke, Canister and Practice. The separate-loading ammunition (separate projectile and charge)

makes the job of loader a lot easier, and also enables the projectiles and charges to be stowed separately, which is considerably safer. When the HESH round hits the target, it is compressed on to the armour, so that, when the charge explodes, shock waves cause the inner surface of the armour to fracture and break up, pieces of the armour then flaking off and flying round the fighting compartment. The APDS round consists of a sub-calibre projectile with a sabot (a light, section "sleeve" that fits round the projectile and fills the bore of the gun) around it: when the round leaves the barrel of the gun, the sabot splits up and falls off, the projectile then travelling at a very high velocity until it strikes the target and pushes its way through the armour.

The Chieftain is fitted with a full range of night-vision equipment including an infra-red searchlight, mounted on the left side of the turret. An NBC pack is fitted in the rear of the turret. This takes in contaminated air, which is then passed through filters before it enters the fighting compartment as clean air.

The Chieftain can ford streams to a depth of 3·5ft (1·066m) without preparation. Deep fording kits have been developed but are not standard issue. The tank can be fitted with an hydraulically operated dozer blade if required. There are two special variants of the Chieftain, the FV4204 Armoured Recovery Vehicle and the FV4205 Bridgelayer. The latter was the first model to enter service and is built at the Royal Ordnance Factory at Leeds. This has a crew of three and weighs just over 53 tons (53,851kg). Two types of bridge can be fitted: the No 8 bridge to span ditches up to 74·79ft (22·8m) in width, and the No 9 bridge to span gaps of up to 40ft (12·2m). The bridgelayer takes three to five minutes to lay the bridge and 10 minutes to recover it. The Chieftain ARV has now replaced the Centurion ARV. The vehicle has a crew of four and a combat weight of 52 tons (52,835kg). Two winches are fitted, one with a capacity of 30 tons (30,482kg) and the other with a capacity of 3 tons (3,048kg). When the spade at the front of the vehicle is lowered, the main winch has a maximum capacity of 90 tons (91,445kg). Armament

Above: A British Army Chieftain shows its paces in a mobility demonstration on rough ground.

Below: A Chieftain with dozer blade installed kicks up the dust; top speed is 30mph (48km/h).

consists of a cupola-mounted 7·62mm machine-gun and smoke dischargers.

In 1974 Iran ordered 125 Shir 1s and 1,225 Shir 2s. The Shir 1, which was already in production at the time of the collapse of the Shah's regime, is basically the Chieftain Mk5/5(P) with a new powerpack consisting of a Rolls-Royce CV12 TCA 12 cylinder 1200hp diesel, David Brown TN37 transmission, and an Airscrew Howden cooling system. The Shir 2 has the same powerpack as the Shir 1 but in addition has Chobham armour. By the late 1978 six Shir 2s had been built.

In 1979 Jordan announced an order for 278 tanks to be delivered in the early 1980s. These are designated Khalid, and are based on

late production standard Chieftain, but with substantial changes to the fire control and automotive systems.

All current Chieftains and Khalids will be able to fire the new Armour-Piercing Fin Stabilized Discarding Sabot (APFSDS) round, which has just entered production. A new high pressure 120mm gun is also under development.

The Chieftain design has proved to be controversial, giving rise to many arguments. However, the main drawback in service has been the poor engine performance; that apart, its major users have been very satisfied, in particular those who use it in battle.

Below: Direct hit by a Chieftain of the 4th/7th Royal Dragoon Guards during a night exercise.

Leopard 1

Origin: Federal Republic of Germany.
Type: MBT.
Crew: 4.
Armament: One 105mm gun; one 7·62mm machine-gun coaxial with main armament; one 7·62mm machine-gun on roof; 4 smoke dischargers on each side of the turret.
Dimensions: Length (including main armament) 31·3ft (9·543m); length (hull) 23·26ft (7·09m); width 10·66ft (3·25m); height 9·1ft (2·764m).
Combat weight: 93,474lb (42,400kg).
Engine: MTU MB 838 Ca.M500 10-cylinder multi-fuel engine developing 830hp at 2,200rpm.
Performance: Road speed 40mph (65km/h); range 373 miles (600km); vertical obstacle 3·77ft (1·15m); trench 9·84ft (3m); gradient 60 per cent.

Development: Without doubt, the Leopard 1 MBT built by Germany has been one of the most successful tanks to be developed since World War II. In the 1950s it was hoped that Germany and France would produce a common tank, but like so many programmes of this type nothing came of it. Prototypes of a new German tank were built by two German consortiums, known as Group A and Group B. At an early stage, however, it was decided to drop the Group B series and continue only with that of Group A. In 1963 it was decided to place this tank in production and the production contract was awarded to Krauss-Maffei of Munich, and the first production Leopard 1 MBT was completed in September 1965.

The Leopard 1 tank has a crew of four with the driver in the front of the hull on the right and the other

three crew members in the turret. The engine and transmission are at the rear of the hull: the complete engine can be taken out in well under 30 minutes, which is a great advantage in battle conditions.

The main armament is the 105mm L7 series gun manufactured at the Royal Ordnance Factory in Nottingham, England. A 7·62mm MG3 machine-gun is mounted coaxially with the main armament and there is a similar machine-gun on the roof of the tank for anti-aircraft defence. Four smoke dischargers are mounted each side of the turret. Sixty rounds of 105mm and 5,500 rounds of machine-gun ammunition are carried. Standard equipment on the Leopard includes night-vision equipment, an NBC system and a crew heater. The vehicle can ford to a maximum depth of 7·38ft (2·25m) without preparation or 13·12ft (4m) with the aid of a schnorkel.

Since the Leopard entered service it has been constantly updated and the most recent modifications include a stabilization system for the main armament, thermal sleeve for the gun barrel, new tracks and passive rather than infra-red vision equipment for the driver and commander. Final production model for the German Army was the Leopard 1A4, which has a new all-welded turret of spaced armour and integrated fire control system.

The Leopard chassis has been the basis for a whole family of variants sharing many common components, some of them (eg the *Gepard*) being manufactured by Krauss-Maffei and others by the MaK company of Kiel. These variants include the Armoured Recovery Vehicle (ARV), the Armoured Engineer Vehicle (AEV), bridgelayer (Biber) and the *Gepard* anti-aircraft weapon system.

The production line for Leopard 1 was reopened recently to meet Greek and Turkish orders. When these are met, the deliveries of the MBT version will have been: Australia (90), Belgium (334), Canada (114), Denmark (120), West Germany (2,437), Greece (106), Italy (920), Netherlands (468), Norway (78), and Turkey (77). This has been a major success by any standard.

Top: German Army Leopard 1 with camouflaged turret during annual Reforger exercises.

Above: Belgian Leopard 1A3s on the range; the badge indicates a Canadian Army competition.

Below: Leopard 1A4 at speed. Mobility has always been the outstanding feature of this tank.

Leopard 2

Origin: Federal Republic of Germany.
Type: MBT.
Crew: 4.
Armament: One 120mm gun; one 7·62mm MG3 machine-gun coaxial with main armament; one 7·62mm MG anti-aircraft machine-gun; 8 smoke dischargers on each side of turret.
Dimensions: Length (including main armament) 31·72ft (9·668m); length (hull) 25·32ft (7·72m); width 12·23ft (3·73m); height 9ft (2·73m).
Combat weight: 121,275lbs (55,000kg).
Engine: MTU MB 873 Ka-501 12-cylinder water-cooled multi-fuel engine developing 1,500hp at 2,600rpm
Performance: Road speed 44·7mph (72km/h); range 341 miles (550km); vertical obstacle 3·93ft (1·2m); trench 9·84ft (3m); gradient 60 per cent.

Development: The Leopard 2 MBT has its origins in a project started in the 1960s. At this time the Germans and the Americans were still working on the MBT-70 programme, so this project had a very low priority. Once the MBT-70 was cancelled in January 1970, the Germans pushed ahead with the Leopard 2, and 17 prototypes were completed by 1974. These prototypes were built by the manufacturers of the Leopard 1, Krauss-Maffei, with the assistance

of many other German companies.

Without doubt, the Leopard 2 is one of the most advanced tanks in the world and the Germans have succeeded in designing a tank with high success in all three areas of tank design: mobility, firepower and armour protection. In the past most tanks have only been able to achieve two of these objectives at once. A good example is the British Chieftain, which has an excellent gun and good armour, but poor mobility; the French AMX-30 is at the other end of the scale and has good mobility, an adequate gun but rather thin armour.

The layout of the Leopard 2 is conventional, with the driver at the front, turret with commander, gunner and loader in the centre, and the engine and transmission at the rear. The engine was in fact originally developed for the MBT-70. The complete powerpack can be removed in about 15 minutes for repair or replacement. At first it was widely believed that the Leopard 2's armour was of the spaced type, but late in 1976 it was revealed that it used the British-developed Chobham armour.

Above right: Leopard 2 grew out of the joint US-German MBT-70 project initiated in the 1960s.

Right: In main armament, Leopard 2 followed the Soviet trend toward smooth-bore tank guns.

M1 Abrams

Origin: USA.
Type: MBT.
Crew: 4.
Armament: One 105mm M68 gun; one 7·62mm machine-gun coaxial with main armament; one 0·5in machine-gun on commander's cupola; one M240 7·62mm machine-gun on loader's hatch.
Dimensions: Length (gun forward) 32·0ft (9·766m); length (hull) 25·97ft (7·918m); width 11·99ft (3·655m); height 7·8ft (2·375m).
Combat weight: 120,000lb (54,432kg).
Engine: Avco Lycoming AGT-T 1500 HP-C turbine developing 1,500hp.
Performance: Road speed 45mph (72·4km/h); range 295 miles (475km); vertical obstacle 4·1ft (1·244m); trench 9ft (2·743m); gradient 60 per cent.

Development: In June 1973 contracts were awarded to both the Chrysler Corporation (which builds the M60 series) and the Detroit Diesel Allison Division of the General Motors Corporation (which built the MBT-70) to build prototypes of a new tank designated M1, and later named the Abrams tank. These tanks were handed over to the US Army for trials in February 1976. In November 1976 it was announced after a four-month delay that the Chrysler tanks would be placed in production.

Production, which commenced at

the Lima Army Tank Plant in Lima, Ohio, in 1979, with the first vehicles being completed the following year, is now also under way at the Detroit Arsenal Tank plant, which, like Lima, is now operated by the Land Systems Division of General Dynamics, which took over Chrysler Defense Incorporated in 1982. By late 1984 some 2,000 M1s had been built and the tank is now entering service at an increasingly rapid rate. The first units to field the M1 were the three armoured battalions of 3rd Infantry Division (Mechanized), who proudly gave the tank its European debut in "Exercise Reforger" in August 1982.

The US Army has a requirement for some 7,058 M1 by the end of Fiscal Year 1989. From 1985 it is expected that the 105mm M68 rifled tank gun will be replaced by the 120mm Rheinmetall smooth-bore gun, which is being produced under the designation XM256; this will fire both West German and American ammunition, although there have been more problems in adapting the turret to take the West German gun than had been anticipated.

The M1 has a hull and turret of the new British Chobham armour, which is claimed to make the tank immune to attack from all shaped-charge warheads and to give dramatically increased protection against other anti-tank rounds, including kinetic energy (i.e., APDS and APFSDS). It has a crew of four;

the driver at the front, the commander and gunner on the right of the turret, and the loader on the left.

The main armament consists of a standard 105mm gun developed in Britain and produced under licence in the United States, and a 7·62mm machine-gun is mounted coaxially with the main armament. A 0·5in machine-gun is mounted at the commander's station and a 7·62mm machine-gun at the loader's station. Ammunition supply consists of 55 rounds of 105mm, 1,000 rounds of 12·7mm and 11,400 rounds of 7·62mm. Mounted each side of the turret is a bank of 6 British-designed smoke dischargers. The main armament can be aimed and fired on the move. The gunner first selects the target, uses the laser rangefinder to get its range and then depresses the firing switch. The computer makes the calculations and adjustments re-

quired to ensure a hit.

The fuel tanks are separated from the crew compartment by armoured bulkheads and sliding doors are provided for the ammunition stowage areas. Blow-out panels in both ensure that an explosion is channelled outward. The suspension is of torsion-bar type with rotary shock absorbers. The tank can travel across country at a speed of 30mph (48km/h) and accelerate from 0 to 20mph (0 to 32km/h) in seven seconds, and this will make the M1 a difficult tank to engage on the battlefield.

The M1 is powered by a turbine developed by Avco Lycoming, running on a variety of fuels including

Above: After exhaustive testing the US Army went for gas turbine propulsion in its M1 Abrams, plus the old 105mm M68 gun.

This gives superior protection against attack from all known projectiles. It is of the laminate type, and consists of layers of steel and ceramics.

The suspension system is of the torsion-bar type with dampers. It has seven road wheels, with the drive sprocket at the rear and the idler at the front, and there are four track-return rollers.

The first prototypes were armed with a 105mm gun of the smooth-bore type, developed by Rheinmetall, but later prototypes had the 120mm smooth-bore gun. The 120mm gun fires two basic types of fin-stabilized ammunition (in which small fins unfold from the rear of the round just after it has left the barrel), and this means that the barrel does not need to be rifled. The anti-tank round is of the Armour-Piercing Discarding Sabot type, and has an effective range of well over 2,405yd (2,200m); at this range it will penetrate a standard NATO heavy tank target. The second round is also fin-stabilized and is designed for use against field fortifications and other battlefield targets. The cartridge case is semi-combustible and only the cartridge stub, which is made of conventional steel, remains after the round has been fired. The job of the loader is eased by the use of the hydraulically-assisted loading mechanism. The gun has an elevation of +20° and a depression of −9°. A standard 7·62mm MG3 machine-gun is mounted coaxially with the main armament. A 7·62mm

MG3 machine-gun is installed on the loader's hatch for use in the anti-aircraft role. 42 rounds of 120mm and 2,000 rounds of 7·62mm ammunition are carried. Eight smoke dischargers are mounted each side of the turret, although production vehicles may well have eight on each side.

A very advanced fire-control system is fitted, which includes a combined laser and stereoscopic rangefinder and the gun is fully stabilized, enabling it to be laid and fired on the move with a high probability of the round hitting the target. Standard equipment includes infra-red and passive night-vision equipment, an NBC system and heaters for both the driver's and fighting compartments. The

Leopard 2 can ford streams to a depth of 2·59ft (0·8m) without preparation, and with the aid of a schnorkel can deep ford to a depth of 13·12ft (4m).

In 1976 a modified version of the tank was delivered to the United States for trials, designated the Leopard 2 (AV), the letters standing for Austere Version. This had many modifications requested by the United States, but Chrysler won the contract for the M1 in November 1976.

The West German Army has ordered 1,800 Leopard 2 MBTs, of which 990 will be built by Krauss-Maffei and the remaining 810 by MaK of Kiel. First production tanks were delivered in October 1979 and production will continue until 1986.

Above: Smoke dischargers, periscopes, commander's sight and MG3 ring-mounted above the loader's hatch adorn the otherwise severely functional square-sided turret fitted to the Leopard 2.

In 1979 the Netherlands Army placed an order for 445 Leopard 2 MBTs for delivery between 1982 and 1986. The Leopard 2 has also been selected for the Swiss Army, but other sales to European armies seem unlikely as it is either too heavy, or too expensive, or because the armies have only just re-equipped with another new MBT (eg, Leopard 1).

Whatever happens, however, this highly efficient and effective MBT will serve a number of armies into the 21st Century.

---

petrol, diesel and jet fuel. All the driver has to do is adjust a dial in his compartment. According to the manufacturers, the engine will not require an overhaul until the tank has travelled between 12,000 to 18,000 miles (19,312 to 28,968km), a great advance over existing tank engines. This engine is coupled to an Allison X-1100 transmission with 4 forward and 2 reverse gears. Great emphasis has been placed on reliability and maintenance, and it is claimed that the complete engine can be removed for replacement in under 30 minutes. The M1 is provided with an NBC system and a full range of night-vision equipment for the commander, gunner and driver. One hazard has, however, become apparent since the tank entered service: the exhaust gases, ejected over the rear of the vehicle, can melt the paint of another vehicle too close behind. Consequently, any M1 on public roads needs another vehicle behind it to keep unsuspecting civilian traffic well away!

Those Europeans who criticize the Americans for failing to make the "two-way street" a reality need look no farther than the M1. This epitome of the US Army's might has British armour, main gun, and smoke dischargers, and a Belgian 7·62mm machine-gun, while later versions will convert to a West German main gun.

Right: Like Chieftain, the M1 Abrams has a slab-sided turret clad in compound armour: future versions will have a 120mm gun.

M48A5

Origin: USA.
Type: MBT.
Crew: 4.
Armament: One 105mm M68; one 0·3in M1919A4E1 machine-gun coaxial with the main armament (some have a 7·62mm M73 MG); one 0·5in machine-gun in commander's cupola.
Dimensions: Length (including main armament) 28·5ft (8·686m); length (hull) 22·57ft (6·882m); width 11·91ft (3·631m); height (including cupola) 10·25ft (3·124m).
Combat weight: 108,000lb (48,989kg).
Engine: Continental AVDS-1790-2A 12-cylinder air-cooled diesel developing 750hp at 2,400rpm.
Performance: Road speed 30mph (48km/h); range 288 miles (463km); vertical obstacle 3ft (0·915m); trench 8·5ft (2·59m); gradient 60 per cent.

Development: In October 1950 Detroit Arsenal started design work

on a new medium tank armed with a 90mm gun. This design study was completed two months later and in December 1950 Chrysler was given a contract to complete the design work and build six prototypes under the designation T48. Production started in 1952 and first deliveries were made to the US Army the following year.

The hull of the M48 is of cast armour construction, as is the turret. The driver is seated at the front of the hull with the other three crew members located in the turret, with the commander and gunner on the right and the loader on the left. The engine and transmission are at the rear of the hull, and are separated from the fighting compartment by a fireproof bulkhead. The suspension is of the torsion-bar type and consists of six road wheels, with the drive sprocket at the rear and idler at the front. Depending on the model, there are between three and five track-return rollers, and some models have a small track tensioning

wheel between the sixth road wheel and the drive sprocket.

The main armament consists of a 105mm gun with an elevation of +19° and a depression of −9°, traverse being 360°. A 0·3in M1919A4E1 machine-gun is mounted coaxially with the main ar-

Above: M48A2 in Turkish Army service have been modernised with improved guns and engines.

mament. The cupola can be traversed through 360°, and the machine-gun can be elevated from −10° to +60°.

M60A3

Origin: USA.
Type: MBT.
Crew: 4.
Armament: One 105mm gun; one 7·62mm machine-gun coaxial with main armament; one 0·5in anti-aircraft machine-gun in commander's cupola.
Dimensions: Length (gun forward) 30·95ft (9·436m); length (hull) 22·79ft (6·946m); width 11·91ft (3·631m); height 10·68ft (3·257m).
Combat weight: 114,600lb (51,982kg).
Engine: Continental AVDS-1790-2A 12-cylinder diesel developing 750bhp at 2,400rpm.
Performance: Road speed 30mph (48km/h); range 280 miles (450km); vertical obstacle 3ft (0·914m); trench 8·5ft (2·59m); gradient 60 per cent.

Development: In the 1950s the standard tank of the United States Army was the M48. In 1957 an M48 series tank was fitted with a new engine for trials purposes and this was followed by another three prototypes in 1958. Late in 1958 it was decided to arm the new tank with the British 105mm L7 series gun, to be built in the United States under the designation M68. In 1959 the first production order for the new

tank, now called the M60, was placed with Chrysler, and the type entered production at Detroit Tank Arsenal in late 1959, with the first production tanks being completed the following year.

From late in 1962, the M60 was replaced in production by the M60A1, which had a number of improvements, the most important being the redesigned turret. The M60A1 had a turret and hull of all-cast construction. The driver is seated at the front of the hull with the other three crew members in the turret, commander and gunner on the right and the loader on the left. The engine and transmission are at the rear, the latter having one reverse and two forward ranges. The M60 has torsion-bar suspension and six road wheels, with the idler at the front and the drive sprocket at the rear; there are four track-return rollers.

The 105mm gun has an elevation of +20° and a depression of −10° and traverse is 360°. A 7·62mm M73 machine-gun is mounted coaxially

Right: 105mm gun barrels on a row of M60A3s with turrets reversed.

Below: Gun stabilisation allows the M60A3 to fire on the move.

with the main armament and there is a 0·5in M85 machine-gun in the commander's cupola. Some 60 rounds of 105mm, 900 rounds of 0·5in and 5,950 rounds of 7·62mm ammunition are carried. Infra-red driving lights are fitted as standard and an infra-red/white light is mounted over the main armament. All M60s have an NBC system. The tank can also be fitted with a dozer blade on the front of the hull.

The M60A2 was a special model armed with a 152mm gun/launcher but has now been phased out of service. Current production model is

the M60A3 with numerous improvements including stabilization of main armament, top loading air cleaner fitted, passive searchlight over main armament, new tracks with removable pads, tube over bar suspension, RISE engine, thermal sleeve for main armament, laser rangefinder, passive night vision devices, new MAG 7·62mm MG, smoke dischargers each side of turret, muzzle reference system, engine smoke dischargers and improved personnel heater.

Most M60A1s of the US Army are now being brought up to this new

The M48 can be fitted with a dozer blade, if required, at the front of the hull. All M48s have infra-red driving lights and some an infra-red/white searchlight mounted over the main armament. The type can ford to a depth of 4ft (1·219m) without preparation or 8ft (2·438m) with the aid of a kit.

The first model to enter service was the M48, and this had a simple cupola for the commander, with the machine-gun mounted externally. The second model was the M48C, which was for training use only as it had a mild steel hull. The M48A1 was followed by the M48A2, which had many improvements including a fuel-injection system for the engine and larger capacity fuel tanks. The M48A2C was a slightly modified M48A2. The M48A3 was a significant improvement as this had a diesel engine, which increased the vehicle's operational range considerably, and a number of other modifications including a different fire-control system. Latest model is the M48A5, essentially an M48A1 or M48A2 with modifications including an M68 main 105mm gun, new

tracks, a 7·62mm M60D coaxial machine-gun and a similar weapon on the loader's hatch, plus many other detail modifications. One interesting modification is the fitting of an Israeli-developed low-profile cupola.

Earlier M48A1, M48A2C and M48A3 versions in the US inventory (some 1,809 tanks) have been updated to M48A5 standard and serve with Army National Guard and Reserve units. Some M48A1 and M48A2 chassis are being used for the Sergeant York (qv), pending availability of M48A5s.

Three flamethrower tanks were developed: the M67 (using the M48A1 chassis), the M67A1 (using the M48A2 chassis) and the M67A2 (using the M48A3 chassis). Also in service is an M48 Armoured Vehicle-Launched Bridge. This has a scissors bridge which can be laid over gaps up to 60ft (18·288m) in width.

The main armament consists of a 105mm gun with an elevation of $+19^0$ and a depression of -9^0, traverse being 360^0. A 0·3in M1919A4E1 machine-gun is mounted coaxially with the main arma-

ment. The cupola can be traversed through 360^0, and the machine-gun can be elevated from -10^0 to $+60^0$.

Above: US Army M48A5 on Exercise Team Spirit 84 in South Korea. The 105mm gun brings it up to virtual M60 standard.

standard, with the aim of an M60A3 fleet totalling 7,347 (1,686 from new production and 5,661 from conversion of M60A1s in Army depots). Of these, 3,786 will be the M60A3 TTS version, which has all the improvements listed above, plus a tank thermal sight.

Specialized versions of the M60 series include the M60 armoured vehicle launched bridge and the M728 Combat Engineer Vehicle which is fitted with a bulldozer blade, 152mm demolition gun and an A-frame for lifting obstacles which is pivoted at the front of the

hull. The basic vehicle can also be fitted with roller type mineclearing equipment or a dozer blade.

By 1984 just under 15,000 M60s had been produced and these serve with some 20 armies. Various updating kits are now being offered (eg, by Teledyne) to see the M60 through to the next century.

Right: The M60A2's 152mm gun/launcher can handle conventional ammunition or Shillelagh ATMs.

Below right: Excessive height is the M60's basic drawback.

Merkava Mk2

Origin: Israel.
Type: MBT.
Armament: One 105mm gun; one 7·62mm machine-gun coaxial with main armament; one 7·62mm anti-aircraft fire-control system.
Dimensions: Length (gun forward); 28·3ft (8·63m); length (hull) 24·4ft (7·45m); width 12·1ft (3·7m); height (commander's cupola) 8·75ft (2·66m).
Combat weight: 132,275lb (60,000kg).
Engine: Teledyne Continental AVDS-1790-6A V-12 diesel developing 900hp.
Performance: Road speed 28mph (45km/h); range 249-311 miles (400-500km); vertical obstacle 3·1ft (0·95m); trench 10·5ft (3·2m); gradient 60 per cent.

Development: Israel started to design an indigenous MBT in the late 1960s and, after many years of speculation, she announced in 1977 that she had indeed developed an MBT called the Merkava (Chariot), which would enter service the following year. The layout of the Merkava is unconventional with the engine and transmission at the front, the driver towards the front on the left and fighting compartment at the rear. The engine is an American Teledyne Continental AVDS-1790-6A, which is a more powerful version of that installed in the M60s used in some numbers by the Israeli Army, and this engine is coupled to an Allison CD-850-6B transmission. The suspension and road wheels are similar to those fitted to the Centurions used by the Israeli Army. There are six road wheels with the drive sprocket at the front, idler at the rear and return rollers, and the tops of the tracks are covered by steel covers to protect the suspension from damage from HEAT attack.

The turret has a very small cross-section and a well sloped front and is difficult to hit when the tank is in a hull down position. The commander and gunner are seated on the right and the loader on the left, with both the commander and loader provided with a hatch in the roof.

Main armament consists of the well tried British 105mm L7 series rifled tank gun, which is fitted with a fume extractor and a thermal sleeve. This gun is manufactured under licence in Israel and is also installed in the Israeli Centurion, M48 and M60 tanks. The gun has an elevation of 20° and a depression of $-8·5^O$; when in a non-combat area the gun is held in position by a travelling lock. In addition to firing all of the standard 105mm projectiles the gun fires the new APFSDS projectile developed by Israel Military Industries. The fire-control system of the Merkava has been developed by Elbit Computers Limited and incorporates a ballistic computer, sensors and a laser rangefinder. One 7·62mm machine-gun is mounted coaxial with the main armament and another is mounted on the roof for anti-aircraft defence. Standard equipment includes night vision devices, NBC system and a fire-suppression system.

The initial version (Mark 1) has

now been succeeded in production by the Mark 2, which has yet better armoured protection, stabilized commander's sight, and improved night-vision aids. A Mark 3 version is under development with "doubled" armour protection and improved suspension and powerpack.

**Above: Survival, of paramount importance to the Israeli armed forces, was the motive for the Merkava's unconventional layout.
Below: Not all Israeli tank operations are in desert conditions: Merkava in heavy mud on the Golan Heights.**

OF-40

Origin: Italy.
Type: MBT.
Crew: 4.
Armament: 105mm rifled gun; one 7·62mm machine-gun coaxial with main armament; one 7·62mm AAMG.
Dimensions: Length (gun forwards) 31·66ft (9·65m); length (hull) 22·6ft (6·89m); width 11·5ft (3·51m); height (to commander's sight) 8·79ft (2·68m).
Combat weight: 94,797lb (43,000kg).
Engine: MTU 10-cylinder 4-stroke; 830hp at 2,300rpm.
Performance: Road speed 40mph (65km/h); road range 373 miles (600km); vertical obstacle 3·6ft (1·1m); trench 9·8ft (3m); gradient 60 per cent.

Development: The Italian Army depended for many years upon the US M47 MBT. Italy was then a partner in the European tank consortium but was left high and dry when France decided to build the AMX-30 and Germany the Leopard 1. The Italians bought 300 M60A1 MBTs from the US to meet their immediate needs and then opted for the Leopard 1 as their definitive MBT. Two hundred of the M60A1s and 720 (out of 920) Leopard 1s have been built in Italy.

Oto Melara, Italy's leading partner in both M60A1 and Leopard 1 programmes, has now produced its own design — the OF-40 — intended for the export market. The only order to date is for 18 by the UAE.

The OF-40 is conventional in

design and bears a marked resemblance to the Leopard 1A4. It is armed with an Oto-Melara-designed 105mm gun.

Above: OF-40 Mk 2, with LLLTV camera mounted on the gun mantlet and stabilisation system for the 105mm rifled gun.

Stridsvagn (S) 103

Origin: Sweden.
Type: MBT.
Crew: 3.
Armament: One 105mm gun; one 7·62mm machine-gun on commander's cupola; 2 7·62mm machine-guns on hull top; 8 smoke dischargers.
Dimensions: Length (including armament) 29·2ft (8·9m); length (hull) 23ft (7m) width 11·15ft (3·4m); height (overall) 8·2ft (2·5m).
Combat weight: 85,980lb (39,000kg).
Engines: Rolls-Royce K.60 multi-fuel engine developing 240bhp at 3,650rpm; Boeing 553 gas turbine developing 490shp at 38,000rpm.
Performance: Maximum road speed 31mph (50km/h), water speed 4mph (6km/h); range 242 miles (390km); vertical obstacle 2·95ft (0·9m); trench 7·55ft (2·3m); gradient 60 per cent.

Development: Of all the tanks in service today, the "S" tank is perhaps the most unusual and controversial. Its design dates back to the 1950s and is based on an original idea by Sven Berge of the Swedish Army Ordnance department. The main battle tank of the Swedish Army in the 1960s was to have been a tank called the KPV, armed with a 150mm smooth-bore gun. Two prototypes of this tank were completed by Landsverk, but these were never fitted with their turrets and armament. These, and a number of other tanks including a Sherman and an Ikv-103 assault gun, were then used to test the basic S tank concept.

In 1958 Bofors was awarded a full development contract and the first two prototype S tanks were completed in 1961. These were powered by a gas turbine engine and an eight cylinder petrol engine. Apart from the 105mm gun, they had five 7·62mm machine-guns, one on the commander's cupola and two in a box on each side of the hull firing forwards. Their suspension was also different from later models. First production tanks were completed in 1966, and 300 were eventually built, the last of them being completed in 1971. The S tank (or to give it the correct name, the *Stridsvagn* 103), has a crew of 3 (commander, driver/gunner and radio operator).

The tank is armed with a 105mm rifled tank gun which is fixed to the hull rather than mounted in a turret as in conventional tanks. This has not only enabled the overall height of the tank to be reduced, but has also allowed an automatic loader to be installed. The 105mm gun is a longer version of the famous British L7 series gun and is made in Sweden. The gun is fed from a magazine which holds 50 rounds of ammunition of the following types:

Armour-Piercing Discarding Sabot, High-Explosive Squash-Head, Smoke and High Explosive. The empty cartridge cases are automatically ejected through a hatch in the rear of the hull. The tank can fire between 10 and 15 aimed rounds per minute. Some of the prototypes were fitted with a ·5in ranging machine-gun, but production models have an optical rangefinder, and a laser rangefinder has now been developed. Two 7·62mm machine-guns are mounted in a box on the left of the hull, firing forwards, and there is a single 7·62mm machine-gun on the commander's cupola. The latter can be aimed and fired from within the vehicle. Some 2,750 rounds of 7·62mm machine-gun ammunition are carried. Eight smoke dischargers are provided, and some S tanks have been fitted with Bofors Lyran flare launchers so that they can engage targets at night.

The suspension is of the hydro-pneumatic type, and consists of four road wheels (these are the same as those fitted to the Centurion tank), with the drive sprocket at the front and the idler at the rear, there being two track-return rollers. The gun is laid in elevation by the driver, who can adjust the suspension so that the gun can be elevated to $+12^\circ$ and depressed to -10°. It is aimed in traverse by slewing the tank in its tracks. When the gun is fired, the suspension is locked so as to provide a more stable firing position.

Another unusual feature of the tank is its powerpack, which is mounted in the forward part of the hull. This consists of two engines, a diesel and gas turbine. The diesel is the Rolls-Royce K.60, whilst the gas turbine is of American design but built in Belgium by FN. For normal operations the diesel is used, but in combat, or crossing very rough country, the gas turbine is also used.

The first production models of the S tank (these were designated Strv. 103As) were not fitted with flotation screens, but these are standard on the Strv. 103Bs, and all earlier tanks have now been refitted with them.

The screen is carried collapsed around the top of the hull and takes about 15 minutes to erect. The tank is propelled in the water by its tracks. There are many lakes and rivers in Sweden too deep for schnorkel crossing, so the only practical solution was the fitting of the flotation screen. The tank is provided with infra-red driving lights but does not have an infra-red searchlight. A dozer blade is mounted at the front of the hull for the preparation of fire positions.

The S tank has a very low silhouette compared with other main battle tanks, and its glacis plate is well sloped, giving the maximum amount of protection available. The S tank has been tested by a number of other countires, including Great Britain and the United States, but no

Above: S tanks on the move with infantry support. The type's main drawback is that it cannot fire while on the move. It has to be stopped and aimed at the target by being slewed in its tracks, requiring good crew coordination.

Top: The only really successful tank to adopt an unconventional layout, the S tank shows its remarkably low front profile.

other country has placed a similar design in production. There are no variants of the S tank, although components of the tank are used in the VK 155 self-propelled gun built by Bofors a few years ago, as well as the Bofors 40mm self-propelled anti-aircraft gun.

Right: Sweden's Stridsvagn 103, or S tank, was designed for the defensive role, with low visibility and good protection. A dozer blade enables it to dig itself in, and the gun is laid by adjusting the tank's suspension.

T-64/-72/-80

Origin: Soviet Union.
Type: MBT.
Crew: 3.
Armament: 125mm smooth-bore gun; one 7·62mm coaxial machine-gun; one 12·7mm (or 14·5mm) AAMG.
Dimensions: Length of hull 22·96ft (7m); width 11·48ft (3·5m); height to commander's cupola 7·54ft (2·3m).
Combat weight: 92,590lb (42,000kg).
Engine: Liquid-cooled diesel; 750bhp.
Performance: Maximum road speed 43·5mph (60km/h); range 280 miles (450km). (Data applies to T-80.)

Development: Considerable confusion was caused in the West over the correct designation of the MBTs which followed the T-62. This was eventually resolved and it is now quite clear that there are two distinct designs: the T-64 and the T-72. The latter is a progressive development of the T-64, with revised suspension and a slightly different turret. It is now accepted that T-72 is constructed of a new type of armour, similar in concept to that developed in the United Kingdom and known as "Chobham armour". This of course means that a large part of NATO's anti-tank weaponry could be negated, especially those projectiles and missiles equipped with a "hollow-charge" warhead.

The principal difference between T-72 and T-64 is that the newer tank has six large road-wheels, whereas T-64 has six rather small road-wheels which are quite unlike those on any other Soviet MBT. There may well also be internal differences between the two tanks, but this will not become apparent until examples of each become available for detailed examination by Western experts.

The T-72 is armed with a 125mm main gun, which is fitted with a fume extractor. The gun fires Armour-Piercing Fin-Stabilised Discarding Sabot (APFSDS), High Explosive (HE), or High Explosive Anti-Tank (HEAT) rounds and an Integrated Fire Control System (IFCS) is installed. The IFCS relieves both commander and gunner of some of their tasks as well as significantly increasing the probability of a first-round hit. An automatic loader is fitted and this, as with T-64, enables the crew to be reduced to three men.

This reduction in the number of crewmen is most significant as it has been strenuously resisted in Western armies, who do, of course, have a much more acute manpower problem than the Soviet Army. This means that the Soviets have been able to "save" 95 soldiers in every tank regiment, and this will have helped them to achieve the recent expansion of tank battalions in motor-rifle regiments from 31 to 40 MBTs.

T-72 was put into production in 1974 and entered service with the Soviet Army shortly afterwards. It is now in production in several state armament factories in the USSR, and is also being produced in Poland and Czechoslovakia. Current production is running at over 2,000 per year, which is sufficient to replace the entire tank fleets of both the British and French armies every year! All Soviet Army front-line divisions have now been re-equipped with this excellent MBT, and the other Warsaw Pact armies are in the process of putting it into service.

Not satisfied with this tank, the Soviet Army has had a new MBT under development since the early

Below: The T-64, predecessor of the T-72, has not been produced in such large numbers, and has been less prominently displayed.

Above: Snorkel-equipped T-72 after a submerged river crossing — a technique featured much in Soviet propaganda, but which has limited tactical application.

Above right: The machine gun highlights the T-72's compact dimensions, a characteristic of Soviet tanks made possible by limiting the height of crewmen.

1970s, known in the West as T-80. Current information is, not unnaturally, somewhat scanty, although it is now known that the new tank is generally similar in size and shape to T-72, but with a slightly longer hull. The gun is the same, and the T-80 has "special" armour, although whether this is the same as on T-72 is a matter for conjecture. The USSR has been a world leader in metallurgy for many years and there is no reason why it should lag behind the West in new armours.

One of the rounds available on the T-80 is believed to have a depleted uranium (DU) warhead. Over 1,000 of this new MBT are already in service in the Soviet Army bringing the total MBT force to over 52,000.

One of the most interesting features of Soviet tank design is the way in which the USSR seems to be able to produce MBTs which excite the envy of Western soldiers for a combat weight some 30 per cent less than MBTs they use. T-80 weighs some 41·3tons (42,000kg), while Leopard 2 is a massive 54 tons (55,000kg).

Right: T-80 on parade. Smooth-bore guns, used on Soviet tanks since the 1960s, have only recently been adopted in the West.

Below: World War II tactics and modern tank warfare do not mix, despite the superficial realism of this exercise with T-72s.

TAM

Origin: Argentina.
Type: MBT.
Crew: 4.
Armament: One 105mm gun; one 7·62mm machine gun coaxial with main armament; one 7·62mm anti-aircraft machine gun; 8 smoke discharges.
Dimensions: Length (with gun forwards) 27ft (8·3m); length (hull) 22·22ft (6·775m); width 10·66ft (3·25m); height (without AA MG) 7·94ft (2·42m).
Combat weight: 57,252lb (30,500kg).
Engine: MTU supercharged 6 cylinder diesel developing 710hp at 2,200 rpm.
Performance: Road speed 46mph (75km/h); road range 342 miles (550km); vertical obstacle 3·25ft (1m); trench 8·2ft (2·5m); gradient 65 per cent.

Development: The Argentinian Army has in the past obtained most of its equipment from the United States but recent American policy has led to a drastic curtailment in the supply of arms to many countries, especially those in South America. So in 1974 the Argentinian Army placed a contract with the West German company of Thyssen Henschel for the design and development of the TAM (Tanque Argentino Mediano) medium tank, and a contract was placed at the same time for the design and development of an infantry fighting vehicle to operate with the TAM, called the VCI (Véhiculo Combate Infanteria). Under the terms of the contract three prototypes of both the TAM and the VCI were to be supplied and a factory was to be established in Argentina to undertake production of both vehicles, which would initially be assembled from components supplied from West Germany but in time would be mostly manufactured in Argentina, not only providing

some employment but also saving valuable foreign exchange costs.

Both the TAM and the VCI are based to a large extent on the chassis of the Marder Mechanized Infantry Combat Vehicle, which entered service with the West German Army in 1971. The hull of the TAM is all of welded steel construction with the driver seated at the front of the well-sloped hull on the left with the engine to high right. The all-welded turret is mounted at the rear of the hull with the commander and gunner on the right and the loader on the left. The suspension system is of the torsion bar type and consists of six dual rubber tyred road wheels with the drive sprocket at the front, idler at the rear and three track return rollers. The first, second, fifth and sixth road wheel stations are provided with a hydraulic shock absorber. The basic model has a range on internal fuel tanks of some 342 miles (550km) but to increase the range to 559 miles (900km) two long range fuel tanks can be mounted at the rear of the hull. The basic vehicle can ford to a depth of 4·59ft (1·4m) without any

preparation but with a schnorkel fitted it can ford to a depth of 13·12ft (4m).

Main armament consists of a 105mm gun which can fire fixed APFSDS, HEAT, HE-T, HESH and WP-T rounds, with a total of 50 rounds being carried, and loaded into the TAM via a door in the rear of the hull or via a small circular door in the left side of the turret. A 7·62mm machine gun is mounted coaxial with the main armament and a similar weapon is mounted on the turret roof for anti-aircraft defence; four electrically operated smoke dischargers are fitted either side of the turret. The fire-control system consists of a panoramic sight for the

Above: The TAM is representative of a recent trend toward indigenous tank production.

commander which has a magnification of from x6 to x20, a coincidence rangefinder which is also operated by the commander while the gunner has a sight with a magnification of x8.

Well over 100 are now in service with the Argentine Army and at least one overseas sale (80 to Peru) has been agreed.

Below: Like the companion VCI, the TAM, with turreted 105mm gun, is based on the chassis of the German Marder MICV.

Type 69

Origin: China.
Type: MBT.
Crew: 4.
Armament: 105mm smooth-bore gun; one 7·62mm machine-gun coaxial with main armament; one 7·62mm machine-gun bow-mounted; one 12·7mm AAMG on commander's cupola.
Dimensions: Length (gun forwards) 29·5ft (9m); length (hull) 20·24ft (6·17m); width 10·7ft (3·27m); height (without AAMG) 8·49ft (2·59m).
Combat weight: approx. 83,774lb (38,000kg).
Engine: V-12 liquid-cooled diesel; 520hp at 2,000rpm.
Performance: Road speed 31mph (50km/h); road range 248 miles (400km); vertical obstacle 2·59ft (0·79m); trench 8·79ft (2·68m); gradient 60 per cent.

Development: As with so much of China's current military equipment, Chinese MBT development can be

traced back directly to a small quantity of Soviet-supplied T-54s received in the early 1950s. A Chinese copy was put into production as the Type 59, but with the Sino-Soviet split separate development commenced and models have been tested with British sights and sensors. Production of the Type 59 has ended and the tank serves in great numbers with the PLA and the armies of Albania, Congo, Kampuchea, North Korea, Pakistan, Sudan, Tanzania and Vietnam.

The Type 69 is a development of the Type 59, the main known differences being in armament, fire control and night vision devices. The main gun is a smooth-bore 105mm weapon, possibly developed from the Soviet 115mm U-5TS, at least one example of which was captured by the PLA in border skirmishes in the early 1970s. The Type 69 gun is fully stabilized and has a fume extractor. A laser rangefinder is fitted on the mantlet, and active infra-red

searchlights and gunsights are fitted. The tank is unique in retaining a bow-mounted machine-gun. Type 69 MBTs are being supplied to Iraq, and other export orders may be expected.

Above: The rounded turret form and five large road wheels indicate the Type 69's derivation from the Soviet T-54, though gun, fire control and night vision devices are all more recent.

Type 74

Origin: Japan.
Type: MBT.
Crew: 4.
Armament: One 105mm L7 series gun; one 7·62mm machine-gun coaxial with the main armament; one 12·7mm anti-aircraft machine-gun; 6 smoke dischargers.
Dimensions: Length (gun forward) 30·87ft (9·41m); length (hull) 22·47ft (6·85m); width 10·43ft (3·18m); height (with anti-aircraft machine-gun) 8·77ft (2·675m) at a ground clearance of 2·13ft (0·65m).
Combat weight: 83,776lb (38,000kg).
Engine: Mitsubishi 10ZF Model 22 WT 10-cylinder air-cooled diesel developing 750bhp at 2,200rpm.
Performance: Maximum road speed 33mph (53km/h); range 186 miles (300km); vertical obstacle 3·28ft (1m); trench 8·86ft (2·7m); gradient 60 per cent.

Development: The Japanese realized in the early 1960s that the Type 61 would not meet their requirements for the 1980s, so in 1962 design work commenced on a new main battle tank. The first two prototypes, known as STB-1s, were completed at the Maruko works of Mitsubishi Heavy Industries in late 1969. Further prototypes, the STB-3 and the STB-6, were built before the type was considered ready for production. The vehicle entered production at the new tank plant run by Mitsubishi Heavy Industries at Sagamihara in 1973.

The layout of the tank is conventional, with the driver at the front of the hull on the left and the other three crew members in the turret. The commander and gunner are on the right and the loader is on the left. The engine and transmission are at the rear of the hull. The suspension is of the hydro-pneumatic type and consists of five road wheels, with the drive sprocket at the rear and the idler at the front. There are no track-return rollers. The suspension can be adjusted by the driver to suit the type of ground being crossed and to alter gun elevation (see below).

The Type 74 is armed with the British 105mm L7 series rifled tank gun, built under licence in Japan. A 7·62mm machine-gun is mounted coaxially with the main armament. The main gun has an elevation of −6·5° and a depression of $9·5°$ and, using the hydropneumatic suspension, an elevation of $15°$ and a depression of −12·5° can be obtained. The fire-control system includes a laser rangefinder and a ballistic computer, both of which are produced in Japan. Some 51 rounds of 105mm ammunition are carried. Prototypes had an automatic loader, but this would have cost too much to install in production tanks. A ·5in M2 anti-aircraft machine-gun is mounted on the roof. On the prototypes this could be aimed and fired from within the turret, but this was also found to be too expensive for production vehicles. Three smoke dischargers are mounted on each side of the turret. The tank is provided with infra-red driving lights and there is also an infra-red searchlight to the left of the main armament. The Type 74 can ford to a maximum depth of 3·25ft (1m) without preparation, although a schnorkel enabling it to ford to a depth of 6·5ft (2m) can be fitted. All tanks are provided with an NBC system.

In designing the Type 74 MBT the Japanese have sought, and managed, to combine the best features of modern tank design within a weight limit of 37·6 tons (38,000kg). There is only one variant of the Type 74 at the present time, which is the Type 78 Armoured Recovery Vehicle; this

Above: The Type 74's suspension can be adjusted to give ground clearance of 8 to 25in (20-65cm).

is provided with a hydraulically operated crane, winch and a dozer blade at the front of the hull. In all, 850 Type 74s will be produced, before production switches to the new STC tank in 1988. This new MBT has a 120mm smooth-bore gun, sophisticated armour and a very high power/weight ratio.

Below: The 105mm gun has only limited depression, but this can be increased by the suspension.

UDES-19

Origin: Sweden.
Type: MBT.
Crew: 3.
Armament: 120mm smooth-bore gun; one 7·62mm coaxial machine-gun; one 7·62mm AAMG; 2 7·62mm remotely controlled bow-mounted machine-guns.
Dimensions: Not known.
Combat weight: 77,160-88,183lb (35,000-40,000kg).
Engine: Not known, but twith approximately 1,000bhp.

Development: The Swedish Army currently has 350 Centurion MBTs (all with 105mm gun) and 300 Stor 103Bs (S-tanks). This latter vehicle illustrated the Swedish designers' ability produce radical solutions to contemporary tactical problems, but there have been serious doubts as to its ability to perform well on the modern European battlefield. The British Army was sufficiently interested to lease a squadron's-worth of S-tanks in the early 1970s, and even produced a fixed-gun prototype of its own, but has not pursued the concept further.

Although the shortcomings of the S-tank have now been recognized — for example, its inability to fire on the move — the Swedish Army is now considering some equally radical solutions to its requirements for a successor to be fielded in the 1990s. Of the three known proposals, it is believed the favoured one is the UDES-19, in which the 120mm smooth-bore gun is mounted externally on a low hull with a three-man crew of commander, gunner and driver. The fixed, external magazine contains 34 rounds and an articulated arm raises the individual round to the gun, regardless of its position. The concept is being tested on a West German Marder MICV chassis, using an externally mounted 105mm L7 series gun.

As with the S-tank other armies will watch the Swedish experiments

with great interest. Many are agreed that the present MBT concept has reached its limits, but none is clear on just where to go next.

Above: This mounting of a 120mm gun on a Marder chassis is one of the configurations being evaluated in Sweden.

Vickers Mk3

Origin: UK.
Type MBT.
Crew: 4.
Armament: One 105mm gun; one 7·62mm machine-gun coaxial with main armament; one 7·62mm anti-aircraft machine-gun; one 12·7mm ranging machine-gun; 12 smoke dischargers.
Dimensions: Length (including main armament 32·11ft (9·788m); length (hull) 24·8ft (7·56m); width 10ft 5in (3·168m); height (to commander's cupola) 10·16ft (3·099m).
Combat weight: 85,098lb (38,600kg).
Engine: GM 12V-7IT turbo-supercharged diesel; 720bhp at 2,500rpm.
Performance: Road speed 31mph (50km/h); range 373 miles (600km); vertical obstacle 3ft (0·914m), trench 8ft (2·438m); gradient 60 per cent.

Development In the 1950s it was decided to set up a tank plant in India and teams were sent abroad to select a design which would meet the requirements of the Indian Army. The Vickers design was successful and in August 1961 a licens-

ing contract was signed. Two prototypes were completed in 1963, one being retained by Vickers and the other being sent to India in 1964. Meanwhile plans were being drawn up for a factory to be built near Madras. Vickers delivered some complete tanks to India before the first Indian tank was completed early in 1969.

These first tanks had many components from England, but over the years the Indian content of the tank has steadily increased and today the Indians build over 90 per cent of the tank themselves. Some 1,300 are in service and the tank gave a good account of itself in the last Indian-Pakistan conflict. The Indians call the tank *Vijayanta* (Victorious).

In designing the tank, Vickers sought to strike the best balance between armour, mobility and firepower within the limits of a tank weighing 38 tons (38,610kg). The layout of the tank is conventional. The driver is seated at the front of the hull on the right with ammunition stowage to his left, and the other three crew members are located in the turret: the commander and gunner to the right and the loader to the left. The engine and

transmission are at the rear of the hull. The engine and transmission are the same as those used in the Chieftain MBT. The suspension is of the torsion-bar type and consists of six road wheels with the drive sprocket at the rear and the idler at the front, there being three track-return rollers.

The Vickers MBT is armed with the standard 105mm L7 series rifled tank gun, this having an elevation of +20° and a depression of −7°, traverse being 360°. A 7·62mm machine-gun is mounted coaxially with the main armament and a

Above: The 105mm-gunned Mk 3, produced for the Kenyan Army, is also available in 155mm SP and twin 35mm AA configurations.

similar weapon is mounted on the commander's cupola. Six smoke dischargers are mounted each side of the turret. Some 44 rounds of 105mm and 3,000 rounds of 7·62mm machine-gun ammunition are carried.

The main armament is aimed with the aid of the ranging machine-gun method, which has been used so successfully in the Centurion tank

---

SELF-PROPELLED ARTILLERY

122mm M-1974

Origin: Soviet Union.
Type: Self-propelled howitzer.
Crew: 4.
Armament: One 122mm howitzer.
Dimensions: Length (hull) 23·95ft (7·3m); width 9·35ft (2·85m); height 7·87ft (2·4m).
Combat weight: 35,273lb (16,000kg).
Engine: YaMZ-238V V-8 water-cooled diesel; 240hp.
Performance: Road speed 37·3mph (60km/h); range 310 miles (500km); vertical obstacle 3·6ft (1·1m); gradient 60 per cent.

Development: The second of the new range of SPs to appear was the 122mm M-1974. This is also a straightforward combination of the 122mm D-30 (*qv*) elevating mass with a new turret, and mounted on a modified PT-76 chassis (there is an additional roadwheel, making seven in all).

The main armament consists of a modified version of the 122mm D-30 towed howitzer and is fitted with a double-baffle muzzle brake and a fume extractor. This fires an HE projectile weighing 48·06lb (21·8kg) to a maximum range of 16,738yd (15,300m). The weapon can also fire a spin-stabilized HEAT projectile which has a range of 1,094yd (1,000m) and can penetrate 18in (460mm) of armour at an incidence of 0 degrees.

The turret is small and very low compared with Western SPs and it appears highly likely that an automatic loader is fitted. Forty

rounds of ammunition are carried, normally a mixture of HE and anti-tank.

The M-1974 is fully amphibious, being propelled in the water by its tracks at a speed of 3mph (4·5km/h) and, unlike many other Soviet amphibious AFVs, a trim vane is not fitted. The vehicle is also fitted with an NBC system.

The M-1974 (Soviet designation — 2S1) is being issued in the Soviet Army as follows: artillery division, 36; motor rifle division, 36; tank division 72. It is also widely used in the non-Soviet Warsaw Pact armies.

Above: Mechanics at work on the engine of an M-1974 SP gun; top speed with the 240hp available is 37mph (60km/h) on roads.

Below: Combining the D-30 towed howitzer and the PT-76 chassis, the M-1974's very compact turret indicates an automatic loader.

with the 105mm gun. The gunner lines up with the target and fires a burst from the 12·7mm ranging machine-gun, and can follow the burst as the rounds are all tracer. If they hit the target, he knows that the gun is correctly aimed and he can then fire the main armament. Some 600 rounds of ranging machine-gun ammunition are carried.

Two types of main calibre ammunition are used: HESH (High Explosive Squash Head) and APDS (Armour-Piercing Discarding Sabot). A GEC-Marconi stabilization system is fitted, and this enables the gun to be aimed and fired whilst the vehicle is moving.

The model of the tank used by India and Kuwait is the Vickers MBT Mk. 1, which has the Leyland L60 engine. There was to have been a Mk.2, with 4 launchers for the British Aircraft Corporation Swingfire ATGW. Vickers Elswick facility is currently producing the Vickers Main Battle Tank Mk 3 for Kenya. This model has a redesigned turret with a cast front which gives increased ballistic protection, Barr and Stroud laser rangefinder, new commander's cupola which enables him to load, aim and fire his 7·62mm GPMG from within the turret, and a General Motors 12V-71T turbo-

charged diesel which develops 720bhp at 2500rpm. Optional equipment for the Mk 3 include passive night-vision equipment, deep wading and flotation equipment, full air filtration and pressurization, heater, fire-control computer, contra-rotating gear for the commander's cupola, fire-detection equipment and the replacement of the coaxial 7·62mm MG by a 12·7mm MG.

In all 76 Mk 3s have been produced for Kenya and 36 are currently in production for Nigeria. A further private venture MBT by Vickers —

Above: Mk 3 mobility demonstration. A high power-to-weight ratio translates into good agility.

the Valiant — which has Chobham armour and 120mm gun, has been developed and is ready for production, but with no orders so far.

152mm M-1973

Origin: Soviet Union.
Type: Self-propelled howitzer.
Crew 6.
Armament: One 152mm howitzer; one 7·62mm AAMG.
Dimensions: Length (gun forwards) 25·52ft (7·78m); length (hull) 23·43ft (7·14m); width 10·5ft (3·2m); height 8·92ft (2·72m).
Combat weight: 50,705lb (23,000kg).
Engine: V-12 diesel; 520hp.
Performance: Road speed 34·2mph (55km/h); range 186 miles (300km); verticle obstacle 3·6ft (1·1m); trench 9·19ft (2·8m); gradient 60 per cent.

Development: The Soviet Union continued to use exclusively wheeled artillery for many years after Western armies had begun the process of converting to self-propelled weapons. This was somewhat surprising in view of the Soviet Army's doctrinal emphasis upon rapid and flowing advance, for which towed artillery is less than ideal. Nor can the innate conservatism of Soviet artillerymen be blamed, since they have been so innovative in other fields. Whatever the reason, they are now making up for lost time and have produced four sturdy, effective and relatively uncomplicated self-propelled weapons, the 152mm M-1973, the 122mm M-1974 and, most recently, the 152mm M-1982 gun and the 203mm SP gun.

The first to appear was the M-1973, which was produced by taking the 152mm D-20 elevating mass, mounting it in a large turret, and utilizing an existing chassis (which appears to be identical to that used by the SA-4 Ganef). The only noticeable modification is that the gun tube is fitted with a fume-extractor to keep the turret clear of toxic gases. Unlike the majority of Soviet AFVs, the M-1973 is not amphibious.

The M-1973 is being issued to the army on a scale of 18 per division, and it is believed that all first-echelon tank divisions have now been complete re-equipped. In addition, at least some motor rifle divisions also have the M-1973.

Main armament consists of a 152mm gun/howitzer which fires an HE projectile weighing 96·131lb (43·6kg) to a maximum range of 26,256yd (24,000m), but unconfirmed reports speak of a rocket-assisted projectile with a range of 40,748yd (37,000m). In common with all other Soviet artillery weapons the M-1973 also has an anti-armour round; this weighs 107·6lb (48·8kg) which will penetrate 5in (130mm) of armour at 1,094yd (1,000m). A total of 40 rounds of ammunition are carried, and the normal maximum rate of fire is 4 rounds per minute; sustained rate is 2 rounds per minute. A nuclear shell has been developed for this gun with a 2KT warhead and this represents a significant increase in Soviet artillery capability.

The newer M-1982 152mm SP gun has a very long tube in an open mounting on a minelayer chassis.

Above right: M-1973 crossing a light bridge. Artillery support is vital in a tank offensive.

Right: The M-1973's 152mm gun mount features twin recuperators, muzzle brake and fume extractor.

152mm DANA

Origin: Czechoslovakia.
Type: Self-propelled gun.
Crew: 6.
Armament: One 152mm howitzer; one 12·7mm AAMG.
Dimensions: Length (gun forward) 34·12ft (10·4m); length (hull) 29·1ft (8·87m); width 9·74ft (2·97m); height 11·56ft (3·525m).
Combat weight: 50,705lb (23,000kg).
Engine: Tatra T3-930-51 V-12; 345hp at 2,200rpm.
Performance: Road speed 50mph (80km/h); range 621 miles (1,000km); trench 6·56ft (2m); gradient 60 per cent.

Development: Towed artillery weapons are cheap, light and, with their wheeled tractors, reliable and economical in their use of fuel, but little protection is provided for the gun crew (and none at all from NBC weapons) and the guns are difficult to move about the gun position. Tracked self-propelled (SP) weapons are complex, expensive, more difficult to maintain, and with much higher fuel consumption, but they do provide good protection, can move at will, and are much quicker in and out of action.

Modern technology has provided a possible midway solution, by mounting a turreted howitzer on a truck chassis; two models are currently in service (152mm DANA, Czechoslovakia; 155mm G-6, South Africa) but there may well be more in the future.

First seen in 1980, the DANA is mounted on the well-proven Tatra 815 chassis, with an armoured cab and a rear-mounted engine. The gun is the Soviet 152mm, in a large square turret which has a long open slot in its roof. An automatic loader is fitted and a crane on the turret roof raises palletized ammunition from resupply vehicles. Gun elevation is a very useful +60°/−3° but turret traverse is limited to the forward arc only. Three stabilizers are lowered to the ground when the gun is in action.

The DANA is an imaginative and effective design, as is to be expected from the Czech armaments industry. Its mobility will be quite adequate to support infantry divisions in terrain with reasonable roads and firm cross-country going.

Above: The DANA's gun turret has a separate fighting compartment each side of the mounting, and an automatic loader.

Below: The fundamental advantage of wheeled SP mountings is their low cost compared with equivalent tracked vehicles.

155mm Palmaria

Origin: Italy.
Type: Self-propelled howitzer.
Crew: 5.
Armament: One 155mm howitzer; one 7·62mm AAMG.
Dimensions: Length (gun forward) 37·64ft (11·474m); length (hull) 24·3ft (7·4m); width 7·71ft (2·35m); height 9·43ft (2·874m).
Combat weight: 101,411lb (46,000kg).
Engine: 4-stroke multi-fuel engine; 750hp.
Performance: Road speed 37·3mph (60km/h); range 250 miles (400km); gradient 60 per cent.

Development: With the switch to self-propelled tracked artillery, the European armies have tended to adopt complex and expensive designs: eg, SP-70, GCT, M110. Seeing the need for a simpler, cheaper but still effective SP for the more sophisticated Third World armies, OTO Melara have developed the 155mm Palmaria.

This weapon system utilizes the chassis of the OF-40 MBT (qv) (but with a different engine), with a large aluminium turret housing the OTO Melara-designed 155mm gun and automatic loader. Design was started in 1977, the prototype was running in 1981 and production started in 1982, with first service deliveries that year — a record which puts the 18-year development cycle of the SP-70 (qv) to shame.

The standard HE round has a range of 26,246yd (24,000m) and a Rocket Assisted Projectile (RAP) can reach out to 32,800yd (30,000m). Libya has ordered 200 of this well-designed weapon, and Nigeria 25.

Below: The Palmaria turret has full 360° traverse and maximum elevation of +70° through −4°.

155mm SP-70

Origin: International.
Type: Self-propelled howitzer.
Crew: 5.
Armament: One 155mm howitzer; one 7·62mm machine-gun.
Dimensions: Length (with gun forward) 33·59ft (10·24m); length (hull) 25·06ft (7·64m).
Combat weight: 95,952lb (43,524kg).
Engine: MTU MB-871 8-cylinder turbocharged diesel; 1,500hp at 2,600rpm.
Performance: Classified.

Development: In the 1960s, three NATO countries, Britain, West Germany and the United States all agreed that they required a new 155mm towed howitzer to replace weapons dating back to World War II. Eventually the United States went on to develop a towed howitzer under the designation of the XM198. Britain and West Germany went ahead and developed the 155mm FH-70, which, unlike the American M198, has an auxiliary power unit which enables it to propel itself around the battery position. For the FH-70 project Britain was the project leader and in 1970 Italy joined the project as an equal partner. There are three production lines for the FH-70, one in each country, with components being supplied by one country to the other two: Britain builds the carriage, West Germany the ordnance and Italy the cradle. First production FH-70s were delivered in 1978.

In 1973 development of the self-propelled version of the FH-70 commenced under the designation of the SP-70 with West Germany the project leader. A total of 12 prototypes of the SP-70 are being built: West Germany is responsible for the ordnance (Rheinmetall), powerback (MTU) and chassis (MaK), Italy responsible for the cradle, recoil system, elevating and balancing equipment (OTO-Melara), and Bri-

tain responsible for the turret, ammunition-handling system and the sighting system (Royal Armament Research and Development Establishment designed the turret and the prototypes were completed by the Royal Ordnance Factory at Leeds).

Trials with the prototypes are expected to continue until the mid-1980s and it is not expected that the SP-70 will enter service until the late 1980s at the earliest. The British Army will replace its 105mm Abbot and 155mm M109s with the SP-70, while in the West German and Italian armies it will replace M109s. Main improvements over the M109 will be a much increased range and a high rate of fire, necessary against fast-moving enemy tanks.

Prototypes of the SP-70 are based on automotive and suspension components of the Leopard 1 MBT, which is in widespread use by NATO forces, including West Germany and Italy. The hull of the SP-70 is of all-welded aluminium construction with the driver's compartment in the front, turret in the centre and the engine and transmission at the rear. The suspension system is of the torsion bar type and consists of seven dual rubber tyred road wheels with the idler at the front, drive sprocket at the rear and four return rollers.

The 155mm weapon is mounted in the forward part of the turret and has a large double-baffle muzzle brake and a fume extractor, and the balancing cylinders either side of the ordnance are housed in armoured housings. Turret traverse and gun elevation are powered and manual controls are provided for emergency use. To enable a high rate of fire to be achieved SP-70 is fitted with an automatic loading system; no details of the rate of fire have been released but it is probable that between six and eight rounds a minute can be fired. Once the ammunition supply has been expended, SP-70 would

move to a predetermined position for ammunition resupply, with ammunition loaded through two doors in the rear of the turret.

The ammunition system of the SP-70 is identical to that of the FH-70 and consists of three projectiles and a charge system. The three projectiles weigh 94·8lb (43·5kg) each and are HE, base ejection smoke (DM105) and illuminating (DM106); they can be fired to a maximum range of 21,880yd (20,000m). In addition all standard NATO 155mm projectiles can be fired, including the recent Martin Marietta Copperhead Cannon Launched Guided Projectile. Under development is a Rocket Assisted Projectile which will have a maximum range of 32,760yd (30,000m). The charge system has eight zones and is divid-

Above: The SP-70 is a long-running programme with some way to go: this early prototype is shown with the turret reversed.

ed into three separate cartridges, zones 1 – 2, 3 – 7, and 8.

This programme is undoubtedly being successful and the final result will be an excellent – if heavy and expensive weapon system. On the other hand, the development period has been extremely lengthy, as is the norm with international partners, with some years to go yet before the first SP-70 reaches a field unit.

Below: Current form of the SP-70, a collaborative programme involving a German gun and Italian mechanism mounted in a British turret on a German chassis.

AMX-GCT

Origin: France.
Type: Self-propelled howitzer.
Crew: 4.
Armament: One 155mm gun; one 7·62mm anti-aircraft machine-gun; 4 smoke dischargers.
Dimensions: Length (with gun forward) 33·46ft (10·2m); length (hull) 21·98ft (6·7m); width 10·33ft (3·15m); height (without anti-aircraft MG) 10·83ft (3·3m).
Combat weight: 92,610lb (42,000kg).
Engine: Hispano-Suiza HS-110 12-cylinder multi-fuel engine developing 720hp at 2,400rpm.
Performance: Road speed 37mph (60km/h); range 280 miles (450km); vertical obstacle 3·05ft (0·93m); trench 6·23ft (1·9m); gradient 60 per cent.

Development: At the present time the standard self-propelled artillery of the French Army consists of 105mm and 155mm weapons on modified AMX-13 type chassis. It was decided some years ago that both weapons would be replaced by a new 155mm weapon since the current weapon of this calibre, the Mk.F3, has a number of drawbacks: the gun cannot be traversed through a full 360°, the gun is on an open mount with no protection for the crew against small arms fire and NBC attack, and it has to be supported in action by a modified AMX armoured personnel carrier for the rest of the crew and the ammunition for the gun. The four main requirements laid down by the French Army were: mobility similar to that of a main battle tank, ability to engage targets quickly through a full 360° at all ranges, high rate of fire with effective ammunition, and full protection for the crew from both NBC attack and small arms fire.

The first prototype of the GCT (*Grande Cadence de Tir*) was completed in 1973; further models followed two years later, and the type is now in production for the French Army. The GCT consists of a slightly modified AMX-30 main battle tank chassis with a new turret of all-welded steel construction. The crew of four consists of the commander, driver and two gunners (one of the gunners is in charge of the fire-control system and elevation and traverse of the main armament, while the other prepares the charges and controls the loading of the gun).

The main armament consists of a 155mm gun with a double baffle muzzle-brake, capable of an elevation of +66° and a depression of −4°, traverse being a full 360°. Elevation and traverse are hydraulic, with manual controls in case of hydraulic failure. The gun is fully automatic and can fire eight rounds in one minute. A total of 42 projectiles and their separate bagged charges is carried in the rear of the turret, arranged in seven racks of six for both projectiles and bags. The propelling charges are contained in combustible cases so that the crew does not have to worry about empty cases littering the floor of the turret. A typical ammunition load would

consist of 36 High Explosive and six Smoke rounds. Large doors are provided in the rear of the turret for reloading purposes, and it takes 3 men about 30 minutes to reload the ammunition. Types of ammunition fired included High Explosive. Smoke and Illuminating, of both French and American manufacture. The HE round has a maximum range of 26,256yd (24,000m), although a rocket-assisted round with a range of 34,461yd (31,500m), is now being developed.

A 7·62mm anti-aircraft machine-gun is mounted on top of the turret, with traverse through a full 360° and elevation limits from −20° to +50°. Some 2,000 rounds of 7·62mm ammunition are carried. In addition there are two smoke dischargers on each side of the turret.

The GCT is provided with an NBC system, and night-vision equipment can be fitted if required. The vehicle can ford to a depth of 6·83ft (2·1m) without preparation. As the crew may well have to remain in the vehicle for up to 24 hours at a time, a bunk has been installed in the turret to allow one member of the crew to rest.

While the introduction of the GCT has increased the effectiveness of the French artillery arm, it is considered by some to be too expensive and too heavy when compared with other self-propelled guns such as the American M109. On the credit side, the ability to fire a large number of rounds in a short space of time is of vital importance on the battlefield of the 1980s. This is because once an SP gun has fired one round, enemy

Below: The ammunition compartment of the GCT accommodates 42 rounds plus separate charges ready for automatic loading.

Above: On its modified AMX-30 chassis the GCT can fire eight rounds and be on its way before counter-battery fire arrives.

gun-locating radars will start to pinpoint its exact position, and within a few minutes the enemy will be returning fire. The role of the GCT will be to fire a burst of 8 rounds and then move to a new firing position before the enemy counter-fire arrives.

The French Army has 190 GCTs on order and deployment is well under way. Some 51 have been ordered by Saudi Arabia, and a further 85 for Iraq.

ASU-85

Origin: Soviet Union.
Type: Airborne fire support vehicle/tank destroyer.
Crew: 4.
Armament: Improved SD-44 85mm gun; one 7·62mm MG coaxial with main armament; one pintle-mounted 12·7mm AAMG.
Dimensions: Length (including gun) 27·85ft (8·49m); length (hull) 19·7ft (6m); width 9·19ft (2·8m); height (hull roof) 6·89ft (2·1m).
Combat weight: 34,171lb (15,500kg).
Engine: Model V-6 6-cylinder in-line water-cooled diesel; 280hp at 1,800rpm.
Performance: Road speed 28mph (45km/h); range 161 miles (260km); vertical obstacle 3·6ft (1·1m); gradient 70 per cent.

Development: One of the major requirements for an airborne force is to have its own fire support available

"on-the-spot" and as soon as possible after the initial landings. The Soviet Army has paid particular attention to this requirement, producing a range of small, light and yet highly effective fire support vehicles such as the BMD, and the ASU-85.

Much heavier and tougher than the earlier ASU-57, this formidable vehicle became possible with the advent of the Mi-6 Hook and Mi-10 Harke helicopters, and — for fixed-wing dropping — high-capacity, multi-chute and retrorocket systems. The ASU-85 was first seen in 1962 and is widely used by the Soviet, Polish and East German airborne divisions. The chassis is based on the ubiquitous PT-76, but is not amphibious. The gun has 12 degree traverse and fires up to 4 rounds per minute. A total of 40 rounds is carried, including HE, APHE and HVAP, giving the vehicle a capability against area targets, people, and

armoured vehicles. It is believed that the ASU-85 is fitted with an NBC system. It also has various night-vision devices, although these are still of the active infra-red type; other target-acquisition and ranging aids may well have been retrofitted.

Another weapon developed specifically for the parachute forces is the RPU-14 140mm multiple rocket-launcher. The 16-tube

Above: ASU-85s of the Chernigov Red Banner Division move off in line after debarking from their An-12 "Cub" transports.

weapon is used exclusively by the Airborne Troops, with 18 held in the artillery battalion of the airborne division. First seen in 1967, the launcher uses the same chassis as the M-1943 57mm anti-tank gun.

Infanterikanonvagn 91

Origin: Sweden.
Type: Tank destroyer.
Crew: 4.
Armament: One 90mm gun; one 7·62mm machine-gun coaxial with main armament; one 7·62mm anti-aircraft machine-gun; 12 smoke dischargers.
Dimensions: Length (with gun forwards) 29ft (8·845m); length (hull) 20·14ft (6·14m); width 9·83ft (3m); height 7·73ft (2·355m).
Combat weight: 35,941lb (15,300kg).
Engine: Volvo-Penta TD 120 A 6-cylinder turbo-charged diesel developing 350hp at 2200rpm.
Performance: Road speed 43mph (69km/h); range 342 miles (550km); vertical obstacle 2·62ft (0·8m); trench 9·19ft (2·8m); gradient 50 per cent.

Development: In 1968 Hägglund and Söner was awarded a develop-

ment contract for a new vehicle to replace the Strv 74 light tank, Ikv-102 and Ikv-103 infantry cannons and the Pansarvarnskanonvagn m/63 self-propelled gun. The first of three prototypes of this vehicle, called the Infanterikanonvagn 91 (or Ikv-91 for short), was completed in 1969; pre-production vehicles were ready in 1974 and the first production model was completed late in 1975.

As in previous desijgns, Hägglund has used standard automotive components wherever possible in the vehicle, which have not only enabled costs to be kept to a minimum but also meant that spare parts can be obtained from normal commercial sources. Some of the components of the Ikv-91 are the same as those used in earlier vehicles developed by the company.

The hull of the Ikv-91 is all of welded steel construction and is

divided up into three compartments, driver's at the front, fighting compartment in the centre and engine compartment at the rear. Main armamanet of Ikv-91 is a Bofors-designed low pressure gun which fires fin-stabilized high explosive and high explosive anti-tank rounds; a total of 59 rounds of ammunition are carried for the main armament. The main armament has an elevation of $+15^\circ$ and a depression of -10° and the turret can be traversed through a full 360°; gun elevation and turret traverse are powered with manual controls for emergency use. The gunner's optical sight incorporates a laser rangefinder to give a high probability of a first-round hit. A 7·62mm machine-gun is mounted coaxially with the main armament and a similar weapon is mounted at the loader's station for use in the anti-aircraft role. Six electrically operated smoke dischargers are mounted either side of the turret.

As there are many lakes in

Sweden, it was necessary that the Ikv-91 should be fully amphibious. Before entering the water a trim vane is erected at the front of the vehicle (this folds back on to the glacis plate when not in use) and low screens are raised around the air inlets and the exhaust and air outlets and four bilge pumps switched on. When afloat the Ikv-91 is propelled by its tracks at a maximum speed of 4·34 miles (7km/h). The vehicle is provided with an NBC system but has no night-vision equipment at the present time.

The Swedish Army has recently ordered a TOW-armed ATGW carrier, based on the chassis of the obsolete Ikv-103 infantry SP cannon. In addition, a revolutionary type of articulated 120mm gun armed tank destroyer is under development.

Below: The Ikv-91, designed to operate with anti-tank units in almost any terrain, has good cross-country performance and full amphibious capability.

Jagdpanzer Kanone/Rakete

Origin: Federal Republic of Germany.
Type: Tank destroyer.
Crew: 4.
Armament: One 90mm gun, one 7·62mm MG3 machine-gun coaxial with main armament; one 7·62mm MG3 anti-aircraft machine-gun, 8 smoke dischargers.
Dimensions: Length (including armament) 28·7ft (8·75m); length (hull) 20·46ft (6·238m); width 9·78ft (2·98m); height (without anti-aircraft machine-gun) 6·84ft (2·085m).
Combat weight: 60,627lb (27,500kg).
Engine: Daimler-Benz Model MB 837 Aa 8-cylinder water-cooled diesel developing 500hp at 2,200rpm.
Performance: Road speed 43·5mph (70km/h); range 249 miles (400km); vertical obstacle 2·46ft (0·75m); trench 6·56ft (2m); gradient 60 per cent.

Development: The Jagdpanzer Kanone (Jpz.4-5 for short) is a member of a range of vehicles developed for the German Army from the late 1950s, the other two members of the family which reached production being the Jagdpanzer Rakete and the Marder MICV. Production began in 1965: 375 were built by Hensche and a similar number by Hanomag, production being completed in 1967.

The primary role of the Jagdpanzer Kanone is to hunt and destroy enemy tanks. It relies on its low silhouette and speed for its survival — it has a very high road and cross-country speed, and can be driven at the same speed backwards and forwards. The hull of the vehicle is of all-welded steel construction

Right: Reminiscent of World War II tank killers, the Jpz.4-5 has inadequate armament for modern anti-tank requirements.

Below: Like most of the 370 built, this Jadpanzer Rakete has been given a HOT missile in place of the original SS-11.

with the maximum armour thickness of 2in (50mm) being concentrated at the front. The fighting compartment is at the front of the hull, with the engine and transmission at the rear. The suspension is of the torsion-bar type, and consists of five road wheels with the idler at the front and drive sprocket at the rear. There are three track-return rollers on each side. The crew of four consists of the commander, gunner, loader and driver.

The 90mm gun is mounted in the front of the hull and is slightly offset to the right. It has a maximum effective range of 2,187yd (2,000m), and a maximum rate of fire of 12 rounds per minute can be achieved. A total of 51 rounds of 90mm and 4,000 rounds of 7·62mm ammunition is carried. An infra-red searchlight is

mounted over the main armament, and this moves in elevation and traverse with the gun.

The Jpz.4-5 is fitted with an NBC system, and can ford streams to a depth of 4·58ft (1·4m) without preparation. A wading kit is also available. This can be fitted quickly, and allows the vehicle to ford to a depth of 6·92ft (2·1m). The Belgian Army has 80 Jpz.4-5s of a slightly different design, these being assembled in Belgium from components supplied by Germany.

The Jagdpanzer Rakete has an almost identical hull to the Jagdpanzer Kanone and has been designed to operate with the latter vehicle in order to give long-range anti-tank support. It has two launchers for French SS-11 anti-tank missiles, a total of 14 missiles (minimum range 547yd or 500m and maximum range of 3,280yd or 3,000) being carried. A total of 370

was built for the German Army between 1967 and 1968. In addition this model has a bow-mounted machine-gun and eight smoke dischargers.

Out of 370 Jagdpazner Raketen, 317 have been refitted with the Euromissile HOT (High-subsonic Optically-guided Tube-launched) missile system. This missile has a number of advantages over the SS-11 missile, including a minimum range of 82yd (75m), a maximum range of 4,374yd (4,000m) and simpler loading procedures. It is also much more accurate, and the aimer merely has to keep the target in his sight, which has a magnification of x7, in order to achieve a hit.

In addition, 162 Jagdpanzer Kanone have been converted to take the US TOW ATGW between 1983 and 1985 for the West German Army. This may well be the end of the SP anti-tank gun concept, which has survived from World War II.

M109A2/A3

Origin: USA.
Type: Self-propelled howitzer.
Crew: 6.
Armament: One 155mm howitzer; one ·5in (12·7mm) Browning anti-aircraft machine-gun.
Dimensions: Length (including armament) 29·92ft (9·12m); length (hull) 20·3ft (6·19m); width 10·83ft (3·295m); height (including anti-aircraft machine-gun) 10·76ft (3·28m).
Combat weight: 55,000lb (24,948kg).
Ground pressure: 10·95lb/in^2 (0·77kg/cm^2).
Engine: Detroit Diesel Model 8V71T 8-cylinder turbocharged diesel developing 405bhp at 2,300rpm.
Performance: Road speed 35mph (56km/h); range 217 miles (349km); vertical obstacle 1·75ft (0·533m); trench 6ft (1·828m); gradient 60 per cent.

Development: The first production models of the M109 were completed in 1962, and some 3,700 examples have now been built (of which about 1,800 are in US Army service), making the M109 the most widely used self-propelled howitzer in the world.

It has a hull of all-welded aluminium construction, providing the crew with protection from small arms fire. The driver is seated at the front of the hull on the left, with the engine to his right. The other five crew members are the commander, gunner and three ammunition members, all located in the turret at the rear of the hull. There is a large door in the rear of the hull for ammunition resupply purposes. Hatches are also provided in the sides and rear of the turret. There are two hatches in the roof of the turret, the commander's hatch being on the right. A 0·5in (12·7mm) Browning machine-gun is mounted on this for anti-aircraft defence. The suspension is of the torsion-bar type and consists of seven road wheels, with the drive sprockets at the front and the idler at the rear, and there are no track-return rollers.

The 155mm howitzer has an elevation of +75° and a depression of −3°, and the turret can be traversed through 360°. Elevation and traverse are powered, with manual controls for emergency use. The weapon can fire a variety of ammunition, including HE, tactical nuclear, illuminating, smoke and chemical rounds. Rate of fire is four rounds per minute for three minutes, followed by one round per minute for the next hour. A total of 28 rounds of separate-loading ammunition is carried, as well as 500 rounds of machine-gun ammunition.

The second model to enter service was the M109A1, identical with the M109 apart from a much longer barrel, provided with a fume extractor as well as a muzzle-brake. The fume extractor removes propellant gases from the barrel after a round has been fired and thus prevents fumes from entering the fighting compartment. The M109A2 has an improved shell rammer and recoil mechanism, the M178 modified gun mount, and other more minor improvements. The M109A3 is the M109A1 fitted with the M178 gun mount and with the same performance capabilities as the production M109A2.

The M109 fires a round to a maximum range of 16,070yd (14,700m); the M109A1 fires to a maximum range of 19,685yd (18,000m). Rocket assisted projectiles (M549A1) increase the maximum range to 26,250yd (24,000m). A new nuclear round (M785) is now under development: it is ballistically compatible with the M549A1 RAP and will utilize the same protective container as the M758 8-in round (see *M110A2 entry*).

The M109 can ford streams to a maximum depth of 6ft (1·828m). A special amphibious kit has been developed for the vehicle but this is not widely used. It consists of nine inflatable airbags, normally carried by a truck. Four of these are fitted to each side of the hull and the last to

Below: The main distinguishing features of the M109A2 involve the gun loading and recoil mechanisms and a new mounting.

the front of the hull. The vehicle is then propelled in the water by its tracks at a maximum speed of 4mph (6·4km/h). The M109 is provided with infra-red driving lights and some vehicles also have an NBC system. To keep the M109 supplied with ammunition in the field Bowen-McLaughlin-York have recently developed the M992 Field Artillery Ammunition Support Vehicle.

Above: Cleaning the secondary armament of an M109 during winter exercises, with paint scheme and camouflage netting appropriate to the landscape.

Below: Twenty years after the first production examples were completed, this M109A1 represents the most widely used SP howitzer in current service.

M110A2

Origin: USA.
Type: Self-propelled howitzer.
Crew: 5 + 8.
Armament: One 8in (203mm) howitzer.
Dimensions: Length (including gun and spade in travelling position) 35·2ft (10·731mm); length (hull) 18·76ft (5·72m); width 10·33ft (3·149m); height 10·33ft (3·143m).
Combat weight: 62,600lb (28,350kg).
Engine: Detroit Diesel Model 8V-7T 8-cylinder turbo-charged diesel developing 405bhp at 2,300rpm.
Performance: Road speed 34mph (54·7km/h); range 325 miles (523km); vertical obstacle 3·33ft (1·016m); trench 7·75ft (2·362m); gradient 60 per cent.

Development: In 1956 the United States Army issued a requirement for a range of self-propelled artillery which would be air-transportable. The Pacific Car and Foundry Company of Washington was awarded the development contract and from 1958 built three different self-propelled weapons on the same chassis. These were the T235 (175mm gun), which became the M107, the T236 (203mm howitzer), which became the M110, and the T245 (155mm gun), which was subsequently dropped from the range. These prototypes were powered by a petrol engine, but it was soon decided to replace this by a diesel engine as this could give the vehicles a much greater range of action. The M107 is no longer in service with the US Army; all have been rebuilt to M110A2 configuration. The M110A2 is also in production by Bowen-McLaughlin-York

Company, and when present orders have been completed the US Army will have a total inventory of over 1,000.

The hull is of all-welded steel construction with the driver at the front on the left with the engine to his right. The gun is mounted towards the rear of the hull. The suspension is of the torsion-bar type and consists of five road wheels, with the fifth road wheel acting as the idler; the drive sprocket is at the front. Five crew are carried on the gun (driver, commander and three gun crew), the other eight crew members following in an M548 tracked vehicle (this is based on the M113 APC chassis), which also carries the ammunition, as only two ready rounds are carried on the M110 itself. The 203mm howitzer has an elevation of +65° and a depression of −2°, traverse being 30° left and 30° right. Elevation and traverse are both hydraulic, although there are manual controls for use in an emergency.

The M110 fires an HE projectile to a maximum range of 26,575yd (24,300m), and other types of projectile that can be fired include HE carrying 104 HE grenades, HE carrying 195 grenades, Agent GB or VX and tactical nuclear. A large hydraulically-operated spade is mounted at the rear of the hull and is lowered into position before the gun opens fire, and the suspension can also be locked when the gun is fired to provide a more stable firing platform. The gun can officially fire one round per two minutes, but a well trained crew can fire one round per minute for short periods. As the projectile is very heavy, an hydraulic hoist is provided to position the pro-

Above: With the chassis braced by the large hydraulically-operated spade and a round ready in the hoist, the crew of an M110 prepare to load the weapon.

Right: With an appropriate motto on the howitzer barrel, the crew of a camouflaged M110 wait for the action to start during Exercise Team Spirit '84.

jectile on the ramming tray; the round is then pushed into the breech hydraulically before the charge is pushed home, the breechlock closed and the weapon is then fired.

The M110 can ford streams to a maximum depth of 3·5ft (1·066m) but has no amphibious capability. Infra-red driving lights are fitted as standard but the type does not have an NBC system.

All M110s in US Army service, and in an increasing number of NATO countries as well, have been brought up to M110A2 configuration. The M110A1 has a new and longer barrel, while the M110A2 is identical to the M110A1 but has a double baffle muzzle brake. The M110A1 can fire up to charge eight

while the M110A2 can fire up to charge nine. The M110A1/M110A2 can fire all of the rounds of the M110 but in addition binary, high-explosive Rocket Assisted Projectile (M650), and the improved conventional munition which contains 195 M42 grenades. The latter two have a maximum range, with charge nine of 32,800yd (30,000km).

The M110A2 also fires the M753 rocket-assisted tactical nuclear round, which entered production in FY 1981. The M753 will be available in two versions: the first as a normal nuclear round; the second as an "Enhanced Radiation" version. These nuclear rounds are packed in very sophisticated containers to prevent unauthorized use and are sub-

Panzerjager SK105

Origin: Austria.
Type: Tank destroyer.
Crew: 3.
Armament: One 105mm gun; one 7·62mm machine-gun coaxial with main armament; 3 smoke dischargers either side of turret.
Dimensions: Length (with gun forwards) 25·47ft (7·763m); length (hull) 18·3ft (5·58m); width 8·2ft (2·5m); height 8·23ft (2·51m).
Combat weight: 38,587lb (17,500kg).
Engine: Steyr 7FA turbo-charged 6-cylinder diesel developing 320hp at 2300rpm.
Performance: Road speed 40·4mph (65km/h); range 323 miles (520km); vertical obstacle 2·62ft (0·8m); trench 7·9ft (2·41m); gradient 75 per cent.

Development: In 1965 Saurer-Werke commenced the development of this well armed and highly mobile tank destroyer to meet the requirements of the Austrian Army. The chassis uses many components of an earlier range of APCs but its layout is quite different with the driver's compartment at the front, turret in the centre and the

engine and transmission at the rear. The hull is all of welded construction and provides the crew with protection from small arms fire and shell splinters.

The FL-12 turret is made under licence in Austria from the French company Fives-Lille-Cail and is identical to that fitted to the AMX-13

light tank and the Brazilian EE-17 (6 x 6) tank destroyer. This turret is of the oscillating type with the 105mm gun fixed in the upper half which in turn pivots on the lower part. The gun can be elevated from −6° to +13° and the turret traversed through a full 360° in 12 to 15 seconds.

The 105mm gun is fed from two revolver type magazines in the turret bustle, each of which holds six

rounds of ammunition. Empty cartridge cases are ejected from the turret through a small trap door in the turret rear. The 2 magazines have enabled the crew to be reduced to three men — commander, gunner and driver — and also allow a high

Below: A spent cartridge case is ejected from the magazine-fed breech of a Panzerjager SK 105 as another round is loaded.

ject to very stringent controls. The ER rounds will not be deployed outside the USA except in emergency.

One major shortcoming of the M110 design has always been its lack of protection for the gun crew: it is virtually the only modern self-propelled gun to suffer such a deficiency. The US Army plans to rectify this by fitting a Crew Ballistic Shelter (CBS), a high, square gun housing that will improve survivability against small arms and shell fragments by some 33 per cent and will also provide collective NBC protection.

One of the problems with heavy artillery of this type is keeping the guns supplied with sufficient ammunition. The weapon is currently supported by an M548 tracked vehicle, but this is to be supplanted in the near future by the M992 Field Artillery Ammunition Support Vehicle (FAASV). M992 has a large armoured housing on an M109 chassis, giving protected transportation for 48 8in (203mm) rounds, which are passed automatically to the gun.

The M110/A1/A2 series serves in 17 armies.

rate of fire to be achieved for a short period; on the other hand once the 12 rounds have been fired at least one of the crew has to leave the vehicle to carry out manual reloading of the two magazines. A total of 44 rounds of 105mm ammunition are carried.

Mounted coaxial to the right of the main armament is a 7·62mm MG42/49 machine-gun, and mounted on either side of the turret are three electrically operated smoke dischargers; a total of 2000 rounds of 7·62mm ammunition are carried. Recently most vehicles have been fitted with a laser rangefinder mounted externally on the turret roof and above this has been mounted an infra-red/white-light searchlight. The Kürassier K, as the vehicle is often called, has no NBC system and no deep fording capability.

This neat and sensible tank destroyer is in service with the armies of Austria, Argentina, Bolivia, Greece, Morocco, Nigeria and Tunisia.

Right: Along with the French 105mm gun, the SK 105's FL-12 turret mounts a coaxial machine gun; the laser rangefinder and IR searchlight are recent additions.

TOWED ARTILLERY

105mm Light Gun

Origin: UK.
Type: Towed field gun.
Calibre: 105mm.
Crew: 4.
Combat weight: 4,096lb (1,858kg).
Dimensions: Length (firing) 23ft (7·01m); length (travelling) 16ft (4·88m); width (firing) 5·84ft (1·78m); width (travelling) 5·84ft (1·78m); height (firing) 8·6ft (2·63m); height (travelling) 4·5ft (1·37m).
Elevation: $-5·5°/+70°$.
Traverse: $11°$.
Range: 18,810yd (17,200m).

Development: The 105mm Light Gun, which was used with great success in the 1982 South Atlantic War, began life in 1965, to replace the 105mm Pack Howitzer then in use with the British Army, and to give greater range, more stability

and better cross-country performance. The design was accepted in 1973 and entered service in 1975. It is now in service with the UK, Brunei, Ireland, Kenya, Malawi, Oman and UAE, and is on order for Australia and being tried out by Switzerland.

The gun is very lightweight, the result of ingenious design and modern manufacturing techniques. Two barrels are available: one (L118) fires British Abbott ammunition, and the other (L119) fires American M1 ammunition. The barrel rotates forward over the trail for towing and can be removed and replaced in 30 minutes with one simple tool.

Right: Loaders stand ready with the next round and its cased charge as a Light Gun is fired. Rate of fire is 6rds/min.

122m D-30

Origin: Soviet Union.
Type: Towed field howitzer.
Calibre: 122mm.
Crew: 7.
Combat weight: Firing 6,944lb (3,150kg).
Dimensions: Length (travelling) 17·71ft (5·4m); width (travelling) 6·39ft (1·95m).
Elevation: $-7°$ to $+70°$.
Traverse: $360°$.
Projectile mass: (HE) 48·1lb (21·8kg).
Muzzle velocity: (HE) 2,264ft/s (690m/s).
Maximum range: (HE) 16,732yd (15,300m).

Development: This howitzer of 122mm (4·8in) calibre, typifies the dramatically advanced and effective design of the latest Soviet artillery. It is towed by a large lunette lug under or just behind the muzzle brake, with its trails folded under the barrel. To fire, the crew of seven rapidly unhitches, lowers the central firing jack (lifting the wheels off the ground)

and swings the outer trails through $120°$ on each side. The gun can then be aimed immediately to any point of the compass. The barrel is carried under a prominent recoil system, has a semi-auto vertically sliding wedge breechblock, and fires cased but variable charge, separate-loading ammunition. In addition to conventional or chemical shells, it fires a fin-stabilized non-rotating HEAT shell from its rifled barrel, giving it a formidable direct fire capability against armour.

The D-30 is the basic field gun of the Soviet Army, and is used throughout the Warsaw Pact. It is in service with some 35 armies around the world and is probably still in production in the USSR. It is also in production in Egypt, and a British firm is building the prototype of an SP version mounted on the chassis of a Combat Engineer Tractor.

Below: D-30s about to fire. Such exposed and concentrated batteries make tempting targets.

155mm FH-70

Origin: International.
Type: Towed field howitzer.
Calibre: 155mm.
Crew: 8.
Combat weight: 20,503lb (9,300kg).
Dimensions: Length (firing) 40·8ft (12·43m); length (travelling) 32·2ft (9·8m); width (firing) 32·2ft (9·8m); width (travelling) 7·2ft (2·2m); height (firing) 7·2ft (2·2m); height (travelling) 8·4ft (2·56m);
Ground clearance: 2·9ft (0·3m).
Elevation: $-5°/+70°$.
Traverse: $56°$ (total).
Range: Standard round 26,247yd (24,000m); RAP 32,810yd (30,000m).

Development: The FH-70 is one of the best examples of a soundly conceived and well executed international collaborative project. The original partners were West Ger-

many and the UK, which, in the early 1960s, agreed on a requirement for a 155mm towed howitzer, fitted with an auxiliary power unit (APU). The MOU was signed in 1968, and the first 6 prototypes completed in 1969-70, Italy joined the project in 1970, and the first trials battery was formed in 1975. Production for the three main partners was completed in 1982, with the FRG taking 216, Italy 164 and the UK 71; Saudi Arabia has ordered some 200-300.

FH-70 is a good design and has proved very satisfactory in service. The elevating mass (ie, the ordnance, cradle and loading system) is used in the SP-70 project (qv).

Below: The FH-70 fires all NATO standard 155mm ammunition, plus a special extended range projectile; a rate of fire of 2rds/min can be sustained for 1hr.

155mm TR

Origin: France.
Type: Towed gun.
Calibre: 155mm.
Crew: 8.
Combat weight: 23,478lb (10,650kg).
Dimensions: Length (firing) 32·8ft (10m); length (travelling) 27·1ft (8·25m); width (firing) 27·56ft (8·4m); width (travelling) 10·1ft (3·09m); height (firing) 5·4ft (1·65m).
Ground clearance: 1·64ft (0·5m).
Elevation: $+66^0/-5^0$.
Traverse: $65^0(27^0$L, 38^0R).
Range: Standard round 26,247yd (24,000m); RAP 36,100yd (33,000m).
Development: Manufactured by GIAT, the 155mm TR is similar in concept to the FH-70 (*qv*) and is intended for use by Motorized Infantry Divisions of the French Army. Eight prototypes were built and troop

trials carried out in 1983. Production started in 1984 and 180 guns have been ordered for the French Army. The 155mm TR is a neat and effective design, and is fitted with an auxiliary power unit (APU) to give limited mobility on the gun position, similar to that fitted on other weapons, like FH-70. The Norwegian Army has shown interest in the 155mm TR but no order has yet been placed.

One of the weaknesses in the Western system is that this project, the FH-70 and the M-198 have been developed in parallel, to similar timescales and to meet very similar requirements.

Right: French equivalent of the FH-70, the TR has comparable performance and fires a similarly extensive range of shells, including "intelligent" projectiles.

M198

Origin: USA.
Type: Towed field howitzer.
Calibre: 155mm.
Crew: 10.
Combat weight: 15,795lb (7,165kg).
Dimensions: Length (firing) 37·08ft (11·302m); length (travelling) 23·25ft (7·086m); width 6firing) 28ft (8·534m); width (travelling) 9·15ft (2·79m); height (firing) 28ft (8·534m); width height (travelling) 9·92ft (3·023m).
Ground clearance: 13in (0·33m).
Elevation: $-4·2^0$ to $+72^0$.
Traverse: $22·5^0$ left and right 360^0 with speed traverse.
Range: 32,808yd (30,000m) with RAP; 24,060yd (22,000m) with conventional round.

Development: In the late 1960s, Rock Island Arsenal started work on a new 155mm howitzer to replace the M114, and this was given the development designation of the XM198. The first two prototypes were followed by eight further prototypes, and during extensive trials these weapons fired over 45,000 rounds of ammunition. The M198 is now in production at Rock Island; the Army has a requirement for 435 M198s while the Marine Corps requires 282. It has also been adopted by a number of other countries, including Australia, India, Greece, Pakistan, Thailand and Saudi Arabia. The M198 is used by airborne, airmobile and infantry divisions. Other divisions will continue to use self-propelled artillery.

The weapon will be deployed in battalions of 18 guns, each battery having 6 weapons. The M198 is normally towed by a 6 x 6 5-ton truck or a tracked M548 cargo carrier, the latter being a member of the M113 family of tracked vehicles. It can also be carried under a Boeing CH-47 Chinook, but (and this is a most important drawback) its 5-ton prime mover cannot. A further problem is that the carriage is some 5in (127mm) too wide to fit into the

Low-Attitude Parachute Extraction System (LAPES) rails in USAF C-130 aircraft.

When in the travelling position, the barrel is swung through 180^0 so that it rests over the trails. This reduces the overall length of the weapon. When in the firing position, the trails are opened out and the suspension system is raised so that the weapon rests on a non-anchoring firing platform. A hydraulic ram cylinder and a 24in (0·609m) diameter float mounted in the bottom carriage at the on-carriage traverse centreline provides for rapid shift of the carriage to ensure 360^0 traverse. This enables the weapon to be quickly laid on to a new target.

The weapon has a recoil system of the hydropneumatic type and the barrel is provided with a double-baffle muzzle brake. The M198 uses separate loading ammunition (eg a projectile and a separate propelling charge) and can fire an HE round to a maximum range of 22,000m, or out to 30,000m with a Rocket Assisted Projectile. The latter is basically a conventional HE shell with a small rocket motor fitted at the rear to increase the range of the shell. The weapon will also be the primary user of the new Cannon Launched Guided Projectile (or Copperhead) round. Nuclear and Improved Conventional Munitions, as well as rounds at present used with the M114, can also be fired. It will also be able to fire the range of ammunition developed for the FH70. Maximum rate of fire is four rounds per minute for the first three minutes, followed by two rounds per minute thereafter. A thermal warning device is provided so that the gun crew know when the barrel is becoming too hot.

Although a great improvement on its predecessors, the M198 has suffered from a number of problems. The desired range and accuracy requirements have been achieved at the expense of mobility and size.

Further, unit price has increased from an original estimate of $184,000 to $421,000 — although the M198 is by no means the only weapon system to suffer such problems.

Above: An M198 at its maximum elevation of 72°. This howitzer uses rocket-assisted shells to reach 18.6 miles (30km) ranges, and also fires the Copperhead laser-homing anti-tank projectile.

AIR DEFENCE WEAPONS

Blowpipe/Javelin

Origin: UK.
Type: Portable air defence missile system.
Dimensions: (Missile) length 54·7in (139cm); body diameter 3in (7·6cm).
Launch weight: (system) 48·3lb (21·9kg).
Propulsion: 2-stage booster-accelerator solid-propellant rocket motor.
Guidance: Radio command/optical tracking.
Range: 3 to 3·7 miles (5 to 6km).
Flight-speed: Supersonic.
Warhead: HE with proximity fuse.

Development: Blowpipe is a lightweight, self-contained SAM system used by 14 armed forces in 10 countries. The system comprises two units: a sealed launch canister containing the missile, and the aiming unit. To prepare for action the aiming unit is clipped to the launch canister (which is treated as an ammunition round) and then put on the operator's shoulder. The missile is ejected from the tube by the first-stage motor and, when well clear of the operator, the second-stage ignites. The operator "gathers" the missile and then steers it to the target by a thumb-stick which transmits signals by a radio link to the missile. This system worked well in the South Atlantic war in 1982, a number of successes being recorded.

Now under development is the Javelin system, using the Blowpipe missile, but with a more powerful motor which significantly reduces the time of flight. A semi-automatic command line-of-sight system (SACLOS) means that the operator need only keep the cross-wires in the sight on the target. Javelin is man-portable, but installations are also under development for use on both soft-skinned and armoured vehicles.

Above: The lightweight multiple launcher for Blowpipe or its SACLOS-guided derivative, Javelin, is a man-portable system comprising three missiles in launch tubes and an aiming unit.

Crotale/Shahine

Origin: France.
Type: Self-propelled air defence missile system.
Dimensions: Length 9·48ft (2·89m); diameter 5·9in (0·15m); wingspan 17·7in (0·54m).
Launch weight: 187lb (85kg).
Propulsion: Single-stage solid propellant rocket motor.
Guidance: Radar or optical.
Range: Maximum slant range 7·45 miles (12,000m); altitude limits 1,640 to 27,900ft (500-8,500m).
Flight speed: Mach 2·3.
Warhead: HE fragmentation with IR proximity fuse.

Development: This system began in 1964 as a South African-funded development by the French firm of Thomson-CSF. Named Cactus, the system was delivered in 1971-73. The same system (now renamed Crotale) has subsequently been ordered by Chile, Egypt, France, Libya, Pakistan, Saudi Arabia and the UAE.

A Crotale battery comprises one acquisition unit and two to three firing units. Both units use a specially-developed 4 x 4 vehicle. The acquisition unit is responsible for target surveillance, identification (IFF) and designation; it uses a Thomson-CSF Mirador radar with a range of 11·2 miles (18km). The missile vehicle mounts four R440 missiles, together with a command transmitter, integrated TV tracking mode, optical tracker and computer.

An improved system, Shahine, has been developed for Saudi Arabia. The firing unit mounts six R460 missiles (developed R440). The radar unit is, as with Crotale, mounted in a separate vehicle. Both units use the AMX-30 MBT chassis. Saudi Arabia ordered 36 units, which have been delivered, and production has now ended.

Above: Launch of a Crotale from the fire unit; the associated acquisition and coordination unit mounts pulse-Doppler surveillance radar and data processing equipment on a similar vehicle.

Gepard/CA1

Origin: Federal Republic of Germany.
Type: Self-propelled air defence gun system.
Crew: 3.
Dimensions: Length 25·4ft (7·73m); width 11ft (3·37m); height (turret roof) 9·9ft (3·01m); height (radar raised) 13·2ft (4·03m).
Combat weight: 104,280lb (47,300kg).
Armament: Two Oerlikon 35mm KDA cannon.
Ranges: (approx): 3,280 to 4,374yd (3,000 to 4,000m).

Development: The need for improvement in air defences became apparent in the late 1950s and for some years there was an intense argument over the relative merits of guns and missiles in this role. One of the first countries to develop a really up-to-date air defence gun was West Germany, which in 1965, decided to develop an all-weather, self-propelled, autonomous system. A Swiss system based on an Oerlikon twin 35mm gun installation mounted on a Leopard 1 chassis was selected, named Gepard. Some 420 have been built for the Bundeswehr, 55 for the Belgian Army and 95 of the slightly different CA1 for the Dutch Army.

The system is based on two Oerlikon 35mm KDA cannon, each with a cyclic rate of fire of 550 rounds per minute. For each gun 310 rounds of anti-aircraft ammunition are carried, together with 20 armour-piercing rounds for anti-tank use. The Gepard has Siemens tracking and search radars, but the CA1 has Hollandse Signalapparaten radars.

Right: Gepard's twin KDA cannon have a combined rate of fire of 1,100rds/min; the CA1 for the Netherlands has Dutch radars.

Patriot, MIM-104

Origin: USA.
Type: Self-propelled air defence missile system.
Dimensions: Length 209in (5·31m); body diameter 16in (40·6cm); span 36in (92cm).
Launch weight: 2,200lb (998kg).
Propulsion: Thiokol TX-486 single-thrust solid motor.
Guidance: Phased-array radar command and semi-active homing.
Range: About 37 nautical miles (68·6km).
Flight speed: About Mach 3.
Warhead: Choice of nuclear or conventional blast/frag.

Development: Originally known as SAM-D, this planned successor to both Nike Hercules and Hawk has had an extremely lengthy gestation. Key element in the Patriot system is a phased-array radar, which performs all the functions of surveillance, acquisition, track/engage and missile guidance. The launcher carries 4 missiles each in its shipping container, from which it blasts upon launch. Launchers, spare missiles boxes, radars, computers, power supplies and other items can be towed or self-propelled. Patriot is claimed to be effective against all aircraft or attack missiles even in the presence of clutter or intense jamming or other ECM.

Fundamental reasons for the serious delay and cost-escalation have been the complexity of the system, the 1974 slowdown to demonstrate TVM (track via missile) radar guidance, and inflation. Unquestionably the system is impressive, but often its complication and cost impress in the wrong way and the number of systems to be procured has been repeatedly revised downwards.

The authorized development programme was officially completed in 1980, when low-rate production was authorized. In 1983 production was cautiously being stepped up and the first operational units were formed in mid-1983. The US Army plans to have 81 Patriot batteries, for which its hardware requirements are 103 fire units and 6,200 missiles.

Below: Patriot's test programme was completed in 1981, with deployment in Germany, Belgium and the Netherlands planned for 1984-1988.

Rapier and Tracked Rapier

Origin: UK.
Type: Self-propelled air defence missile system.
Dimensions: Length 7·35ft (2·24m); diameter 5·23in (0·133m); wingspan 15in (0·381m).
Launch weight: 93·9lb (42·6kg).
Propulsion: 2-stage solid-propellant rocket motor.
Guidance: Command to line-of-sight.
Range: Slant range 4 miles (6·5km); altitude limits 164 to 9,840ft (50 to 3,000m).
Flight speed: Mach 2+.
Warhead: Semi-armour-piercing with crush fuse and HE.

Development: The original Rapier model consisted of a wheeled fire-unit towed behind a 1-tonne vehicle, and mounting four missiles. The fire unit included a surveillance radar, an IFF coder/decoder and a microwave command transmitter. Intended for use against fast, low-flying targets, Rapier is designed actually to impact the target, and has repeatedly shown its ability to do so, both in peacetime tests and in combat. The requirement to operate in bad weather and at night led to the development of the DV181 (Blindfire) radar.

The next and inevitable development was the Tracked Rapier, originally developed by BAe with the former Shah's Iran in mind. Using the M548 cargo-carying derivative of the US Army's MII3 APC. The crew of three travel in an aluminium-armoured cab, and behind this is an air-conditioner and a power unit protected from missile efflux by a blast shield. The firing-unit is mounted on a turntable on the rear of the vehicle, with eight missiles at instant readiness.

Well over 14,000 missiles have been produced so far in what promises to be a very long production run. Towed Rapier is already in service with, or on order for, Australia, Brunei, Iran, Oman, Qatar, Singapore, Switzerland, Turkey, UAE, the United States Air Force, Zambia, and the British RAF Regiment. Tracked Rapier has so far only been ordered by the British Army.

The latest development is Rapier Laserfire, a palletized installation which can fit on any flat-platform 3- or 4-tonne vehicle, and has an automatic laser tracker.

Above: Tracked Rapier. The dome between the missiles houses surveillance radar and IFF, with the command antenna above it.

Below: Towed Rapier in operation with the Canadian 12th Air Defense Regiment during an exercise in West Germany.

Roland

Origin: International.
Type: Self-propelled air defence missile system.
Crew: 3.
Dimensions: Length 7·87ft (2·4m); diameter 6·3in (0·16m); wingspan 1·64ft (0·5m).
Launch weight: 146·6lb (66·5kg).
Propulsion: Two-stage solid-propellant rocket motor.
Guidance: Optical or radar.
Range: Slant range 3·9 miles (6·3km); altitude limits 66 to 16,400ft (20 to 5,000m).
Flight speed: Mach 1·6.
Warhead: HE with impact and proximity fuses.

Development: In the mid-1960s France and Germany started development of an air defence system, with two modes: clear-weather (Roland 1) and all-weather (Roland 2). The system entered service in 1980.

The French Army has ordered 205 firing units and 9,103 missiles, and uses the AMX-30 MBT chassis. West Germany requires 344 firing units and 12,526 missiles, and uses the Marder APC chassis. Roland is in service with the armies of Argentina, Brazil, France, Iraq, West Germany and the USA, and is on order for Nigeria, Jordan and Venezuela. The US Army selected Roland in 1975 to be mounted on the M109 chassis, with large orders promised, but progressive cutbacks have ended with 27 firing units.

The missile can be used in both clear and all-weather conditions, and has always been intended for use in a gun/missile mix. The Argentine Army deployed a Roland unit to Port Stanley in the South Atlantic War, and it fired 8 out of its 10 missiles, but only one possible success (a Sea Harrier) resulted.

Right: Sabots are discarded as a US Army Roland is launched.

SA-7 Grail

Origin: Soviet Union.
Type: Portable air defence missile system.
Dimensions: Length 53·15in (1·35m); diameter 2·75in (70mm).
Launch weight: (missile alone) 20·3lb (9·2kg).
Range: Up to 6·25 miles (10km).
Flight speed: About Mach 1·5.
Warhead: Smooth case frag.

Development: Originally called Strela (arrow) in the West, this simple infantry weapon was originally very similar to the American Redeye, and suffered from all the latter's deficiencies. These included inability of the uncooled PbS IR seeker to lock on to any heat source other than the nozzle of a departing attacker — with the single exception of most helicopters which could be hit from the side or even ahead, if the jetpipe projected enough to give a target.

The basic missile is a tube with dual-thrust solid motor, steered by canard fins. The operator merely aims the launch tube at the target with an open sight, takes the first pressure on the trigger, waits until the resulting red light turns green (indicating the seeker has locked on) and applies the full trigger pressure. The boost charge fires and burns out before the missile clears the tube. At a safe distance the sustainer ignites and accelerates the missile to about Mach 1·5.

The 5·5lb (2·5kg) warhead has a smooth fragmentation casing and both graze and impact fuses. This is lethal only against small aircraft, and in the Yom Kippur war almost half the A-4s hit by SA-7s returned to base. Height limit is still widely given as 4,921ft (1,500m), but in 1974 a Hunter over Oman was hit at 11,500ft (3,505m) above ground level.

An improved missile has been in production since 1972 with augmented propulsion, IR filter to screen out decoys, and much better guidance believed to house a cryogenic cooler in a prominent launcher nose ring.

There are probably 50,000 missiles and nearly as many launchers, large numbers of them in the hands of terrorists all over the world. Users include Angola, Bulgaria, Cuba, Czechoslovakia, East Germany, Egypt, Ethiopia, Hungary, Iraq, North Korea, Kuwait, Libya, Mozambique, Peru, Philippines (Muslim guerrillas), Poland, Rumania, Soviet Union, Syria, Vietnam and PDR of the Yemen. A small-ship version is SA-N-7.

Right: SA-7 Grail in the hands of a Soviet artilleryman. Many thousands more are in service with more than a score of users, including terrorist organisations.

Below: An SA-7 is launched from the rear hatch of a BMP. The booster charge has burnt out before the missile left the tube, with relatively little backblast.

SA-8 Gecko

Origin: Soviet Union.
Type: Self-propelled air defence missile system.
Dimensions: Length 10·5ft (3·2m); diameter 8·25in (210mm); span 25·2in (640mm).
Launch weight: About 419lb (190kg).
Propulsion: Single-stage, solid-propellant rocket motor.
Guidance: Command guidance; semi-active radar or IR terminal guidance.
Range: 1,750 to 13,123yd (1,600 to 12,000m).
Flight speed: Mach 2.
Warhead: Fragmentation; proximity fuse.

Development: A surprise in the 7 November 1975 Red Square parade was a dozen completely new vehicles each carrying quadruple launchers for this advanced and highly mobile system which was rather incorrectly called "the Soviet Roland", and it was almost certainly derived from SA-N-4.

Despite its great size the 6 x 6 amphibious vehicle is air-portable in an An-22, and carries missiles ready to fire. Inside the body, or hull, are an estimated eight further missiles, enough for two reloads. Towards the rear of the vehicle is the rotatable and elevating quad launcher, surmounted by a folding surveillance radar probably operating in the J-band. Between this installation and the cab is a large guidance group comprising a central target-tracking radar, two missile guidance-beam radars, two command-link horns for gathering, an optical tracker and an LLTV and telescopic sight.

All the radars have flat-fronted Cassegrain aerials, the main set being a J-band (13-15 GHz) tracker with a range of about 15·5 miles (25km). Each guidance aerial has a similar but smaller geometry, with limited azimuth movement; below each is the command link horn. After careful study, semi-active radar homing has been judged unlikely and it is believed all SA-8 missiles have IR homing.

The missile has small fixed tail fins, small nose canard controls, a radar beacon and external flare. The dual-thrust solid motor gives very high acceleration to a burn-out speed greater than Mach 2, the average speed in a typical interception being about Mach 1·5. It is believed missiles are fired in pairs, with very short time-interval, the left and right missile-tracking and command systems operating on different spreads with frequency-agility in the I-band to counter ECM

and jamming by the target, with TV tracking as a back-up. The warhead weighs about 110lb (50kg) and has a proximity fuse.

A new version with six missiles appeared in 1980 and is designated SA-8B; the missiles are mounted in containers. SA-8A/B systems are deployed in five four-vehicle

Above: SA-8 mounts four missiles along with search, acquisition, tracking and guidance radars on a six-wheeled amphibious vehicle; SA-8B has six missiles.

batteries in each Soviet tank and motor-rifle division. Other users include Iraq, Jordan and Syria.

SA-10

Origin: Soviet Union.
Type: Surface-to-air missile system.
Dimensions: Overall length 22·9ft (7m); diameter 18·7in (0·48m).
Performance: Range 62 miles (100km); effective altitude low-to-high.
(All data estimated).

Development: The SA-10 SAM system has been operational since 1980. By late 1984 US sources estimated that there were some 350 launchers in service, each mounting four missiles, deployed at some 70 sites. The Transporter-Erector-Launcher (TEL) vehicle is based on the MAZ-543 8 x 8 chassis, but has only one narrow cab for the driver with a large cabin behind it, presumably for power and electronic equipment. The four missiles are in closed canisters (as with SA-X-12) and are launched vertically. A second vehicle, also on a MAZ-543 chassis, mounts a phased-array radar. Some reports suggest that the SA-10 may have an anti-strategic missile capability, in which case these missile batteries could become part of an ABM system giving area coverage.

Below: US DoD representation of the mobile SA-10 system, with planar array radar and TEL units on wheeled chassis.

SA-X-12

Origin: Soviet Union.
Type: Tactical, mobile surface-to-air missile system.
Specifications: None yet known (see text).

Development: First revealed in the US annual publication "Soviet Military Power" in 1984, the SA-X-12 is a highly mobile, tactical, surface-to-air missile system which is probably due to enter service in 1985/86. A launch battery consists of four key types of vehicle. The missile transporter-erector-launcher (TEL) mounts two ready-to-fire missiles in closed canisters hinged about the rear end of the vehicle roof. The TEL also has a guidance-

Below: Mobile SA-12 system, with two TELs, two planar array radars, and resupply and crew/command vehicles, all on tracked chassis.

radar mounted on an erectable mast. The second vehicle is a surveillance-tracking radar, with a large antenna, similar in appearance and size to that used for the US Patriot system. Third is a command-post vehicle and last a missile resupply vehicle with a crane and two rear-mounted devices for carrying four missiles and transferring them direct to the TEL. All these four vehicles are track-mounted, and would appear to be derivations of the "Ganef" vehicle.

The SA-X-12 appears to be intended either to replace or supplement the SA-4 SAM in frontal AD missile brigades. The system is clearly capable of intercepting high-altitude/high performance aircraft, but may also have a capability against both ballistic battlefield missiles (eg, Lance) and even INF missiles (eg, Pershing).

SA-13

Origin: Soviet Union.
Type: Self-propelled air defence missile system.
Dimensions: Length (approx) 6·56ft (2m); diameter 4·72in (12cm).
Launch weight: 66lb (30kg).
Propulsion: Solid propellant rocket motor.
Guidance: Passive infra-red terminal guidance.
Range: Maximum clear-weather slant range 4·97 miles (8km); altitude limits 164 to 32,800ft (50 to 10,000m).
Flight speed: Approx Mach 1·5 to 2.

Development: Mounted on a vehicle based on the MT-LB APC, the SA-13 first appeared in the late 1970s. It was deployed in GSFG in 1980, replacing SA-9 (Gaskin) on a one-for-one basis in motor-rifle and tank regiments. The missiles are mounted in canisters on a pylon; one version has four missiles and another six. A range-only radar is mounted on the pylon head.

The Soviet Army has developed a sophisticated (and expensive) mix of air defence weapons including: company/battalion, SA-7A and SA-14 shoulder-fired SAMs; regiment, SA-9 and SA-13 vehicle-mounted SAMs plus ZSU-23-4 self-propelled gun; division, SA-6A/B, SA-8A/B and SA-11 vehicle-mounted SAMs; front/army SA-4A/B vehicle-mounted SAM.

Right: The SA-13 replacement for SA-9 mounts missiles and radar on a derivative of the MT-LB.

Sergeant York, M998 DIVAD

Origin: USA.
Type: Self-propelled air defence gun system.
Crew: 3.
Dimensions: Length (hull) 23·35ft (7·12m); length (guns forward) 25·16ft (7·67m); width 11·9ft (3·63m); height (turret roof) 11·22ft (3·42m); height (antenna) 15·1ft (4·61m).
Combat weight: 119,997lb (54,431kg).
Armament: 2 x 40mm L/70 guns.
Ranges: (approx) Slant 3,280yd (3,000m); ground role 4,374yd (4,000m).

Development: For many years NATO armies have envied the Soviet ZSU-23-4 self-propelled air-defence gun, in which a proven 23mm quad MG was married to a valve-technology radar and mounted on an existing chassis to produce one of the most effective current air-defence systems. The US Army seems at last to have heeded this object lesson in cost-effectiveness with the Sergeant York DIVAD.

The US Army's main problem lay in deciding just what kind of forward air-defence system it needed: gun, missile or a mix? Once it was decided that a gun was required, there followed many studies on calibre, type and number of tubes per chassis. The process began in 1962 and it was not until 1977 that the Army asked industry to submit proposals, using the M48A5 chassis and as much "mature, off-the-shelf" equipment as possible. The resulting competition was won in 1981 for Ford, who combined a twin Bofors L/70 installation with a radar developed from that used on the F-16 fighter, on the M48 chassis. First delivery took place in 1983 and the first battery will be fielded in 1985. Total requirement is 618 fire units.

The Bofors L/70 gun, in wide service throughout NATO, is both effective and extremely reliable. Two combat rounds will be used: a pre-fragmented proximity-fused HE round for use against aircraft; a point-detonating round for use against ground targets. Maximum slant range against aircraft is about 9,843ft (3,000m) and maximum ground range is some 4,370yd (4,000m). 502 rounds are carried for each gun and combined rate of fire is some 620 rounds per minute.

The radar automatically determines target type, assigns priorities and updates the fire control unit, which automatically aligns the turret and guns. The Identification Friend-or-Foe (IFF) has 90 per cent commonality with that fitted to the Stinger SAM (qv). Air defence engagements on the move are possible, although accuracy in this case is open to question.

The turret is intended to be mounted on the M48A5 chassis, although most early production will be on M48A1 or M48A2 chassis, re-engined and with new transmissions. This is because all M48A5 chassis are currently required for active US Army units, for the National Guard, or for operational reserve stocks.

Above: Components of the DIVAD system include the M48A5 tank chassis, and radar developed from the F-16 fighter's APG-66.

Below: The twin 40mm L/70 guns can engage targets either automatically or under crew control using optical sights.

Stinger, FIM-92A

Origin: USA.
Type: Portable air-defence missile system.
Dimensions: (Missile) length 60in (152cm); body diameter 2·75in (7cm); span 5·5in (14cm).
Launch weight: 24lb (10·9kg); whole package 35lb (15·8kg).
Propulsion: Atlantic Research dual-thrust solid.
Guidance: Passive IR homing (see text).
Range: In excess of 3·1 miles (5km).
Flight speed: About Mach 2.
Warhead: Smooth-case frag.

Development: Designed in the mid-1960s as a much-needed replacement for Redeye, Stinger has had a long and troubled development but is at last in service. An im-

proved IR seeker gives all-aspect guidance, the wavelength of less than 4·4 microns being matched to an exhaust plume rather than hot metal, and IFF is incorporated (so that the operator does not have to rely on correct visual identification of oncoming supersonic aircraft).

In FY 1981 the first 1,144 missiles for the inventory were delivered at $70·1m, and totals for 1982 and 1983 were respetively 2,544 at $193·4m and 2,256 at $214·6m. Total requirements for the US Army and Marine Corps are currently some 17,000 fire units and 31,484 missiles.

An improvement programme called Stinger-POST (Passive Optical Seeker Technique) is now in hand. This operates both the ultra-violet and infra-red spectra, the combination of the two frequency bands giv-

ing improved discrimination, longer detection range, and greater ECCM options. The Soviet counterparts are SA-7 and its replacement, SA-14.

Above: Stinger was developed to provide an all-aspect attack capability, with in-built IFF and greater resistance to ECM.

ZSU-23-4

Origin: Soviet Union.
Type: Self-propelled air defence gun system.
Dimensions: length 20·67ft (6,300mm); width 9·67ft (2,950mm); height (radar stowed) 7·38ft (2,250mm).
Engine: V-6 six-in-line water-cooled diesel, 240hp.
Combat weight: 13·78 tons (14,000kg).
Armament: Quadruple ZU-23 23mm anti-aircraft, 1,000 rounds.
Speed: 27mph (44km/h).
Range: 162 miles (260km).

Development: Extremely dangerous to aircraft out to slant range of 6,600ft (2,000m), the ZSU-23-4 is a neat package of firepower with its own microwave target-acquisition and fire-control radar, and crew of 4 in an NBC-sealed chassis derived from the amphibious PT-76. Each gun has a cyclic firing rate of 800 to 1,000rds/min, and with liquid-cooled barrels can actually sustain this rate. The crew of four comprises commander, driver, radar observer and gunner, and there is plenty of room in the large but thin-skinned turret. Gun travel is unrestricted in traverse, and from − 7° to +80°.

First seen in 1955, this vehicle is used throughout Warsaw Pact armies where it is popularly named Shilka. ZSU-23-4 was tested under battle conditions in the 1973 Arab-Israeli War, where it proved to be one of the most effective low-level defence systems. However, its radar suffers from "clutter" when trying to deal with targets below 200ft (60m).

The ZSU-23-4 is in many ways an archetypical piece of modern Soviet equipment, which marries a quadruple 23mm cannon derived from the well-proven ZU-23 to a Gainful chassis. Added to this is a valve-technology fire control and target-acquisition system. These simple steps have produced a cheap weapon system which is very effective and much respected by any pilot who might have to fly against it.

ZSU-23-4 is issued on a scale of

four to each motor rifle and tank regiment, giving 16 per division. They are used to protect columns on the line of march, and would normally be expected to operate in pairs. The practical rate of fire is about 200 rounds per minute per barrel, fired in 50 round (per barrel) bursts. A successor has been reported to be under development (ZSU-X), but it can be safely assumed that this weapon system will continue to serve with the Soviet Army for many years to come.

Right: GDR Army crews at work on their ZSU-23-4s. The weapon is used throughout the Warsaw Pact and in 15 other countries.

Below: ZSU-23-4s on parade. Combined rate of fire is 3,400rds/min, with a practical limit of four 3.5sec bursts per minute of 200 rounds each.

INFANTRY ANTI-TANK WEAPONS

AT-4 Spigot

Origin: Soviet Union.
Type: Wire-guided SACLOS infantry anti-tank weapon.
Dimensions: Launcher length 4·26ft (1·3m), height 26in (660mm). Missile length 30in (118mm), diameter 4·72in (120mm).
Weight: Launcher 88lb (40kg); missile 22 to 26lb (10 to 12kg).
Propulsion: Two-stage rocket motor.
Guidance: Semi-automatic command line-of-sight (SACLOS).
Range: 2,187 to 2,734yd (2,000 to 2,500m).
Flight speed: 590 to 656ft/sec (180 to 200m/s).

Warhead: Shaped charge.

Development: Code-named "Spigot" by NATO, the AT-4 is a high-performance infantry, anti-tank missile fired from a tube, and is similar in many respects to the Euromissile MILAN. The AT-4 has been in service with the Soviet and other Warsaw Pact armies for some ten years but photographs became available in the West only in about 1980.

Control is Semi-Automatic Command Line-of-Sight (SACLOS) and guidance is by means of electric signals which pass down the twin wires deployed from the missile during flight. The missile takes 12·5 seconds to reach its target at maximum range.

Above: AT-4 in the firing position. The raised sighting and tracking head enables the operator to remain under cover.

AT-5 Spandrel

Origin: Soviet Union.
Type: Wire-guided SACLOS vehicle-mounted anti-tank missile.
Dimensions: Launcher length 3·94ft (1·2m); diameter 5·3in (13·5m).
Weight: Missile 22 to 26lb (10 to 12kg).
Propulsion: Two-stage rocket motor.
Guidance: Command to line-of-sight; optical tracking.
Range: 4,374yd (4,000m) (estimated).
Flight speed: 492 to 820ft/sec (150 to 250m/s).
Warhead: Shaped charge.

Development: Allotted the NATO reporting name of "Spandrel", the AT-5 is a tube-launched system first seen on BRDM-2 armoured cars in the Red Square Parade on 7 November 1977. Each vehicle has five launch tubes in a row, mounted on a trainable turntable on the vehicle's roof. A further ten reloads are carried inside the vehicle, and these are fitted through the hatch in the roof immediately behind the turntable.

The tube resembles that of AT-4 "Spigot" (and that of the MILAN) and has a blow-out front closure and flared tail through which passes the efflux from the boost charge. This bursts the missile out of the launch-tube prior to ignition of its own motor. Folding wings, SACLOS guidance via trailing wires and general similarity to MILAN seem more than just chance. One curious feature, however, is that the missile tubes are always left unpainted, which may be due to a need to minimise the internal temperature of the tube, and this could be intended to protect the propellant.

The Group of Soviet Forces in Germany (GSFG) has replaced almost all its AT-2 "Swatter" and AT-3 "Sagger" anti-tank missile systems with AT-5, giving a significant improvement in anti-tank capability, although like all hollow-charge warheads the AT-5 is unlikely to be effective against tanks using the British Chobham armour. This revolutionary armour, utilising a ceramic sandwich, defeats all such warheads and the next generation of anti-tank missiles will have to be design bureaux have shown remark-their effectiveness and penetration.

Nevertheless, the Soviet Army's design bureaux have shown remakable ingenuity in the past and will eventually find a solution.

Below: Five-round AT-5 launcher on a BRDM-2. The launch tubes are one third larger in diameter than those for the AT-4; five ready rounds and five reloads are carried aboard each vehicle.

Dragon M47 FGM-77A

Origin: USA.
Type: Infantry anti-tank/assault missile.
Dimensions: Length 29·3in (74cm); body diameter 4·5in (11·4cm); fin span 13in (33cm).
Launch weight: 24·4lb (11·1kg).
Propulsion: Recoilless gas-generator thruster in launch tube; sustain propulsion by 60 small side thrusters fired in pairs upon tracker demand.
Guidance: See text.
Range: 200 to 3,300ft (60-100m).
Flight speed: About 230mph (370km/h).
Warhead: Linear shaped charge, 5·4lb (2·45kg).

Development: Dragon was designed as a medium-range complement to TOW (*qv*). In service since 1971, Dragon comes sealed in a glass-fibre launch tube with a fat rear end containing the launch charge. The operator attaches this to his tracker, comprising telescopic sight, IR sensor and electronics box. When the missile is fired, its 3 curved fins flick open and start the missile spinning. The operator holds the sight on the target and the tracker automatically commands the missile to the line of sight by firing appropriate pairs of side thrusters. The launch tube is thrown away and a fresh one attached to the tracker. The Army and Marine Corps use the basic Dragon, while developments involve night sights and laser guidance.

The Dragon system is not without its problems. Perhaps the most important is that the missile body diameter of 4·5in (11·4cm) sets the limit on the size of the warhead. The effectiveness of a shaped charge warhead is a function of its diameter, and at least 6in (15cm) is likely to be needed to counter the new armours coming into service on the latest Soviet tanks. In addition, the missile is slow; this aggravates the difficulties of the operator, who must hold his breath throughout the flight of the missile. The operator is also adjured to grasp the launch-tube tightly, for if he does not his shoulder may rise at the moment of launch, thus sending the missile into the ground. Finally, the rocket thrusters have been found to deteriorate in storage, and many need replacement.

Initial plans for a Dragon replacement centred on a programme designated IMAAWS, but this was halted in 1980. The new programme is called Rattler and was scheduled to start development in FY84.

Below: Test-firing a Dragon AT missile. Its limitations are referred to above, but apparently it is "user-friendly"!

LAW80

Origin: UK.
Type: Portable anti-tank weapon system.
Propulsion: Single-stage rocket motor.
Dimensions (launcher): Length (folded) 3·28ft (1m); length (gun) 4·9ft (1·5m); calibre 3·7in (94mm).
Launch weight: 19·4lb (8·8kg).
Range: 22 to 547yd (20 to 500m).

Development: The Light Anti-armour Weapon 80 (LAW80) shows a novel approach to the problem of giving an anti-tank capability to soldiers on any part of the battlefield. The British Army requirement was for a weapon able to defeat a modern tank at ranges out to 547yd (500m), but to be simple and accurate, and usable with minimal training by soldiers of both combat and supporting arms.

The answer to the first is a high-explosive anti-tank (HEAT) hollow-charge warhead of the greatest possible diameter. It is in the second area, however, that the ingenuity is applied, since sighting and ranging are achieved by means of a built-in spotting rifle, complete with a fire-loaded magazine of five rounds of ammunition. These rounds are ballistically matched to the main projectile and have a tracer tail and flash head. This simple solution has proved a great success on trials, and the LAW80 is now in full production for the British Army, the Royal Air Force Regiment, and the Royal Marines.

Right: Preparing a LAW80 for action. The launcher includes a spotting rifle firing tracer rounds to improve accuracy.

MILAN

Origin: France.
Type: Wire-guided SACLOS infantry anti-tank weapon.
Dimensions: Launcher length 35·4in (900mm); height 25·6in (650mm); width 16·5in (420mm). Missile length 30·2in (769mm); diameter 3·54in (90mm), warhead diameter 4·52in (265mm); wingspan 10in (265mm).
Weights: Launcher 36lb (16·4kg); missile 14·7lb (6·65kg).
Propulsion: Two-stage rocket motor.
Guidance: Semi-automatic command line-of-sight (SACLOS), using jet spoiler in sustainer motor rocket exhaust.
Range: 27 to 2187yd (25 to 2,000m).
Flight speed: On launch 246ft/sec (75m/s); at 2,187yd (2,000m) 656ft/sec (200m/s).
Warhead: Shaped charge 6·57lb (2·98kg).

Development: The Missile d'Infanterie Leger Anti-char (MILAN) (infantry missile anti-tank) was jointly developed by Nord in France and Messerschmitt-Bolkow-Blohm in the Federal Republic of Germany. To produce the successful system a joint company known as Euromissile was set up, and some 158,000 had been produced by January 1982 for use in 21 armies. Systems for the British Army are produced under licence by British Aerospace.

The basic system comprises a missile in a sealed tube, two of which can be carried by one infantryman; and a firing-post (comprising an optical sight, Infra-red tracker and – for infantry – a tripod mount) carried by a second man. To launch the missile the operator removes the rear cap from the tube, gets the tank target in the optical cross-hairs and squeezes the firing-button. A stack of thin DB discs in the tube impart 75g for 0·01 seconds to propel the missile from the tube, at the same time blowing through the rear nozzle and propelling the tube rearwards off the firing-post. Once the missile is at a safe distance from the firer the motor ignites, and a 1·5 second boost is followed by an 11 second sustainer burn, which accelerates the missile constantly until burnout at a range of some 6,562ft (2,000m). Thus, average speed increases with range, and time of flight to burnout is just 12·5 seconds.

MILAN has a night-firing capability by adding the MIRA thermal imaging device, and this has been adopted by the German, French and British armies. This device weighs some 15·4lb (7kg) and is mounted on the standard firing post. The TI device can detect targets out to a range of 3,280yd (3,000m), and is proving to be an invaluable addition to the infantryman's armoury.

As the missile leaves its launch tube four small wings spring open, rolling the missile to achieve stability, and holding it at a level some 20in (0·5m) above the line of sight. The operator simply has to hold the cross-hairs on the target, a great improvement on the first-generation anti-tank missiles where the operator had to control the missile throughout its flight and literally fly it onto the target.

MILAN penetrates at least 3·3in (850mm) of solid armour at 90° incidence with a very high hit probability.

MILAN is widely used on vehicles as well as by the infantry, although no air-launched version has ever been developed. Customers include Belgium, France, Germany, Greece, South Africa, Spain, Syria, Turkey and the UK, and total production will be well over 200,000 missiles.

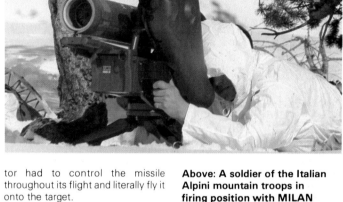

Above: A soldier of the Italian Alpini mountain troops in firing position with MILAN launcher. Italy is one of a total of 32 countries using this very effective and extremely versatile weapon system.

Below: The Mira thermal imaging system attaches to the standard tracking head of the MILAN launcher to provide night firing capability. The device is in service with the British, French and West German armies.

RPG-7/RPG-16

Origin: Soviet Union.
Type: Portable anti-tank rocket launcher.
Propulsion: Single-stage rocket motor firing 32·8ft (10m) after tube exit.
Dimensions (projectile); Length 25·45in (64·6cm); diameter 3·15in (8cm). **(Launcher):** Length 38·9in (99cm); calibre 1·57in (4·0cm).
Weight: (Launcher) 15·4lb (7kg); (projectile) 4·96lb (2·25kg).
Range: (Moving target) 328yd (300m); (stationary target) 547yd (500m).
Velocity: (Muzzle) 984ft (300m/s).

Development: Standard anti-armour weapon of Soviet infantry, the RPG-7V replaced an earlier weapon derived from the World War II German Panzerfaust, which merely fired the hollow-charge projectile from a shoulder-rested tube. RPG-7V fires a new projectile which, a few metres beyond the muzzle, ignites an intenal rocket to give shorter flight-time, flatter trajectory and better accuracy. The HEAT or HE warhead has improved fusing, the HEAT round penetrating to

12·6in (320mm) of armour. The optical sight is frequently supplemented by the NSP-2 (IR) night sight. As the grenade is fired from the tube four large fins toward the rear flick out. At the very end of the weapon small offset fins give slow rate of roll to maintain stability.

This type of anti-tank grenade has been developed as a direct result of the Soviet experiences in World War II, where battles frequently became so desperate that grenade-armed tank-killing parties were essential to success. Whether such tactics would stand any chance against modern tanks, especially those fitted with "Chobham-type" armour, is questionable.

There have been various reports of improvements to the missile, including a new two-stage rocket intended to increase range.

A folding version, designated RPG-7D, was originally issued to airborne troops, but is now on general issue throughout the Warsaw Pact. A new weapon — the RPG-16 — appears to be a development of the RPG-7, although the launch tube is fitted with a bipod on the muzzle,

suggesting a substantial weight increase.

The variety and capability of Soviet Army anti-tank weapons is impressive and indicates what a threat they consider NATO armour to be.

Above: A Soviet Army private takes up position with his RPG-7 rocket launcher to cover an infantry advance; modern tank armour may have rendered such weapons ineffective.

RPG-18

Origin: Soviet Union.
Type: Infantry/all arms anti-tank weapon.
Dimensions: Complete weapon length 27·75in (705mm), open 39·3in (1,000mm); diameter 2·75in (70mm). Missile calibre 2·5in (64mm).
Weight: System 8·81lb (4kg); missile 5·5lb (2·5kg).
Propulsion: Rocket motor, all-burnt on launch.
Range: Effective 218yd (200m); Maximum 382yd (350m).
Muzzle velocity: 374ft/sec (114m/s).
Warhead: Shaped charge, 2·5in (64mm) diameter.

Development: As the threat from tanks and armoured personnel carriers has increased, all armies have been seeking to improve the protection of their infantry, particularly at section/squad level. The Soviet Army has for many years had some excellent anti-tank weapons in its RPG series, starting with the RPG-2, and with its latest manifestation being the RPG-16. The RPG-18, however, follows a new line of development and bears a remarkable

resemblance to the US Army's M72, even to the extent of a series of simple, illustrated instructions on the side of the launch tube.

The launcher is made from a light alloy tube, which is 27·75in (705mm) long in the carrying mode, and 39·4in (1,000mm) long when open and ready for firing. Prior to firing, two rubber end-caps must be removed and two simple plastic sights raised. The missile has a small rocket motor which is all-burnt prior to leaving the launch tube. Maximum effective range is some 218yd (200m) and the warhead will penetrate about 10in (250mm) of armour plate.

This is a neat and handy little weapon with a useful performance against armoured vehicles. As the RPG-16 has such a good capability, however, it is difficult to see why the Soviet Army felt that it needed another weapon system for use in the same area. A possible explanation may be that the RPG-16 operator needs some training while the RPG-18 needs none at all.

Below: The RPG-18 launcher with its missile. The launcher is extended before firing.

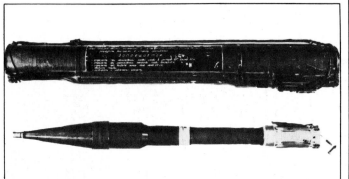

Strim LRAC 89

Origin: France.
Type: Portable anti-tank rocket launcher.
Propulsion: Single-stage, constant-pressure propulsive charge.
Dimensions (launcher): Length (ready to fire) 5·25ft (1·6m); calibre 3·5in (89mm). **(Projectile):** Length 23·6in (600mm); diameter 3·5in (89mm).
Launch weight (projectile): 4,85lb (2·2kg).
Range: (anti-tank) 656yd (600m); (anti-personnel) 1,093yd (1,000m).

Development: The Lance-Roquette Anti-Char de 89mm (LRAC 89) is a portable, tube-launched anti-tank rocket system. The glassfibre/plastic launch-tube has a bipod and is fitted with adjustable shoulder rest and hand-grip. The rocket is issued in a glassfibre/resin carrying tube and the second crewman needs only to remove the front cap and attach this tube to the rear of the launcher. The rear cover is left on to protect

the rocket until ready for use and when it is removed the firing circuit is completed.

The warhead is a hollow-charge, which can meet the NATO requirement of penetrating a single 4·7in (120mm) plate at 65 degrees, and double plate of 2·7in (40mm) followed by one of the 4·3in (110mm) both at 60 degrees and 5·9in (150mm) apart. A special round has been developed for anti-personnel/anti-vehicle use. This has a dual-purpose warhead composed of moulded steel balls which become about 1,600 projectiles and are lethal up to some 66ft (20m) from the point of burst. Two smoke rockets are also available and also a parachute illuminating flare. The LRAC 89 is in wide-scale use with the French Army.

Below: Strim in firing position. The rocket reaches its effective range limit of 400m (1,300ft) in just over 1.5 sec; a three-man team would be normal, but one-man operation is also possible.

TOW 2, BGM-71

Origin: USA.
Type: Heavy anti-tank missile.
Propulsion: Hercules K41 boost (0·05s) and sustain (1s) motors.
Dimensions: Length 55·1in (140cm); body diameter 6in (15·2cm); span (wings extended) 13·5in (34·3cm).
Launch weight: (BGM-71A) 47·4lb (21·5kg).
Range: 213 to 4,000yd (65 to 3,750m).
Flight speed: 625mph (1,003km/h).
Warhead: (BGM-71A) Picatinny Arsenal 8·6lb (3·9kg) shaped-charge with 5·3lb (2·4kg) explosive.

Development: The TOW (Tube-launched, Optically-tracked, Wire-guided) missile is likely to set an all-time record in the field of guided-missile production. Prime contractor Hughes Aircraft began work in 1965 to replace the 106mm recoilless rifle. The missile's basic infantry form is supplied in a sealed tube which is clipped to the launcher. The missile tube is attached to the rear of the launch tube, the target sighted and the round fired. The boost charge pops the missile from the tube, firing through lateral nozzles amidships. The 4 wings indexed at 45° spring open forwards, and the 4 tail controls flip open rearwards. Guidance commands are generated by the optical sensor in the sight, which continuously measures the position of a light source in the missile relative to the LOS and sends steering commands along twin wires. These drive the helium-pressure actuators working the four tail controls in pairs for pitch and yaw. In 1976 production switched to ER (Extended-Range) TOW with the guidance wires

lengthened from 9,842ft (3,000m) to the figure given. Sight field of view reduces from 6° for gathering to 1·5° for smoothing and 0·25° for tracking.

TOW reached IOC in 1970, was used in Vietnam and the 1973 Middle East war, and has since been produced at a higher rate than any other known missile. The M65 airborne TOW system equips the standard American attack helicopter, the AH-1S TowCobra and the Marines' twin-engine AH-1J and -1T Improved SeaCobra, each with a TSU (Telescopic Sight Unit) and two quad launchers. Other countries use TOW systems on the BO 105, Lynx, A109, A129, 500MD and other attack helicopters.

In late 1981 production began of the Improved TOW, with a new warhead triggered by a long probe, extended after launch to give 15in (381mm) stand-off distance for greater armour penetration. The shaped-chargehead, with LX-14 filling and a dual-angle deformable liner, is also being retrofitted to many existing rounds.

Hughes is now producing TOW 2, which has several I-TOW improvements, plus a new warhead with the same diameter as the rest of the missile with a mass of 13lb (5·9kg) and an even longer (21·25in, 540mm) extensible probe, calculated to defeat all tanks of the 1990s. Flight performance is maintained by a new double-base motor giving about 30 per cent greater total impulse, and the command guidance link has been hardened.

Below: TOW 2, with 6in (152mm) warhead and probe extended, after launch from an M2; side exhausts avoid guidance wires.

BATTLEFIELD ROCKETS AND MISSILES

122mm BM-21

Origin: Soviet Union.
Type: Multiple rocket system.
Weight: One rocket, 101lb (45·9kg); launcher, 7,718lb (3500kg); vehicle, launcher and 40 rounds, 11·3 tons (11,500kg).
Length: Rocket 8·99ft (2,740mm); vehicle, 24·1ft (7,350mm).
Calibre: 4·8in (122mm).
Engine: (Vehicle) ZIL-375 vee-8 gasoline, 175hp.
Speed: (Vehicle) 47mph (75km/h).
Launcher: Elevation 0° to +50°.
Traverse: +120°.
Time to reload: 10 minutes.
Maximum range: (Rocket) over 16,395yd (15,000m).

Development: An important multi-rocket system which first appeared in November 1964, the BM-21 uses a smaller-calibre rocket than any other of its era, and can thus fire a greater quantity (40). It is the first rocket system carried by the outstanding Ural-375 truck, which among other attributes has exceptional cross-country capability. The BM-21 is used by the Soviet ground forces and by those of several Warsaw Pact and other nations, totalling at least 27 armies.

BM-21 is an especially significant asset to the Soviet artillery. The rockets are fired in salvo, or "rippled" in sequence or selected individually, always with the vehicle

parked obliquely to the target to avoid blast damage to the unprotected cab. Each launcher delivers its full 40-round salvo in less than 30 seconds, producing 0·73 tons (760kg) of HE on the target. Reloading of the Soviet Army's BM-21 takes some 10 to 15 minutes, but the Czechoslovak Army has developed its own version carrying a palletized reload, which can load a second salvo in about one minute.

Multiple rocket launchers such as this can be used to bombard an area, delivering a devastating concentration of fire at critical moments in the battle. They are also particularly suitable for delivering chemical agents, and such projectiles are known to have been produced.

Further Soviet multiple rocket systems appear regularly. The BM-24 is a 240mm system with 12 rounds mounted on a ZIL-157 6 x 6 truck. BM-25 is a 250mm missile, six of which are mounted on a KrAZ-214 6 x 6 chassis. The latest is the BM-27, a 220mm rocket with a range of 21·7 to 24·9 miles (35 to 40km) and mounted on a ZIL-135 8 x 8 truck. This entered service in 1978-9 but little is known of it yet in the West.

These Soviet rocket systems are extremely effective and can bring down heavy fire in a very short time.

Above: Ripple fire from a BM-21 battery. The rockets, each with a 19kg (41.9lb) warhead, can be fired individually or in salvo.

Below: BM-21s in the firing position; three six-launcher batteries form the rocket artillery battalion of a Soviet division.

122mm M1972 MRS

Origin: Czechoslovakia.
Type: Multiple rocket system.
Length: Long rocket 10·6ft (3·23m); short rocket 6·2ft (1·9m).
Launch weight: Long rocket 170lb (77kg); short rocket 97lb (43·8kg).
Launcher elevation: +50⁰/0⁰.
Launcher traverse: 240⁰.
Reload time: 2-3 minutes.
Maximum range: Long rocket 22,420yd (20,500m); short rocket 12,030yd (11,000m).

Development: In a number of cases the Czechoslovak army industry has taken a basically Soviet idea and produced a greatly superior solution; it did this with the OT-64 8 x 8 APC and has also done it with this 122mm M1972 MRS. It has taken the Soviet

BM-21 122mm rocket launcher unit and mounted it on the rear of the excellent Tatra 813 (8 x 8) truck. It has also mounted a pack of 40 missiles just in front of the launcher unit, thus reducing reloading time from 10 minutes on the Soviet vehicle to just two to three minutes. Some vehicles are fitted with a BZT dozer blade for clearing obstacles and preparing fire positions.

This very effective system is in service with the armies of Czechoslovakia, East Germany and Libya.

Right: The M1972 multiple rocket system mounts the BM-21 launcher plus palletised reloads on a Tatra 813 truck.

FROG-7

Origin: Soviet Union.
Type: Artillery rocket.
Dimensions: Length 29·53ft (9·0m); diameter 23·6in (600mm).
Launch weight: About 5,511lb (2,500kg).
Range: 9·3 miles (15km) to 40·4 miles (65km).

Development: FROG is the NATO acronym for "Free Rocket Over Ground" and is used to designate a series of tactical missiles starting with the FROG-1 which entered service in 1957. FROG-1 and -2 have long since left service, but FROG-3 is believed still to serve with some second echelon units. Mounted on a modified PT-76 chassis FROG-3 has a range of some 25 miles (40km) with a 992lb (250kg) warhead. FROG-4 and -5 differed mainly in the diameter of the missile, and may have had slightly greater ranges than the -3. FROG-6 is a dummy rocket used in training.

FROG-7 was the next in the series. First seen in 1967 it has — like all FROGs from -3 onwards — a central sustainer and a ring of 20

peripheral boost nozzles; there is, however, only one propulsion stage. The airframe is cleaner, the fins larger, and motor performance higher than previous missiles, while the launcher is simpler, with quicker elevation. There are thought to be speed brakes for range adjustment, but details of the necessary radar (doppler) tracking and radio command system are not known.

The carrier vehicle is the ZIL-135 wheeled prime mover, with an onboard crane for rapid reload. Cross-country performance is as good as the PT-76, except for the lack of an amphibious capability.

Successor to FROG-7 is the SS-21, which may already have been deployed in the Group of Soviet Forces Germany. No performance details have yet been published. FROGs (and presumably SS-21s) are held in motor rifle division rocket battalions, which comprise 2 batteries each of 2 launchers.

Below: East German Army FROG-7s on their TELs, with GAZ-66 command post vehicle.

Lance, MGM-52C

Origin: USA.
Type: Battlefield missile.
Dimensions: Length 20·24ft (6·17m); body diameter 22in (56cm).
Launch weight: 2,833 to 3,376lb (1,285-1,527kg) depending on warhead.
Propulsion: Rocketdyne P8E-9 storable-liquid 2-part motor with infinitely throttleable sustainer portion.
Guidance: Simplified inertial.
Range: 45 to 75 miles (70-120km) depending on warhead.
Flight speed: Mach 3.
Warhead: M234 nuclear 468lb (212kg, 10KT), W-70-4 ER/RB (neutron) or Honeywell M251, 1,000lb (454kg) HE cluster.

Development: In service since 1972, this neat rocket replaced the earlier Honest John rocket and Sergeant ballistic missile, with very great gains in reduced system weight, cost and bulk and increases in accuracy and mobility. Usual vehicle is the M752 (M113 family) amphibious tracked launcher, with the M688 carrying two extra missiles and a loading hoist. For air-dropped operations a lightweight towed launcher can be used. In-flight guidance accuracy, with the precisely con-

trolled sustainer and spin-stabilization, is already highly satisfactory, but a future missile could have DME (Distance Measuring Equipment) command guidance. The US Army has 8 battalions, 6 of which are deployed in Europe with 6 launchers each; the 2 remaining battalions are at Fort Sill, Okla. Lance production lasted from 1971 to 1980, during which time 2,133 missiles were built. Lance is in service with the armies of Belgium, FRG, Italy, Israel, Netherlands and the UK.

Lance is the most powerful long-range missile currently under the direct control of the tactical ground commander. Its importance lies in its potential for breaking up Warsaw Pact second and third echelon forces before they can be committed.

A successor in this vital mission is under development as the Corps Support Weapon System (CSWS). This is intended to carry an even wider variety of payloads over ranges up to 124 miles (200km), using simpler support equipment and requiring fewer men.

Below: As well as tactical nuclear warheads, Lance can carry chemicals, terminally-guided submunitions or scatterable mines.

MLRS

Origin: USA.
Type: Multiple rocket system.
Dimensions: Rocket length 13ft (3·96m); diameter 8·94in (227mm).
Launch weight: Rocket 600lb (272kg).
Propulsion: Atlantic Research solid rocket motor.
Range: Over 18·6 miles (30km).
Flight speed: Just supersonic.
Warhead: Dispenses payload of sub-munitions, initially 644 standard M42 bomblets.

Development: Known from 1972 until 1979 as the GSRS (General Support Rocket System), the MLRS (Multiple Launch Rocket System) entered service with 1st Infantry Division (Mechanized) at Fort Riley, Kansas, in 1983. It has the same battlefield mobility as armoured formations, being carried on a tracked vehicle with a trainable and elevating launcher; this can be rapidly loaded with two six-round containers without the crew of three leaving their cab. Each box houses six preloaded tubes with a 10-year shelf life. The crew can ripple-fire from two to 12 rounds in less than one minute, the fire control re-aiming after each shot. The rocket is highly accurate and is intended to carry any of three types of sub-munition, M42

shaped-charge grenade-size, scatterable anti-armour mine, or guided sub-missiles. In the future a binary chemical warhead may also be developed. Each launcher load of 12 missiles is said to "place almost 8,000 sub-munitions in an area the size of four (US) football fields". The first production system was delivered to the Army in early 1982, by which time $317 million had been voted for the first 112 vehicles and 6,210 rockets. Production is intended to rise to 5,000 rounds per month in a programme costing an estimated $4·2 billion.

The carrying vehicle is designated a Self-Propelled Launcher Loader (SPLL) and is based on the M2 IFV (*qv*). The SPLL weighs some 50,000lb (22,680kg) fully loaded and is air-portable in a C-141 StarLifter. It can travel at 40mph (64km/h) and can ford a depth of 40in (1·01m), but is not amphibious.

One of the major problems with high rate-of-fire rocket systems is that of resupply, and MLRS is no exception. Each battery of nine launchers will have its own ammunition platoon of 18 resupply vehicles and trailers, and there will be many more farther back in the logistic system.

In mechanized and armoured divisions there will be one MLRS battery in the general support battalion

(with two batteries of M110A2), while light divisions will have an independent battery. There will be an MLRS battalion of three batteries with each corps.

The MLRS is also the subject of a major NATO programme involving France, Italy, the Federal Republic of Germany, and the United Kingdom.

Above: A full load of 12 MLRS rockets can be ripple-fired in under 1min, with automatic re-aiming after each launch.

Below: The tracked launcher/ loader carries driver, gunner and commander, and the crew can stop, fire and depart the scene without leaving the cab.

Pershing, MGM-31

Origin: USA.
Type: Battlefield missile.
Dimensions: Length 34·5ft (10·51m); body diameter 40in (1·01m); fin span about 80in (2·02m).
Launch weight: About 10,150lb (4,600kg).
Propulsion: Two Thiokol solid motors in tandem, first stage XM105, second stage XM106.
Guidance: Army-developed inertial made by Eclipse-Pioneer (Bendix).
Range: 100 to 460 miles (160-740km).
Flight speed: Mach 8 at burnout.
Warhead: Nuclear, usually W-50 of approximately 400KT.

Development: Originally deployed in 1962 on XM474 tracked vehicles as Pershing 1, the standard US Army long-range missile system has now been modified to 1a standard, carried on 4 vehicles based on the M656 5-ton truck. All are transportable in a C-130. In 1976 the 3 battalions with the US 7th Army in Europe were updated with the ARS (Azimuth Reference System), allowing them quickly to use unsurveyed launch sites, and the SLA (Sequential Launch Adapter) allowing one launch control centre easily to fire three missiles. To replenish inventory losses caused by practice firings, 66 additional Pershing 1a missiles were manufactured in 1978-80.

Deployment of Pershing 1a in mid-1983 totalled 108 launchers with US Army Europe and 72 with West German forces (these latter being operated by the Luftwaffe, not the Army). The US has started to replace its Pershing 1as with Pershing IIs on a one-for-one basis, but the intentions of the West German government are not yet known.

Pershing II has been studied since 1969 and has been in full development since 1974. It mates the ex-isting vehicle with Goodyear Radag (Radar area-correlation guidance) in the new nose of the missile. As the forebody plunges down towards its target the small active radar scans the ground at 120rpm and correlates the returns with stored target imagery. The terminal guidance corrects the trajectory by means of new delta control surfaces, giving c.e.p. expected to be within 120ft (36m). As a result a lighter and less-destructive warhead (reported to be based on the B61 bomb of some 15KT) can be used, which extends maximum range. The Pershing II is fitted with an "earth-penetrator" device which enables the nuclear warhead to "burrow" deep underground before exploding. This is clearly intended for use against buried facilities such as HQs and communications centres.

Development of Pershing II was envisaged as being relatively simple and cheap, but it has, in the event, proved both complicated and expensive. One problem has been with the rocket motors, which are entirely new to obtain the greatly increased range. A further complication was the sudden elevation of the Pershing II programme into a major international issue, with the fielding of the missiles becoming a test of US determination. The problems were further compounded by repeated failures in the test programme, but the final test was a success and full production has gone ahead. First fielding took place in December 1984 and all 108 are due to be deployed by December 1985.

The reason for the furore over Pershing II stems from its quite exceptional accuracy. The "hard-target kill potential" of a nuclear

Below: Pershing II test launch. The new model uses the same two-stage solid-propellant rocket motor as earlier versions.

warhead is derived from the formula: $(\text{Raw Yield})^{2/3} - (\text{c.e.p.})^2$. This means that the effect can be increased by two methods. In the first, the raw yield is increased, but this not only leads to a larger warhead, and thus a larger missile, but also the rate of increase in effect decreases as the raw yield is increased, ie, there is a law of diminishing returns. The other method of achieving a greater effect is to increase the accuracy (ie, decrease the c.e.p.), and as the effect increases by the square of the c.e.p., this is far more efficacious. Hence,

Above: The new re-entry vehicle with improved guidance carries a W85 warhead with a 15kT yield, compared with the 1a's 400kT.

the c.e.p. of 120ft (36m) has led to a warhead of much less raw yield but very much greater effect than that fitted to the Pershing 1a. This seems to place a number of hard targets in the western USSR in danger.

Below: Soldiers of the 56th Brigade, US Army, carry out field assembly of Pershing II sections after their arrival in Germany.

Pluton

Origin: France.
Type: Battlefield missile.
Dimensions: Length 25·1ft (7·64m); body diameter 25·6in (65cm).
Launch weight: 5,341lb (2,423kg).
Propulsion: Dual-thrust solid-propellant SEP rocket motor.
Guidance: Simplified inertial.
Range: 12 to 74·6 miles (20 to 120km) depending on warhead.
Warhead: 15KT or 25KT.

Development: Pluton is a very neat and highly capable battlefield support missile system, entirely French in design. The launch vehicle is based on the AMX-30 tank chassis, which transports and fires the missile from a container.

The missile has a simplified inertial guidance system and can be fitted with 2 warheads: one, of 25KT yield, is devised from the AN-52 free-fall weapon used by Mirage IIIs and Jaguars, and the second is a 15KT weapon.

Operational with five regiments of the French Army, Pluton's planned deployment of 120 systems was eventually reduced to only 30, though several times this number of missiles were manufactured, to provide reloads. Pluton will eventually be replaced by another mobile system, Hades, which is expected to have a nuclear warhead of between 10 and 25KT, and a range of at least 215 miles (350km). The missiles would be housed in canisters which would be elevated for firing at the rear of their semi-trailer transporter.

There could be 190 missiles in service, beginning in 1992.

Right: Pluton is carried on and launched from AMX-30 tank chassis; each Pluton regiment has six launchers plus command vehicles, and real-time target data can be provided by RPVs.

SS-12

Origin: Soviet Union.
Type: Battlefield missile.
Dimensions: Length 37·73ft (11·5m); diameter 43in (1·1m).
Launch weight: Probably about 17,636lb (8,000kg).
Warhead: 1 x 1MT nuclear.
Range: Estimated at up to 560 miles (900km).

Development: First reported in Western literature in 1967, this mobile ballistic missile almost comes into the strategic category, because it can menace Western Europe from WP soil and is universally agreed to have a warhead in a megaton range. Yet in many ways it is similar to Scud-B; it is little different in length, rides on an erector/launcher mounted on an MAZ-543, and almost certainly has similar strap-down simplified inertial guidance. One of the few obvious differences, apart from the much greater missile diameter, is that the erector/launcher is in the form of a ribbed container, split into upper and lower halves, which protects the weapon from the weather while it is travelling. It is possibly shock-mounted, and the container may even offer limited protection against nuclear attack.

Though there are clear illustrations of the complete weapon system on the march, or elevating for firing, the missile itself remains almost unknown, and the data are little more than the best guesses of Western intelligence. It is reasonable to assume that there is a single rocket engine burning storable RFNA/UDMH. Steering may be by refractory jet-deflector vanes, but a later method would be desirable for maximum range. The Soviet Ground Forces enjoy a wealth of superb purpose-designed vehicles, and the MAZ-543 transporter/launcher is one of the best. A beautiful exercise in packaging, it is powerful, highly mobile on rough ground, air-conditioned for extremes of heat or cold, and has automatic regulation of tyre pressure from the driver's cab on the left side. The right front cab is the launch-control station, as in the Scud-B system. The rest of the launch crew sit in the second row of seats in line with the rear doors on each side. Some related vehicles are amphibious.

Like all Soviet tactical missiles SS-12 is intended for "shoot and scoot" operation. But it is too large for snappy reloading and in any case this needs the services of one, if not two, additional vehicles. Resupply missiles are carried in their own ribbed casings, with propellant tanks empty, and even with fast pressurized-gas transfer the fuelling process must take about a quarter of an hour. The likelihood is that the Soviet Ground Forces already have a detailed itinerary of pre-surveying firing sites offering good concealment throughout Western Europe. So far as is known, this powerful thermonuclear weapon serves only with the Soviet Union.

The successor to SS-12 is the SS-22 which has recently been deployed in East Germany and in the western Military Districts of the USSR. No pictures are yet available, but it is reported that SS-22 is more accurate than SS-12, with a longer range and more powerful warhead. It is also believed to be solid-fuelled.

Above right: An SS-12 Scaleboard is raised to launch position by its MAZ-543 transporter/erector/launcher. The driver's cab is on the left, with the control station on the right.

Right: There are believed to be some 120 SS-12s in service, with 70 in the Western and the remainder in the Far Eastern TVDs (theatres of military operations); their replacement by SS-22s is currently under way.

ARMOURED PERSONNEL CARRIERS

4K 7FA G127

Origin: Austria.
Type: Armoured personnel carrier.
Dimensions: Length 19·2ft (5·87m); width 8·2ft (2·5m); height 5·5ft (1·69m).
Combat weight: 32,628lb (14,800kg).
Engine: Steyr 7FA 6-cylinder liquid-cooled, 4-stroke turbocharged diesel; 320hp at 2,300rpm.
Performance: Maximum speed 40mph (63·6km/h); road range 323 miles (520km); gradient 75 per cent; vertical obstacle 2·6ft (0·8m); trench 6·9ft (2·1m).
Crew: Driver, commander, section of 8 infantrymen.

Development: The Saurer-Werke of Austria developed a successful range of APCs during the 1950s and 1960s, which entered service with the Austrian Army in some numbers. The chassis was then used as the basis for the 4K 7FA SK 105 tank destroyer, armed with a 105mm gun mounted in a French oscillating turret. The Kürassier K, as this vehicle is called, is in service with the Austrian Army, as well as a number of others in Africa and South America.

Steyr have taken the 4K 7FA APC, improved its armour protection and fitted the engine and transmission of the SK 105, to produce the 4K 7FA G 127. This is now in service in Austria, Greece, Morocco, Nigeria and Tunisia.

Above right: The 4K 7FA G127 is a sound, well-designed "battle taxi"; the MG3 is normally stowed in the crew compartment.

Right: With no turret or heavy weapon, the G127's main armament is a 12·7mm machine gun on a simple ring mounting.

AIFV

Origin: USA.
Type: Infantry fighting vehicle.
Crew: 3 + 7.
Armament: One 25mm Oerlikon KBA-BO2 cannon; one 7·62mm MAG machine-gun coaxial with main armament.
Dimensions: Length 17·25ft (5·26m); width 9·25ft (2·82m); height (roof) 6·56ft (2·0m); height (turret) 8·59ft (2·62m).
Combat weight: 30,174lb (13,687kg).
Engine: Detroit Diesel 6V53T V6 liquid-cooled diesel; 264hp at 2,800rpm.
Performance: Road speed 38mph (61·2km/h); water speed 3·9mph (6·3km/h); road range 304 miles (490km); trench 5·35ft (1·63m); gradient 60 per cent.

Development: The Armoured Infantry Fighting Vehicle (AIFV) stems from a 1967 US Army requirement for a MICV, for which FMC produced an improved M113 designated XM765. This was rejected by the US Army, but FMC carried on development, in a company-funded programme, which resulted in the AIFV. The first order came from the Netherlands (1,601 AIFV; 119 Improved TOW version), followed by Belgium (514) and the Philippines.

The hull of the AIFV is made of aluminium armour with an appliqué layer of spaced, laminated steel armour, the two combining to give excellent protection. The weapon station includes a 25mm Oerlikon cannon and a coaxial 7·62mm machine-gun. The crew compartment accommodates 7 infantrymen, and there are 5 firing ports. There are the usual special-to-role versions: command-post, TOW missile vehicle, mortar prime mover, recovery vehicle and cargo carrier.

Possessing considerable commonality with the M113, the AIFV still represents a major advance without the complexity and vast expense of vehicles such as the M2.

Right: A development of the ubiquitous M113, the FMC AIFV adds laminate armour and trades internal space for an EWS with 25mm and 7·62mm guns.

Below: The EWS (enclosed weapon station) can be removed and the vehicle adapted to other roles: the TOW missile version mounts a twin TOW launcher and 7·62mm machine gun in its place.

AMX-10P

Origin: France.
Type: APC.
Crew: 3 + 8.
Armament: One 20mm M693 cannon; one coaxial 7·62mm MG.
Dimensions: Length 18·96ft (5·78m); width 9·12ft (2·78m); height (hull top) 6·3ft (1·92m); (overall) 8·43ft (2·57m).
Combat weight: 31,305lb (14,200kg).
Engine: Hispano-Suiza HS-115 V-8 diesel; 280hp at 3,000rpm.
Performance: Road speed 40mph (65km/h); water speed 4·4mph (7km/h); range 372 miles (600km); vertical obstacle 2·3ft (0·76m); gradient 60 per cent.

Development: The French Army was one of the first to produce a reasonable MICV, the AMX VCI (Vehicule de Combat d'Infantrie), which entered production in 1957. The VCI was based on the AMX-13 tank chassis, carried a crew of 3 and an infantry section of 10 and was armed with either a 20mm cannon or a 12·7mm MG. Successor to the VCI is the AMX-10P MICV, which entered service in 1973, since when over 2,000 have been produced for the French Army and overseas customers.

The basic infantry vehicle is made of welded aluminium and is armed with a 20mm cannon mounted in a Toucan II turret. The 8-strong infantry section sits in the rear compartment; they have individual periscopes but do not have firing ports. There are numerous versions of the AMX-10, including command posts, ambulances, repair vehicles, artillery observation vehicles, as well as various fire support vehicles fitted with HOT ATGW or a 90mm gun. The AMX-10 is in service with at least 10 armies.

Below: The AMX-10P's 20mm M693 cannon has dual feed, enabling HE or AP rounds to be selected as the situation requires.

BMD

Origin: Soviet Union.
Type: Airborne infantry fighting vehicle.
Crew: 3+6.
Armament: One 73mm Model 2A28 smooth-bore gun; one 7·62mm MG coaxial with main armament; 2 fixed bar-mounted 7·62 MGs; one "Sagger" ATGW.
Dimensions: Length 17·7ft (5·4m); width 8·62ft (2·63m); height (variable suspension) 5·3 to 6·5ft (1·62 to 1·87m).
Combat weight: 14,770lb (6,700kg).
Engine: Type 5D-20 V-6 liquid-cooled diesel; 240hp.
Performance: Road speed 43·5mph (70km/h); water speed 6mph (10km/h); range 199 miles (320km); vertical obstacle 2·6ft (0·8m); gradient 60 per cent.

Development: Two of the major requirements of a parachute-landed force are the ability to move small groups of men rapidly around the battlefield, and as strong an anti-tank defence as possible. The BMD is a very neat attempt to answer the first of these, and to add a reasonable contribution to the second. First seen in the November 1973 parade in Red Square this trim little air-droppable fire support vehicle is yet another of the "quart-in-a-pint-pot" designs which seem to flow from the Soviet State arsenals. Though such aircraft as the Antonov AN-22 (NATO codename "Cock") could easily carry the BMP APC, it was judged that the same capability could be built into a smaller and lighter APC capable of being airlifted in greater numbers, and more readily dropped by parachute.

At first styled "M-1970" in the West, the correct Soviet designation is "Boyevaya Mashina Desantnaya" (BMD). The vehicle has a crew of 3, and carries 6 parachute soldiers in open seats in the back. It is armed with the same turret as the much larger BMP-1, mounting a 73mm low-pressure gun, a 7·62mm coaxial machine-gun, and a "Sagger" ATGW on a launcher rail. One remarkable feature is the mounting of two fixed machine guns in the front corners of the hull. It is also fully amphibious, using hydrojet propulsion. It is a truly remarkable achievement to design all this capability into a vehicle weighing no more than 6·59 tons (6·7 tonnes).

BMDs were widely used in the Soviet invasion of Afghanistan and would be used in any major airborne operation. Its roles include bold reconnaissance immediately following a landing, rapid movement away from the DZs (especially to capture key targets), direct support of infantry assaults, and anti-tank defence.

Above: The compact but potently armed BMD can provide instant support for airborne operations.

Below: BMDs rigged for air dropping; on the ground they can carry six plus three crew.

BMP-1/-2

Origin: Soviet Union.
Type: MICV.
Crew: 3 + 8.
Armament: One 30mm cannon;
one 7·62mm coaxial machine-gun;
one AT-5 (Spandrel) ATGW.
Dimensions: Length 22·1ft
(6·74m); width 9·64ft (2·94m);
height overall 7·1ft (2·15m).
Combat weight (approx): 33,069lb
(15,000kg).
Engine: 6-cylinder in-line water-
cooled diesel; 300hp at 2,000rpm.
Performance: Road speed
49·7mph (80km/h); water speed
3·7 to 5mph (6 to 8km/h); range
310 miles (500km); vertical
obstacle 2·6ft (0·8m); gradient 60
per cent.

Development: When it was first
seen by Western observers in 1967,
the BMP-1 was thought to be exactly what the West's own armies
needed: a true Mechanised Infantry
Combat Vehicle (MICV). It was
significantly smaller than the West's
own APCs, but with much greater
firepower. The eight troops have
multiple periscopes and can, at least
in theory, fire on the move. There is
a vehicle crew of three: commander
(who is also commander of the dis-
mounted infantry section), driver,
and gunner. Both crew and
passengers have nuclear/biological/
chemical warfare protection in the
pressurized hull, and air filters are fitted as standard.

The 73mm low-pressure gun has a
smooth bore and fires fin-stabilized
HEAT rounds, the automatic loader
giving a firing rate of eight rounds
per minute. The missile launcher
above the gun carries one round
ready to fire, and three more rounds
are carried inside the vehicle. In addition, one of the infantrymen inside
the vehicle normally carries an SA-7
Grail SAM.

The driver is seated at the front of
the hull on the left, with the vehicle
commander to his rear, the engine
being mounted to the driver's right.
BMP is fully amphibious, being propelled in the water by its tracks. A
full range of night-vision equipment
is fitted, although only the old-
fashioned active infra-red types have
been seen so far, which is, of
course, easily detected on a modern
battlefield.

A number of variants have been
developed. One minor modification
to the basic vehicle was that a
sharper bow has been fitted to all
versions appearing since 1970, clearly designed to improve the handling
characteristics in water. In 1975 a
variant was seen with the troop
compartment replaced by a rear-
mounted turret mounting a battlefield radar, known as BMP-SON,
of which no clear picture is yet
available in the West. A reconnaissance version is now rapidly
replacing the ageing PT-76. The
radar version is armed with only a
machine-gun; the recce version
lacks an ATGW, but carries observation equipment as well as its 73mm
gun in an enlarged turret. The recce
version is designated BMP-R.

The Soviet Army received a rude

shock when the BMP was used in
combat for the first time in the 1973
Yom Kippur war. Used by the Egyp-
tians exactly as taught in the Soviet
tactical text-books, the result frequently verged on disaster. It was
demonstrated beyond doubt that
the idea of a MICV which would
charge on to enemy positions with
all armament blazing away — including the infantrymen's rifles firing
through ports — was very nice in
theory but totally unworkable in
practice.

There followed a very open, extremely frank, and valuable discussion in Soviet military journals, as a
result of which the whole concept of
the use of BMP was revised. Soviet
doctrine now dictates that the whole
concept of the use of APCs will involve the infantry dismounting some
220 to 330yd (200 to 300m) short of
the objective and completing the
final phase of the assault on foot,
covered by fire from artillery, tanks
and the BMPs. It is, incidentally, interesting to note that the West
learned a great deal from the Soviet
discussions on the role of MICVs
and APCs, and probably avoided
some very expensive mistakes as a
result!

The long-awaited successor to
BMP-1 appeared in a Moscow
parade in November 1982, although
it had been in service for at least 3
years prior to that. The most
noticeable difference is the new turret, mounting a 30mm cannon,
which apears to have a dual role
against both ground and air (eg.
helicopter) targets. The AT-5 (Spandrel) ATGW is much more effective
than the Sagger fitted to BMP-1. The
BMP-2 is a sensible development of the BMP-1 and seems to incorporate most of the lessons learned with the earlier vehicle.

**Below: BMP-1s in snow. BMP-2
has several changes, including
30mm cannon and AT-5 ATGW.**

**Above: A BMP-1 halted at the
shore line to discharge its
infantry squad during a river-
crossing operation.**

**Below: BMP-1 is in wide-scale
use; this reconnaissance version
is in service with the Polish
Army.**

BTR-70

Origin: Soviet Union.
Type: APC.
Crew: 2 + 14.
Armament: One 14·5mm KPV HMG; one coaxial 7·62mm PKT MG.
Dimensions: Length 25·75ft (7·85m); width 9·22ft (2·81m); height (hull top) 7·1ft (2·15m); height (turret top) 7·97ft (2·43m).
Combat weight (approx): 23,148lb (10,500kg).
Engines: Two diesel engines.
Performance: Not known.

Development: First seen in November 1961, the BTR-60 family of armoured personnel carriers is impressive and is widely used in Warsaw Pact forces and has been exported to at least 30 other countries. The large hull is boat-shaped for good swimming and to deflect hostile fire. It runs on eight land wheels, all powered and with power steering on the front four. Tyre pressures are centrally controlled at all times. The twin rear engines can be switched to drive waterjets.

The basic BTR-60P has an open top or canvas hood, and carries two crew plus 16 troops. Typical armament is a 12·7mm and from one to three 7·62mm SGMB or PK. The BTR-60PK has an armoured roof, carries 16 passengers and has a single 12·7 or 7·62mm gun. The PB has a turret with coaxial 14·5mm KPVT and 7·62 PKT (the same turret as on the BTR-40P-2) and carries 14 troops. There is a special version for platoon and other commanders, with extra communications (BTR-60PU). BTR-60P is the standard amphibious APC for the Soviet marines.

Following some doubt in the West as to the Soviet Army's intentions with wheeled APCs it is now known that a new design has entered service — the BTR-70. Like the BTR-60 this is an 8 x 8 design, and was first seen by Western observers at the November military parade in Moscow's Red Square. BTR-70 has a longer hull than BTR-60 with a redesigned engine compartment which changes the rear-end shape slightly. There is also a rather more marked gap between the front and rear pairs of the wheels. Apart from these differences, however, BTR-70 seems to be very similar to BTR-60PB, with the same turret and armament, although it is possible that

Above: Despite the success of the BMP tracked IFVs the BTR-60 8x8 wheeled APC has been developed into BTR-70 seen here.

the newer vehicle may have diesel engines in place of the two petrol engines in BTR-60. The unfortunate infantrymen still have to debus over the sides of the vehicle, a hazard which the Soviet Army appears to consider to be an acceptable risk.

M2/M3

Origin: USA.
Type: MICV.
Crew: 3 + 6.
Armament: One 25mm Hughes Chain Gun; one 7·62mm machine-gun coaxial with main armament; twin launcher for Hughes TOW ATGW.
Dimensions: Length 21·16ft (6·45m); width 10·5ft (3·20m); height 9·74ft (2·97).
Combat weight: M2, 49,969lb (22,666kg); M3 49,298lb (22,362kg).
Engine: Cummins VTA-903T water-cooled 4-cycle diesel developing 506bhp.
Performance: Road speed 41mph (66km/h); water speed 4·5mph (7·2km/h); range 300 miles (384km); vertical obstacle 3ft (0·91m); trench 8·33ft (2·54m); gradient 60 per cent.

Development: The United States Army has had a requirement for an MICV for well over 15 years. The first American MICV was the XM701, developed in the early 1960s on the M107/M110 self-propelled gun chassis. This proved unsatisfactory during trials. The Americans then tried to modify the current M113 to meet the MICV role: a variety of different models was built and tested, but again these vehicles failed to meet the army requirement. As a result of a competition held in 1972, the FMC Corporation, which still builds the M113A2, was awarded a contract to design an MICV designated the XM723.

The XM723 did not meet the requirements of the US Army and further development, based on the same chassis, resulted in the Fighting Vehicle System (FVS), which comprised two vehicles, the XM2 Infantry Fighting Vehicle and the XM3 Cavalry Fighting Vehicle. These were eventually accepted for service as the M2 and M3 Bradley Fighting Vehicles. The US Army has a requirement for some 6,882 M2/M3 vehicles, and three battalions of M2s were formed in 1983, the first at Ford Hood, Texas.

The primary task of the M2 in the eyes of the US Army is to enable infantry to fight from under armour whenever practicable, and to be able to both observe and to use their weapons from inside the vehicle. The M2 will replace some, but not all, of the current M113 APCs, as the latter are more than adequate for many roles on the battlefield. The M2 has three major advances over the existing M113 APC. First, the IFV has greater mobility and better cross-country speed, enabling it to keep up with the M1 MBT when acting as part of the tank/infantry team. The tank provides long-range firepower while the IFV provides firepower against softer, close-in targets. The M2's infantry also assist tanks by locating and destroying enemy anti-tank weapons.

The hull of the M2 is of all-welded aluminium construction with an ap-

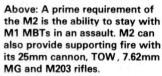

Above: A prime requirement of the M2 is the ability to stay with M1 MBTs in an assault. M2 can also provide supporting fire with its 25mm cannon, TOW, 7·62mm MG and M203 rifles.

Below: An M2 infantry squad can deploy an M60 MG, five M16A1 rifles (two with M203 40mm grenade launchers) three M72 LAWS and up to three Dragon ATGW launchers.

pliqué layer of steel armour welded to the hull front, upper sides and rear for added protection. The hull sides also have a thin layer of steel armour, the space between the aluminium and steel being filled with foam to increase the buoyancy of the vehicle. The armoured protection of the IFV is claimed to be effective against Soviet 14·5mm armour-piercing rounds and 155mm air-burst shell splinters.

The driver is seated at the front of the vehicle on the left, with the engine to his right. The two-man turret is in the centre of the hull and the personnel compartment is at the rear. Personnel entry is effected through a large power-operated ramp in the hull rear. The two-man power-operated turret is fully stabilized and is armed with a 25mm Hughes Chain Gun and a coaxial 7·62mm machine-gun. The weapons can be elevated to +60° and depressed to −10°, turret traverse being 360°. Mounted on the left side of the turret is a twin launcher for the Hughes TOW ATGW. A total of 900 rounds of 25mm, 2,340 rounds of 7·62mm and seven TOW missiles are carried. The troop compartment is provided with six firing ports (two in each side and two at the rear) for the 5·56mm M231 weapon. The M231 is a specially developed version of the M16, cut-down and sealed in a ball mount. It is somewhat ironic that the outcome of a requirement for the infantry to be able to use their weapons from inside the vehicle should be an additional and specialized rifle.

Three M72A2 light anti-tank weapons are also carried. The M2 is fully amphibious, although a flotation screen is required, and is propelled in the water by its tracks. An NBC system is fitted, as is a full range of night vision equipment.

Some 3,300 M3 Cavalry Fighting Vehicles are to be purchased to replace M60s and M113s in armoured

cavalry units and in the scout platoons of mechanized infantry and tank battalions. The M3 is outwardly identical with the M2: the major differences lie in the internal stowage and the layout of the crew compartment. The M3 carries twice the number of stowed 25mm rounds and 10 stored TOW missiles. Only 2 cavalrymen are housed in the rear compartment and the firing ports are not used.

The chassis of the M2/M3 is also used as the basis for the Vought Multiple Launch Rocket System and the Armoured, Forward-Area, Rearm Vehicle (AFARV), which has been designed to supply MBTs with ammunition when they are in the battlefield area.

Right: The M2's 25mm Chain Gun can defeat the front armour of 19 of the Warsaw Pact's current range of 25 armoured vehicles, but against tanks TOW is needed, preferably used in flank attacks; the M2 carries five reloads.

M113 series

Origin: USA.
Type: APC.
Crew: 2 + 11.
Armament: One Browning 0·5in (12·7mm) machine-gun.
Dimensions: Length 15·95ft (4·863m); width 8·8ft (2·686m); height 8·2ft (2·5m).
Combat weight: 25,000lb (11,341kg).
Engine: General Motors Model 6V53 6-cylinder water-cooled diesel developing 215bhp at 2,800rpm.
Performance: Road speed 42mph (67·6km/h); water speed 3·6mph (5·8km/h); range 300 miles (483km); vertical obstacle 2ft (0·61m); trench 5·5ft (1·68m); gradient 60 per cent.

Development: In the early 1950s the standard United States Army APC was the M75, followed in 1954 by the M59. Neither of these was satisfactory and in 1954 foundations were laid for a new series of vehicles. In 1958 prototypes of the T113 (aluminium hull) and T117 (steel hull) armoured personnel carriers were built. A modified version of the T113, the T113E1, was cleared for production in mid-1959 and production commenced at the FMC plant at San Jose, California,

in 1960. It is also built in Italy by Oto Melara, which has produced a further 4,000 for the Italian Army and for export. In 1964 the M113 was replaced in production by the M113A1, identical with the earlier model but for a diesel rather than a petrol engine.

The M113A1 had a larger radius of action than the earlier vehicle. The M113 had the distinction of being the first armoured fighting vehicle of aluminium construction to enter production. The driver is seated at the front of the hull on the left, with the engine to his right. The commander's hatch is in the centre of the roof and the personnel compartment is at the rear of the hull. The infantry enter and leave via a large ramp in the hull rear, although there is also a roof hatch over the troop compartment. The basic vehicle is normally armed with a pintle-mounted Browning 0·5in machine-gun, which has 2,000 rounds of ammunition. The M113 is fully amphibious and is propelled in the water by its tracks. Infra-red driving lights are fitted as standard. FMC has developed a wide variety of kits for the basic vehicle including an ambulance kit, NBC kit, heater kit, dozer-blade kit, various shields for

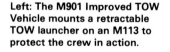

Left: The M901 Improved TOW Vehicle mounts a retractable TOW launcher on an M113 to protect the crew in action.

Above: Another M113 derivative, the M548 cargo carrier, forms the basis for the M730 quad Chaparral SAM launcher.

machine-guns and so on.

The current production model is the M113A2 which is essentially an M113A1 with improved engine cooling and improved suspension. Most US Army M113 and M113A1 vehicles are now being brought up to M113A2 standard.

There are more variants of the M113 family than any other fighting vehicle in service today, and there is room here to mention only some of the more important models. The

M577 is the command model, with a much higher roof and no armament. There are 2 mortar carriers: the M125 with an 81mm mortar, and the M106 with a 107mm mortar. The flame thrower model is known as the M132A1, and is not used outside the United States Army. The M806 is the recovery model, and this is provided with a winch in the rear of the vehicle and spades at the rear. The anti-aircraft model is known as the Vulcan Air Defense System or

Marder

Origin: Federal Republic of Germany.
Type: MICV.
Crew: 4 + 6.
Armament: One Rheinmetall Mk 20 Rh 202 20mm cannon; one coaxial 7·62mm MG3 MG; one remote-controlled 7·62mm MG3 MG.
Dimensions: Length 22·3ft (6·79m); width 10·6ft (3·34m); height (hull top) 6·23ft (1·9m); height (turret top) 9·4ft (2·86m).
Combat weight: 62,169lb (28,200kg).
Engine: MTU MB 833 Ea-500 6-cylinder liquid-cooled diesel; 600hp at 2,200rpm.
Performance: Road speed 46·6mph (75km/h); range 323 miles (520km); vertical obstacle 3·28ft (1m); gradient 60 per cent.

Development: The Marder MICV was designed and produced under the directions of men who had fought in World War II, especially on the Eastern Front and who thus had first-hand experience of dealing with the Soviet Army. A series of prototype MICVs appeared throughout the 1960s, and in 1969 the order was finally placed with Rheinstahl for the vehicle now designated Marder, of which some 3,111 were produced between 1971 and 1975. Further chassis were produced for the Roland 2 SAM system. Crew for the MICV consists of the commander, driver, two gunners and six infantrymen who enter and exit via a ramp at the rear.

Marder is constructed of steel armour and, with a combat weight of 27·75tons (28,200kg), is by far the heaviest MICV/APC in service. Armament comprises a 20mm cannon and two 7·62mm MGs, and there are also four spherical firing ports for the infantrymen. Few variants of Marder have appeared, as it is a very expensive platform. How-

ever, Thyssen Henschel have developed the TAM medium tank and the VCTP MICV for the Argentine Army, both based on the Marder.

Above: The German Army's **FlaRakPz mounts a Roland II SAM system on the chassis of the Marder combat vehicle.**

Above: Whereas the lavishly **equipped MICVs now in service are substantial weapon systems in their own right, the M113, standard Western APC since the 1960s, acts as a more basic personnel delivery system.**

Below: Based on the M113A1, **with modified suspension and a raised cab, the M548 cargo carrier is used for a variety of load-carrying tasks. The 0.5in MG is provided with 300 rounds of ammunition.**

M163; this is armed with a 6-barrelled 20mm General Electric cannon. The M548 tracked cargo carrier is based on an M113 chassis, can carry 5 tons (5,080kg) of cargo and is fully amphibious. There are many models of the M548, including the M727, which carries 3 Hawk surface-to-air missiles, and the M730, which carries 4 Chaparral short-range surface-to-air missiles. Yet another version, the M752, carries the Lance tactical missile system, whilst the M688 carries 2 spare missiles.

One recent model is the M901 Improved TOW Vehicle (ITV), with a retractable launcher that carries 2

Hughes TOW ATGWs in the ready-to-launch position. Almost 2,000 of these vehicles have been ordered by the US Army. The latest model to be ordered is the Surface-Launched Unit Fuel-Air Explosive (SLUFAE) launcher, which is an unguided rocket system based on the M548 chassis.

The M113 series and its derivatives will remain in service with the US and foreign armies for many years to come. Like the "Jeep" of World War II, it is cheap, simple to manufacture, easy to maintain, and effective in use. The US Army may well one day wish that the same applied to the M2/M3 series.

MCV-80

Origin: UK.
Type: APC.
Crew: 2 + 8.
Armament: One 30mm Rarden cannon; one 7·62mm Chain Gun coaxial with main armament.
Dimensions: Length 20·8ft (6·34m); width 9·94ft (3·03m); height (hull roof) 6·3ft (1·93m); height (turret) 8·99ft (2·74m).
Combat weight: 52,910lb (24,000kg).
Engine: Rolls-Royce CV8 V-8 diesel; 550hp at 2,300rpm.
Performance: Road speed 46·6mph (75km/h); road range 310 miles (500km); trench 8·2ft (2·5m); gradient 60 per cent.

Development: Like the US Army, the British introduced a simple, box-like APC (FV432) in the 1960s, and then spent some 15 years arguing about a successor. The basis of the argument was whether the requirement was for a "battle taxi" which delivered the infantry to the crucial point fast and in relative safety, or whether it was for a Mechanised Infantry Combat Vehicle (MICV) which enabled the infantry to fight from inside, only emerging in an at-

tack when the vehicle had arrived at the objective. The MICV concept is epitomized by the German Marder and the Soviet BMP *(qqv)* and undoubtedly some British prototypes of such a vehicle were tested.

The MCV-80 has been developed by GKN Sankey; detailed design work started in 1977 and three prototypes were running in 1980. The vehicle has a 2-man turret mounting a 30mm Rarden cannon and a 7·62mm Chain Gun. The 8 men of the infantry section do not, however, have firing ports and the MCV-80 has thus ended up as a more heavily armed APC, rather than a MICV.

The British Army has a require-

ment for 1,800 to 2,000 MCV-80s. Some early models were tested on Exercise Lionheart in 1984 and were reported to have been very satisfactory. There will be a number of variants, including command vehicles, mortar carriers, repair and recovery vehicles, and a mechanised engineer vehicle. GKN Sankey is also examining a version called APC90 which is shorter, has one less road wheel, lighter and costs 25 per cent less. No costs for MCV80 have been announced, but it is likely to be very expensive.

Below: An MCV-80 keeps pace with a Challenger MBT during trials in wintry conditions.

MT-LB

Origin: Soviet Union.
Type: APC.
Crew: 2 + 11.
Armament: One 7·62mm PKT MG.
Dimensions: Length 21·2ft (6·45m); width 9·35ft (2·85m); height (turret top) 6·14ft (1·87m).
Combat weight: 26,234lb (11,900kg).
Engine: YaMZ 238V, V-8 cylinder diesel; 240hp at 2,100rpm.
Performance: Road speed 38mph (61·5km/h); water speed 3-3·7mph (5-6km/h); range 310 miles (500km); vertical obstacle 2·3ft (0·7m); gradient 60 per cent.

Development: This vehicle was first seen in 1970 and was designated by the West as the "M-1970" APC. It is now known that the correct designation is "MT-LB". The hull is of all-welded steel construction with the engine between the crew and passenger compartments. There are two rear doors (as with BMP) and hatches over the passenger compartment. The sole vehicle weapon is one 7·62mm machine-gun, mounted in a turret identical to that on the BTR-60P. The vehicle is fully amphibious, being propelled in the water by its tracks. Although slightly shorter and narrower than the BMP,

the MT-LB does, in fact, carry two more passengers.

It is of considerable interest that the Soviet Army has felt the need to develop another type of APC when it already had a superb vehicle available in the shape of the BMP. The MT-LB has not been produced in such large numbers as the BMP, although it is known to be in service with the Soviet and East German armies. Presumably, therefore, the Soviet Army wanted a less sophisticated and more adaptable vehicle, and considered that the additional expense of another development and production programme would be worth it in the long term. BMP has only been seen as an APC, command vehicle or radar vehicle, whereas the MT-LB serves as APC (3 crew plus 10 infantrymen), command vehicle, artillery command post, artillery tractor, cargo carrier, minelayer and radio vehicle. It is clearly a useful and versatile machine, and is in service with the armies of the USSR, Bulgaria, East Germany, Hungary, Poland and Yugoslavia.

In addition, the US Army has acquired some MT-LBs (presumably from the Middle East) for training its "Red Army opposite forces".

Above: Anti-tank grenadier, machine-gunner and rifleman deploy from their MT-LB APC.

Below: E. German Army MT-LBs towing T-12 100 guns and carrying crew and ammunition.

LAV-25

Origin: Switzerland.
Type: Light armoured vehicle/APC.
Crew: 3 + 6.
Armament: M242 25mm
Bushmaster automatic cannon;
M240 7·62mm coaxial MG.
Dimensions: Length 21ft (6·4m);
width 7·22ft (2·2m) height 8·2ft
(2·5m).
Weights: Empty 19,850lb
(9,004kg); combat loaded 27,559lb
(12,501kg).
Engine: Detroit Diesel 6V-53T,
6-cylinder turbocharged diesel;
275hp at 2,800rpm.
Performance: Maximum road
speed 63mph (101km/h); road
range 485 miles (781km);
swimming speed 6mph (9·65km/h);
gradient 70 per cent; side slope 35
per cent.

Development: In 1981-82 the US
Army's Tank-Automotive Command
(TACom) carried out a series of tests
for a light armoured vehicle (LAV) to
be procured for the US Army and

US Marine Corps. Four vehicles
were tested: the British Alvis
Scorpion-Stormer, the Swiss-
designed but Canadian-produced
MOWAG ''Piranha'', and the
Cadillac-Cage V-150 and V-300. The
Scorpion-Stormer series is tracked;
the remainder are wheeled. Both the
Army and Marine Corps required
these vehicles for employment with
the Rapid Deployment Force (RDF),
although the Army's original requirement was reduced from 2,315 to 680
in 1982, and is now totally in doubt.
Indeed, in mid-1983 the Congress
denied funds for Army procurement
of the LAV because it was considered that the requirement had not
been properly justified. This is not to
say, however, that the Army will not
get its LAV in the long run.
The US Army's basic requirement
is for a Mobile Protected Gun-Near
Term (MPG-N), which differs
substantially from the LAV required
by the USMC. The Army plans to
equip its light divisions with 2 MPG

battalions, each with 41 LAVs and
40 HMMWVs (Hummer), for reconnaissance, fire support, and anti-
tank defence. These battalions appear — on paper, at least — to be
too lightly equipped, even for the
RDF.
The Original Swiss Piranha vehicle
is a 6 x 6 vehicle and serves in
Nigeria and Sierra Leone. The 6 x 6
version is also produced in Canada

**Above: With turret-mounted
25mm gun and coaxial machine
gun, the LAV-25 will be used by
USMC Rapid deployment Force.**

as the Grizzly (APC) Cougar (76mm
fire support) and Husky (recovery
vehicle); some 491 have been produced. The LAV-25 for the USA has
a slightly longer hull and is an 8 x 8
vehicle.

LVTP-7

Origin: USA.
Type: Amphibious APC.
Crew: 3 + 25.
Armament: One M85 0·5in
(12·7mm) machine-gun.
Dimensions: Length 26·1ft
(7·94m); width 10·73ft (3·27m);
height (overall) 10·69ft (3·26m);
height (turret roof) 10·24ft
(3·12m).
Combat weight: 52,768·9lb
(23,936kg).
Engine: Cummins VT400
8-cylinder, water-cooled diesel;
400hp at 2,800rpm.
Performance: Road speed 45mph
(72·42km/h); water speed 8·4mph
(13·5km/h); range 300 miles
(482km); gradient 60 per cent.

Development: In 1964, the Marines
issued a requirement for a new
LVTP and the FMC Corporation was
selected to build 17 prototypes. The
first of these was completed in 1967
under the designation of LVTP-X12.
Trials were carried out in Alaska,
Panama and various other Marine installations, and in 1970 FMC was
awarded a production contract for
942 vehicles. The first production
LVTP-7 was completed in August
1971 and production continued until
September 1974.
The role of the LVTP-7 is to
transport marines from ships off
shore to the beach, and, if required,
to carry them inland to their objective.
The hull of the LVTP-7 is of all-
welded aluminium construction and
varies in thickness from 7 to 45mm.
The engine and transmission are at
the front of the hull and can be
removed as a complete unit if required. The driver is seated at the
front, on the left, with the commander to his rear. The LVTP-7 is
armed with a turret-mounted M85
0·5in machine-gun. This is mounted
on the right side and has an eleva-

tion of +60° and a depression of
−15°; traverse is a full 360° and a
total of 2,000 rounds of ammunition
is carried. The personnel compartment is at the rear of the hull, where
the 25 marines are provided with
bench type seats which can be
quickly stowed so that the vehicle
can be used as an ambulance or
cargo carrier. The usual method of
entry and exit is via a large ramp at
the rear of the hull.
The LVTP-7 is propelled in the
water by two water-jets, one in each
side of the hull towards the rear.
These are driven by propeller shafts
from the transmission. Basically
pumps draw water from above the
track, and this is then discharged to
the rear of the vehicle. Deflectors at
the rear of each unit divert the
water-jet stream for steering, stopp-
ing and reversing.
There are two special versions of
the LVTP-7 in service. The first of
these is the LVTR-7. This is used to
repair disabled vehicles, for which a
wide range of equipment is carried,
including an hydraulic crane and
winch. The second model is the
LVTC-7, a special command model,
with additional radios and other
equipment. Planned engineer and
howitzer versions were not produced.
All LVTP-7s are now being rebuilt
to the LVTP-7A1 standard and fitted
with new Cummins VT400 8-cylinder
diesel and many other im-
provements, including smoke
generators, passive night vision
equipment, improved fire suppression system, installation of PLARS,
improved crew/troop compartment
ventilation and improved electric
weapon station. In addition, vehicles
are now being produced to the
LVTP-7A1 standard.
LVTP-7s also serve with the armed forces of Argentina, Italy, Spain,
South Korea, Thailand and
Venezuela.

**Above: The LVTP-7 carries 25
marines in addition to the three-
man crew; each side of the rear
ramp are the square-section
water jet outlets, with curved
deflectors for steering.**

**Below: An LVTP-7 climbs ashore
on a Turkish beach during a US
Marine Corps exercise. The front
of the hull is protected by
welded aluminium armour to a
thickness of 45mm (1¾in).**

Ratel IFV

Origin: Republic of South Africa.
Type: Infantry fighting vehicle.
Crew: 3 + 7.
Armament: One 20mm cannon; one coaxial 7·62mm MG; 2 pintle-mounted 7·62mm AAMG.
Dimensions: Length 23·65ft (7·21m); width 8·26ft (2·52m); height (overall) 9·58ft (21·92m); height (hull) 6·92ft (2·11m).
Combat weight: 40,784lb (18,500kg).
Engine: D3256 BTXF 6-cylinder in-line diesel; 282hp at 2,200rpm.
Performance: Road speed 65·25mph (105km/h); range 621 miles (1,000km); vertical obstacle 1·15ft (0·35m); gradient 60 per cent.

Development: The Ratel 6 x 6 wheeled infantry fighting vehicle (IFV) has been developed in the Republic of South Africa to meet the country's special requirements. These include good road range and speed, protection, minimal maintenance, and good cross-country mobility. This last requirement is, however, in the context of vast expanses of desert and scrub, rather than the close, broken European type of terrain.

The Ratel appeared in July 1974 and entered service in 1977. It now serves in the South African Army in large numbers and has seen much action in that country's many small conflicts. The hull is all-welded steel, with the driver in the centre of the front of the vehicle. The infantry section of 6 men sit on bench seats in the centre of the passenger compartment, and have both vision blocks and firing ports.

The IFV has a 20mm cannon in a turret mounted immediately behind the drive. The Ratel 90 Fire Support Vehicle has a 90mm gun in the same type of turret. There is also a 60mm mortar version, a command vehicle and a logistic support vehicle on an 8 x 8 chassis.

Above: The Ratel 90 fire support variant of the basic wheeled APC carries a 90mm gun with coaxial and pintle-mounted AA machine guns.

RECONNAISSANCE VEHICLES

AMX-10RC

Origin: France.
Type: Reconnaissance vehicle.
Crew: 4.
Armament: One 105mm gun; one 7·62mm MG coaxial with main armament; 2 smoke dischargers.
Dimensions: Length (gun forward) 30ft (9·15m); length (hull) 20·8ft (6·35m); width 9·68ft (2·95m); height (overall) 8·8ft (2·68m).
Combat weight: 34,832lb (15,800kg).
Engine: Hispano-Suiza HS-115 water-cooled 8-cylinder diesel; 260hp at 3000rpm.
Performance: Road speed 53mph (85km/h); water speed 4·5mph (7·2km/h); range 497 miles (800km); vertical obstacle 2·3ft (0·7m); gradient 60 per cent.

Development: The Panhard EBR 8 x 8 heavy armoured car served in the French Army from 1951 onwards but in 1970 the French initiated a requirement for a replacement. The vehicle was designed by the Atelier de Construction d'Issy-les-Molineaux (AMX), and is designated the AMX-10RC. It utilizes some of the components of the AMX-10 APC (eg, the automotive system), and mounts a 105mm gun in a large turret. So far 264 have been ordered for the French Army, and Morocco has ordered 108.

The AMX-10RC has a 6 x 6 wheeled configuration and a hydro-pneumatic suspension, its height being variable between 8·26in (210mm) and 18·5in (470mm) for land travel, which is extremely useful for traversing very rough terrain. A special setting of 23·6in (600mm) is for amphibious operations. Unusually for a wheeled vehicle (and one so fast), the AMX-10RC is skid-steered (ie, as with a tracked vehicle).

Above: Along with the 105mm gun, the AMX-10RC is equipped with comprehensive fire control, including a low-light TV system.

Below: For off-road operations in the snow, this AMX-10RC has its suspension raised to give maximum ground clearance.

Type 73

Origin: Japan.
Type: APC.
Crew: 3 + 9.
Armament: One M2 HB 0·5in (12·7mm) MG; one bow-mounted 7·62mm MG.
Dimensions: Length 19·03ft (5·8m); width 9·19ft (2·8m); height (including MG) 7·22ft (2·2m); height (hull) 5·57ft (1·7m).
Combat weight: 29,320lb (13,300kg).
Engine: Mitsubishi 4ZF air-cooled 2-stroke V4 diesel; 300hp at 2,200rpm.
Performance: Road speed 43·5mph (70km/h); range 186 miles (300km); vertical obstacle 2·3ft (0·7m); gradient 60 per cent.

Development: The first Japanese Ground Self-Defense Force (JG SDF) APC was the 8U-60, which entered service in 1960. The simple,

sturdy and reliable vehicle is a conventional APC, similar in concept to the M113 and will remain in service at least until the late 1980s.

The development of the next APC — Type 73 — began in 1967. It is another conventional design, although, like the Type 60, it has a bow-mounted MG, an unusual feature in modern APCs. The section is 9 strong; 1 mans the 12·7mm MG and 8 sit on inward-facing, fold-up seats. There are 6 T-shaped firing ports: 2 on each side and 2 in the rear door. These have close-fitting covers and are not visible from any distance. The basic Type 73 is not amphibious, but a special kit is available to achieve this. The only variant, strangely, is a self-propelled ground-wind measuring unit, used with the 130mm Type 75 SP multiple rocket launcher.

A new vehicle is now under

development to replace the Type 60 APC. This Mechanised Infantry Combat Vehicle will be armed with a Hughes 25mm Chain Gun and the MAT ATGW.

Above: Carrying a crew of three plus nine-man infantry section, the Type 73 is armed with 12·7mm and 7·62mm MGs and features infra-red driving and fighting aids.

BRDM-2

Origin: Soviet Union.
Type: Reconnaissance vehicle.
Crew: 4.
Armament: One 14·5mm KPVT machine-gun; one 7·62mm PKT coaxial machine-gun (but see below for variants).
Dimensions: Length 18·86ft (5·75m); width 7·71ft (2·35m); height (overall) 7·58ft (2·31m).
Combat weight: 15,432lb (7,000kg).
Engine: GAZ-41 V-8 water-cooled petrol engine; 140hp at 3,400rpm.
Performance: Road speed 62mph (100km/h); water speed 6·2mph (10km/h); range 466 miles (750km); gradient 60 per cent.

Development: The standard reconnaissance vehicles in the Soviet Army in the early 1950s were the elderly BA-64 armoured car and the BTR-40, which had appeared in 1948. Both had major drawbacks in that they were not amphibious and they lacked an adequate cross-country performance; this made them quite unsatisfactory to the army commanders as they did not fit in with their tactical concepts. In the late 1950s, therefore, a new vehicle entered production: the BRDM-1, also designated BTR-40P.

This vehicle has a hull of all-welded steel which provides protection from small-arms fire. All four wheels are powered and a central tyre-pressure regulation system is provided. One most unusual feature is that there are two belly-wheels on each side, which can be lowered by the driver when crossing rough country. The vehicle is fully amphibious, water propulsion being a single water-jet in the rear of the hull. There are many versions, some carrying Snapper, Swatter or Sagger anti-tank guided-weapons (ATGW). Other models include BRDM-U, a command vehicle, and BRDM-1rkh, is used to mark lanes through NBC-contaminated areas.

In 1966 a further vehicle appeared, known variously as BRDM-2, BTR-40PB, or BTR-40P-2. This has a

modified hull, a more powerful engine, a gun turret similar to that fitted on BTR-60PB, and a land-navigation system. Again, many different version have been developed, including one armed with Sagger, and a new air-defence vehicle mounting four launch-tubes for modified SA-7 Grail SAMs, or, in the latest version, for SA-9 (Gaskin) SAMS. A major difference between BRDM-1 and -2 is that the former has the engine in front and the crew compartment in the rear, while the latter has the crew in the front.

There are many thousands of both types in service with the Soviet Army and with other Warsaw Pact nations. They will normally be found in the reconnaissance units at battalion, regiment and divisional level, with the -rkh versions serving in chemical recce companies.

Above: BRDM-2, showing the boat-shaped hull front for swimming, 14·5mm KPVT turret and crew compartment hatches.

Below: Sagger-armed BRDM-2s, with retractable launchers each housing six AT-3 missiles, carry out observation tasks.

Scorpion

Origin: UK.
Type: Reconnaissance vehicle.
Crew: 3.
Armament: One 76mm gun; one 7·62mm MG coaxial with main armament, 8 smoke dischargers.
Dimensions: Length (hull) 15·73ft (4·794m); width 7·3ft (2·235m); height (overall) 6·9ft (2·1m).
Combat weight: 17,798lb (8,073kg).
Engine: Jaguar J60 4·2 litre 6-cylinder petrol engine; 190hp at 4,750rpm.
Performance: Road speed 50mph (80·5km/h); range 400 miles (644km); vertical obstacle 1·64ft (0·5m); gradient 60 per cent.

Development: The very successful Scorpion family stems from a 1950s

British Army requirement for a reconnaissance vehicle which would also have a fire-support and anti-tank capability. The result was the Combat Vehicle Reconnaissance (Tracked) which was put into production by the firm of Alvis. (A wheeled vehicle — the Fox — was also produced, but has been less successful.) A second Scorpion production line was set up at the British Leyland factory at Malines in Belgium. Current estimated orders total 3,300.

The basic vehicle has a cast aluminium hull and a crew of three. It is powered by a 4·2 litre Jaguar petrol engine, although a diesel-engined version is now on offer. There are numerous variants, including: Scorpion reconnaissance

MORTARS

81mm L16 ML

Origin: UK.
Type: Portable infantry mortar.
Dimensions: Length of barrel 50·4in (128m); calibre 81mm; baseplate diameter 21·5in (546mm).
Weight: Barrel 27·07lb (12·28kg); mounting 26lb (11·8kg); baseplate 25lb (11·36kg).
Range: Charge 8 minimum 2,296yd (2,100m); maximum 6,189yd (5,660m).
Ammunition: L15A3 HE bomb, L31E3 HE bomb, L19A4 smoke bomb, L20 illuminating bomb.

Development: The British Army used the famous 3in (76·2mm) mortar for many years as its standard infantry mortar, where it earned a high reputation as the "battalion commander's artillery". Eventually it was decided to replace the weapon and the result was the 81mm L16 ML mortar, which is proving to be as much of a success as its predecessor.

The mortar consists of a barrel, a K-mounting, a base-plate and a sight. The forged steel barrel is very simple and has cooling fins on the bottom half. A threaded breech-plug screws into the end. The K-mount steadies the barrel and is of an interesting design which was adopted to achieve simplicity and lightness. The base-plate and sight are Canadian; and the sight includes a tritium source for night firing.

Considerable development effort has been put into the ammunition, in order to achieve the maximum possible range and lethality. The current standard round is the L15A3 with which a maximum range of 6,189 yd (5,660m) can be obtained with charge 8. However, a new round, the L31, has a range of 6,315yd (5,775m) and is also easier to load. There are also smoke bombs, illuminating bombs and a practice bomb.

The mortar can be broken down into three "man-portable" loads of 25, 26 and 27lb (11·3, 11·8 and 12·25kg), although it is unlikely to be

used in this manner for very long. The mortar can also be carried in a Land Rover or other suitable field-car, but its optimum mounting is in an APC (eg, FV 432), where it is fired through the hatch. This gives the mortar much flexibility in rapid movement from one launch site to another.

The L16 mortar is in use with the British Army and the Royal Marines, but has also sold well abroad. Current users include Austria, Bahrain, Canada, Guyana, India, Kenya, Malawi, Malaysia, New Zealand, Nigeria, Norway, Oman, Qatar, United Arab Emirates and North Yemen. Its most important overseas order has recently been announced, with the acceptance of the L16 for US Army service as the M252, with procurement quantities of 5,600 mortars and 2·6 million rounds of ammunition being mentioned. These will be produced at Royal Ordnance Factories in the UK.

Mortars are a very effective method of bringing heavy fire to bear both rapidly and accurately, while

their light weight, simplicity of operation and rugged construction make them ideal for the infantry. Most armies use them in large numbers, although only the Soviet and Warsaw Pact armies go in for the really large calibre weapons. In Western armies even the 4·2in (107mm) mortars have ceased to be used in all but a few, and the 81mm is now supreme. Of the 81mm mortars now in service there can be no doubt that the British L16, and especially with the L31

Above: Soldiers in NBC gear prepare to fire an L16 from its turntable mount in an FV432.

round, is the best and is likely to remain so for many years to come, especially following the crucial American order.

Below: Evasive action. The L16's heavy barrel enables 15rds/min to be maintained indefinitely, with the barrel at 540°C (1,000°F).

vehicle (76mm gun, see data above). Scorpion 90 light tank (90mm Cockerill gun); Striker anti-tank vehikcle (five Swingfire ATGW); Spartan APC (7·62mm MG); Stormer APC (7·62mm); Samaritan ambulance; Sultan command vehicle (7·62mm MG); Samson recovery vehicle (7·62mm MG); Scimitar reconnaissance vehicle (30mm Rarden cannon); Streaker high mobility road carrier.

Left: Scorpion on manoeuvres. The basic Combat Vehicle Reconnaissance (Tracked) has spawned a whole family of AFVs.

Right: A Swingfire anti-tank missile is launched from an FV102 Striker. The missile-armed variant of Scorpion carries a five-round launcher, with five reloads stowed in the hull.

M-160

Origin: Soviet Union.
Type: Wheeled heavy mortar.
Dimensions: Travelling order, length 15·9ft (4·86m); width 6·7ft (2·3m); height 5·5ft (1·69m). Barrel: length 14·9ft (4·55m); calibre 160mm.
Weight: Travelling, 3,240lb (1,470kg); firing 2,866lb (1,300kg).
Range: Minimum 820yd (750m); maximum 880yd (8,040m).
Crew: 7.

Development: The Soviet Army has always been keen on employing mortars, using them in vast numbers. They have also developed some very large calibre weapons, including 6·3in (160mm) and even 9·5in (240mm), whereas the largest calibre found in a Western army is 4·72in (120mm). The largest calibre Soviet mortars have barrels of such length that they must be "broken" like a shotgun in order to load the bomb, rather than drop the bomb down the tube as is done in all other mortars. One such breech-loading mortar is

the 6·3in (160mm) M-160, which succeeded the well tried and tested 160mm M-1943 in the late 1960s.

The M-160 has a barrel 14·9ft (4,550mm) long and weighs 3,240lb (1,470kg) in travelling order. It has a maximum range of 8,792yd (8,040m) and its crew of seven can maintain a rate of fire of 2 to 3 rounds per minute.

The Soviet Army normally deploys its mortar batteries in a straight line, as it does its gun batteries. Wherever feasible these heavy mortars are employed on battery tasks as part of the higher level artillery fire plan, but they can be taken under local control if necessary.

Unlike Western armies the Soviets do not appear to use their mortars from APCs, and neither of the major APC types has a suitable roof hatch.

Below: Muzzle flash illuminates this view of a 160mm M-160 mortar in action. The 91.5lb (41·5kg) HE bomb can be fired to a range of 5 miles (8km).

M-240

Origin: Soviet Union.
Type: Wheeled super-heavy mortar.
Dimensions: Travelling order, length 21·3ft (6·51m); width 8·1ft (2·49m); height 7·2ft (2·21m). Barrel: length 17·6ft (5·38m); calibre 240mm.
Weight: Travelling 8,600lb (3,900kg); firing 7,960lb (3,610kg).
Range: Minimum 1,640yd (1,500m); maximum 10,608yd (9,700m).
Crew: 8.

Development: This huge weapon is one of the largest mortars ever to enter full service with a major army. The bomb weighs 220lb (100kg) and requires four soldiers to load it into the vertically-sliding breech, which is surprisingly high off the ground. This

mortar was first seen by Western eyes in the Red Square parade in November 1953, hence its alternative designation, M-1953. The M-240 was at one time reported to be nuclear-capable, but this has never been confirmed.

A new self-propelled 240mm mortar has been reported from East Germany. This may have an automatic loader, which, coupled with the inherent mobility of an SP mount, would make it a highly effective weapon, able to bring down very heavy fire and with very quick reaction times.

Below: Typical straight-line deployment pattern of a battery of 240mm M-240s, the biggest mortar in service. The barrel length demands breech loading.

M29A1

Origin: USA.
Type: Mortar.
Calibre: 81mm.
Weight of barrel: 27·99lb (12·7kg).
Weight of baseplate: 24·91lb (11·3kg).
Weight of bipod: 40lb (18·15kg).
Total weight with sight: 115lb (52·2kg).
Elevation: +40⁰ to +85⁰.
Traverse: 4⁰ left and 4⁰ right.
Maximum range: 5,140yd (4,700m).
Rate of fire: 30rpm for 1 minute; 4-12rpm sustained.

Development: In service with the US Army and some Allied countries, the 81mm M29 mortar is the standard medium mortar of the US Army and is in service in two basic models, infantry and self-propelled. The standard infantry model can be disassembled into components, each of which can be carried by one man—baseplate, barrel, mount and sight. The exterior of the barrel is helically grooved both to reduce weight and to dissipate heat when a high rate of fire is being achieved.

The mortar is also mounted in the rear of a modified member of the M113 APC family called the M125A1. In this vehicle the mortar is mounted on a turntable and this enables it to be traversed quickly through 360⁰ to be laid on to a new target. A total of 114 81mm mortar bombs are carried in the vehicle.

The mortar can fire a variety of mortar bombs, including HE (the M374 bomb has a maximum range of 5,025yd (4,595m)), white phosphorus (the M375 bomb has a maximum range of 5,180yd (4,737m)), and illuminating (the M301 bomb has a maximum range of 3,444yd (3,150m)). Sustained rates of fire range from 4 to 12 rounds, depending on ammunition used.

The 81mm M29 has been replaced in certain US Army units by the new M224 60mm lightweight Company Mortar.

Above: The M29A1 version of the US Army's standard 81mm mortar introduced a new barrel for higher rates of sustained fire.

MACHINE GUNS

M60

Origin: USA.
Type: General-purpose machine-gun.
Calibre: 7·62mm.
Length overall: 43·5in (1,105mm).
Length of barrel: 22in (560mm).
Weight (LMG version); 23·2lb (10·51kg).
Range (effective); 1,094yd (1,000m); (with tripod) 1,969yd (1,800m).
Rate of fire (cyclic); 550rpm.
Muzzle velocity: 2,805ft/s (855m/s).

Development: The M60 is the standard GPMG of the US Army and has now replaced the older 0·30 Browning machine gun. The weapon was developed by the Bridge Tool and Die Works and the Inland Division of General Motors Corporation, under the direction of Springfield Armory. Production of the M60 commenced in 1959 by the Maremont Corporation of Saco, Maine.

The M60 is gas-operated, air-cooled and is normally used with a 100-round belt of ammunition. To avoid overheating the barrel is normally changed after 500 rounds have been fired. Its foresight is of the fixed blade type and its rearsight is of the U-notch type and is graduated from about 656ft to 3,937ft (200 to 1200m) in about 328ft (100m) steps. The weapon is provided with a stock, carrying handle and a built-in tripod. The M60 can also be used on an M122 tripod mount, M4 pedestal mount and M142 gun mount for vehicles. Other versions include the M60C remote for helicopters, M60D pintle mount for vehicles and helicopters and the M60E2 internal model for AFVs.

The M60 will remain in service with the US Army for many years. It is sturdy, reliable and highly effective.

Above: The M60's rather flimsy carrying handle is disregarded by this paratrooper of the 101st Airborne in Sinai in 1983.

Below: M60 with blank firing attachment; fixing bipod and gas cylinder to the barrel was a basic flaw in the design.

M224

Origin: USA.
Type: Lightweight company mortar.
Calibre: 60mm.
Total weight: 46lb (20·9kg); (hand-held with M8 baseplate) 17lb (7·7kg).
Maximum range: 3,828yd (3,500m).

Development: During the Vietnam campaign it was found that the standard 81mm M29 mortar was too heavy for the infantry to transport in rough terrain, even when it is assembled into its three main components. In its place the old 60mm M19 mortar was used, but this had a short range.

The M224 has been developed to replace the 81mm M29 mortar in non-mechanized infantry, airmobile and airborne units at company level, and is also issued to the US Marine Corps. The weapon comprises a lightweight finned barrel, sight, M7

baseplate and bipod, although if required it can also be used with the lightweight M8 baseplate, in which case it is hand-held. The complete mortar weighs only 46lb (20·9kg) compared to the 81mm mortar which weighs 115lb (52kg). The M224 fires an HE bomb which provides a substantial portion of the lethality of the 81mm mortar with a waterproof "horseshoe" snap-off, propellant increments, and the M734 multi-option fuse. This new fuse is set by hand and gives delayed detonation, impact, near-surface burst (0-3ft, 0-0·9m), or proximity burst (3-13ft, 0·9-3·96m).

The mortar can be used in conjunction with the AN/GVS-5 hand-held laser rangefinder which can range up to 10,936yd (10,000m) to an accuracy of ±10·936 (±10m). this enables the mortar to engage a direct-fire target without firing a ranging bomb first. The M224 fires a variety of mortar bombs to a max-

imum range of 3,828yd (3,000m). The Army has ordered 1,590 of these mortars, the Marine Corps 698.

Above: Lightweight replacement for the M29 at company level, the M224 can be hand-held or bipod-mounted for firing.

MG3

Origin: Federal Republic of Germany.
Type: General-purpose machine-gun.
Calibre: 7·62mm.
Length overall: (with butt) 48·3in (1,225mm); (without butt) 43·2in (1,097mm).
Length of barrel: 22·3in (565mm).
Weight with bipod: 24·4lb (11·05kg).
Range (effective): (with bipod) 875yd (800m); (with tripod) 2,400yd (2,200m).
Rate of fire (cyclic): 700-1,300rpm.
Muzzle velocity: 2,690ft/s (820m/s).

Development: One of the most famous of all the small arms used in World War II was the 7·92mm MG42, which first saw combat in May 1942. From then to the end of

the war it was encountered on every front, and was treated with great respect by all who had to face it. When the West Germans entered NATO, they decided to update the MG42 and convert it to take the NATO standard 7·62mm round. This was manufactured by Rheinmetall as the MG 42/59, but was designated MG1 by the Bundeswehr. In addition, some original MG42s were rebarrelled to take the 7·62mm round; these are designated MG2. The current version — MG3 — went into service in 1968. The MG1 and MG3 are used by at least 11 armies, and are produced in West Germany, Italy, Spain, Turkey, Portugal and Pakistan.

Below: Bundeswehr infantryman with MG3, current production derivative of the WWII MG42.

PK series

Origin: Soviet Union.
Type: General-purpose machine-gun.
Calibre: 7·62mm.
Length overall: 45·7in (1,173mm).
Length of barrel: 25·9in (658mm).
Weight: 19·8lb (9·0kg).
Range (effective): 1,093yd (1,000m).
Rate of fire (cyclic): 650rpm.
Muzzle velocity: 2,755ft/s (840m/s).

Development: Though a hotch-potch of other weapons (mostly the Kalashnikov AK-47), the PK family is an excellent series of weapons which can be described as the first Soviet GPMGs (general-purpose MGs). Unlike almost all other Soviet rifle-calibre weapons except the sniper's rifle, it fires the long rimmed cartridge with over twice the propellant charge of the standard kind. It

is a fully automatic gas-operated gun with Kalashnikov rotating bolt, Goryunov cartridge extractor and barrel-change, and Degtyarev feed system and trigger. The PKS is the PK on a light tripod for sustained or AA firing. The PKT is a solenoid-operated version without sights, stock or trigger mechanism for use in armoured vehicles. The PKM is the latest service version with unfluted barrel and hinged butt rest, weighing only 8·39kg (18^1/2lb); on a tripod it becomes the PKMS. The PKB has stock and trigger replaced by a butterfly trigger for pintle mounting on armoured vehicles (but the standard PK and PKM can be fired from, say, BMPs).

Below: Standard PK light MG in action with troops of the Soviet (upper) and Polish armies; several variants exist.

M249

Origin: USA.
Type: Squad automatic weapon (SAW).
Calibre: 5·56mm.
Lengths: Overall 39·4in (100cm); barrel 18·5in (47cm).
Weights: Empty 15·5lb (7·03kg); with 200-round magazine 22lb (9·97kg).
Effective range: 1,421yd (1,300m).
Rate of fire: 750rpm.
Muzzle velocity: 3,033ft/s (924m/s).

Development: The SAW idea was conceived in 1966, but it has taken a long time to reach service. When the M16 was issued to infantry squads, all infantrymen had an automatic weapon, but with a maximum effective range of some 330yd (300m) only. It was considered that each fire team in the squad needed a weapon of greater all-round capability than the M16, but obviously not a weapon as heavy or as sophisticated as the M60. The SAW meets this requirement, and will be issued on a scale of one per fire team, ie, two per squad. The SAW may also replace some M60s in non-infantry units.

The M249 SAW is a development of the Belgian Fabrique Nationale (FN) "Minimi". Current orders are being met from the FN factory, but it is intended to set up a production line in the USA. Current requirements are for 26,000 M249s for the Army and 9,000 for the Marine Corps in a five-year programme, but further orders will doubtless follow.

The M249 is very smooth in operation and displays a reliability that is considered exceptional in light machine-guns. Fully combat ready, with a magazine of 200 rounds, bipod, sling, and cleaning kit, the M249 weighs 22lb (9·97kg), which is still 1lb (0·4kg) less than an empty M60 machine-gun. The M855 ball round fired from the M249 will penetgrate a US steel helmet at a range of 1,421yd (1,300m). Overall, the M249 is superior to the Soviet PKM 7·62mm (despite this being bigger and heavier, and with a smaller mag), and the RPK 5·45mm (bigger, lighter, smaller mag).

Above: Bridging the firepower gap between M16 and M60, the M249 light machine gun is based on the Belgian Minimi; the 5·56mm ammunition is supplied in 100- or 200-round boxes.

ASSAULT RIFLES

5.45mm AK-74

Origin: Soviet Union.
Type: Assault rifle.
Weight: (With loaded magazine) 7·94lb (3·6kg).
Length: (AK-74) 36·6in (930mm); (AKS-74 folded) 27·2in (690mm).
Ammunition: 5·45mm.
Muzzle velocity: 2,854ft/sec (870m/sec).
Effective range: About 547yd (500m).

Development: The success of the United States Armalite AR-15 (M-16) rifle firing the 5·56mm round in the 1960s led most armies to review their rifle designs. In the West some very curious weapons have appeared together with sound designs such as the British 4·85mm Individual Weapon, and there was intense competition to win the order for the next NATO standard rifle. The USSR appeared, however, to be satisfied with the AK series, which gave excellent service both in the Warsaw Pact armies and with many revolutionary forces.

In the mid-1970s there was considerable curiosity in the West as to whether Soviet small arms designers would follow the tendency towards a smaller calibre, but for a long time there was no evidence of any activity. Then suddenly, in the usual Russian way, Soviet parachute troops appeared on a Moscow parade carrying a totally new weapon which had passed through design, tests, troop trials and into production without a word leaking to the West.

The basic rifle is the AK-74, and is of 5·45mm calibre. It is a Kalashnikov design and is an evolutionary, smaller calibre version of the famous AK-47. One of the most interesting features is the muzzle brake, which is extremely effective, reducing recoil, even when firing a long burst, to negligble proportions. This leads to great accuracy, although the muzzle flash is some three times that of other rifles.

The basic AK-74 has a wooden butt and furniture, but a special version (AKS-74) is made for the parachute forces.

Above: Soviet paratroops on parade in Red Square, Moscow, with AKS-74 folding-butt versions of the assault rifle.

Below: Motorised infantry on an exercise with AK-74s, whose accurate burst capability is well suited to such tactics.

HK33E

Origin: Federal Republic of Germany.
Type: 5·56mm assault rifle.
Dimensions: Length (fixed butt) 36·2in (920mm); (retracted butt) 29in (735mm); barrel 15·3in (390mm).
Weights: (Fixed butt) 8·0lb (3·65kg); (retracting butt) 8·7lb (3·98kg).
Rate of fire: 750 rounds per minute.
Muzzle velocity: 3,018ft/sec (920m/s).
Range, effective: 437yd (400m).

(Specifications apply to rifle; carbine is lighter, shorter).

Development: One of the most successful of current small arms manufacturers, Heckler and Koch have produced a number of rifles which have sold all over the world. One of these is the HK 33E, which is in production in West Germany and Thailand, and is in service with a number of armies in Southeast Asia.

The rifle is a scaled-down version of the standard G3 7·62mm rifle.

The HK 33E has a long and rather involved history. Its origins lie with a World War II German rifle which was redesigned in the 1950s in Spain to become the CETME rifle. That weapon was taken as the starting point for the West German Army's first post-War rifle — the Heckler and Koch G3. The G3 did not use gas for its operation, as does virtually every other rifle, but relies on delayed blowback. In this the breech is not fully locked, but is held closed until the pressure drops to a safe level, when rollers are forced out of the recesses they have been forced into during the breech block's forward travel. The residual gas pressure then forces the breech back taking the spent case with it.

The HK 33 was a logical development of the G3, using a 5·56mm x 45 rounds. Another rifle based on the G3 is the G41, but this was developed to fire the new standard NATO 5·56mm x 45 round. Heckler and Koch, long one of the more innovative small arms firms, are also working on the 4·7mm G11 rifle, a "space-age" weapon in appearance, which is designed to fire a 4·7mm caseless round.

Below: HK 33A2 assault rifle; other versions are the shortened A1, A3 with telescoping butt and ZF with telescopic sight.

FA MAS

Origin: France.
Type: Rifle.
Calibre: 5·56mm.
Length overall: 29·8in (757mm).
Length of barrel: 19·2in (488mm).
Weight: (including 25-round loaded magazine) 9·32lb (4·23kg).
Range: (effective) 328yd (300m).
Rate of fire: (cyclic) 900-1,000rpm.
Muzzle velocity: 3,249ft/sec (960m/s).

Development: The 5·56mm FA MAS is in full-scale production at the St Etienne arsenal and will have re-equipped the entire French Army by 1990. With the usual French flair, it is of very unusual appearance, but is a compact, reliable and effective weapon. Particular thought has been given to making the weapon equally easy to use for left-handed or right-handed firers with relatively simple changes from one to the other.

The rifle is of "bullpup" design, and one of the most striking visual features is the large plastic carrying handle, which also provides protection for the rear and foresights. The cocking handle is placed centrally under the carrying handle, and thus is easy to use with either hand. A bipod is permanently fitted, but is normally folded back out of the way; it weighs only 0·37lb (0·17kg). The bipod is used in the LMG role and also when the weapon is being used for grenade-launching.

Below: FA MAS rifle with grenade attached and forward sight erected for indirect fire; the rifle needs to be braced against a solid surface in this mode.

Above: The distinctive handle holds the bipod and detachable sight for grenade fire, as well as protecting the rifle sights and return spring cylinder.

Below: Light and compact, the FA MAS is a versatile weapon able to deliver single shots, three-round bursts and full automatic fire as required.

G3A3

Origin: Federal Republic of Germany.
Type: Rifle.
Calibre: 7·62mm.
Length overall: (fixed butt) 40·35in (1025mm).
Length of barrel: 17·7in (450mm).
Weight: (including 20-round loaded magazine) 11·35lb (5·15kg).
Range: (effective) 437yd (400m).
Rate of fire: (cyclic) 500 to 600rpm.

Muzzle velocity: 2,560 to 2,625ft/s (780 to 800m/s).

Development: The Heckler and Koch 7·62mm G3 rifle is based on a weapon produced by some expatriate German designers in Spain in the early 1950s. Design rights were transferred first to a Dutch firm and then to Heckler and Koch, who started to produce the rifle as a replacement for the FN FAL in

Bundeswehr service. Since then the G3 has been produced under licence in some 14 countries and equips the armed forces of at least 50 countries.

The G3 is of somewhat unusual design in that it works on delayed blowback, a method in which the breech is never fully locked in the strict sense of the word. Other designers had attempted this but had always hit insoluble problems, especially with the extraction; the ingenious German designers found the answers, however, and the G3 is

a reliable and effective weapon.

Major variants of the standard G3A3 are a sniping rifle with a telescopic sight (G3A3ZF) and paratroop rifle with a retracting butt (G3A4).

Below: G3 A3ZF sniper's rifle with telescopic sight. Like the standard G3 A3, this has a plastic butt stock and handguard; a paratroop version is also produced with a retractable butt. Along with the MG3 machine gun, they use the 7·62mm x 51 cartridge.

Galil ARM

Origin: Israel.
Type: Rifle.
Calibre: 5·56mm.
Length overall: 38·54in (979mm).
Length of barrel: 18·54in (471mm).
Weight: (including 35-round loaded magazine): 10·82lb (4·91kg).
Range: (effective) 547yd (500m).
Rate of fire: (cyclic): 650rpm.
Muzzle velocity: 3,215ft/s (980m/s).

Development: The Israeli Army probably has more combat experience than any other army in the post-World War II period, and this, coupled with a skilled and imaginative arms industry, has led to some interesting weapons systems. In the late 1950s and early 1960s the

Israelis re-equipped with the 7·62mm FN FAL, but their experiences in the 1967 Six-Day War led to a decision to change to 5·56mm calibre. A number of designs were subjected to an intensive series of tests, but the Israeli-designed Galil proved the eventual winner.

The Galil has a more than passing similarity to the Soviet AK-47 (of which the Israelis have captured large numbers), but with many Israeli refinements, and it is in every way a superior weapon. There are three versions. The ARM is an assault rifle and light machine-gun, with a folding metal stock, bipod and carrying handle. The AR is similar but without the bipod. Finally, the SAR has a short barrel.

The Israeli Army uses the 5·56mm

Galil in some numbers, although other rifles are also in service. A 7·62mm version of the Galil is made for export.

Above: The Galil weapon system includes rifle, bipod, grenades and accessories; the 12-round mag is for grenade cartridges.

L70E3

Origin: UK.
Type: Individual weapon system.
Calibre: 5·56mm.
Length overall: 30·3in (770mm).
Length of barrel: 20·4in (518mm).
Weight: (including 30-round loaded magazine): 11·11lb (5·04kg).
Rate of fire: (cyclic) 650-800rpm.
Muzzle velocity: 2,952ft/s (900m/s).

Development: After centuries of using its own design of rifle, the British Army found itself equipped from the 1950s to the 1980s with a foreign weapon: the Belgian 7·62mm L1A1 Self-Loading Rifle. With the move from 7·62mm calibre to something smaller came the opportunity to move back into the design field, and a new rifle was

entered in the NATO competition based on the 'bullpup' principle tested in the 0·28mm (7·11mm) EM2 rifle of the 1950s. The lesson of that earlier NATO competition was well learnt, however, and the new rifle was designed from the start to be barrelled for either the British-preferred 4·85mm, or for the US-backed 5·56mm calibre. Thus, even though the Belgian SS109 5·56mm cartridge was the winner, the British were able to adopt their own weapon.

The new weapon weighs 11·11lb (5·04kg) with a magazine of 30 rounds, compared with 11lb (5kg) for an L1A1 with a magazine of 20 rounds. Rather more significant is the reduction in length from 45in (1,143mm) to 30·3 (770mm), which makes a lot of different to a modern

M16A1/A2

Origin: USA.
Type: Rifle.
Calibre: 5·56mm.
Length overall: (with flash suppressor) 38·9in (99cm).
Length of barrel: 19·9in (50·8cm);
Weight: (including 30-round loaded magazine) 8·2lb (3·72kg).
Range: (accurate lethal) 874yd (800m).
Rate of fire: 700-950rpm (cyclic); 150-200rpm (automatic); 45-65rpm (semi-automatic).
Muzzle velocity: 3,280ft/s (1,000m/s).

Development: The M16 (previously the AR-15) was designed by Eugene Stone and was a development of the earlier 7·62mm AR-10 assault rifle. It was first adopted by the US Air Force, and later the US Army adopted the weapon for use in Vietnam. When first used in combat, numerous faults became apparent and most of these were traced to a lack of training and poor maintenance. Since then the M16 has replaced the 7·62mm M14 as the standard rifle of the United States forces. To date over 5,000,000 have been manufactured, most by Colt Firearms, and the weapon is also made under licence in Singapore, South Korea and the Philippines. Twenty-one armies use the M16.

The weapon is gas-operated and the user can select either full automatic or semi-automatic. Both 20- and 30-round magazines can be fitted, as can a bipod, bayonet, telescope and night sight. The weapon can also be fitted with the M203 40mm grenade launcher, and this fires a variety of 40mm grenades to a maximum range of 382yd (350m). The M203 has now replaced the M79 grenade launcher on a one-for-one basis. The M231 is a special

model which can be fired from within the M2 Bradley Infantry Fighting Vehicle.

There has been consistent dissatisfaction with the M16A1 in the US Army, and even more so in the other main user — the US Marine Corps. One of the major complaints is its lack of effectiveness at ranges above 340yd (300m), which has come to a head with the increased emphasis on desert warfare with the Rapid Deployment Force (RDF). This, combined with the high average age of current stocks, led to a major review in 1981.

As a result, a "product improved" weapon (M16A2) is now in production. A major feature is a stiffer and heavier barrel, utilizing one-turn-in-7in (17·8cm) rifling — as opposed to one-turn-in-12in (30·5cm) — to enable the new standad NATO 5·56mm (0·218in) round to be fired. Other features are a three-round burst capability to replace the current full-automatic, an adjustable rearsight, and a modified flash eliminator. The opportunity has also been taken to introduce tougher "furniture", ie, butt-stock, pistol grip, and handguard.

A programme is in hand to examine new technologies for incorporation in a possible future weapon. These include controlled-burst fire and multiple projectiles (eg, flechettes). This is designated the Advanced Combat Rifle programme.

Above: An M16 fitted with the Multiple Integrated Laser Engagement (MILE) system in use during a training exercise.

Below: M16 with M203 pump-action grenade launcher; this can fire 40mm HE, buck shot, smoke and illuminating grenades.

Right: Externally almost identical to the original model, the M16A2 incorporates a number of improvements whose overall effect is to nearly double the weapon's effective range while greatly improving reliability.

soldier who spends much of his time travelling in an APC, truck or helicopter. Most important, however, is that the 'bullpup' design makes the weapon much easier to handle, and therefore reduces training problems and increases battle accuracy.

The Individual Weapon (IW) will replace the rifle and sub-machine gun, and the Light Support Weapon (LSW) (basically the IW with a bipod) will replace the GPMG. The overall programme is designated Small Arms for the 1980s (SA80).

Left: Alternative firing modes using the L70E3 Individual Weapon; with a bipod the rifle becomes a light machine gun.

Right: The x4 optical sight enables the weapon to be used in poor light conditions, and is also useful for surveillance.

Sea Weapons

The single most remarkable phenomenon in naval weapons systems is the increase in dependence upon the electromagnetic spectrum as sensors, command, control and communications systems, and weapons themselves become more sophisticated, and more dependent upon electronics for their operation. Submarines are growing in absolute quantities and are operated by an ever-increasing number of navies, although only one more navy (China) has moved into the nuclear-submarine arena in the past decade.

Aircraft carriers remain the visible symbol of maritime might. The large, conventional take-off and landing (CTOL) carrier is still supreme, with the US carrier fleet expected to be joined later in the decade by the new Soviet nuclear-powered carrier, and in the early 1990s by the French *Charles de Gaulle.* Other navies, seeing the vital nature of afloat air support but unable to afford such large ships, have looked to the V/STOL fixed- and rotary-winged carriers (eg, British *Illustrious,* Italian *Garibaldi,* Spanish *Asturias*) to give at least a limited answer.

Among other surface combatants the reappearance of the large capital ship is surprising, as the US Iowa class battleships are reactivated in reply to the new Soviet Kirov class battlecruisers. Such a ship represents a large concentration of manpower and weaponry in one hull and will be a very high-value target in any future naval conflict. It is at the somewhat lower level of destroyers and frigates that the major growth is occurring, however, as every nation with any maritime pretensions seeks some means of imposing its national will at sea.

The growing sophistication of these ships and their weapons systems is, however, leading to an appalling escalation in costs: the US Aegis cruisers ordered in FY83 will cost $3,200 million for the three, while the smaller, less complex British HMS *Boxer* costs £120 million at 1981 prices. This spiral is facing all but the two largest navies with a major dilemma as they try to maintain effective and balanced fleets.

Above: The 16 in guns of the battleship USS *New Jersey* in action off the coast of Lebanon in September 1984.

AIRCRAFT CARRIERS

Clémenceau Class

Origin: France.
Type: Aircraft carrier.
Number: 2 ships.
Displacement: 27,307 tons standard; 32,780 tons full load.
Dimensions: Length overall 870ft (265m); beam 168ft (51·2m); draught 28·1ft (8·6m).
Propulsion: Geared steam turbines, 126,000shp; 2 shafts; 32kt.
Armament: 8 single 3·9in (100mm) DP guns.
Aircraft: 16 Super Étendard; 3 Étendard IVP; 7 Breguet Alizé; 2 Aérospatiale Alouette III.
Complement: 1,338.

Development: Clémenceau and Foch were commissioned in the early 1960s and incorporated all the major advances made in carrier operating techniques during the immediate postwar period. The flight deck is angled at 8° to the ship's major axis. The forward aircraft lift is offset to starboard and the after lift is positioned on the deck-edge to clear the flight deck and to increase hangar capacity.

The attack aircraft is the Dassault Super Étendard, which can carry anti-ship (ASM) missiles such as the Exocet, as was demonstrated so dramatically by the Argentine naval

air arm during the South Atlantic war in 1982. The small size and light construction of the ships, together with the limited capacity of their lifts and catapults, have made it difficult to find a replacement for the Crusader fighter aircraft formerly carried. The ASW aircraft is the Breguet Alizé; ten are carried, but these are now somewhat aged, although no replacement is in sight. A further limitation on the effectiveness of these ships is the lack of integral airborne early warning (AEW) aircraft, and a French carrier group would thus suffer the same problems as the Royal Navy task force experienced in the South Atlantic in 1982.

The French Navy plans to replace these two aircraft carriers with new nuclear-powered carriers in the 1990s. These will be enormously expensive for such a relatively small navy and will take up a disproportionate amount of the available resources, but they will nevertheless be a valuable addition to NATO's maritime capabilities.

Below: The BAe Harrier demonstrator prepares for takeoff during trials aboard Foch; in the event, the French navy preferred to stick with non-STOVL aircraft.

Above: The aircraft carrier Clemenceau during a naval revue with the crew on parade and a pair of Super Frelon helicopters on the flight deck.

Below: Clémenceau under way, with a mix of Super Etendard, Etendard IVP, Crusader and an Alizé fixed wing aircraft, plus two Alouette helicopters.

Giuseppe Garibaldi Class

Origin: Italy.
Type: Aircraft carrier.
Number: 1 ship.
Displacement: 10,100 tons standard; 13,370 tons full load.
Dimensions: Length overall 591ft (180·1m); beam 100ft (30·5m); draught 22ft (6·7m).
Propulsion: COGAG; 4 LM2500 gas turbines; 80,000shp; 2 shafts; 30kt.
Armament: 4 Otomat Mk 2 SSM launchers; 2 octuple Albatros SAM systems; 3 twin Breda 40mm guns; 2 triple Mk 32 torpedo tubes.
Aircraft: 18 SH-3D Sea King.
Complement: 550.

Development: Following a series of helicopter-carrying cruisers, the Italian Navy has finally decided to produce a full-blown aircraft carrier, although the project has been a far from easy one and there may be further problems ahead. Nevertheless, *Giuseppe Garibaldi* was launched on 4 June 1983 and is scheduled to join the fleet in 1985. She is designed for ASW operations and will carry eighteen SH-3D Sea King helicopters in place of the AB 204/212s which serve aboard earlier ships. The hangar, which is located centrally, is 361ft (110m) long and 20ft (6m) high, with a maximum width of 49ft (15m). Twelve Sea Kings can be struck down in the hangar, which is divided into three sections by fire curtains, but the other aircraft must remain on the flight deck. Six helicopter spots are marked on the deck.

Short-range air defence and close-in anti-missile systems are fitted, and these are all of Italian design and manufacture, as are the Selenia surveillance, 3-D tracking and fire control radars. A bow sonar is fitted; this is a Raytheon DE-1167, manufactured in Italy under licence by Elsag.

Giuseppe Garibaldi's most striking feature is a 91ft (28m) 6° ski-jump, which has no feasible helicopter application and is quite clearly intended for use by Sea Harrier V/STOL aircraft. This whole problem has been the subject of a major interservice row between the Italian Navy and Air Force: the Navy desperately wants fixed-wing aircraft at sea but is not allowed to operate them, while the Air Force is adamant in its refusal to provide a sea-going component. There the matter rests for the moment, but meanwhile the Italian Navy has built itself a very useful light aircraft carrier, optimised for the ASW role. The design makes an interesting comparison with the Spanish *Príncipe de Asturias* (qv).

Right and above right: The *Giuseppe Garibaldi* under construction at Italcantieri's Monfalcone yard in October 1982, and immediately after her launch in June 1983.

Invincible Class

Origin: UK.
Type: Light aircraft carrier.
Number: 3 ships.
Displacement: 16,256 tons standard; 19,812 tons full load.
Dimensions: Length overall 677ft (206·3m); beam 105ft (32m); draught 24ft (7·3m).
Propulsion: COGAG; 4 Rolls-Royce Olympus TM3B gas turbines, 112,000shp; 2 shafts; 28kt.
Armament: 1 twin Sea Dart launcher (22 missiles); 2 Phalanx CIWS.
Aircraft: 5 Sea Harrier FRS.1; 9 Sea King HAS.5; 2 Sea King AEW.
Complement: 1,000 (+ 320 air group).

Below; HMS *Illustrious* passes her sister *Ark Royal,* the latter still fitting out at the Swan Hunter yard where both ships were built, on the Tyne in 1982.

Development: It is indisputable that without the air cover provided by HMS *Invincible* and the older *Hermes* the British task force could never have succeeded in re-establishing British rule over the Falkland Islands in the South Atlantic war of 1982. In fact, the British were very lucky to have these carriers at all, especially the *Invincible,* which has had a very chequered history and was very nearly cancelled on several occasions. Initially this class was to operate large ASW helicopters, but late in the design process provision was made for operating Sea Harrier aircraft to intercept hostile reconnaissance and ASW patrol aircraft. A

Right: *Invincible* under way. Note the slightly angled flight deck offset to port to clear the Sea Dart launcher in the centre of the foredeck.

final change in 1976-77 required the ships to be able to act as commando carriers as well!

The hangar has a 'dumb-bell' shape, with a narrow centre-section and wide bays at each end, dictated by the large exhaust uptakes for the gas turbines to starboard and imposing some constraints on the movement of fixed-wing aircraft within it. Unlike earlier RN carriers, the Invincible class has an open forecastle. The 600ft (182·9m) × 44ft (13·4m) flight deck is offset to port and angled slightly to avoid the Sea Dart launcher which is fairly central on the foredeck. *Invincible* and *Illustrious* have a 7° ski-jump at the forward end of the flight deck, which enables the Sea Harrier to increase its payload by some 1,500lb (680kg). *Ark Royal*, however, has a 12° ski-jump, which means that the Sea Dart launcher has had to be moved. Following experience in the Falkland Islands, US Vulcan/Phalanx close-in defence weapons have been installed, one

beside the Sea Dart launcher forward and the second on the after end of the flight deck.

A further lesson from the Falklands has led to the inclusion of two Sea King AEW helicopters in the air wing. During that conflict *Invincible* operated between eight and twelve Sea Harrier/Harrier GR.3 aircraft in addition to her complement of Mk5 Sea Kings. Her sister ship *Illustrious*, which relieved her in the summer of 1982, operated ten Sea Harriers while on patrol in the South Atlantic. The limited hangar capacity means that a permanent deck park must be operated.

For the NATO EASTLANT role, which remains their primary mission, the Invincibles are equipped with a sophisticated ASW command centre and first-rate communications. They are also fitted with Type 184 hull sonar but have no shipborne anti-submarine weapons. The planned sale of *Invincible* to Australia has been cancelled.

Above: The profusion of antennas and radomes highlight the comprehensive sensor and communications fit on *Illustrious*.

Below: Sea Harriers and a Sea King on *Illustrious'* flight deck. Note the Vulcan/Phalanx on the after end of the deck.

Kiev Class

Origin: Soviet Union.
Type: ASW aircraft carrier.
Number: 4 ships.
Displacement: 36,000 tons standard; 42,000 tons full load.
Dimensions: Length overall 899ft (274m); beam 157·5ft (48m); draught 33ft (10m).
Propulsion: Geared steam turbines, 180,000shp; 4 shafts; 32kt.
Armament: 8 SS-N-12 launchers; 2 twin SA-N-3 launchers; 2 twin SA-N-4 launchers; 2 twin 3in (76·2mm) DP guns; 8 30mm Gatling CIWS; 1 twin SUW-N-1 launcher; 2 RBU 6000 launchers; 2 quintuple 21in (533mm) torpedoe tubes.
Aircraft: 12 Yak-36MP Forger; 18–21 Ka-25 Hormone-A/B.
Complement: 2,500.

Development: The primary role of these splendid warships is ASW, and for this they carry two squadrons of 15–18 Kamov Ka-25 Hormone-A helicopters, an SUW-N-1 launcher forward and the usual complement of ASW mortars and torpedo tubes. Principal shipborne ASW sensors are a low-frequency sonar mounted in the bow, and a variable-depth sonar streamed from the stern.

The principal air defence system, the SA-N-3, is the same as on the Moskvas, but the layout of the Kievs' launchers — one forward and one aft of the bridge island — is obviously more satisfactory. The other major air defence weapons are similarly deployed: a twin 76mm gun mounting on the forecastle and a second on the after end of the superstructure; an SA-N-4 'bin' on the port side of the forecastle and another on the starboard side of the island; and groups of paired Gatling CIWS at each of the four quadrants.

The long, narrow hangar accom-modates the two squadrons of helicopters and a squadron of Yakovlev Yak-36 Forger VTOL air-craft. The flight deck is angled at 4° but does not have the 'ski-jump' which has made such a difference to the performance of the Sea Harriers on the British Invincible class ships, probably because the Forger cannot perform rolling short take-offs.

Kiev and *Novorosiiysk* are de-ployed with the Northern Fleet and *Minsk* with the Pacific Fleet; the fourth of class, *Kharkov*, joined the latter when she completed her sea trials late in 1984. In the event of hostilities the Kievs would probably be employed in support of Soviet submarines in their respective areas, against NATO submarine and sur-face threats. This could involve the protection of the SSBN bastions in the Barents Sea in the west and in the Sea of Okhotsk in the east, and also offensive forays against NATO ASW barrier forces in such critical areas as the GIUK gap.

The next class is already under construction at the Nikdayev yard on the Black Sea. The lead ship, *Kre-mlin* (75,000 tons), will enter service in about 1994.

Above right: *Kiev* on her first deployment to the Soviet Northern Fleet. A "Forger" and four 'Hormones'' are on deck.

Right: Along with a compre-hensive array of ASW aircraft and weapons, *Kiev* carries four twin SS-N-12 launchers on the forecastle.

Below: *Minsk's* profile is domi-nated by the massive island superstructure, which carries the air search and fire control radars for the weapons systems.

Kitty Hawk Class

Origin: USA.
Type: Multi-role aircraft carrier.
Number: 4 ships.
Displacement: 60,100–61,000 tons standard; 80,800–82,000 tons full load.
Dimensions: Length overall 1,048–1,073ft (319·4–327m); beam 250–268ft (76·2–81·7m); draught 36ft (11m).
Propulsion: Geared steam turbines, 280,000shp; 4 shafts; 30kt.
Armament: 2 twin Terrier Mk 10 launchers in CV 64; 2 octuple NATO Sea Sparrow Mk 29 launchers in CV 63; 3 octuple NATO Sea Sparrow Mk 29 launchers in CV 66–67; 3 Phalanx CIWS.
Aircraft: 24 F-14A Tomcat; 24 A-7E Corsair; 10 A-6E Intruder + 4 KA-6D; 4 E-2C Hawkeye; 4 EA-6B Prowler; 10 S-3A Viking; 6 SH-3H Sea King.
Complement: 2,879–2,990 + 2,500 (air wing).

Development: Although there are significant differences between the first pair completed and the last two vessels, these four carriers are generally grouped together because of their common propulsion system and flight deck layout.

Kitty Hawk (CV 63) and *Constellation* (CV 64) were ordered as improved Forrestals, incorporating a number of important modifications. The flight deck showed a slight increase in area, and the layout of the lifts was revised to improve aircraft-handling arrangements. The single port-side lift, which on the Forrestals was located at the forward end of the flight deck — and was therefore unusable during the landing operations — was repositioned at the after end of the overhang, outside the line of the angled deck. The respective positions of the centre lift on the starboard side and the island structure were reversed, so that two lifts were available to serve the forward catapults. A further improved feature of the lifts was that they were no longer strictly rectangular, but had an additional angled section at their forward end which enabled longer aircraft to be accommodated. The new arrangement proved so successful that it was adopted by all subsequent US carriers.

Right: *Kitty Hawk* underway in the Pacific, the four E-2Cs and a single A-7 dwarfed by the massive area of the flightdeck.

Nimitz Class

Origin: USA.
Type: Multi-role aircraft carrier.
Number: 3 ships + 1 building + 2 ordered.
Displacement: 81,600 tons standard; 91,400 tons full load.
Dimensions: Length overall 1,092ft (332·8m); beam 251ft (76·5m); draught 37ft (11·3m).
Propulsion: Nuclear; 2A4W reactors 260,000shp; 4 shafts; 30kt.
Armament: 3 octuple Sea Sparrow Mk 25 launchers in CVN 68–69; 3 octuple NATO Sea Sparrow Mk 29 launchers in CVN 70–71; 3 Phalanx CIWS in CVN 70–71.
Aircraft: 24 F-14A Tomcat; 24 A-7E Corsair; 10 A-6E Intruder + 4 KA-6D; 4 E-2C Hawkeye; 4 EA-6B Prowler; 10 S-3A Viking; 6 SH-3H Sea King.
Complement: 3,073–3,151 + 2,625 (air wing).

Development: The Nimitz class was originally envisaged as a replacement for the Midway class. The development of more advanced nuclear reactors made nuclear propulsion an increasingly attractive option, and the high initial cost associated with nuclear propulsion was accepted in return for the proven benefits of high endurance and reduced life-cycle costs. The two A4W reactors which power the Nimitz class each produce approximately 130,000shp compared with only 35,000shp for each of the eight A2W reactors installed in *Enterprise*. Moreover, the uranium cores need replacing less frequently than those originally used in *Enterprise*, giving a full 13-year period between refuellings.

The reduction in the number of reactors from eight to two also allowed for major improvements in the internal arrangements below hangar deck level. Whereas in *Enterprise* the entire centre section of the ship is occupied by machinery rooms, with the aviation fuel compartments and the missile magazines pushed out towards the end of the ship, in *Nimitz* the propulsion machinery is divided into two separate units, with the magazines between them and forward of them. The improved layout has resulted in an increase of 20 per cent in aviation fuel capacity and a similar increase in the volume available for munitions and stores.

Flight deck layout is almost identical to that of *John F. Kennedy*. At hangar deck level, however, there has been a significant increase in the provision of maintenance workshops and spare parts stowage. Maintenance shops have all but taken over the large sponson which supports the flight deck, and at the after end of the hangar there is a large bay for aero-engine maintenance and testing. The increased competition for internal volume even in a ship of this size is illustrated by the need to accommodate some 6,300 men (including air group) — the original Forrestal design on which *Nimitz* and her sisters are based provided for 3,800!

Sensor provision and defensive weapons are on a par with *John F. Kennedy*, although the third ship, *Carl Vinson*, has NATO Sea Sparrow and Phalanx in place of the BPDMS launchers of the earliest ships, which will be similarly fitted in the near future. *Vinson* is also fitted with an ASW control centre and specialised maintenance facilities for the S-3 Viking; these will also be installed in *Nimitz* and *Eisenhower* at future refits.

Delays in construction caused by shipyard problems resulted in rocketing costs, and in the late 1970s the Carter administration attempted, unsuccessfully, to block authorisation of funds for the construction of a fourth carrier in favour of the smaller, conventionally-powered CVV design. The CVV was never popular with the US Navy, however, and the Reagan administration is now committed to the continuation of the CVN programme, and two ships beyond *Theodore Roosevelt* are currently projected. *Nimitz* and *Eisenhower* serve in the Atlantic and *Vinson* in the Pacific; *Theodore Roosevelt* will join the fleet in 1987.

Right: The huge dimensions of the Nimitz class are clearly shown here; flightdeck width is no less than 252ft (76.8m).

Below: *Nimitz* at sea. US planning goal is 15 deployable carrier battle groups by 1990.

America (CV 66), the third ship of the class, was completed after a gap of four years and therefore incorporated a number of further modifications. She has a narrower smokestack and is fitted with an SQS-23 sonar — the only US carrier so equipped.

The first three ships were all completed with two Mk 10 launchers for Terrier missiles. The need to accommodate SPG-55 missile guidance radars in addition to a growing number of air surveillance antennae led to the adoption of a separate lattice mast abaft the island in this and subsequent classes.

In 1963 it was decided that the new carrier due to be laid down in FY 1964 would be nuclear-powered, but Congress baulked at the cost, and the ship was finally laid down as a conventionally-powered carrier of a modified Kitty Hawk design. John F. Kennedy (CV 67) can be distinguished externally from her near-sisters by her canted stack — designed to keep the corrosive exhaust gases clear of the flight deck — and by the shape of the forward end of the angled deck. She also abandoned the Terrier missile system, which consumed valuable space and merely duplicated similar area defence systems aboard the carrier escorts, in favour of the Basic Point Defense Missile

System (BPDMS). The earlier three vessels are at present being similarly modified, the Terrier launchers being removed and replaced by a combination of NATO Sea Sparrow and

Gatling-type Phalanx CIWS guns.

Kitty Hawk and Constellation have served since completion in the Pacific. America and John F. Kennedy serve in the Atlantic.

Above: America in the Indian Ocean with airpower on display: A-7Es line the foredeck, with F-14s and E-2Cs forward of the island facing a row of A-6Es.

Príncipe de Asturias

Origin: Spain.
Type: ASW carrier.
Number: 1 ship.
Displacement: 14,700 tons full load.
Dimensions: Length overall 640ft (195·1m); beam 81ft (24·7m); draught 29·8ft (9·1m).
Propulsion: COGAG; 2 LM2500 gas turbines; 46,600shp; 1 shaft; 26kt.
Armament: 4 Meroka CIWS.
Aircraft: 17 mixed V/STOL and helicopters.
Complement: 793 (inc. air group).

Development: *Príncipe de Asturias* has been built to replace the elderly, ex-US *Dédalo*. The design is based on that of the ill-fated US Navy Sea Control Ship of the early 1970s to provide ASW and air superiority missions in low-threat areas. A simple design with single-shaft propulsion and an austere electronic fit, *Príncipe de Asturias* is built around her aviation facilities, which include a full-width hangar occupying the after two-thirds of the ship. A feature not in the original SCS design is the full 12° ski-jump which will enable the AV-8S Matadors to take off with maximum payload. The only fixed armament is four Spanish-designed Meroka 20mm close-in weapons systems. *Príncipe de Asturias* was laid down in 1979 and launched on 22 May 1982.

The air wing will comprise 6–8 AV-8S Matadors, 6–8 SH-3 Sea Kings for the ASW role, and 4–8 troop-carrying helicopters. As with *Dédalo*, the Spanish Navy has shown what can be done on a limited budget if sufficient thought and sound judgement is applied. The new ship may be unsophisticated by the standards of the larger navies, but the fact is that she will give the Spanish Navy a seaborne air capability which it would otherwise lack.

Right: Launch of the Spanish Navy's *Príncipe de Asturias*, highlighting the ski jump for AV-8S operations.

BATTLESHIPS

Kirov Class

Origin: Soviet Union.
Type: Battlecruiser.
Number: 2 ships + 2 building.
Displacement: 22,000 tons standard; 25,000 full load.
Dimensions: Length overall 814ft (248m); beam 75ft (23m); draught 25ft (8m).
Propulsion: CONAS; 160,000shp; 2 shafts; 32kt.
Armament: 20 SS-N-19 launchers; 12 SA-N-6 launchers; 2 SA-N-4 launchers; SA-N-8 launchers (*Frunze*); 2 single 3·9in (100m) guns (*Kirov*); 1 twin 5·1in (130mm) gun (*Frunze*); 8 30mm Gatling CIWS; 1 twin SS-N-14 launcher (*Kirov*); 1 RBU 6000 launcher; 2 RBU 1000 launchers; 2 quintuple 21in (533mm) torpedo tubes.
Aircraft: 3 Ka-25 Hormone-A.
Complement: 900.

Development: *Kirov* is the ultimate realisation of the Soviet Rocket Cruiser concept. Her powerful defensive armament and extensive command facilities enable her to perform her primary mission of challenging NATO surface forces either alone or at the centre of a Surface Action Group. The need for such capabilities, in addition to a large battery of anti-ship missiles, has resulted in a ship more than twice the displacement of any Western cruiser currently in service.

The main armament comprises twenty SS-N-19 anti-ship missiles, with a range of about 250nm (460km). These are housed in an extensive box magazine, 66ft × 40ft (20m × 15m) and perhaps 33ft (10m) deep, just forward of the main superstructures, and are fired vertically from individual silos, each covered by a hinged hatch. The magazine itself may well be armoured. The vertical launch system, which dispenses with reloads, enables saturation missile attacks to be launched against a high-value target such as a Carrier Battle Group.

Forward of the SS-N-19 magazine is the major surface-to-air system, which also employs vertical launch. There are twelve hinged hatches, arranged in three rows of four. The asymmetrical disposition of the three rows suggests a reloading mechanism sited between them with a magazine beneath, possibly for some 60 reloads. The SA-N-6 is thought to be a high-performance missile with a range of at least 27nm (50km) and a speed of Mach 5 to 6. Since the large Top Globe guidance radars are at either end of the ship, it seems likely that an autopilot is employed in the initial phase of flight, the missile being subsequently picked up by the guidance radar and deposited close enough to the target for its homing head to take over. Whatever the exact nature of the guidance system, it would be surprising, in view of the adoption of vertical launch, if a number of missiles could not be kept in the air simultaneously.

The SA-N-6 system is backed up by launchers for the close-range SA-N-4 abreast the forward deckhouse, two single-mounted 100mm guns aft and four groups of 30mm Gatlings in the "quadrants" of the ship. ECM are also exceptionally complete.

Tucked inside the break in the forecastle is a twin-tube launcher for SS-N-14 anti-submarine missiles. Target data are provided by a large low-frequency bow sonar — probably the same model as on *Kiev* — and an independent variable depth sonar. In addition to the customary mortars and torpedo tubes — the latter behind sliding doors in the hull — there are probably three Hormone-A helicopters for longer-range anti-submarine search and attack and to assist with targeting for the SS-N-14 missiles. The Hormone-As, and possibly a further two helicopters of the Hormone-B missile guidance version, are accommodated in a hangar beneath the quarterdeck. A 40ft × 16ft (15m × 5m) lift at the forward end of the quarterdeck descends to form part of the hangar floor, the opening being closed by twin hinged doors. This arrangement increases stowage space in the hangar.

The propulsion system of *Kirov* is thought to be a combined nuclear and steam (CONAS) plant. Two nuclear reactors, each rated at about 35,000shp, give a cruising speed of 24kt, while the steam plant — either in the form of independent geared turbines or of oil-fired superheater — provides the necessary boost to give a maximum speed of more than 30kt. This seems an unduly complex arrangement; it was probably made necessary by the limited power available from Soviet marine reactors.

Kirov is clearly a very powerful ship indeed and incorporates a number of significant technological advances, including the first operational vertical launch system in the world. If she has a weakness it is that she represents such a considerable investment of resources that she constitutes a high-value target on a par with the carriers she is designed to hunt. This may impose restrictions on her deployment.

Right: Stern view of the Soviet battlecruiser *Kirov*, a powerful challenge to NATO surface forces.

Below: *Kirov*'s long forecastle houses the vertical launch tubes for her main missile armament.

Iowa Class

Origin: USA.
Type: Battleship.
Number: 4 ships.
Displacement: 58,000 tons full load.
Dimensions: Length overall 887ft (270m); beam 108ft (33m); draught 38ft (12m).
Propulsion: 4 geared turbines, 212,000shp; 4 shafts; 33kt.
Armament: 8 Tomahawk launchers; 8 Harpoon launchers; 9 16in (406mm), 12 5in (127mm) guns (*New Jersey, Iowa*), 20 5in (127mm) guns (*Missouri, Wisconsin*); 20 40mm quadruple guns (*Missouri*), 4 Phalanx, CIWS (*New Jersey, Wisconsin*).
Complement: 1,606.

Development: These four battleships (*Iowa* (BB 61), *New Jersey* (BB 62), *Missouri* (BB 63) and *Wisconsin* (BB 64)) have had extraordinary careers. The second largest battleships ever built (largest were the Japanese Yamatos), the Iowas were designed to survive ship-to-ship combat against enemy vessels armed with 18in (45·72cm) guns, (i.e. the Yamatos); they were all commissioned in 1943-44 and saw action in the Pacific in World War II. After the war, other navies first mothballed and then scrapped their battleships, leaving only these four US ships as apparent relics of a bygone age.

They saw action again: all four served in the Korean War (1950-53) and *New Jersey* in the Vietnam War, in both conflicts operating in the shore bombardment role. That seemed to be the end of the story, but the rapid expansion of the Soviet Fleet in the 1960s and 1970s, and especially the appearance of the Kirov class (*qv*), led to modernisation and recommissioning of these mighty ships. *New Jersey* re-joined the fleet on 28 September 1982, followed by *Iowa* on 18 April 1984; *Missouri* and *Wisconsin*, currently in dockyard hands, will rejoin in July 1986 and 1988 respectively.

In the modernisation programme all electronics systems have been either upgraded or replaced, and the ships have been given a general refurbishment. Two 5in (127mm) gun mounts have been removed and replaced by two quadruple Tomahawk cruise missile launchers, and eight Harpoon launch canisters have been fitted. Four LAMPS-III helicopters are carried, but these have to be stored on the open quarterdeck as there is no hangar accommodation. A more fundamental refit programme involving the removal of the after 16in (406mm) turret has been cancelled.

These ships are being returned to service with the fleet at a small fraction of the cost of new ships with comparable capabilities. They represent a valuable addition to the navy's strike forces and the Americans must be deeply grateful to the far-sighted men who decided not to follow the lead of others and scrap them.

Above: Trial launch of a Tomahawk cruise missile from one of eight quad box launchers.

Below: Among *New Jersey's* new equipment is the SPS-49 air surveillance radar.

SSNs

Alfa Class

Origin: Soviet Union.
Type: SSN.
Number: 6 boats.
Displacement: 3,500 tons surfaced; 3,800 tons submerged.
Dimensions: Length 260·1ft (79·3m); beam 32·8ft (10m); draught 24·9ft (7·6m).
Propulsion: Nuclear, 45,000shp; 1 shaft; 40 + kt dived.
Armament: 6 21in (533mm) torpedo tubes.
Complement: 60.

Development: The first Alfa class SSN was completed at the Sud-omekh Yard in Leningrad in 1970 and after protracted tests was scrapped in 1974, reportedly as a result of a major leak from the nuclear reactor. The second boat was laid down in 1971, the third in 1976 and the fourth in 1977, and production ended after six had been built.

The Alfa class submarines are much shorter than previous Soviet SSNs, indicating the possibility of a new and much smaller nuclear reactor; some sources suggest that it may be liquid-metal cooled. These boats are very fast and and are reported to have run under NATO

task forces on exercise at such speeds (40kt) that effective counter-action in a combat environment would have been virtually impossible.

The hull is constructed of titanium alloy, this conferring a maximum diving depth of some 3,000ft (914m); indeed, the long construction times may be related to difficulties in fabricating such materials. The long, low fin and the total absence of any protruding devices in any published photographs suggest that great attention has been paid to reducing the noise signature. The hull is coated with Clusterguard, an anechoic substance designed to resist detection.

There is a large low-frequency sonar array in the bow, and the submarine is fitted with six torpedo tubes and twelve reloads. This class is clearly at the very forefront of submarine technology, and is regarded with great respect by NATO forces. A major concern is that the Alfa's apparent top speed is equal to, or possibly even greater than, the speed of NATO torpedoes, which leaves ASW commanders with a crucial problem.

Los Angeles Class

Origin: USA.
Type: SSN.
Number: 31 + 20 building + 7 ordered.
Displacement: 6,000 tons surfaced; 6,900 tons submerged.
Dimensions: Length 360ft (109·7m); beam 33ft (10·1m); draught 32·3ft (9·8m).
Propulsion: Nuclear, 35,000shp; 1 shaft; 30 + kt dived.
Armament: Harpoon and Tomahawk missiles; SUBROC; 4 21in (533mm) torpedo tubes.
Complement: 127.

Development: The first Los Angeles class SSN entered service in 1976; thirty are now in commission, twenty are under construction and a further seven are on order, making this one of the most massive and expensive defence programmes undertaken by any nation. The Los Angeles boats are much larger than any previous US Navy SSN and have a higher submerged speed. They have the very long-range BQQ-5 sonar and the BQS-15 short-range system, and also operate towed arrays. Weapons fits can include SUBROC, Harpoon and Tomahawk, as well as conventional and wire-guided torpedoes. From the 34th boat the Tomahawks will be carried in 15 vertical launch tubes in the bow. New boats also

have a mine-laying capability. Thus, like all other US SSNs, although they ar eprimarily intended to hunt and destroy other submarines and to protect SSBNs, the Los Angeles submarines can also be used without modification to sink surface ships at long range, while Tomahawk will enable them to attack strategic targets well inland. The BQQ-5 sonar is particularly effective, and is reported on one occasion to have enabled an American SSN to track two Soviet Victor class (qv) SSNs simultaneously. New boats are being given an improved under-ice capability to enable them to operate more effecitvely against Soviet SSBNs in Arctic waters.

The Los Angeles class is very sophisticated and each boat is an extremely potent fighting machine; moreover, with a production run of at least 57 boats it must be considered a very successful design. Nevertheless, the class is also becoming very expensive: in 1976 the cost of each boat was estimated at $221·25 million, but the boat bought in FY79 cost $325·6 million and the three in FY84 $663 million each. Not even the USA can continue at that rate.

Right: *City of Corpus Christi,* her tall, slender fin highlighting the enormous size of this class.

Above: The long, low fin with retractable antennas of a surfaced Alfa class SSN.

Below: Short and sleek, the Alfa class boats may be able to outrun Western ASW torpedoes.

Above; Launch of the Los Angeles class submarine *Phoenix* in December 1979.

Below: The Los Angeles class SSN *City of Corpus Christi* running on the surface.

Mike Class

Origin: Soviet Union.
Type: SSN.
Number: 1 boat + 6 building?
Displacement: 8,000 tons surfaced; 9,700 tons submerged.
Dimensions: Length 360ft (110m).
Propulsion: Nuclear, 1 shaft; 40kt dived.
Armament: SS-N-21 missiles; 6 torpedo tubes.

Development: Few details are yet known of the Mike class, one of two new types of Soviet attack submarine revealed in 1984 (the other is the Sierra class, of 6,400 tons displacement). The first Mike is believed to have been launched in May 1983, entering service in late 1984.

The class is almost certainly constructed of titanium, and the fabrication problems met with the Alfa class (qv) have apparently been overcome; indeed, it is now clear that in this field Soviet technology is well ahead of that in the West.

This class follows the tendency towards ever-increasing size, and at just under 10,000 tons submerged the Mike is currently the largest SSN in service in any navy. The torpedo tubes may be larger than the usual 21in (533mm), in order to accommodate the new SS-N-21 missile.

Below: DoD rendition of the massive Mike class Soviet SSN.

Rubis Class (SNA72)

Origin: France.
Type: SSN.
Number: 2 boats + 4 building + 2 ordered.
Displacement: 2,385 tons surfaced; 2,670 tons submerged.
Dimensions: Length 236·5ft (72·1m); beam 24·9ft (7·6m); draught 21ft (6·4m).
Propulsion: Nuclear, 4MW; 1 shaft, 25kt dived.
Armament: Tube-launched SM.39 missiles; 4 21in (533mm) torpedo tubes.
Complement: 66.

Development: France came late on to the nuclear submarine scene and, under strong pressure from President de Gaulle, went straight to SSBNs. Such a massive programme, which for political reasons had to be entirely French in character, took up all the available national resources for many years. It was not until the 1974 naval programme, therefore, that the French Navy was able to turn its attention to SSNs, with the first of the SNA72 class — Rubis — being laid down in December 1976 and launched on 7 July 1979. She joined the fleet in 1982, following extensive trials, and has been joined by Saphir in 1984, with four more to follow in 1986-88 and the final two in 1993-94. The French intend to form two squadrons, both based at Toulon.

The SNA72 class are the smallest SSNs in any navy and are based fairly closely on the Agosta class conventional submarines. Such small size suggests that the French have produced a dramatic development in

Trafalgar Class

Origin: UK.
Type: SSN.
Number: 2 boats + 2 building + 2 ordered.
Displacement: 4,000 tons surfaced; 5,208 tons submerged.
Dimensions: Length 280ft (85·4m); beam 32·1ft (9·8m); draught 26·9ft (8·2m).
Propulsion: Nuclear, 15,000shp; 1 shaft; 30 + kt dived.
Armament: 5 21in (533mm) torpedo tubes; sub-harpoon from mid-1980s.
Complement: 97.

Development: The latest type of British SSN is the Trafalgar class, the first of which was launched on 1 July 1981 and commissioned on 27 May 1983. Six of these boats are on order, with the possibility of at least one more.

The Trafalgar class is a logical development of the Swiftsure design but the hull has been slightly stretched by including one more section in the parallel body, thus increasing overall length from 272ft (82·9m) to 280ft (85·4m). The diameter of the pressure hull remains unchanged at 32·25ft (9·83m), but there is an increase in submerged displacement to 4,920 tons. A new type of reactor core is used, and the machinery is mounted on rafts to insulate it from the hull and thus cut down radiated noise. There are five 21in (533mm) torpedo tubes, with twenty reloads.

The cost of these SSNs is a good indicator of the problem facing the major navies. At 1976 prices the building costs of the Swiftsure were: Swiftsure £37·1 million; Superb £41·3 million; Sceptre £58·9 million; and Spartan £68·9 million. The cost of the fourth Trafalgar submarine, including weapons systems and equipment, will be £185 million!

Right: HMS *Tireless,* third of the Trafalgar class, was launched on 17 March, 1984.

Below: HMS *Trafalgar* during sea trials. The class is equipped with the latest Plessey Marine Sonar.

nuclear reactor design, especially compared with the rather large devices in the Le Redoutable class SSNs.

Armament, sensors and fire control systems are based on those currently in service in the Agosta s. Four 21in (533mm) torpedo tubes are fitted, with ten reloads. From 1985 onwards the SNA72 class will be fitted for the SM.39, an adaptation of the very succesful MM.38 Exocet surface-launched anti-ship missile; like the US Harpoon, SM.39 will be tube-launched from underwater. A follow-on class of SSNs is already on the drawing boards.

Left: Launch of *Saphir*, second of the class, in 1981.

Right: The Rubis class are very compact boats, with nuclear reactors much smaller than those in Le Redoutable SSBNs.

Victor-III Class

Origin: Soviet Union.
Type: SSN.
Number: 18 boats.
Displacement: 5,000 tons surfaced; 6,000 tons submerged.
Dimensions: Length 341·2ft (104m); beam 32·8ft (10m); draught 23·9ft (7·3m).
Propulsion: Nuclear, 30,000shp; 1 shaft; 30kt dived.
Armament: SS-N-15/SS-NX-16 missiles; 6 21in (533mm) torpedo tubes.
Complement: 90.

Development: First seen by Western observers in 1968 the Victor class is a second-generation Soviet nuclear-powered attack submarine. Somewhat shorter than the November class but with as great a beam, sixteen of the first type — Victor-I — were built. These were followed by the Victor-II, which is 15·25ft (4·6m) longer to enable it to carry the SS-N-15 missile.

Only seven of these boats were built before production changed to yet another development, 11·5ft

(3·5m) longer still and with a cylindrical pod mounted on top of the upper rudder. This has subsequently been confirmed as the first Soviet towed sonar array, and it can be assumed that this device will be seen more widely on other attack submarines in future. Other sensors are a large, low-frequency sonar array in the bow, and a medium-frequency array for torpedo control. It is also reported that Victor-III hulls are coated with the Soviet Clusterguard anechoic protection to attenuate the reflections which are returned to searching hostile warships. There are now eighteen Victor-IIIs, and production has ceased; the successor to the class is the Sierra (qv).

On 27 February 1982 the Italian Sauro class submarine *Leonardo da Vinci* detected a Soviet Victor-I at a depth of some 984ft (300m), 25 miles (40km) south-east of the naval base at Taranto. The Italians tracked the Soviet submarine for some 18 hours until it left Italian territorial waters. In reporting this incident the Italian authorities made it clear that this was

by no means the first such incursion by a Soviet submarine. It is also noteworthy that the public announcement was absolutely positive about the nationality *and type* of the target, a rare acknowledgement of the precision of modern underwater identification technology.

Above: A Victor-III appears to flaunt the fin-mounted towed array sonar.

Below: A Victor-III diving. The hull is coated with an anechoic substance to help reduce chance of detection.

SSGNs

Oscar Class

Origin: Soviet Union.
Type: SSGN.
Number: 3 boats + ?building.
Displacement: 10,000 tons surfaced; 14,000 tons submerged.
Dimensions: Length 469ft (143m); beam 60ft (18·3m); draught 36ft (12m).
Propulsion: Nuclear, 40,000shp; 2 shafts; 30 + kt dived.
Armament: 24 SS-N-19 SLCM; 8 21in (533mm) torpedo tubes.
Complement: c130.

Development: Like other recent Soviet submarine classes, the Oscar class is significantly larger than its predecessors, with a submerged displacement of 14,000 tons and a length of 469ft (143m). The principal weapon is the SS-N-19, with 24 missiles housed in vertical launch tubes, mounted twelve per side abreast the long, low fin and outside the pressure hull.

The Oscar class appears to be the Soviet Navy's reaction to the increased range of US carrier groups' defensive measures, as made possible by, for example, the Lockheed S-3 Viking ASW aircraft (qv). Thus the SS-N-19 has a range of 277 miles (445km) compared to the 69 miles (111km) of the Charlie-II's SS-N-9. The Oscar could also act as the advance guard of a Soviet Navy task group clearing the way for SSBNs to transit the Iceland-Faroes 'chokepoint' and move out into the Atlantic.

The first Oscar was launched at Severodvinsk in 1980 and joined the fleet in 1983. The second boat was launched in 1982 and the third in 1983 and the delivery rate is now expected to stabilise at one or two boats per year. The single boat of the Papa class is now thought to have been a test-bed for the Oscar/SS-N-19 combinations.

By 1985 Oscar-class SSGN patrols had become a matter of routine and production was continuing. In addition, the elderly Echo II SSGNs were being updated by replacing the 1960 vintage SS-N-3 missiles by the much more effective SS-N-12.

Above: 14,000-ton Oscar-class cruise missile submarine (SSGN), armed with 24 nuclear-capable SS-N-19 anti-ship missiles with a 340 mile (550km) range. Missiles are targetted on NATO carrier battle groups.

Right: An Oscar-class SSGN running with its sail surfaced and one periscope raised, photographed by a vigilant NATO ASW patrol aircraft, probably in the Barents Sea. Note the hatch-covers for quiet running.

SSKs

Agosta Class

Origin: France.
Type: SSK.
Number: 10 boats
Displacement: 1,450 tons surfaced; 1,725 tons submerged.
Dimensions: Length 221·7ft (67·6m); beam 22·3ft (6·8m); draught 17·7ft (5·4m).
Propulsion: 2 diesel, 3,600bhp; 1 electric, 4,600hp; 1 shaft; 20kt dived.
Armament: 4 21in (533mm) torpedo tubes.
Complement: 52.

Development: *Agosta,* name-ship of the class, joined the French fleet in July 1977 and was followed by three more in 1977-78, completing the French Navy's own order. A further four Agostas have been built in Spain by Bazan and two have been built in France for the Pakistani Navy. The South Africans wanted to order two as well, but this was blocked for political reasons. Egypt was interested at one time in purchasing two Agostas, but no order has ever been announced.

Somewhat larger than the previous Daphnés, the Agosta class is intended for distant-water operation. Only four torpedo tubes are fitted, but there are twenty reloads and special devices for rapid reloading. The tubes are 21in (533mm) in diameter, the first time that the French have abandoned their previous 21·7in (550mm); this is presumably intended to enhance the export potential of the type. ASW equipment includes a passive sonar set (DSUV-2) with 36 microphones, two active sets, and a passive ranging set under a spiky dome on the foredeck.

Considerable attention has been devoted to silent running, and an unusual feature is the fitting of a small 30hp electric motor for very quiet, low-speed movement whilst on patrol.

If the French Navy sticks to its announced intention to concentrate on nuclear-powered submarines in future, the Agosta class could be the last of a distinguished and interesting line of French conventional boats.

Above: The Agosta class; probably the last non-nuclear boats in the French Navy.

Below: Hurmat (S 136) is one of two French-built Agosta class submarines in service with the Pakistan Navy.

Kilo Class

Origin: Soviet Union.
Type: SSK.
Number: 5 boats + ? building.
Displacement: 2,500 tons surfaced; 3,200 tons submerged.
Dimensions: Length 229·6ft (70m); beam 29·5ft (9m); draught 23ft (7m).
Propulsion: Diesel-electric.
Armament: 8 21in (533mm) torpedo tubes.
Complement: 55.

Development: Despite the fact that the Tango class has proved to be a successful SSK and is still in production, a second type of SSK, the Kilo class, is being built in the Komsomolsk yard in the Far East,

specifically, it appears, for the Pacific Fleet; it may be that the Kilo has been designed for a particular operational requirement in the Pacific. Bearing in mind the stealth capabilities of the SSK, one such role could be to operate in the shallow waters of the Sea of Okhotsk and the Yellow Sea to protect the Soviet SSBNs against intruding US Navy SSNs. The Kilo class adopts an Albacore-type hull for the first time in a Soviet SSK and can be expected to be very quiet in operation.

Right: Soviet Navy Kilo class conventionally-powered submarines are currently in production in Pacific yards.

Näcken Class

Origin: Sweden.
Type: SSK.
Number: 3 boats.
Displacement: 1,030 tons surfaced, 1,125 tons submerged.
Dimensions: Length 162·4ft (49·5m); beam 18·4ft (5·6m); draught 18·4ft (5·6m).
Propulsion: diesel, 2,200bhp; 1 electric; 1 shaft; 20kt dived.
Armament: 6 21in (533mm) and 2 16in (406mm) torpedo tubes.
Complement: 19.

Development: Sweden's first submarine was launched in 1904 and since then there has been a succession of small but efficient boats, designed primarily for use in the Baltic. The five submarines of the Sjöormen class joined the Swedish fleet between 1967 and 1969. The design was based, like so many others at that time, on the revolutionary Albacore hull, as were the indexed cruciform after control surfaces. The large fin towers over the bow and the forward control surfaces were mounted on the fin rather than on the bows, in line with

American practice. Endurance was estimated to be some 21 days, and the class was intended for operations in the very tricky Baltic waters. The crew normally numbered 23, although operation with a mere 18 was possible for short periods.

The Sjöormens were followed in the late 1970s by the Näcken class. These boats can operate at depths of up to 984ft (300m) and may well be intended to deploy outside the Baltic in the deep trenches of the Skagerrak. Based on the Sjöormen, the Näcken is a little smaller, and very great attention has been paid to quietness and to effective control at slow speeds.

Successor to the Näcken class will

be the A17 or Västergötland class; four boats are due for delivery in 1987–89. These 1,000-ton boats will be highly automated, with a crew of only 20 officers and ratings. Lead yard is Kockums of Malmö.

Below: Nacken class submarines are manned by a crew of 19, a very small number in comparison with other vessels of similar size.

Tango Class

Origin: Soviet Union.
Type: SSK.
Number: 21 boats.
Displacement: 3,000 tons surfaced; 3,700 tons submerged.
Dimensions: Length 301·8ft (92m); beam 29·5ft (9m); draught 23ft (7m).
Propulsion: 3 diesel, 6,000bhp; electric, 6,000hp; 3 shafts; 16kt dived.
Armament: 8 21in (533mm) torpedo tubes.
Complement: 62.

Development: The Soviet Navy has a vast fleet of conventional submarines: at least 200 are currently in service, with a further 100 in reserve. Some fifty Whiskey-V boats are in service, and it was one of these that ran aground near the Swedish naval base at Karlskrona in 1982. The next class was the Romeo, an improved Whiskey, of which 560 were initially planned although this was cut back to twenty when the nuclear fleet was expanded. Some sixty of the Foxtrot class are in service, built between 1958 and 1971 for the Soviet Navy although small scale production continues for export.

The Tango class was first seen by Western observers at the Sebastopol Naval Review in July 1973, and production has continued ever since at a

rate of some 2–3 boats per year. This class is of advanced design, and it is of interest that the Soviet Union does not seem particularly keen to export it, despite the considerable market for a conventionally-powered submarine of this size and capability. It is clearly regarded as complementary to the SSNs, and could well be intended for use in the extensive shallow waters around the Soviet Union. The continued production of this class, and of the later Kilo as well, makes it clear that the Soviets do not intend to follow the US lead in aiming for an all-nuclear submarine fleet.

The hull of the Tango class has very smooth lines, but one noteworthy feature is that forward of the fin there is a marked rise of some 3ft (0·91m). This will undoubtedly improve seakeeping qualities on the surface, but also suggests a requirement for extra volume in the forward end of the boat, possibly for some new weapon system; it has also been reported that the latest boats have a slightly longer hull to allow for a new weapon to be fitted. The smooth lines of these submarines suggest a small and compact design, but this is deceptive as they are, in fact, the largest conventional submarines currently in production (which continues at a rate of two per year).

Above: Unlike the US Navy, the Soviet Navy continues to build diesel-powered submarines of the Kilo and Tango (above) classes.

Below: The rise of the hull forward of the fin suggests that the Tangos may be able to carry the SS-N-15 ASW.

Type 209 Class

Origin: Federal Republic of Germany.
Type: SSK.
Number: 44 boats + 12 building.
Displacement: 1,100 tons surfaced; 1,210 tons submerged. See text.
Dimensions: Length 178·4ft (54·4m); beam 20·3ft (6·2m); draught 17·9ft (5·5m).
Propulsion: 4 MTU diesel, 7,040kW; 1 Siemens electric, 3,070kW; 1 shaft; 23kt dived.
Armament: 8 21in (533mm) torpedo tubes.
Complement: 33.

Development: Germany has a special place in the history of the submarine, ending World War II with some outstanding designs which, fortunately for the Allies, failed to attain operational status in significant numbers. It was not until 1954 that Germany was allowed to construct the Type 205, a small coastal boat, twelve of which were built for use in the Baltic. Two improved Type 205s were built in Denmark (Narvhalen class), and a further fifteen improved Type 205s, optimised for deep diving, were built in Germany for Norway as Type 207s.

Having completed the Type 205s, design work started on a follow-on class of 450-ton boats, the main concern being with greater battery power to meet the demands of the ever-increasing numbers of electrical and electronic devices but without reducing submerged speed or endurance. Construction of the first boat (U-13) began in November

1969, and the eighteenth and last joined the Bundesmarine in September 1971. Made of special non-magnetic steel, these submarines have served the German Navy well and, so far as is known, have totally avoided the notorious corrosion problems that affected the earlier Type 205s. The opportunity was taken in this class to upgrade the active and passive sonars and the fire-control system, and wire-guided torpedoes were fitted for the first time in a German submarine design.

The British Vickers Shipbuilding Group constructed three submarines for Israel under licence from Ingenieurkontor Lübeck (IKL) but fitted with British weapons systems. Described variously as an adaptation of the Type 206 or as a smaller Type 209, the IKL 540 is optimised for operations in the warmer waters of the eastern Mediterranean. A unique fitting is the fin-mounted SLAM, a quadruple Blowpipe SAM installation, giving the Israeli submarines an anti-aircraft capability.

Raising the Allied-imposed displacement limit on German submarine construction to 1,000 tons led to the design of the very successful Type 209, which has met the need of many navies for a new submarine, conventionally armed and powered but with up-to-date sensors and electronics and with minimal demands on highly skilled manpower. Operators include Greece (eight Type IK 36), Argentina (two IK 68), Peru (two 62 with 35-man crews), Colombia (two IK 78 with 35-man crews) and Turkey (two IK 14 with 35-man

Above: Substantial numbers of Type 209s have been built for export: this example flies the Federal German flag during trials.

Below: The Yildiray, first of three Type 209s built in Turkey, under construction at Gölcük prior to her launch in 1979.

TR 1700

Origin: Federal Republic of Germany.
Type: SSK.
Number: 2 boats + 4 building.
Displacement: 1,700 tons surfaced; 2,535 tons submerged.
Dimensions: Length 216·5ft (66m); beam 23·9ft (7·3m); draught 21·3ft (6·5m).
Propulsion: 4 diesel, 5,900hp; 1 electric, 6,600kW; 1 shaft; 25kt dived.
Armament: 6 21in (533mm) torpedo tubes.
Complement: 34.

Development: Thyssen Nordsee-werke built 15 Type 207 submarines for Norway in the years 1962-67, followed by 10 Type 206 boats in 1970-1975 for the Bundesmarine. This experience of building to IKL designs led to Thyssen designing their own boat, aimed primarily at the export market.

With a submerged displacement of 2,535 tons the TR 1700 is by far the largest post-war German design, but extensive automation reduces the manning requirement to just 34. The battery storage capacity is considerable, and an underwater 'burst' speed of 25kt for one hour is claimed, considerably greater than for any previous conventional submarine. The boat can remain submerged for up to 70 days using the schnorkel, and a torpedo reload time of just 50 seconds is claimed.

So far only six TR 1700s have been ordered, all for the Argentine Navy. The first two, *Santa Cruz* and *San Juan*, were built in West Germany. *Santa Cruz* was launched in September 1982, and both were commissioned in 1984. Following a period of working up in European waters they travelled south and joined the Argentine fleet in late 1984.

The other four boats are being built in Buenos Aires by Astilleros Domecq Garcia, the first of them being laid down in October 1983. They will be commissioned at a rate of one a year from 1986 onwards.

These boats, being much larger and having much greater range than Argentina's present Type 209s (also German-built) will obviously be an important addition to the fleet, and one that will doubtless have a significant effect on the naval balance of power in the South Atlantic, and must cause the Royal Navy to review its plans for the defence of the Falkland Islands.

A modified and stretched version – the TR 1700A – is on offer to Australia for its SSK order.

Below: *Santa Cruz* on trials before transfer with sister *San Juan* to Argentina. They are reckoned to be capable boats.

crews). An improved version (IK 81), longer, and with a larger detection dome, has been ordered by Venezuela (4), Ecuador (2), Turkey (1), Greece (4), Peru (6) and Indonesia (4). Turkey, having received three from Howaldtswerke, is now producing her own; two have been completed so far, and another eight are planned. Similarly, Howaldtswerke is constructing two for India, who will then produce a further two in her own yard at Bombay. Three Type 209s are being built for Brazil, whilst a Chilean order for two was completed in 1984.

The Type 209 is similar in shape and layout to the Type 205 but has increased dimensions, greater battery capacity and more powerful propulsion. The hull is completely smooth, with retractable hydroplanes mounted low on the bows, cruciform after control surfaces and a single screw. Careful hull design and powerful motors result in a staggering underwater 'burst' speed of 23kt. Designed for patrols of up to 50 days, these boats are armed with eight 21in (533mm) torpedo tubes and have a full range of sensors. They are one of the most successful contemporary submarine designs, and during the 1982 South Atlantic War the Royal Navy treated the two Argentine Type 209s with the greatest respect, although it is not clear whether the two boats actually went to sea.

Twelve of the German Navy's 18 Type 206s are to be modernised in 1988-91. A new design for the German Navy, the Type 211, will replace the Type 205s in the early 1990s.

Right: Bow threequarter view of a Type 209 with the Kiel shipyards in the background.

Below: Modest size and conventional power proved an attractive combination, especially for South American navies.

Upholder (Type 2400) Class

Origin: UK.
Type: SSK.
Number: 1 ordered.
Displacement: 2,125 tons surfaced; 2,362 tons submerged.
Dimensions: Length 230·5ft (70·25m); beam 24·9ft (7·6m); draught 24·6ft (7·5m).
Propulsion: 2 diesel, 5,400bhp; 1 electric; 1 shaft; 20 + kt dived.
Armament: 6 21in (533mm) torpedo tubes.
Complement: 44.

Development: It was once intended that the Oberon class would be the last conventional boats for the Royal Navy. It is now, however, accepted that, although the nuclear-powered submarine has many advantages, there is still an operational need for the conventional type as well. There is thus an urgent need to replace the Porpoise and Oberon boats since the hulls are all over 20 years old, and more importantly, the design itself is based on the technology of the early 1950s. The new RN patrol submarine (SSK-01) will be based very closely upon the Vickers Type 2400, so called because its submerged displacement is 2,400 metric tonnes (2,362 tons). This type will be built at a rate of about one per year with the first, *HMS Upholder*, being commissioned in 1987–88. At least ten will be ordered for the Royal Navy and they will be a very welcome addition to the fleet. A unit cost of £50 million has been quoted, but it would seem probable that it will be very much more.

The new class will have a 'teardrop' type hull with cruciform after hydroplanes and retractable forward hydroplanes mounted low on the bow; the RN has never accepted the US mountings on the fin. Many special features have been incorporated into the design to minimise radiated noise and thus achieve a marked reduction in the noise signature. A major factor is the maximum use of labour-saving devices, thus reducing the crew to 46 compared to 69 in the Oberons. A diesel-electric propulsion system, comprising a single-pitch propeller on a shaft directly driven by a twin-armature electric motor, is fitted. On the surface, and when 'snorting', two four-stroke high-speed diesels are used, each driving a 1·25 MW AC generator. The most modern sensors will be fitted, and there will be six torpedo tubes in the bow. These will fire normal torpedoes or Harpoon missiles, or can be used to lay mines.

There is a large overseas market for this type and size of patrol submarine with countries who want to replace their Oberons, Guppies and Balaos, and the Type 2400 is already on offer to Australia and Canada. In the Australian competition the favourites are the British Vickers design and the West German Thyssen TR 1700A, although both will be somewhat modified (longer and with greater displacement) to meet Australian requirements.

Below: Cutaway of the new Type 2400 class submarine, the first of which is expected to commission in the late 1980s.

Yuushio Class

Origin: Japan.
Type: SSK.
Number: 6 boats + 2 building.
Displacement: 2,200 tons surfaced.
Dimensions: Length 249·3ft (76m); beam 32·5ft (9·9m); draught 24·6ft (7·5m).
Propulsion: 2 diesel, 4,200bhp; 1 electric, 7,200hp; 1 shaft; 20kt dived.
Armament: 6 21in (533mm) torpedo tubes.
Complement: 75.

Development: The Japanese Maritime Self-Defence Force (JMSDF) produced its first indigenous postwar submarine design in 1959 and a second, improved class in 1961-62. All these boats have now been stricken, but the culmination of this line — the four boats of the Asahio class — is still in service. This is a neat and workmanlike design, with the now somewhat rare feature of twin 12·7in (322mm) stern torpedo tubes, which are feasible only because of the twin propellers.

The next class — Uzushio — was based on a US Navy design, with an Albacore-type 'teardrop' hull for faster and quieter underwater performance. The hull is of very high quality steel to permit a diving depth of up to 650ft (220m). Seven of the class are in service.

The latest Japanese submarines are those of the Yuushio class, an all-round improvement on the Uzushio class and capable of slightly higher speeds. Both the Uzushio and Yuushio submarines have their torpedo tubes mounted amidships, firing outwards at an angle of 10° to the hull, a feature they share with the US Navy's SSNs. This is done in order to free the entire bow area for a large sonar array. The Uzushio class have pressure hulls of high-tensile steel (NS-63), permitting a diving depth of some 1,970ft (600m). The Yuushios, however, use even more modern steel (NS-90), giving a claimed diving depth of some 3,280ft (1,000m). The first two will be retrospectively fitted with Sub-Harpoon missiles, but the remaining boats of the class are being fitted with these missiles prior to delivery. A noteworthy feature of these submarines is their large complements, much greater for the size of hull than in other navies. (Compare, for instance, the complement allocated for Yuushio Class boats with that for the Type 2400s, described above).

The Yuushios are very advanced, as would be expected from such a technologically capable nation, and would seem to be fully equivalent to SSNs in most features except that of underwater endurance. This may be encouraging the JMSDF to seek a solution to the problem of freeing the non-nuclear submarine from the necessity of having to come up to the surface to 'breathe' at regular intervals.

The role of these Japanese submarines in war would be to defend Japanese waters from incursions by foreign (presumably Soviet) surface and underwater vessels. It may be assumed that this would include patrols in the Sea of Okhotsk, which is a known haven for Soviet SSBNs.

Below: The Yuushio class submarine *Mochishio* running on the surface; later members of the class are equipped to launch Sub-Harpoon cruise missiles.

Zwaardvis/Walrus Classes

Origin: Netherlands.
Type: SSK.
Number: 2 boats + 6 building.
Displacement: 2,350 tons surfaced, 2,640 tons submerged.
Dimensions: Length 217·2ft (66·2m); beam 33·8ft (10·3m); draught 22·3ft (7·1m).
Propulsion: 3 diesel, 4,200bhp, 1 electric; 1 shaft; 20kt dived.
Armament: 6 21in (533mm) torpedo tubes.
Complement: 67.
(Specifications given for Zwaardvis class; see text for Walrus class variations).

Development: The two boats of the Zwaardvis class are among the largest conventional submarines currently in service, matched by the US Barbel, the Soviet Tango and the projected British Upholder classes. The design of the Zwaardvis does, in fact, owe a great deal to the Barbel, with a similar Albacore hull, giving considerable internal depth and making it possible to incorporate two decks. This gives a generally roomy interior, which has great advantages for living conditions. Three diesel generators power the propulsion motor for surface running and two groups of batteries provide underwater power. A single, five-bladed propeller is mounted abaft the cruciform control surfaces, and a 'burst' speed in excess of 20kt has been reported.

The Zwaardvis boats will be supplemented in the late 1980s by two improved submarines designated the Walrus class. The dimensions and silhouettes are virtually identical, but use of the new French 'Marel' high-tension steel will increase diving depth by at least 50 per cent in the new vessels. An updated and automated fire control and command system (codenamed 'Gipsy') will allow a reduction in complement from 65 to 49, a very significant factor in such a small navy. The first two boats will replace *Dolfijn* and *Yeehond* in the late 1980s, and the second pair will replace *Potvis* and *Tonjin* in the early 1990s.

It is planned to fit the six boats of the Walrus class with the new and highly effective GEC Avionics Type 2026 towed array sonar, together with a Hollandse Signaalapparaten processing system and an Ameeco "wet end".

These Dutch submarines are capable of protracted operations in the Atlantic, where they would operate as part of the task groups which the Royal Netherlands Navy has undertaken to provide as part of its NATO commitment. Two 'Improved Zwaardvis' boats will be delivered to Taiwan in 1985-86, although a follow-on order has been blocked by political pressure.

Below: *Tijgerhaai,* one of two Zwaardvis class SSKs in service with the Dutch Navy. Two Walrus class boats launched in 1984/85 are expected to join the fleet in 1988, while construction of a further pair at Rotterdam should be completed in 1992/93, and two more will be started in 1991 for completion in 1996.

CRUISERS

Slava Class

Origin: Soviet Union.
Type: Cruiser.
Number: 2 ships + ? building.
Displacement: 10,500 tons standard; 12,000 full load.
Dimensions: Length overall 613·4ft (187m); beam 65·6ft (20m); draught 25ft (8m).
Propulsion: 4 gas turbines, 120,000shp 2 shafts; 34kt.
Armament: 16 SS-N-12 launchers; 8 SA-N-6 launchers; ? SA-N-7 launchers; 2 twin SA-N-4 launchers; 2 twin 5·1in (130mm) guns; 6 30mm Gatling CIWS; 2 RBU 6000 launchers; 8 21in (533mm) torpedo tubes.
Aircraft: 1 helicopter.
Complement: *c*600.

Development: Built at the Nikolayev North yard on the Black Sea, the first Slava class cruiser was launched in 1979 and commissioned in 1983, followed by the second a year later. The ship is very large by modern standards, the most unusual feature being the sixteen SS-N-12 launchers in eight dual mountings on the forecastle, four either side of the bridge complex. There are also eight vertical launchers (64 missiles) for the SA-N-6 system abaft the engine room uptakes and a twin 5·1in (130mm) gun mounting forward.

In many ways the *Slava* is a smaller version of the *Kirov* (qv), except that it has conventional gas-turbine power.

Below: The four pairs of SS-N-12 launchers on each side of the bridge complex form the primary armament of the cruiser *Slava,* while a twin 130mm gun is mounted on the forecastle.

Bottom: The Ka-25 "Hormone-B" helicopter visible on the stern pad aboard *Slava* is used to provide mid-course guidance for the SS-N-12s; vertical SAM launchers are abaft the uptakes.

Sovremenny Class

Origin: Soviet Union.
Type: Destroyer.
Number: 4 ships + 3 building.
Displacement: 6,500 tons standard; 7,950 full load.
Dimensions: Length overall 510ft (155m); beam 57ft (17m); draught 20ft (6m).
Propulsion: Geared steam turbines, 110,000shp; 2 shafts; 33kt.
Armament: 2 quadruple SS-N-22 launchers; 2 SA-N-7 launchers; 2 twin 5·1in (130m) guns; 4 30mm Gatling CIWS; 2 RBU 1000 launchers; 2 twin 21in (533mm) torpedo tubes.
Complement: 400.

Development: In 1980 the first of a new series of destroyers with a clear anti-surface mission made its appearance in the Baltic. *Sovremenny* ran her sea-trials without her main armament but with most of her surveillance and fire control radars in place. The major weapon system eventually proved to be a new anti-ship missile, the SS-N-22, a development of the SS-N-9.

Immediately aft of the funnel is a telescopic hangar for a Hormone-B missile guidance helicopter. Major guns of a new 5·1in (130mm) model are installed fore and aft, intended for fire support operations. The gun bears a remarkable similarity to the 8in (203mm) gun under development for the US Navy during the late 1970s.

The other major weapon system is a medium-range surface-to-air system with the forward mount between the bridge and the forward gun and the after mount between "V" turret and the helicopter platform. This is the SA-N-7 which has been undergoing trials aboard the Kashin class destroyer *Provorny*. ASW armament is minimal, comprising only a pair of RBU mortars and twin banks of torpedo tubes. There is a bow sonar, but no variable depth sonar.

It is somewhat surprising that the Soviet Navy should have persisted with steam propulsion in this class, especially as its ASW counterpart, the *Udaloy*, is powered by gas turbines.

Above: The Sovremenny class vessels carry surface-to-surface weapon systems to complement the ASW-capable Udaloy class.

Below: A member of the Sovremenny class during sea trials in October 1980, before armament is installed.

Ticonderoga Class

Origin: USA.
Type: Cruiser.
Number: 2 ships + 2 building + 9 ordered.
Displacement: 9,600 tons full load.
Dimensions: Length overall 563ft (171·7m); beam 55ft (17m); draught 31ft (9·4m).
Propulsion: COGAG; 4 LM2500 gas turbines, 80,000shp; 2 shafts; 30kt.
Armament: 2 twin Standard MR SM-2 Mk26 launchers; 8 Harpoon launchers; 2 5in (127mm) guns; 2 Phalanx CIWS; ASROC; 6 12·75in (324mm) torpedo tubes.
Aircraft: 2 LAMPS helicopters.
Complement: 375.

Development: The new missile cruiser *Ticonderoga* is the first operational vessel to be fitted with the Aegis Combat System. It was originally envisaged that this system would be installed in nuclear-powered escorts such as the Strike Cruiser (CSGN) and the CGN 42 variant of the Virginia class, but the enormous cost of Aegis combined with that of nuclear propulsion proved to be prohibitive under the restrictive budgets of the late 1970s. Moreover, two Aegis escorts were required for each of the twelve carrier battle groups, and as not all the carriers concerned were nuclear-powered, it was decided to utilise the growth potential of the fossil-fuelled Spruance design to incorporate the necessary electronics.

The Aegis Combat System was developed to counter the saturation missile attacks which could be expected to form the basis of Soviet anti-carrier tactics during the 1980s. Conventional rotating radars are limited both in data rate and in number of target tracks they can handle, whereas saturation missile attacks require sensors which can react immediately and have a virtually unlimited tracking capacity. The solution adopted in the Aegis system is to mount four fixed planar antennae each covering a sector of 45 degrees on the superstructures of the ship. Each SPY-1 array has more than 4,000 radiating elements that shape and direct multiple beams. Targets satisfying predetermined criteria are evaluated, arranged in sequence of threat and engaged, either automatically or with manual override, by a variety of defensive systems.

At longer ranges air targets will be engaged by the SM-2 missile, fired from one of two Mk 26 launchers. The SM-2 differs from previous missiles in requiring target illumination only in the terminal phase of flight. In the initial and mid-flight phase the missile flies under autopilot towards a predicted interception point with initial guidance data and limited mid-course guidance supplied by the Aegis system. This means that no fewer than 18 missiles can be kept in the air in addition to the four in the terminal phase, and the Mk 99 illuminators switch rapidly from one target to the next under computer control. At closer ranges back-up is provided by the two 5in (127mm) guns, while "last-ditch" self-defence is provided by two Phalanx CIWS guns, assisted by ECM jammers and chaff dispensers.

Ticonderoga and her sisters are designed to serve as flagships, and are equipped with an elaborate Combat Information Centre (CIC) possessing an integral flag function able to accept and co-ordinate data from

Above: USS *Ticonderoga* showing the after pair of SPY-1 antennas, key parts of the Aegis system.

Right: Missile element of the Aegis system is the twin Standard launcher fore and aft.

other ships and aircraft. Twenty-six units are currently projected, and it is envisaged that they will operate in conjunction with specialised ASW destroyers of the Spruance class and a new type of AAW destroyer (DDG-51).

Udaloy Class

Origin: Soviet Union.
Type: Cruiser.
Number: 6 ships.
Displacement: 6,500 tons standard; 8,000 tons full load.
Dimensions: Length overall 531·4ft (162m); beam 59ft (18m); draught 20ft (6m).
Propulsion: 2 gas turbines, 30,000shp; 2 gas turbines, 6,000shp; 35kt.
Armament: 2 quadruple SS-N-14 launchers; 8 SA-N-8 launchers; 2 3·9in (100m) guns; 4 30mm Gatling CIWS; 2 RBU 6000 launchers; 4 21in (533mm) torpedo tubes.
Aircraft: 2 Ka-25 Hormone-B or Ka-27 Helix-A.
Complement: 350.

Development: The *Udaloy* is of great interest because it is optimised for the ASW role and is clearly intended to be the anti-submarine component of a mixed battle group operating at some distance from its base, probably in the northern and central Atlantic. The ASW armament is exceptionally powerful. There are the now standard quadruple SS-N-14 launchers abreast the bridge, two RBU 6000 rocket launchers, and 21in (533mm) torpedo tubes amidships. *Udaloy* has two separate hangars for its pair of Hormone-A ASW helicopters, the first Soviet cruiser or destroyer to be equipped to operate two rather than one aircraft. The landing platform is large and located above the VDS well

on the stern, but the hangar floor is one deck lower with a ramp for moving the aircraft from one level to the other. The sharp rake of the bow suggests a large low-frequency sonar dome fitted below, and this tends to be confirmed by the characteristics of the bow wave. There is a VDS at the stern, streamed over the transom in line with Soviet naval practice.

Other weapons are somewhat limited. There are two 100mm dual-purpose guns in single mountings in the "A" and "B" positions, together with four 30mm Gatling CIWS. Air defence is provided by eight SA-N-8 SAM launchers. Air and surface surveillance radars are not so numerous as on earlier Soviet warships, suggesting greater electronic sophistication.

In view of the similarity in dimensions and displacement between this class and the Sovremenny class, which appeared at the same time but are optimised for the anti-surface role, it is surprising that the two classes do not share a common hull. This certainly would have been the case for a Western navy, but the political and economic constraints upon the Soviet Navy are much less severe and they have been permitted to optimise the hull form as well. The propulsion method is also different, *Sovremenny* having steam turbines and *Udaloy* gas turbines (which are particularly suited for the ASW mission).

These ships are classified by the

Above: *Udaloy* on sea trials. The armament is already aboard, but there are empty platforms for electronic equipment.

Below: Armament of the Udaloy class includes quad SS-N-14 launchers each side of the bridge and SAM tubes fore and aft.

Soviets as *"bol'shoy protivolod-ochny korable"* (large ASW ship, BPK) and by NATO as DDGs, although their size suggests that the designation "cruiser" is much more appropriate. Six vessels of this very capable and interesting class are in service, with more to follow.

Virginia Class

Origin: USA.
Type: Cruiser.
Number: 4 ships.
Displacement: 11,000 tons full load.
Dimensions: Length overall 585ft (178m); beam 63ft (19m); draught 30ft (9m).
Propulsion: Nuclear; 2 D2G reactors, 60,000shp; 2 shafts; 30kt.
Armament: 2 twin Standard MR Mk 26 launchers; 2 quadruple Harpoon launchers being fitted; 2 5in (127mm) guns; ASROC; 6 12·75in (324mm) torpedo tubes.
Aircraft: 2 LAMPS helicopters.
Complement: 473.

Development: Following closely upon the two CGNs of the California class, the *Virginia* incorporated a number of significant modifications. While the basic layout of the class is identical to that of their predecessor, the single-arm Mk 13 launchers of the *California* were superseded by the new Mk 26 twin ASROC launcher forward, and a helicopter hangar was built into the stern.

The magazine layout and missile-handling arrangements of the Mk 26 constitute a break with previous US Navy practice. In earlier missile cruisers and destroyers booster-assisted missiles such as Terrier were stowed in horizontal magazine rings, and the shorter Tartar missiles in cylindrical magazines comprising two concentric rings of vertically stowed missiles. The magazine associated with the Mk 26 launcher, however, has a continuous belt feed system with vertical stowage capable of accommodating a variety of missiles. This means that ship's length is the only limiting factor on the size of the magazine, which is capable of being "stretched" or "contracted" to suit the dimensions of the vessel in which it is to be installed. It has also eliminated the requirement for a separate launcher for ASROC. In the Virginia class ASROC rounds are carried in the forward magazine alongside Standard MR surface-to-air missiles. The elimination of the ASROC launcher and its associated reloading deckhouse has saved 16·4ft (5m) in length compared with *California*.

The installation of an internal helicopter hangar in a ship other than an aircraft carrier is unique in the postwar US Navy. The hangar itself is 42ft × 14ft (12·8 × 4·3m) and is served by a stern elevator covered by a folding hatch.

The electronics outfit is on a par with *California*, with two important differences. The first is the replacement of the SQS-26 sonar by the more advanced solid-state SQS-53,

Above: The Virginia class cruiser *Mississippi*. The twin-arm Mk 26 launchers distinguish her from the earlier California class.

and the older Mk 114ASW FC system by the digital Mk 116. The second is the retention of only the after pair of SPG 51 tracker/illuminators, reducing the number of available

---

DESTROYERS

Broadsword Class

Origin: UK.
Type: Destroyer.
Number: 6 ships + 4 building + 3 ordered.
Displacement: 3,500 tons standard; 4,000 tons full load.
Dimensions: Length overall 430ft (131·2m); beam 48·5ft (14·8m); draught 14ft (4·3m).
Propulsion: 2 Rolls-Royce Olympus gas turbines, 56,000shp; 2 Rolls-Royce Tyne gas turbines, 8,500shp; 2 shafts; 30 + kt.
Armament: 4 Exocet SSM launchers; 2 6-barrelled Seawolf SAM launchers; 2 40mm guns; 2 triple Mk 32 torpedo tubes.
Aircraft: 2 WG13 Lynx.
Complement: 223.

Development: The *Broadsword* (Type 22) class is the successor to the very successful Leander class, the first of these new ships being ordered on 8 February 1974, launched in May 1976 and commissioned on 3 May 1979. The remaining twelve of the class are following at approximately yearly intervals. This class is designed primarily for ASW and has a comprehensive weapons and sensor fit for the role. The ships are also fitted out as group leaders and are fully capable of acting as "Officer in Tactical Command" (OTC) command ships.

The main ASW sensor is the Type 2016 sonar, the roll-stabilised array being mounted in a GRP keel dome in line with RN practice. ASW weapons are two triple Mk 32 torpedo tubes firing Mk 46 torpedoes. Batch I and II ships (the first ten of the class) have only two 40mm guns and their main armament is all-missile, the first RN warship class to attempt this transition. However, the Falklands War proved that this was an error, and Batch III ships have a single 4.5in (115mm) DP gun. The ships also have a hangar and flight deck for two WG13 Lynx ASW helicopters, although only one is normally carried in peacetime.

The Batches II and III ships are being constructed 41ft (12·5m) longer than the earlier ships, bringing overall length up to 471ft (143·6m). The first two Batch II ships are reported to cost an estimated £120 million each (1981 prices), a fearsome price and one which is indicative of the rapidly escalating costs of modern defence equipment.

These ships are amongst the most sophisticated ASW vessels afloat and have already established a good reputation for efficiency and effectiveness. *Broadsword* and *Brilliant* of the class took part in the South Atlantic War in 1982, where their Seawolf missiles proved of great value.

From the 7th ship, lightweight Seawolf will be fitted. This can be in a 4-barrel launcher or in a vertical launch system, promising faster reaction and multiple targetting.

Right: The Type 22 destroyer *Broadsword* was the first all-missile armed Royal Navy warship, with Exocet SSM and Seawolf SAM launchers. Later batches of the class will have a 4.5in gun on the foredeck, as well as the Goalkeeper 30mm CIWS.

channels (including the SPG-60) from five to three. This modification looks forward to the conversion of the ships to fire the SM-2 missile, which requires target illumination only in the terminal phase. The ships are also scheduled to receive Harpoon, Tomahawk, and two Phalanx CIWS guns at future refits.

The original requirement was for eleven ships of this class, which would then combine with earlier CGNs to provide each of the CVANs projected at that time with four nuclear-powered escorts. After only four units of the class had been laid down, however, further orders were suspended while consideration was given first to the Strike Cruiser (CSGN) and then to a modified CGN 38 design with Aegis. Both these projects were abandoned in favour of the conventionally powered CG-47 now under construction.

All four ships of the Virginia class currently serve with the Atlantic Fleet, where they have the job of protecting the carriers *Nimitz* and *Eisenhower.*

Right: The Virginia class cruiser *Texas.* Only four of a projected 11 ships of the class were completed, the last in 1980.

Georges Leygues (C70) Class

Origin: France.
Type: Destroyer.
Number: 4 ships + 3 building.
Displacement: 3,380 tons standard; 4,170 tons full load.
Dimensions: Length overall 455·9ft (139m); beam 45·9ft (14m); draught 18·7ft (5·7m).
Propulsion: Rolls-Royce Olympus gas turbines, 52,000shp; 2 SEMT-Pielstick diesels, 10,400bhp; 2 shafts; 30kt.
Armament: 1 3·9in (100mm) gun; 2 20mm guns; 2 torpedo tubes.
Aircraft: 1 WG13 Lynx.
Complement: 226.

Development: The French Navy has produced a very efficient hull design for the C70 ships which are being produced in two versions — the basic C70 (Georges Leygues) class for ASW, and the C70AA (AA = "anti-airienne") for air defence. Current plans envisage eight of the ASW version and four of the AA, but further orders might well be forthcoming. The first of the class, *Georges Leygues,* was commissioned on 10 December 1979 and four are now with the fleet; the remaining three will commission during the period 1985-89. There was to be an eighth ship, but it was cancelled for economic reasons.

Although these ships are officially classified as "corvettes" by the French Navy, they have been given "D" pennant numbers and would, in fact, be counted as destroyers by any other navy.

The principal ASW sensors are the bow-mounted, low-frequency DUBV-23 sonar and the very closely related DUBV-43 variable-depth sonar. The DUBV-23 array is mounted in a streamlined bulb and performs both search and attack func-

tions. In the DUBV-43 the transducer array is mounted in a "fish" which is streamed over the stern at distances of up to 820ft (250m) and at depths ranging from 33ft (10m) to 656ft (200m). Virtually all elements of the two systems are identical apart from the transducer arrays, and even they consist of essentially similar components.

The main shipborne ASW weapon is the L5 torpedo, of which ten are mounted in fixed launchers. The L5 is an electric-powered 21in (533mm) weapon with an active/passive head and a speed of 35kt. The other main element of the C70's ASW capability is the two WG13 Lynx helicopters which have identical airframes with the British version but are fitted with French avionics and sensors. Other weapons include four Exocet launchers (a further four can be added in war) and 26 Crotale SAMs. There is also one 3·9in (100mm) gun in a DP mounting on the foredeck.

The second batch of four ships is expected to have a number of improvements. Visually, the most noticeable change will be the raising of the bridge by one deck level as the current position (low and rather far forward) has proved unsuitable in bad weather, especially with seas breaking over the bow. The ships will also carry a new sonar, the SS-48, a very sophisticated bow-mounted VDS system which has been some ten years in development. This will be fitted in the fifth and subsequent ships during construction, and will be retrofitted to the earlier four during mid-life refits. It is also hoped to install the new Vampir long-range infra-red surveillance sensor, as well as a new version of the Crotale SAM missile with an additional sea-skimming anti-ship capability.

Above: The Georges Leygues class ASW destroyer *Montcalm,* classed as a corvette though allocated a "D" pennant number.

Below: *Georges Leygues* herself. The basic hull is used for eight ASW and four AA vessels with appropriate weapon/sensor fits.

Iroquois (DDH280) Class

Origin: Canada.
Type: Destroyer.
Number: 4 ships.
Displacement: 3,551 tons standard; 4,700 tons full load.
Dimensions: Length overall 426ft (129·8m); beam 50ft (15·2m); draught 14·5ft (4·4m).
Propulsion: 2 Pratt & Whitney FT4A2 gas turbines, 50,000shp; 2 Pratt & Whitney FT12 AH3 gas turbines, 7,400shp; 2 shafts; 29kt.
Armament: 1 5in (127mm) OTO Melara Compact gun; 2 Sea Sparrow Mk 25 launchers; 2 triple Mk 32 torpedo tubes; 1 triple Limbo ASW mortar.
Aircraft: 2 CHSS-2 Sea King.
Complement: 285.

Development: The Royal Canadian Navy (RCN) has a wealth of experience of ASW operations in the North Atlantic stemming from its major contribution to the Allied effort in the Battle of the Atlantic during World War II. During the war and for some years afterwards the RCN relied upon British designs for its destroyers and frigates (albeit built in Canadian yards) but in 1951 it decided to design its own ships. The result has been a series of unusual-looking vessels, packed with innovations and well suited to their role in the frequently inhospitable environment of the northern seas.

First was the St Laurent class (2,260 tons), comprising six ships commissioned 1956–57. All underwent major refits in the 1970s but are due to be replaced by a new class in the late 1980s. Next into service was the Restigouche class (three ships), followed by the "improved Restigouche" class (four), the latter armed with Sea Sparrow SAMS, two 3in (76mm) guns and ASROC. The design was further developed into the Mackenzie class (four ships) and the Annapolis class (two), which were commissioned in 1962–64.

In the early 1970s the DDH280 class (*Iroquois, Huron, Athabaskan, Algonquin*) appeared. These have a distinctive appearance with a high bridge and hangar, surmounted by a lattice mast and bifurcated funnel. The RCN has always used large helicopters in relation to the size of its hulls, and the DDH280s carry two CHSS-2 Sea Kings. Landing is assisted by the "Beartrap", a Canadian-invented cable device which is attached to the hovering helicopter and which then hauls the aircraft down on to the deck. ASW sensors include a hull-mounted SQS-505 sonar in a 14ft (4·26m) dome and an SQS-505 variable-depth sonar, which is streamed from the stern. ASW weapons include the Mk 10 Limbo mortar and two triple Mk 32 torpedo tubes firing Mk 46 torpedoes.

The next class of destroyers (the Halifax class) is awaited with considerable interest, as the Canadian designers will doubtless have a few more surprises for the naval world. Six ships have been ordered, each equipped with one Sea King helicopter and displacing 4,254 tons full load. They will be armed with 8 Harpoon SSMs, 2 Sea Sparrow SAM systems (2 × 14 missiles), 2 ASW torpedo tubes, a 57mm DP gun and a Phalanx CIWS. The first ship will be laid down in 1985 and completed in 1988.

Above: HMCS *Iroquois* (left), with the earlier Annapolis class *Nipigon* alongside. Canadian destroyers are of distinctive apperance, with such features as the *Iroquois'* very high bridge.

Luda Class

Origin: People's Republic of China.
Type: Destroyer.
Number: 14 ships + 2 building.
Displacement: 3,250 tons standard; 3,900 tons full load.
Dimensions: Length overall 430ft (131m); beam 45ft (13·7m); draught 15ft (4·6m).
Propulsion: Geared turbines, 60,000shp; 2 shafts; 36kt.
Armament: 6 Hai Ying 2 SSM launchers; 2 ASW rocket launchers; 2 twin 5.2in (133mm) guns; 4 twin 37mm guns; 4 twin 25mm guns.
Complement: 215.

Development: The armed forces of the People's Republic of China (PRC) were, in their early days, almost totally dependent upon the USSR for weapons systems design and supply. Steps were quickly taken to establish indigenous production facilities although these, too, simply produced Soviet designs at first. The next stage was Chinese improvements on Soviet designs, but now Chinese designers are at last coming into their own. One of their products in the naval field is the Luda class destroyer, which has some similarity to the Soviet Kotlin class.

One of the most interesting features is the fitting of the SSM missile launcher tubes on traversing mounts, with one triple mount abaft each funnel. There was a plan to fit British Sea Dart SAMs, and a £100 million contract was announced in January 1983; this was, however, subsequently cancelled in a period of Chinese defence procurement retrenchment, owing to the high costs.

There is a major building programme in progress for the Chinese Navy, which already has an SSBN and an SSN at sea. Numerically the largest fleet in Asia, if not yet the most powerful, the Chinese Navy will soon become a force of major strategic significance.

Above: Launch of a Hai Ying 2 (SS-N-2 type) missile from a Chinese Luda class destroyer.

Below: Apart from the two triple missile launchers, the Luda class carry guns and ASW rockets.

Shirane/Haruna Classes

Origin: Japan.
Type: Destroyer.
Number: 2 ships/2 ships.
Displacement: 5,200 tons standard; 6,800 tons full load.
Dimensions: Length overall 521ft (158·8m); beam 57·5ft (17·5m); draught 17·5ft (5·3m).
Propulsion: Geared steam turbines, 70,000shp; 2 shafts; 32kt.
Armament: 1 octuple ASROC Mk 16 launcher; 2 single 5in (127mm) Mk 42 guns; 1 Sea Sparrow Mk 25 launcher; 2 Phalanx CIWS; 2 triple Mk 32 torpedo tubes.
Aircraft: 3 SH-3B Sea King.
Complement: 350.

Development: The Soviet Pacific Fleet is based at Vladivostok, Sovetskaya Gavan and Petropavlovsk-Kamchatskiy, none of which have direct access to the ocean. It is Japan's misfortune to sit astride most of the exits, as well as being the only non-Soviet Navy with access to the Sea of Okhotsk. Mindful of the large submarine component of the Soviet Navy, the Japanese Maritime Self-Defence Force (JMSDF) has built up a naval element with a primary ASW role. Its surface ships are of Japanese design and construction, but tend to use weapons systems and armament of US origin,

either locally manufactured under licence or purchased directly. The showpieces of the fleet are the four helicopter-carrying ASW destroyers of the Haruna and Shirane classes; the former were completed in 1973–74 and the latter in 1980–81.

Both classes have a very large hangar which accommodates three Mitsubishi SH—3B (licence-built Sikorsky Sea King) helicopters. The spacious flight deck extends to the stern, and incorporates the Canadian "Beartrap" haul-down system. Both classes have an ASROC launcher forward of the bridge and Mk 64 torpedo tubes in two triple mountings abreast the bridge. Hull sonars of Japanese design and manufacture are fitted (OQS-3 in *Haruna*, OQS-101 in *Shirane*). *Shirane* currently is the only class to have SQS-35 VDS, but this will be fitted to the earlier pair at their next refit. The Shiranes also have SQR-18 TACTASS towed arrays. Surface armament includes two quick-firing 5in (127mm) Mk 42 guns in all four ships, complemented by BPDMS and Phalanx CIWS in the Shirane class, The *Harunas* are to be fitted with Sea Sparrow at their next refit. Second of the Shirane class is *Kurama,* and second Haruna is *Hiei.*

These are two classes of im-

pressive and capable ships, in the best Japanese maritime traditions. The JMSDF will soon have a surface fleet of 35 destroyers and 18 frigates, all optimised for the ASW role, with a further three destroyers (one of 4,500 tons, one of 3,400 tons, one of 2,000 tons) being ordered in the FY85 budget.

Above: The destroyer *Haruna*, built to a Japanese design but armed with US weapons. Their main role is ASW.

Below: An SH-3B Sea King lands on the *Shirane's* helicopter deck. On the foredeck are the two 5in guns and ASROC launcher.

Spruance/Kidd Classes

Origin: USA.
Type: Destroyer.
Number: 31 ships + 1 building/4 ships.
Displacement: 5,830 tons standard; 7,810 tons full load.
Dimensions: Length overall 563·2ft (171·7m); beam 55·1ft (16·8m); draught 19ft (5·8m).
Propulsion: 4 LM2500 gas turbines, 80,000shp; 2 shafts; 33kt.
Armament: 2 quadruple Harpoon SSM launchers; 1 octuple Sea Sparrow Mk 25 launcher; 2 single 5in (127mm) guns; 2 Phalanx CIWS; 1 ASROC launcher; 2 triple Mk 32 torpedo tubes. See text.
Aircraft: 1 SH-3 Sea King or 2 SH-2D.
Complement: 296.
(Specifications given for Spruance class; see text for Kidd class variations.)

Development: The Spruance class was designed to replace war-built destroyers of the Gearing and Sumner classes. At 7,810 tons full load (twice the displacement of the ships it has replaced) the Spruance epitomises the philosophy which envisages the construction of large hulls with block superstructures (maximising internal volume) fitted with easily maintained machinery and equipped with high-technology sensors and weapons that can be added to or updated by modular replacement at a later stage. The aim is to minimise "platform" costs and maximise expenditure on weapons systems ("payload").

The advanced ASW features of the Spruances are largely hidden. The ASROC launcher, for example, has a magazine containing no fewer than 24 reloads. The large hangar to port of the after funnel accommodates two LAMPS III helicopters, while the triple Mk 32 torpedo tubes are concealed behind sliding doors. The bow sonar is the SQS-53, which can operate in a variety of active and passive modes, including direct path, bottom-bounce and convergence zone. This has proved so successful that the SQS-35 VDS originally scheduled for the class will not now be installed, but the SQR-19 towed array will be fitted to all ships of the class in the late 1980s.

The adoption of an all-gas-turbine propulsion system, which employs paired LM2500 turbines en échelon, is an arrangement which not only gives easy maintenance and reduces manning but also significantly cuts underwater noise emission. The class is fitted with computerised data-handling systems in a well-designed Combat Information Centre (CIC), and also has the Mk 86 Gunfire Control System (GFCS) and the Mk 116 Underwater Fire Control System (UFCS).

The flexibility of the Spruance design is such that it formed the basis for the new Ticonderoga class Aegis cruisers, and for the four ships ordered for the Imperial Iranian Navy (but acquired for the USN in 1979 following the fall of the Shah and now designated the Kidd class). These were completed for the Iranians for the AAW role, having, for example, twin-arm Mk 26 launchers in place of the original ASROC and Sea Sparrow launchers. However, the ASROC missiles can still be fired from the forward Mk 26 launcher, and the SQS-53 sonar is fitted. Thus the Kidds have an ASW capability not too far short of the Spruances and the gap will doubtless be narrowed in the course of future refits, especially when the SQR-19 is fitted.

One additional ship of the Spruance class was ordered in 1979. This unit (DD997) was to have increased hangar and flight-deck space to operate both helicopters and VTOL aircraft, but it has been completed as a standard vessel. The US Navy plans to build a new class, DDG-51, (starting in FY86) to replace the fourteen destroyers of the Sherman and Hull classes.

Below: The Spruance class ASW destroyer *Briscoe.* The uncluttered exterior of these vessels conceals substantial firepower.

Top: USS *Elliot,* another of the 32-ship Spruance class, showing the design's ample space for new sensors and weapons.

Above: The Kidd class destroyer USS *Chandler,* originally ordered by the Imperial Iranian Navy as the *Anoushirvan.*

FRIGATES

Amazon Class

Origin: UK.
Type: Frigate.
Number: 6 ships.
Displacement: 2,750 tons standard; 3,250 full load.
Dimensions: Length overall 384ft (117m); beam 40·5ft (12·3m); draught 19ft (5·8m).
Propulsion: 2 Rolls-Royce Olympus gas turbines, 56,000shp; 2 Rolls-Royce Tyne gas turbines, 8,500shp; 2 shafts; 30kt.
Armament: 4 Exocet SSM launchers; 1 4·5in (114mm) gun; 1 quadruple Seacat launcher; 2 20mm guns; 2 triple torpedo tubes.
Aircraft: 1 WG13 Lynx.
Complement: 175.

Development: Before the first of the "broad-beamed" Leander class frigates had been laid down, Vosper-Thorneycroft received a contract for a new design to be prepared in collaboration with Yarrow. They produced the Type 21 Amazon class, the first being launched in 1971 and commissioned in 1974. These were the first Royal Navy ships to be designed from the outset to be powered solely by gas turbines, and were also the first for many years to be designed by a commercial firm.

The ASW fit includes a Type 184M hull-mounted sonar, and the ASW armament comprises two triple torpedo tubes for Mk 48 torpedoes; the ships also have a flight deck and hangar for the Westland WG13 Lynx ASW helicopter. Other armament includes a 4·5in (114mm) main gun in a DP mounting, four Exocet SSMs, a Seacat SAM launcher and two 20mm Oerlikon cannon. Despite this heavy armament (relative to the hull size), the complement is only 175 officers and ratings.

Each of the Amazons has some 135 tons of aluminium alloy in its structure and the spectacular destruction of the *Ardent* and *Antelope* during the South Atlantic War led to

reports that it was the aluminium that was responsible. This was investigated fully after the war and the chairman of the working party stated that he was "not aware of any evidence to suggest that any ship was lost because of the use of aluminium alloys in its construction. Nor was there any evidence that aluminium or aluminium alloys had burned" (*Financial Times*, 24 December 1982).

These are handsome ships that have served the Royal Navy well, and it is indeed sad that two out of the eight should have been lost in a war. It is also worth noting that no fewer than seven out of the eight in the class were in the Falklands area during the war. Apart from the *Amazon*, the other ships in the class are *Active*, *Ambuscade*, *Arrow*, *Alacrity* and *Avenger*.

Above: The Type 21 Amazon class frigate *Ambuscade* in the English Channel. Two of her sisters, *Ardent* and *Antelope*, were sunk by Skyhawk attacks off the Falkland Islands in 1982.

Below: HMS *Arrow* at sea on exercise. *Arrow* suffered only superficial damage from 30mm cannon shells off the Falklands, though the incident highlighted shortcomings in the radar.

Godavari Class

Origin: India.
Type: Frigate.
Number: 2 ships + 4 building.
Displacement: 3,600 tons full load.
Dimensions: Length overall 396·9ft (121m); beam 46·2ft (14·1m); draught 29·5ft (9m).
Propulsion: 2 geared turbines; 30,000shp; 2 shafts; 27kt.
Armament: 4 SS-N-2C launchers, 1 SA-N-4 launcher; 2 twin 57mm guns; 8 twin 30mm guns; 6 torpedo tubes.
Aircraft: 2 Sea King.
Complement: 313.

Development: With one of the most powerful fleets in Asia, the Indian Navy shows a fascinating mixture of Soviet and Western technology, not only in complete ships but also in weapons systems and sensors aboard them. Thus the surface fleet

comprises three Kashin destroyers and twelve Petya II frigates — all Soviet designs — and twelve frigates either British-built or developed from British designs. The oldest of these frigates still in service are two British-built Whitby class ships, *Talwar* and *Trishul*, much modified in refits and now armed with three Soviet SS-N-2 launchers. These were followed by six Bombay-built frigates based on the British "broad-beamed" Leander but with numerous local modifications.

Latest in this series is the Godavari class, a further and more radical adaptation of the Leander design. The main armament is of Soviet origin (SS-N-2C, SA-N-4), whilst two Westland Sea King ASW helicopters are carried, using the Canadian "Beartrap" haul-down gear. Search, missile and gun-

control radars are all of Soviet design, but the sonar is British. The prominent forecastle of the Leanders has been deleted and one of the major visual features is the very large hangar. It is a major achievement to operate two such large helicopters from a vessel with a 46·2ft (14·1m) beam, indicating a considerable degree of expertise in Indian Naval Air Arm pilots.

Above: The Indian Navy has successfully combined warships of both Soviet and Western origin, and the Godavari class frigates take the process even further, mounting Soviet missile armament (and fire control radars) and a Sea King helicopter on a hull derived from the British Leander, with British sonar but indigenous radars.

Kortenaer Class

Origin: Netherlands.
Type: Frigate.
Number: 12 ships + 2 air defence versions + 13 ordered.
Displacement: 3,050 tons standard; 3,630 tons full load.
Dimensions: Length overall 428ft (130m); beam 47·2ft (14·4m); draught 14·3ft (4·4m).
Propulsion: 2 Rolls-Royce Olympus gas turbines; 50,000shp; 2 Rolls-Royce Olympus gas turbines, 8,000shp; 2 shafts; 30kt.
Armament: 1 quadruple Harpoon SSM launcher; 1 octuple Sea Sparrow SAM launcher; 2 single 3in (76mm) guns; 2 twin Mk 32 torpedo tubes.
Aircraft: 2 WG13 Lynx.
Complement: 167.

Development: Following failure to agree on the specification for the successor to the Leander class *(Van Speijk* in Dutch service) the British and the Dutch decided to go their separate ways. The result for the Dutch Navy is the Kortenaer class of which twelve have so far been ordered, the last two being laid down as flagships, with extra air defences and without the hangar, although the flight deck is being retained. The

Right: The frigate *Kortenaer,* lead ship of a class ordered by three navies in addition to the Dutch.

ASW fit includes a bow-mounted SQS-505 sonar, two Lynx helicopters (only one is normally carried in peacetime) and two twin Mk 32 torpedo tubes for Mk 46 torpedoes.

The West German Navy has ordered six ships of a modified Kortenaer design to replace the old ships of the Köln and Fletcher classes. These ships (the Bremen class) are intended for ASW missions in the Baltic, and all six have been commissioned. The Greek Navy has ordered four, with the possibility of a fifth to follow later; the first two units of the Elli class were built in Holland and are now in service, but the second pair will be built locally in the Hellenic Shipyards at Scaramanga, although the keels have not yet been laid down. Portugal has also confirmed an order for three ships, one to be built in Holland and the other two in local yards.

Thus with confirmed orders for 21 of these well-designed, well-equipped and highly respected ships already received, the Dutch are justified in their belief that they have produced a "winner", which has more claim than most to being the NATA "standard frigate".

Krivak-I/II Classes

Origin: Soviet Union.
Type: Escort ship.
Number: 32 ships.
Displacement: 3,300 tons standard; 3,800 tons full load.
Dimensions: Length overall 405ft (123m); beam 46ft (14m); draught 16ft (5m).
Propulsion: COGAG; 72,000shp; 2 shafts; 32kt.
Armament: 2 twin SA-N-4 launchers; 2 twin 76mm guns in Krivak-I; 2 single 3·9in (100mm) guns in Krivak-II; 4 SS-N-14 launchers; 2 RBU 6000 launchers; 2 quadruple 21in (533mm) torpedo tubes.
Complement: 220.

Development: The Krivak was designed as a "2nd-rate" counterpart to the "1st-rate" Kresta-II and Kara classes, with which it initially

shared the same BPK classification. Although the class followed on from the Kashin in terms of construction dates, the Krivak is smaller, has an altogether more sophisticated ASW outfit, lacks an area defence SAM system — arguably the main armament of the Kashin — and is easier to build The latter factor made it possible to allocate construction to the smaller Baltic and Black Sea shipyards, leaving the slipways of the traditional naval yards free for the construction of larger units.

The major ASW system is the SS-N-14 missile, fired from a bulky quadruple launcher forward. This is backed up by RBU 6000 mortars immediately forward of the bridge and torpedo tubes amidships. The bow sonar is probably the same model as that fitted in the Kresta-II and the Kanin conversions; there is, in addi-

Oliver Hazard Perry Class

Origin: USA.
Type: Frigate.
Number: 35 ships + 14 building + 3 ordered.
Displacement: 3,605 tons full load.
Dimensions: Length overall 445ft (135·6m); beam 45ft (13·7m); draught 14·8ft (4·5m).
Propulsion: 2 gas turbines, 41,000shp; 1 shaft; 29kt.
Armament: 1 Harpoon/Standard Mk 13 launcher; 1 3in (76mm) gun; 1 Phalanx CIWS; 2 triple Mk 32 torpedo tubes.
Aircraft: 2 SH-2.
Complement: 210.

Development: The Oliver Hazard Perry (FFG7) class originated in the "Patrol Frigate" programme which was to constitute the "low" end of a high/low mix, providing large numbers of cheap escorts with reduced capabilities. These were to balance the sophisticated and very costly specialist ASW and AAW ships whose primary mission was to protect carriers. Strict limitations were placed on cost, displacement and manpower, and the FFG7 was to be built in small yards, keeping construction techniques simple, making the maximum use of flat panels and

Above: The *Oliver Hazard Perry,* lead ship of the low-cost class of frigates designed to complement the Spruance class.

Below: The single Mk 13 Tartar launcher on the foredeck and the SPS-49 surveillance radar are prominent features of the FFG7s.

Above: Krivak-I photographed in the Baltic by an RAF Nimrod; gun armament comprises two twin 76mm mountings forward of the VDS.

Left: A Krivak-I in the English Channel. The quadruple SS-N-14 launcher on the foredeck houses anti-submarine missiles.

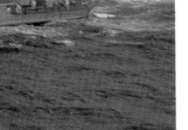

Gatling; ECM provision is also minimal compared with the "1st-rate" BPKs.

The Krivaks are fitted with four gas turbines, two for cruising (24,000shp) and two for boost (48,000shp). The ships thus have rapid acceleration from a cold start, coupled with good endurance (4,600 miles at 20kt). The eleven Krivak-IIs differ from the earlier vessels in having two 100mm guns in single mounts and the break to the quarterdeck further aft.

The Krivak has never been the heavily-armed super-destroyer that Western commentators have claimed, nor was it designed as such. The complement of anti-submarine missiles is small by Western standards; there is no shipborne helicopter to provide target data and independent ASW at longer ranges; AAW and ECM capability is inadequate for open-ocean operations; and the propulsion system, while easy to maintain, probably leaves the Krivaks short on endurance. They are, nevertheless, clearly adequate for their intended mission, for production has continued at a rate of three to four ships per year over the past twelve years and shows no sign of coming to an end.

tion, an independent variable depth sonar on the stern.

Only close-range air defence is provided, in the form of SA-N-4 "bins" fore and aft and a pair of 76mm mountings (later ships, designated Krivak-II, have single 100mm). The Krivak is unusual in its generation in having no "last-ditch" anti-missile system such as the 30mm

Left: A Krivak-II class destroyer photographed in the Indian Ocean by a US Navy reconnaissance aircraft. The Krivak-IIs have two single 100mm guns aft, along with the stern-mounted housing for the VDS.

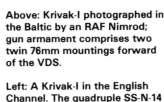

bulkheads, and ensuring that passageways kept deliberately straight. The hull structure is prefabricated in modules of 35, 100, 200 or 400 tons, allowing the shipyards to select the most convenient size.

Like the previous frigate classes, the Perry has only one screw, but the layout is much more compact as the result of using gas turbines. Two LM2500s (the same model as in the Spruances) are located side by side in a single engine room, and small retractable propulsion pods are fitted just aft of the sonar dome to provide emergency "get-you-home" power as well as help docking. Each of these pods has a 325hp engine, and the two together can propel the ship at a speed of some 10kt.

The FFG7 has a Mk 13 launcher forward for Standard (MR) SAMs and Harpoon ASMs and an OTO Melara 3in (76mm) gun on top of the superstructure. ASROC is not fitted, but there is a large hangar aft for two LAMPS helicopters. The SQS-56 sonar is hull-mounted inside a rubber dome; it is a new austere type, much less sophisticated than the SQS-26. It was planned, however, that the FFG7 would operate in company with other frigates equipped with the

SQS-26 and would receive target information from their sensors via data links. Later in the decade it is intended to fit the SQR-19 towed-array sonar in all active FFG7 frigates.

The FFG7 has been tailored to accommodate only those systems envisaged in the near future, including the SH-60 LAMPS-III, fin stabilisers, the Link 11 data transfer system and a single Phalanx CIWS. Once these have been installed, however, there remains only a further 50-ton margin for additional equipment.

Four Perrys have been built in the USA for the Royal Australian Navy, and three are being built by Bazan at Ferrol for the Spanish Navy.

Right: *Oliver Hazard Perry* **on sea trials. The large double hangar accommodates two SH-2 LAMPS helicopters, and later members of the class will be lengthened and given extra equipment to handle the new SH-60B Seahawk LAMPS III.**

Far right: *Oliver Hazard Perry* **undergoes a mine-resistance test, an aspect of naval warfare which NATO navies are taking increasingly seriously.**

AMPHIBIOUS VESSELS

Ivan Rogov Class

Origin: Soviet Union.
Type: Landing ship.
Number: 2 ships + 1 building.
Displacement: 11,000 tons standard; 14,000 tons full load.
Dimensions: Length overall 522ft (159m); beam 80ft (25m); draught 21/28ft (6·5/8·5m).
Propulsion: COGAG; 24,000shp; 2 shafts; 26kt.
Armament: 1 twin SA-N-4 launcher; 1 twin 3in (76·2m) gun; 4 30mm Gatling CIWS; 1 20-barrelled rocket launcher.
Aircraft: 5 Ka-25 Hormone-A.
Complement: 400.

Development: Although *Ivan Rogov* carries the same BDK designation as the Alligator class, her displacement is more than twice that of her predecessor. More importantly, the design represents a significant break with previous Soviet amphibious ships in that it incorporates a hangar for helicopters and a docking-well for amphibious landing craft — both features of Western amphibious construction but never before adopted by the Soviet Navy. The result is a ship capable not only

of direct beach assault but also of both "horizontal" and "vertical" landing operations.

The lower part of the ship is built around a continuous tank deck, with workshops and accommodation to the sides and the traditional bow doors and ramp of an LST. Capacity is estimated at 10 tanks, plus 30 APCs and other vehicles. The vehicle load can be increased by utilising the midships section of the upper deck, access to which is gained by lowering a hinged ramp located immediately aft of the break in the forecastle.

At the after end of the ship the tank deck leads down into a docking well some 98ft (30m) long and 66ft (20m) wide, closed by a large stern gate. An unusual feature of the docking well is that the height of the deckhead rises towards the stern. This enables the *Rogov* to operate ACVs of the Gus or Lebed classes, both of which have tall tail rudders. Two ACVs can be accommodated

Below: *Ivan Rogov's* **bulky superstructure accommodates a battalion of naval infantry.**

side by side, with an alternative loading of the new Ondatra class LCMs, which are thought to have been designed specifically for *Rogov* and her successors.

The upper part of the ship is dominated by a massive block superstructure which extends to the sides of the ship. Apart from carrying the major defensive weapon systems and sensors, the superstructure has troop accommodation four decks high for a full battalion of Naval Infantry (522 officers and men) and also contains a large hangar for Ka-25 Hormone helicopters which appear to be fitted to serve in both the ASW and the troop-carrying roles. The hangar itself runs between the funnel uptakes, opening out into a wide bay offset to starboard at its after end. The sloping hangar floor is a continuation of the raised helicopter platform aft, and at its forward end leads down on to the main section of the upper deck via a fixed ramp. Helicopter take-off and landing spots, each with its own flying control cabin, are marked out aft of the break in the forecastle and above the stern.

Above: The forward and after landing pads for *Ivan Rogov's* **"Hormones" are clearly visible.**

AAW capabilities are exceptionally complete for an amphibious vessel, and comprise an SA-N-4 missile launcher at the after end of the superstructure, a twin 76mm mounting on the forecastle, and 30mm Gatlings on either side of the foremast. Forward of the main superstructure block, to starboard, is a tall, narrow deckhouse, on top of which is a rocket launcher.

The construction of *Ivan Rogov* marks a significant advance in the long range amphibious capabilities of the Soviet Navy. However, the attempt to put the capabilities of the LPH, LPD and LST into one hull does not appear to be entirely successful. The docking well is small by Western standards, and the ability to land heavy vehicles except by direct beaching is strictly limited.

Right: Stern view of *Ivan Rogov* **with the hangar doors open. Note the stern ramp for ACV and landing craft access.**

Ropucha Class

Origin: Soviet Union.
Type: Landing ship.
Number: 17 ships.
Displacement: 3,450 tons standard; 4,400 tons full load.
Dimensions: Length overall 360ft (110m); beam 49ft (15m); draught 12ft (3·5m).
Propulsion: Diesel, 10,000bhp; 2 shafts; 17kt.
Armament: 2 twin 57mm guns.
Complement: 95.

Development: Built, like the Polnocny class, in Poland, the *Ropucha* is altogether larger and more capable than its predecessor. The main improvements, apart from the increase in vehicle capacity resulting from the increase in size, lie in the powerful modern anti-aircraft armament and the extensive troop accommodation.

The tank deck is continuous, with bow and stern ramps provided. Above the squared-off bow section is a long, sliding, hatch cover for alongside loading or off-loading of vehicles and supplies.

The conventional LST hull-form is surmounted by an exceptionally long superstructure with accommodation for large numbers of troops. At either end of the superstructure are twin 57mm automatic mountings, controlled by a single Muff Cob director. Air search and navigation radars are carried atop a tall lattice foremast. Forward of the bridge there appears to be provision for the installation of an SA-N-4 "bin" launcher — although the associated Pop Group guidance radar is not fitted.

The relatively slow building rate of the Ropucha class may indicate greater Soviet emphasis on large high-speed ACVs for short-range assault.

Below: Stern view of a Ropucha. Note the stern gate, the split funnels and the forecastle hatch.

Tarawa Class

Origin: USA.
Type: Amphibious assault ship.
Number: 5 ships.
Displacement: 39,300 tons full load.
Dimensions: Length overall 820ft (249·9m); beam 126ft (38·4m); draught 26ft (7·9m).
Propulsion: Geared steam turbines; 70,000shp; 2 shafts; 24kt.
Armament: 2 octuple Sea Sparrow Mk25 launchers; 3 single 5in (127mm) guns.
Aircraft: 30 helicopters/VTOL aircraft (AV-8A, CH-53D, CH-46D, AH-1T, UH-1N; see text).
Complement: 900 (+ 2,000 troops).

Development: The Tarawa class LHAs are large amphibious vessels combining in a single hull capabilities which previously had required a number of separate specialist types. They incorporate both an aircraft hangar and a docking well for landing craft, and can therefore employ "vertical" or "horizontal" assault techniques. The hangar is located in the after part of the ship directly above the docking well; both are 268ft (81·7m) in length and 78ft (23·8) wide, and the hangar has a 20ft (6·1m) overhead to enable the latest heavy-lift helicopters to be accommodated. A typical loading of helicopters would include 9–12 CH-46 Sea Knights, six CH-53D Sea Stallions, four AH-1 Sea Cobra gunships and 2–4 UH-1 utility helicopters. Detachments of AV-8A Harriers are frequently embarked, and in a recent NATO exercise no fewer than 11 such aircraft operated from the deck of *Nassau*. The essentially defensive armament includes three lightweight 5in (127mm) guns for fire support operations.

The docking well can accommodate four LCUs, and is served by an elaborate cargo transfer system employing a central conveyor belt and 11 overhead monorail cars. Forward of the docking well is a "multi-storey car park" for tanks, guns, trucks and LVTP-7 amphibious personnel carriers. Above the vehicle decks there is accommodation for both the Commander Amphibious Task Group (CATG) and the Landing Force Commander (FLC) and their respective staffs. These ships generally serve as flagships of Marine amphibious squadrons.

The next amphibious assault ship class will be the 40,000-ton LHD-1 Multipurpose Amphibious Assault Ship, which will be similar in general design to the Tarawa class. The lead ship was authorised in FY84 and it is planned to construct 10 or 11 by the late 1990s. Those ships will be known as the Wasp class. The first is due to commission in 1989, and will be very slightly bigger than the Tarawa class, while being more flexible, accommodating up to 42 Sea Knight helicopters, or 30 plus 6 to 8 AV-8B Harriers, or 20 Harriers and 4 to 6 Sea Hawks when in her secondary role as sea control ship. The ship could support a mix of attack and utlity/transport helicopters, and will also carry three LCAC air-cushion vehicles.

Above: The Tarawa class ships have large docking wells which can accommodate four LCUs for loading by conveyor belt and monorail.

Right: Designed to act as flagships of Marine amphibious squadrons, *Tarawa* and her sisters have flight decks big enough for substantial numbers of large transport helicopters, or for AV-8A detachments.

SMALL COMBATANTS

Pegasus Class

Origin: USA.
Type: Patrol hydrofoil.
Number: 6 craft.
Displacement: 231 tons full load.
Dimensions: Length overall 132ft (40m); beam 28ft (8·6m); draught 6ft (1·9m).
Propulsion: (Hullborne) 2 Mercedes-Benz diesels, 1,600bhp, 12kt; (foilborne) 1 General Electric gas turbine, 18,000hp, 40kt.
Armament: 2 quadruple Harpoon launchers; 1 3in (76mm) Mk 75 gun.
Complement: 21.

Development: The PHM was one of the four new designs in the "low" programme advocated in Zumwalt's Project 60. It was envisaged that squadrons of these fast patrol craft would be deployed at the various choke-points — in particular those in the Mediterranean and the NW Pacific — through which the surface units of the Soviet Navy needed to pass in order to reach open waters. High speed and a heavy armament of anti-ship missiles would enable the PHM to make rapid interceptions, and the relatively low unit cost meant that large numbers could be bought.

The Italian and Federal German Navies, with similar requirements in the Mediterranean and the Baltic respectively, participated in the development of the design. The Germans planned to built 12 units of their own in addition to the 30 originally projected for the US Navy.

Technical problems with the hydrofoil system resulted in cost increases, and opponents of the PHM programme, pointing to the limited capabilities of the design, tried to obtain cancellation of all except the lead vessel. Congress insisted, however, on the construction of the six units for which funds had already been authorised.

The propulsion system of the PHM comprises separate diesels driving two waterjets for hullborne operation and a single gas turbine for high-speed foilborne operation.

In order to fit in with the requirements of the NATO navies the OTO-Melara 76mm gun and a Dutch fire control system were adopted. The Mk 94 GFCS on *Pegasus* was bought direct from HSA but the Mk 92 systems on the other five are being manufactured under licence. The original anti-ship missile armament has been doubled, with two quadruple mounts replacing the four singles first envisaged.

Below: *Pegasus* during a test run off San Diego. These vessels can make 40kt, but their operational limitations and the escalating cost of the propulsion system led Italy and West Germany to reconsider, while the US Navy built only six.

Below: Harpoon launch from the patrol hydrofoil Pegasus, originally envisaged as the first of a very large class.

Pauk Class

Origin: Soviet Union.
Type: Fast attack craft.
Number: 12 craft + ?
Displacement: 580 tons full load.
Dimensions: Length overall 187ft (57m); beam 34·4ft (10·5m); draught 6·6ft (2m).
Propulsion: Diesels, 16,000bhp; 2 shafts; 28–34kt.
Armament: 2 RBU 1200 launchers; 1 3in (76mm) gun; 1 30mm Gatling CIWS; 1 SA-N-5 launcher; 4 16in (406mm) torpedo tubes.
Complement: 80.

Development: The Soviet Navy's Pauk design is especially interesting as being one of the smallest specialis-ed anti-submarine ships. First seen by Western observers in 1980, Pauk is intended to be the replacement for the ageing Poti class and it would ap-pear that, unusually for them, the Soviets have adopted the hull of the Tarantul class missile ship rather than develop a new one.

The ASW sensor and weapon fit is probably as comprehensive as could possibly be installed in a hull of this size. There is a prominent housing for a dipping sonar on the transom, making the Pauk the smallest ship to carry such a device; there may well also be a hull-mounted sonar in the bow position, but this is as yet un-confirmed. Main ASW weapons are four single 15·7in (400mm) electric-powered acoustic homing torpe-does, mounted amidships. There are also two RBU 1200 250mm ASW mortars for close-in attack and, for good measure, two six-round depth-charge racks mounted at the stern on either side of the VDS housing.

The Soviet predilection for ever larger guns is followed on the Pauk which has a single 76mm in a dual-purpose mounting on the forecastle and well clear of the superstructure, giving it an excellent field of fire. For close-in air defence there is an ADG6-30 six barrelled 30mm Gatling on the after superstructure, together with an SA-N-5 Grail SAM launcher below it on the quarterdeck. These air defence weapons systems are controlled by a single Bass Tilt direc-tor, mounted on a pedestal at the after end of the bridge structure. The propulsion system is all-diesel, ex-haust outlets being located in the hull sides. With an assessed 12,000shp, these should give a maximum speed in the region of 26kt.

As in so many other classes, the Soviet naval architects have manag-ed to pack a great deal into a small hull, and this class represents a substantial addition to the Soviet short-range ASW forces. Western ship designers could well take note of this class, which will undoubtedly be built in large numbers.

Below: A Pauk class fast attack craft in the Baltic. Note the prominent housing for the vari-able depth sonar on the stern.

Sarancha

Origin: Soviet Union.
Type: Missile boat.
Number: 1 boat.
Displacement: 280 tons standard; 320 tons full load.
Dimensions: Length overall 148ft (45m); beam 36ft (11m); draught 9ft (2·8m).
Propulsion: COGAG; 24,000shp; 2 shafts; 45kt.
Armament: 2 twin SS-N-9 launchers; 1 twin SA-N-4 launcher; 1 30mm Gatling CIWS.
Complement: Not known.

Development: The Sarancha appears to be an R&D vessel for a hydrofoil successor to the Nanuchka. The adoption of a combination of foils and gas-turbines suggests an emphasis on quick reaction times rather than endurance. Instead of patrolling the periphery of Soviet sea-space to prevent incursions by hostile surface units, the Sarancha would presumably be deployed to strategically placed harbours to make high-speed sorties to intercept contacts made by aerial reconnaissance.

Evidence that such a mission is envisaged for the Sarancha rests in its medium-range missile armament. Two pairs of elevating launchers for SS-N-9 missiles are fitted on either side of the bridge, and these would require relay aircraft for guidance beyond horizon range. The Sarancha is also fitted, like the Nanuchkas, with an SA-N-4 air defence system, and it has a 30mm Gatling with a cyclic rate of fire probably up to 3000 rounds per minute at the after end of the superstructure.

The Sarancha has only one-quarter the displacement of the Nanuchkas and may have proven too small for the area defence role. It would also be incapable of the long-range deployments frequently undertaken by the Nanuchkas: construction of the latter class, therefore, continued at the same Petrovsky yard as the Saranchas throughout the late 1970s. Nevertheless, the Sarancha is a fascinating design, showing once again the Soviet ability to put a lot of weaponry and sensors on a small hull.

AIR CUSHION VEHICLES

Aist Class

Origin: Soviet Union.
Type: Air-cushion landing craft.
Number: 16 ships.
Weight: 220 tons full load.
Dimensions: Length overall 157ft (48m); beam 57ft (18m).
Propulsion: Two gas turbines, 24,000bhp; 65kt.
Armament: 2 twin 30mm guns.

Development: The Aist is the Soviet Navy's first large military hovercraft. Unlike the experimental ACVs currently undergoing trials for the US Navy, it is not designed to be accommodated in the docking wells of larger landing ships, but is intended for independent high-speed assault operations over relatively short distances. Such a craft would clearly be extremely useful in the Baltic, where it would be invaluable to defensive minefields laid by the Danish and Federal German navies to protect their respective coastlines.

The Aist is powered by two marinised gas turbines, each rated at 12–14,000hp. The turbines drive four axial lift fans and four propeller units, giving a maximum speed estimated at 65kt. A continuous tank deck with ramps at either end can accommodate two MBTs (T-62 or T-72) or five PT-76 amphibious tanks. A reduction in the number of vehicles enables up to 150 troops to be carried.

Above the vehicle hangar is a navigating bridge with good all-round vision and a lattice mast carrying the navigational radar. Twin

30mm mountings are fitted side by side at the forward end of the hangar, and the Drum Tilt FC director is installed atop the bridge. As well as providing anti-aircraft defence, the guns can be used for the suppression of shore defences.

Right: The Soviet Navy's first large military hovercraft is used for short-range independent amphibious assault transport and can carry a mix of light tanks and troops over a range of up to 350 miles (560km +) at 65kt (125km/h).

Right: Bow view of an *Aist*, showing the twin 30mm guns each side of the bridge and the Drum Tilt fire-control radome.

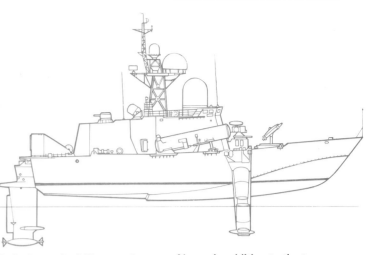

Left: *Sarancha* foilborne, showing the twin exhausts for the 2-shaft COGAG gas turbines, twin launcher for the SS-N-9 "Siren" surface-to-surface missiles and the large dome above the bridge for the Band Stand air surveillance radar which may also be used to track the SS-N-9s after launch so that guidance commands can be transmitted.

Above: In addition to the two pairs of SS-N-9s (positioned each side of the bridge) the *Sarancha* mounts an SA-N-4 SAM launcher on the foredeck and a 30mm Gatling-type close-in weapon aft on the superstructure. A hemispherical Fish Bowl radome is mounted above the Band Stand, with the smaller Bass Tilt at the foot of the mast.

LCAC

Origin: USA.
Type: Landing Craft Air-Cushion (LCAC).
Number: 108 on order
Weight: 170 tons mission weight.
Dimensions: Length overall 88ft (26·8m) beam 47ft (14·4m).
Propulsion: Four gas-turbines; 12,280 bhp; 50 knots.
Armament: None.
Cargo: 60 tons.
Complement: 5.

Development: As in so many other Western countries the USA has spent many years examining the whole air-cushion concept, starting with the deployment of virtually standard British machines in South Vietnam in the late 1960s. In the early 1970s attention began to concentrate on the use of ACVs in the ship-to-shore logistics role and two trials craft were built: one by Aerojet-General (designated JEFF-A) and a second by Bell Aerosystems (JEFF-B). Following extensive (and protracted) trials of these, Bell was awarded a contract to produce a new craft, which combines the best features of the two designs.

The new Landing Craft Air-Cushion (LCAC) is now in full production and 108 are on order, of which 54 will be based at Little Creek, Virginia, to serve the Atlantic Fleet, and 54 at Camp Pendleton, California, to serve the Pacific Fleet. The first production LCAC is already completed and undergoing trials.

The LCAC is a pure logistics machine and is designed to carry vehicles and stores from amphibious warfare ships standing offshore onto and over the beaches. It is not intended for anything other than very limited overground travel, although it could obviously penetrate deep inland using rivers or crossing swamps. The engines, control cabin and services are all located in narrow superstructures running down either side of the craft, leaving a clear and unobstructed deck space available for the payload. This 1,800 sq ft (167·23m²) can take up to 60 tons of cargo, which enables it to carry a Main Battle Tank (MBT).

Another rapidly developing role for ACVs is that of minehunting, where their low weight and acoustic signatures, and virtual immunity to underwater explosions gives them many advantages over a conventional ship type of minehunter/sweeper. The LCAC is thus being examined to assess its suitablity for this role, especially in support of an amphibious force en route to a landing operation.

Above: The first production LCAC en route to its delivery to the US Navy on 14 December, 1984, by Bell Aerospace Textron.

Below: LCAC-1 showing its main features, but its only payload on this occasion is a solitary transportable container.

ASW WEAPON SYSTEMS

AUSTRALIA

IKARA

Ikara is a system in which a guided missile is used to carry a torpedo to the vicinity of a submarine target, thus cutting down dramatically on the "dead time" between the acquisition of the target and the arrival of the weapon. The missile is launched from a surface warship and flies at high subsonic speed out to a maximum range of some 12 miles (20km). A data link back to the ship ensures that the missile flies to the computer-predicted optimum launch point, where the torpedo is released, makes a parachute descent to the sea and then carries out a normal homing attack on the target. Mk 44 and Mk 46 torpedoes are among those which can be carried, and the system is in service with the Australian, Brazilian and Royal Navies.

FRANCE

L5 MULTIPURPOSE TORPEDO

This torpedo is in service with the French and Belgian Navies. It is powered by silver-zinc batteries and has a speed of some 35kt, which suggests that its capability against modern, fast, Soviet submarines (some of which are capable of 40kt + speeds) is marginal. It has an active/passive head capable of homing attacks, either direct or in programmed search.

ITALY

TYPE A 184 TORPEDO

This wire-guided torpedo can be launched from surface ships or submarines against either surface or submarine targets. It has a range of some 8.7 miles (14km) and a speed of some 35kt. The torpedo is controlled from the launching ship until its own acoustic sensors acquire the target, when it is allowed to carry out a normal homing attack. The A 184 is in service with the Italian Navy.

SOVIET UNION

ASW ROCKET LAUNCHERS

Virtually all Soviet surface warships carry at least one ASW rocket launcher, of which there are several varieties. The rockets are fired in a predetermined pattern from a multiple launcher which is remotely trained and elevated. Current models include the RBU 1800 (a 5-barrelled 250mm system used in older ships); RBU 2500 (16-barrelled 250mm system, fitted in older cruisers and destroyers and in some small escorts); RBU 4500A (6-barrelled 300mm system with automatic loading); RBU 6000 (300mm system with barrels arranged in a circular fashion, range about 6,560yd (6,000m), fired in paired sequence, fitted in many modern warships); and RBU 1200 (6-barrelled system, fitted on the quarters of larger warships, manual reloading, may have an anti-torpedo role). The RBU-2500 rocket has an estimated weight of 397 to 440lb (180 to 200kg).

SS-N-14 (SILEX)

The SS-N-14 is fitted on most modern, large Soviet surface warships. The system appears similar in concept to the Ikara (qv) in that a missile carries a torpedo to the vicinity of a target where it is dropped to carry out a normal search and homing attack. The Soviet missile is 24.6ft (7.6m) long and is fired from either a twin-arm launcher (e.g. *Moskva*) or from a quadruple bin (e.g. *Udaloy*). The flight profile shows a height of about 2,500ft (750m) and a speed of mach 0.95, for a maximum range of 34 miles (55km). SS-N-14 is believed to have an alternative nuclear warhead, whilst the homing torpedo version may also have an anti-ship capability.

SS-N-15/16

The SS-N-15 is fitted to Soviet attack submarines of the Alfa, Papa, Tango, Kilo and Victor-III classes, and may well be fitted to others as well. It is an ASW system similar to the US SUBROC in which an underwater-launched missile travels to the surface, follows an airborne flight path and then releases a depth bomb (SS-N-15) or a homing torpedo (SS-N-16). Maximum range is estimated to be about 34 miles (55km).

SWEDEN

BOFORS TYPE 375 ROCKET SYSTEM

Developed in Sweden, this rocket system is in service with at least eight navies. The missile weighs some 551lb (250kg) and carries 220lb (100kg) of TNT or 176lb (80kg) of hexotonal. The launcher has either two or four tubes and is automatically reloaded from an operating room below; fuses are set automatically at proximity, time or impact. The missile has a rocket motor and follows a flat trajectory, thus minimising time of flight. Maximum range is approximately 4,000yd (3,657m).

UK

STINGRAY

Stingray can be launched from helicopters, fixed-wing aircraft and surface ships, and is now in service with the Royal Navy. It is an autonomous acoustic-homing torpedo and is claimed to be equally effective in shallow and deep water. An onboard computer can make its own tactical decisions during the course of an attack. The torpedo has an endurance of 8 minutes at 45kt.

USA

ADVANCED LIGHTWEIGHT TORPEDO (ALWT)

The latest Soviet submarines are not only faster and capable of diving deeper than previous types, but are also being constructed of new, stronger materials. In addition, they are being coated with anechoic paint or even (it has been suggested) anechoic tiles to reduce the reflective signature. Thus although in-service torpedoes can be improved to a certain extent, the challenge is now such that a totally new torpedo is needed; the current US programme is called ALWT. This is expected to have a similar size and weight to the Mk 46, although every effort is being made to increase the vital element of speed. A stored chemical energy propulsion system (SCEPS) is used, and it is intended to use a directed-energy warhead to achieve the required penetration of target hulls.

ASROC

ASROC consists of a nuclear depth bomb (approx 1kT) or a Mk 46 torpedo attached to a solid-propellant rocket motor. It is fired either from a box-shaped 8-cell launcher or from the Mk 10 Terrier launcher. On launch, the missile follows a ballistic trajectory and the rocket motor is jettisoned at a predetermined point. If the payload is a torpedo it descends by parachute to the surface, where its homing head and motor are activated; a depth bomb free-drops and is detonated at a set depth. Range is estimated to be 1.25-6.2 miles (2-10km). This very successful system is in service on some 240 ships of twelve navies. A vertically launched version is now under development.

ASW-SOW

The Anti-Submarine Warfare Stand-Off Weapon (ASW-SOW) is designed to replace the SUBROC system (qv); it may also be available for launch from surface ships. The missile will be stored in a canister in a standard torpedo tube and on launch the entire canister will leave the submarine and travel to the surface, where the missile motor will ignite, the canister will detach itself and the missile proceed towards the target. The payload will be the new Advanc-ed Lightweight Torpedo (ALWT), a Mk 46 torpedo (surface ships) or a nuclear depth bomb.

MK 46 TORPEDO

This torpedo is used by at least 21 navies, and is a deep-diving, high-speed device which can be launched from surface warships, helicopters or fixed-wing aircraft; it can also be carried by the ASROC and Ikara systems (qv). The Mk 46 has an active/passive acoustic homing sensor for its role of submarine attack. The increased speed of Soviet SSNs, allied to the new Clusterguard paint which seriously attenuates the acoustic response, has given rise to the Near-Term Improvement Program (NEAR-TIP) which will be both applied to new-production Mk 46s and retrofitted to in-service torpedoes. The Mk 46 is also used in the CAPTOR system, in which an encapsulated torpedo (hence "CAPTOR") is moored in deep water and then launched on detection of a suitable hostile target. These mines can be delivered by surface ships, aircraft or submarines, and add a significant new dimension to mine warfare.

SUBROC

SUBROC is a nuclear missile (1kT warhead) designed for use against hostile SSBNs. It is launched from a standard torpedo tube and after a short underwater journey it rises to the surface and becomes airborne until it returns to the water and the warhead sinks to a set depth before detonating. Range is about 35 miles (56km) and airborne speed is in excess of Mach 1; estimated lethal radius of the W55 warhead is 3-5 miles (5-8km). Some USN SSNs are equipped to carry SUBROC; each carries 4-6 missiles. SUBROC is scheduled to be replaced by ASW-SOW (qv) in the late 1980s.

Above: A SUBROC missile breaks the surface after launch from a submarine torpedo tube; the missile can travel up to 35 miles (56km) at supersonic speed.

Below: Whereas SUBROC has only a nuclear depth charge, ASROC, seen here leaving USS *Brooke,* can carry a homing tor-pedo as an alternative warhead.

Top right: Quadruple SS-N-14 "Silex" launcher on a Kresta II ASW cruiser; like ASROC, the missile can carry nuclear or homing torpedo warheads.

Above right: A rocket-propelled depth bomb is fired from one of the two RBU 6000 launchers on a Kashin class cruiser; two RBU 1000s are also carried.

Above: The Australian Ikara system uses a guided missile to carry a homing torpedo up to 12 miles (20km) before releasing it for a parachute descent.

Below: Launch of a Mk 46 torpedo from a surface ship: the weapon can also be air-launched, as well as forming the payload of both Ikara and ASROC systems.

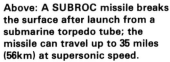

NAVAL GUNS

FRANCE

100mm DUAL-PURPOSE MODEL 1968-II
Calibre: 100mm.
Number of barrels: One.
Rate of fire: 80 rounds per minute.
Associated radar: Various.
Remarks: This completely automatic gun is intended to be used against both surface and air targets, including sea-skimming missiles. It serves on ships of the navies of France, the Federal Republic of Germany, Belgium, Portugal, Argentina and Greece.

ITALY

OTO 76/62 COMPACT
Calibre: 76mm.
Number of barrels: One.
Rate of fire: 10–85 rounds per minute.
Associated radar: Various.
Remarks: This very successful weapon is in service with 35 navies, including the US Navy, and is intended for use as a dual-purpose anti-aircraft and anti-ship gun in vessels down to the size of motor gunboats. Eighty rounds can be fired without reloading. The very neat shield is made of fibreglass and is proof against nuclear fallout.

127/54 GUN MOUNTING
Calibre: 127mm.
Number of barrels: One.
Rate of fire: 45 rounds per minute.
Associated radar: Various.
Remarks: Another OTO Melara design, this gun is in service with at least six navies and is installed on destroyers and frigates. It is completely automated, and the reloading, feeding and firing operations are controlled by just one man.

SOVIET UNION

30MM CLOSE-IN WEAPON SYSTEM (CIWS)
Calibre: 30mm.
Number of barrels: Six.
Rate of fire: 3,000 rounds per minute.
Associated radar: Bass Tilt.
Remarks: This is an exceptionally neat and functional design, intended for use against sea-skimming missiles and "smart" munitions. It is mounted on ships varying in size from fast attack craft to the Kirov class battlecruisers.

TWIN 30MM AA GUN
Calibre: 30mm.
Number of barrels: Two.
Rate of fire: 500 rounds per minute per barrel.
Associated radar: Drum Tilt or Muff Cob.
Remarks: A fully automatic, remote controlled weapon in a very neat mounting. In service since 1960 this mounting is fitted in many Soviet warships, especially the smaller types.

TWIN 130MM DUAL-PURPOSE
Calibre: 130mm.
Number of barrels: Two.
Rate of fire: 35 rounds per minute.
Associated radar: Kite Screech.
Remarks: This new mounting has appeared on a number of the recent Soviet ships. There are two turrets on the Sovremenny class, one on the Slava class, and two on the second ship of the Kirov class.

SINGLE 100MM DUAL-PURPOSE
Calibre: 100mm.
Number of barrels: One.
Associated radar: Kite Screech.
Remarks: This new mounting has appeared on several recent Soviet ships, including Udaloy (two turrets) and the first of the Kirov class battlecruisers (two turrets). The mounting is totally enclosed and operation is automatic. First use of the gun was in the Krivak-II frigates, which have two turrets in "X" and "Y" positions.

TWIN 76MM DUAL-PURPOSE
Calibre: 76mm.
Number of barrels: Two.
Rate of fire: 60 rounds per minute.
Associated radar: Owl Screech.
Remarks: First seen in the West in 1962, this gun is now mounted in many classes of Soviet ship, including Kiev class carriers, Kynda and Kara cruisers, Kashin and Modified Kildin destroyers, and at least four types of frigate.

UK

SINGLE 4·5IN GUN MOUNTING MK 8
Calibre: 113mm.
Number of barrels: One.
Rate of fire: 25 rounds per minute.
Associated radar: Various.
Remarks: The ordnance in this mount was developed from that used on the British Army's Abbott field gun. The naval mounting is fully automatic and is housed in a neat protective shield. In Royal Navy service it is fitted to the Types 82, 42, 22 and 21 ships, and it also serves on board vessels of the Argentine, Brazilian, Iranian, Libyan and Thai Navies.

USA

MULTI-BARREL 20MM/76 CALIBRE MK 15
Calibre: 20mm.
Number of barrels: Six.
Rate of fire: 3,000 rounds per minute.
Associated radar: Pulse doppler search, acquisition, tracking and electronic spotting radar.
Remarks: The 1982 naval war in the Falklands showed only too clearly the danger posed to expensive warships by air-and ship-launched missiles. Therefore, even more attention is being devoted to last-ditch anti-missile defences such as this Vulcan/Phalanx Close-in Weapon system which uses the M61A1 Gatling gun with an extremely high-speed computer-aided fire control system. The system destroys missiles by firing high-kinetic-energy 20mm projectiles at them causing them to explode before hitting the target ship. The system is in service with vessels of the US Navy and Britain's Royal Navy, and is also in production for Australia, Japan and Saudi Arabia.

SINGLE 5IN/54 CALIBRE MK 42
Calibre: 5in (127mm).
Number of barrels: One.
Rate of fire: 20-40 rounds per minute.
Associated radar: Various.
Remarks: The Mk 42 Mod 7 has automatic ammunition feed mechanisms, the principal manual function being in setting up the cylindrical loading drums. Total crew for the Mod 7 is 14 (of whom 4 are in the turret) but this is reduced to 12 in the Mod 10. The mounting is in service with the US, Australian, German, Japanese and Spanish Navies. A further development, the Lightweight Mod 9, is fitted to the DE 1052 class frigates of the US Navy.

SINGLE 5IN/54 CALIBRE MK 45
Calibre: 5in (127mm).
Number of barrels: One.
Rate of fire: 20 rounds per minute.
Associated radar: Various.
Remarks: Successor to the Mk 42

Above: Full automation allows the potent OTO 76/62 Compact to be mounted on vessels as small as the Italian Navy's 62.5t Sparviero class hydrofoils.

Right: The pair of single 100mm dual-purpose guns aboard the Soviet battlecruiser *Kirov* have been replaced on later members of the class by a twin 130mm mounting.

Far right: The forward twin 30mm/L65 mounting on an ex-Soviet Osa I class fast attack craft of the East German navy. A similar mounting is located aft, and four single SS-N-2 launchers are also carried.

(above), the Mk 45 mounting is probably the lightest for its calibre in the world. It is fully automated and has a crew of six, none of whom are in the turret. The system is fitted in the Aegis cruisers, the Virginia and California class CGNs and the Kidd class destroyers.

TRIPLE 16IN TURRET
Calibre: 16in (406mm).
Number of barrels: Three.
Rate of fire: 2 rounds per minute.
Associated radar: Various.
Remarks: The only remaining battleship guns, these weapons are all mounted on the US Navy's Iowa class, the four ships of which are now rejoining the fleet. Although intended mainly for use in a shore bombardment role the guns would obviously be used against other warships if the opportunity arose; their effect against modern unarmoured ships would be devastating. However, it is rather improbable that they would be offered such an chance.

Right: The 16in guns carried by the Iowa class can reach a range of 25 miles (40km) with a 2,700lb (1,225kg) projectile.

Below: Despite the prevalence of guided missiles, guns such as this 5in/54cal Mk45 aboard the USS *Lockwood* remain important for rapid fire over short ranges.

SHIPBORNE MISSILES

Surface-to-air missiles

Albatros

Origin: Italy.
Propulsion: Solid.
Dimensions: Length 12ft 1·7in (3·7m); diameter 8·0in (203mm); span (moving wings) 31·5in (800mm), (fixed fins) 25·2in (640mm).
Launch weight: 485lb (220kg).
Range: Up to 62 miles (100km).
Warhead: HE.

Development: This system has been developed from Mk1 form (1968-73) which was intended to provide fire control for guns and RIM-7H Sparrow SAMs; the Mk2 uses Selenia's Aspide AAM. It has SARH guidance with CW illumination of the target, and is said to be capable against all anti-ship aircraft or missiles including those descending from above or skimming the waves. A lightweight version, with four-box launcher is offered for small vessels down to 200 tons.

Albatros is in service with the Italian Navy and 13 others.

SA-N-3 Goblet

Origin: Soviet Union.
Propulsion: Solid.
Dimensions: Length 20ft (6·1m).
Launch weight: 1,190lb (540kg).

Range: 20nm (35km).
Warhead: HE.

Development: Installed aboard most major Soviet surface combatants completed in the 1970s, Goblet first appeared on the Moskva and Kresta II class vessels. It is operated with twin launchers; early models had four reload hatches, later models only two; the magazine holds 22 missiles. Guidance is by command, and the missile is associated with the "High Lights" fire control radar.

SA-N-4

Origin: Soviet Union.
Propulsion: Solid.
Length: Length 11·5ft (3·5m).
Range: 6nm (10km).
Warhead: HE.

Development: Installed as primary anti-air system in some small combatants (such as Grisha and Nanuchka craft), and as secondary system in major combatants completed in the 1970s (including Kiev, Kirov and Slava class vessels), this missile uses a twin "pop-up" launcher in a cylindrical bin containing 18 reloads, and is associated with the

"Pop Group" fire control radar. There is thought to be a similarity with the SA-8 Gecko land mobile missile system, which has a near-identical radar group.

SA-N-6

Origin: Soviet Union.

Development: Very few details are known about this missile system, which is reported to be part of the armament of Kirov class battle-cruisers and Slava class guided missile cruisers, both of which have appeared in the 1980s. The SA-N-6 is thought to be a naval version of the SA-10, which is reportedly a 20·5ft (6·25m) long land mobile SAM system that entered service in 1985. In Soviet surface warships a vertical launch system is reported, from 12 hatches on the foredeck of Kirov, for instance.

The 1985 edition of the US DoD's "Soviet Military Power" stated that, "The Kirov is outfitted with an array of air defense weapons, including 96 long-range SA-N-6 missiles and, on the second unit (Frunze), provisions for 128 SA-NX-9 shorter range SAMs," and that Slava has 64 SA-N-6 missiles.

SA-N-7

Origin: Soviet Union.

Development: This is a new naval SAM said to resemble the SA-11 short-range land mobile SAM which is reported to have a range of from 1·8 miles to 17 miles (3-28km) and be capable of engaging targets flying at from 100 to 50,000ft (30 to 14,000m). The 1985 DoD's "Soviet Military Power" credits the latest Sovremenny DDG as carrying 44 SA-N-7s, and this vessel has many fire control/target illuminating radars — so the missile could have a multiple-target role.

Seacat

Origin: UK.
Propulsion: Solid.
Dimensions: Length 4·8ft (1·48m); diameter 7·5in (190·5mm); span 25·6in (650mm).
Launch weight: 143lb (65kg).
Range: Up to 4 miles (6·5km).
Warhead: HE.

Development: This Royal Navy SAM, which is also used by many other navies in various forms, stems from the SX-A5 research SAM of the

1950s, which in turn was based on the Malkara anti-tank weapon. It is also associated with the land-mobile Tigercat and Hellcat ASM. It is probably the most cost-effective weapon in its class, most customers expecting to retain it in service until the late 1980s.

Seacat is delivered and stored in glassfibre containers which fit closely around the wings, and is small enough for loading to be manual if necessary. It has radio command guidance which enables it to be integrated with almost any form of sighting and fire control system. One system uses a quad launcher slaved to a director bin manned by the two operators, one of whom rotates the launcher to the target azimuth while the other searches with powerful, binoculars whose elevation is followed by the missiles. When a target is acquired within range a missile is fired and the operator guides the missile to the target via a joystick. Further options are a lightweight triple launcher and a one-man sight/director for small craft. Operation against anti-ship missiles is also a possibility.

NATO Seasparrow

Origin: USA.
Propulsion: Solid.
Dimensions: Length 12ft (3·6m); diameter 7·8in (200mm); span 3·4in (1·02m).
Launch weight: 441lb (200kg).
Range: 15 miles (25km).
Warhead: HE.

Far left, top: Seasparrow ship-to-air missile launching from *Enterprise*.

Far left, centre: Twin launchers for SS-N-12 missiles on the Soviet carrier *Kiev*. The launcher for SA-N-3 can also be seen.

Far left, bottom: Rare photo of an SA-N-4 launch from a Koni of the East German Navy.

Top left: A Standard MR being launched from a single overhead launcher at Point Mugu Pacific Missile Test Center. There are surface-to-air and surface-to-surface versions of the same basic missile system, and other variants which differ in guidance systems and range.

Left: Italy's Otomat surface-to-surface missile is launched from *Sparviero* at speed.

Below: Launch of the UK-built Seawolf SAM, which is capable of anti-missile operation.

Development: NATO Seasparrow (RIM-7H) is a development of the US Navy's Seasparrow (RIM-7E) which is based on the air-launched Sparrow. RIM-7E still forms the basis of the USN's Basic Point Defense System (BPDMS), serving aboard many DDGs, FFGs and CGNs, though it is being superseded by the RIM-7H version. This is an Improved PDMS in CVs, Spruance and other class DDs, some LCCs, LPHs and LHAs, and large replenishment vessels. It is also the subject of a massive international co-production programme involving several NATO nations, and serves aboard frigates of Belgium, Canada, Denmark, Italy, Netherlands, Norway, Spain and West Germany. Some NATO nations, including Turkey, have land-based Seasparrow sites; and Japan has also purchased the system.

NATO Seasparrow entered full-scale production in 1973 and there are now over 100 systems installed. The RIM-7H version was also used in Italy's Albatros system.

Basically, RIM-7H is a high-performance version of the Sparrow AAM with folding fins. It is fired from a lightweight 8-round box launcher, is reloaded manually, has automatic tracking, uses continuous wave (CW) semi-active homing guidance and the Mk91 fire control radar system.

RIM-7M entered USN and other NATO service in 1983. This is a vertical launch system, whose missile has a Jet Vane Control unit added, which controls the initial tip-over, orientation and course control, after which the normal guidance control takes over.

Seawolf

Origin: UK.
Propulsion: Solid.
Dimensions: Length 6·5ft (1·9m); diameter 7·1in (180mm); span 22in (559mm).
Launch weight: 180lb (82kg).
Range: 4 miles (6·4km).
Warhead: HE.

Development: Seawolf is installed aboard Royal Navy Type 22 frigates and was retrofitted to the last ten Leander class frigates in place of Seacat. It is a lightweight point defence system, launched from a six-box launcher, and is effective against missiles which dive on the ship or skim the wavetops.

The basic RN system, GWS.25, uses a Marconi Type 965 L-band PD radar to detect incoming targets; the entire sequence is automatic and takes milliseconds. Range, bearing and velocity are fed to a computer which forms an unambiguous track, and initiates threat evaluation and IFF interrogation. If these processes (which take up to 5 seconds in all) result in an immediate-threat designation, an appropriate launcher/tracker is selected and aimed in azimuth. In a fraction of a second the tracker searches up and down on the designated azimuth and its computer calculates the aim-off necessary to direct the selected missile into the gathering beam. Guidance is by command line of sight (CLOS), resulting in a small missile.

The system can fire up to three barrels in rapid sucession and steer all to the same target simultaneously. In development firings, Seawolf has actually struck a 4·5in Mk8 shell, and in 1983 a Seawolf intercepted and destroyed an MM 38 Exocet anti-ship missile in a trial.

There are numerous alternatives to the GWS.25, including extremely lightweight systems for smaller vessels, a four-barrel launcher, and a vertical launch system.

Standard

Origin: USA.
Propulsion: Solid.
Dimensions: (MR) 14·5ft (4·47m); (ER) 26ft (7·98m); diameter (both) 13·5in (343mm); span 36in (914mm).
Launch weight: (MR) 1,280lb (581kg); (ER) 2,350lb (1,066kg).
Range: 17 miles (27·8km).
Warhead: HE.

Development: The Standard family began in the 1960s as two missiles: MR (Medium Range) RIM-66 to replace Tartar on USN DDGs, FFGs and new CGNs; and ER (Extended Range) RIM-67 to replace Terrier on CGs. The family is related to the Standard air-launched ARM (Anti-Radar Missile).

The ER version uses the MK10 twin launcher (40 to 60 missiles; 80 in CGN 9), and the SPG-55 fire control radar with the Mk76 MFCS. The smaller MR is the same missile but does not have a booster; it uses the Mk11 twin launcher (42 missiles), a single Mk13 launcher (40 missiles), a single Mk22 launcher (16 missiles) or a twin Mk26 launcher (22-44 missiles). Fire control is by SPG-51 radar with Mk74 MFCS. Both missiles have semi-active homing guidance, and both have fully automatic magazine and loading.

The Standard missile has been continuously developed in various ship- and air-launched forms, such as adaptation as an interim SSM (awaiting deployment of Harpoon), most recent developments including an altitude record-breaking "kill" of a target drone by SM-2 version with Thiokol booster motor and probable deployment of a nuclear warhead in Standard SM-2 in the 1990s, following studies underway now.

Well over 10,000 missiles have been delivered to the USN and other navies, with procurement funds alotted for many more thousands.

Surface-to-surface missiles

Otomat

Origin: Italy/France.
Propulsion: Twin boost rockets; Turbomeca TR281 Arbizon turbojets for cruise.
Dimensions: Length 15ft 9¾in (4·82m); diameter 18·1in (460mm); span 47in (1,190mm).
Launch weight: 1,697lb (770kg).
Range: Mk1 3·75-37 miles (6-60km); Mk2 up to about 62 miles (100km).
Warhead: 463lb (210kg) HE.

Development: This cruise missile with cruciform wings and rear control fins can be fired from a wide variety of platforms including ships of FPB size upwards, aircraft and land mobile platforms. On ships it is fired from a box container slaved to a fixed launch table. Target data are fed from any source such as helicopter data-link, and launch direction can be up to 180° off-target. Mk2 is a seaskimmer all the way to the target, and a Teseo version incorporates a radio link for mid-course guidance by an airborne platform.

SS-N-9 Siren

Origin: Soviet Union.
Propulsion: Solid.
Length: (Estimated) 30ft (9·2m).
Range: 60-90nm (110-170km).

Development: Though no pictures of the actual missile, or official details, have been released, it is thought Siren arms the Soviet Nanuchka class boats (in triple elevating launchers), which first appeared in 1969. It is also said to be fitted in paired launchers on Sarancha class hydrofoils, is used with "Band Stand" fire control radar, and needs relay aircraft beyond horizon range.

Possible use of SS-N-9 from vertical launch apparatus aboard Papa class submarines has also been suggested.

SS-N-12 Sandbox

Origin: Soviet Union.
Dimensions: Length up to 38ft (11·5m); diameter 35in (900mm); wingspan 8·5ft (2·6m).
Range: 250nm (460km).

Development: Sandbox is thought to be the replacement for SS-N-3 and to have been in service since 1976. This missile is said to be the one housed in eight launcher/containers aboard Kiev class aircraft carriers, and also possibly on new Soviet RKRs currently under construction in the Black Sea. Mid-course guidance by airborne platform is considered probable, especially if launched by Echo-class submarines, which is also thought possible.

SS-N-19

Origin: Soviet Union.
Length: 33ft (10m).
Range: 200nm (370km).

Development: This missile is thought to be contained in vertical-launch hatches aboard the Kirov-class battle cruisers, and also to be the likely armament of Oscar-class submarines; both vessels were first seen in 1980. No photographs or official details have been released; data above are guesses, and other guestimates have put the range as high as 310 miles (500km). It is further surmised that nuclear warheads would be carried, and that mid-course guidance could be supplied by Hormone helicopters or the Kirovs themselves.

Appendix

STRATEGIC WEAPONS: Air-launched Missiles

L: length   D: diameter   S: span
N/A: Not available or not applicable

Weapon	Type	Origin	Dimensions	Warhead	Propulsion/speed	Range	Remarks
AS-3	ASM	USSR	L 48ft 11in (14·9m); D 72·8in (1·85m); S 30ft 0in (9·14m)	Nuclear or thermonuclear	1 Tumanskii R-11 or R-13 afterburning turbojet; Mach 1·8	400 miles (650km)	NATO "Kangaroo"; 1 carried by Tu-20/-95 "Bear-B/-C"; operational since 1960; possibly inertial or pre-programmed guidance.
SRAM, AGM-69A	Short-range attack missile	USA	L 15ft 10in (4·83m); D 17¾ in (44·45cm); S 15in (38·1cm) each of 3 fins	200kT W-69 thermonuclear	Two-pulse solid motor, Mach 2·8-3·2	35-105 miles (56-169km) depending on attack profile	20 carried by B-52G/H, 6 by FB-111A; Thiokol HTPB-propellant motor being fitted; obsolescent.

STRATEGIC WEAPONS: Ballistic Missile Submarines

L: length   B: beam   D: draught
N/A: Not available or not applicable

Weapon	Type	Origin	Dimensions	Weaponry	Propulsion/speed (surfaced/submerged)	Range	Remarks
Golf I/II/III/IV/V classes	SSB (SS)	USSR	L 351ft 5in (98·0m); B 27ft 11in (8·5m); D 21ft 0in (6·4m)	3xSS-N-5(II); SS-N-8 (III); SS-N-6 (IV); SS-NX-? (V)	3-shaft diesel/electric, 6,000/12,000hp = 17/14kt	22,700 miles (36,500km)	21 launched 1959-63; 3 Golf I (Mod), 13 Golf II, 1 Golf III/IV/V in service; Golf I (Mod)s no missiles but additional antennae.
Lafayette class	SSBN	USA	L 425ft 0in (129·5m); B 33ft 0in (10·1m); D 31ft 6in (9·6m)	16 Poseidon C-3 (19 boats); 16 Trident C-4 (12 boats)	1-shaft nuclear, 15,000shp = 20/30kt	N/A	31 launched 1962-66; last 12 sometimes known as Benjamin Franklin class; also armed with torpedoes and Subroc.

STRATEGIC WEAPONS: ICBMs, IRBMs, MRBMs

L: length   D: diameter   c: circa   est: estimated
N/A: Not available or not applicable

Weapon	Type	Origin	Dimensions	Warhead	Propulsion	Range	Remarks
CSS-1	MRBM	China	N/A	20mT nuclear (est)	Liquid fuel rocket	c.680 miles (1,000km)	Similar to Soviet SS-3; operational since 1970; 100 deployed.
CSS-2	IRBM	China	N/A	N/A	Single-stage liquid fuel rocket	1,550-2,450 miles (2,500-4,000km)	Operational since 1971; about 70 in service by 1980.
CSS-X-4	ICBM	China	L c.32ft 10in (10m); D c.9ft 10in (3m)	5mT nuclear (est)	2-stage liquid fuel rocket	8,000 miles (13,00km) (est)	Two test launches May 1980; small numbers deployed subsequently.
SS-4	MRBM	USSR	L c.68ft 11in (21m); D c.5ft 3in (1·6m)	1mT nuclear or HE	Single-stage liquid fuel rocket	930-1,120 miles (1,300-1,800km)	NATO "Sandal"; first deployed operationally 1958; over 200 in service.
SS-5	IRBM	USSR	L c.82ft 0in (25m); D c.7ft 10in (2·4m)	1mT nuclear	Single-stage liquid fuel rocket	2,200 miles (3,500km) est.	NATO "Skean"; operational since 1961; most now replaced by SS-20.
SS-9	ICBM	USSR	L c.114ft 10in (35m); D 9ft 10in (3m)	20-25mT nuclear	3-stage liquid fuel rocket	7,460 miles (12,000km)	NATO "Scarp"; 288 deployed in late 1960s; replaced by SS-18; Mods 3 and 4 test only.
SS-11	ICBM	USSR	L c.65ft 7in (20m); D c.8ft 2in (2·5m)	200kT nuclear	2-stage liquid-fuel rocket	6,200-8,000 miles (10,000-13,000km)	NATO "Sego"; Mods 1, 2, 3, and 4 with variations in warhead; operational since 1970; 520 in service by 1984.
SS-13	ICBM	USSR	L 65ft 7in (20m); D 5ft 7in (1·7m)	1x1mT or 3 MRV nuclear	3-stage solid fuel rocket	4,970 miles (8,000km)	NATO "Savage"; only 60 deployed because of poor accuracy, unreliable propulsion; to be replaced by SS-16.
SS-14	IRBM	USSR	L 36ft 0in (11m); D 4ft 7in (1·4m)	Thermonuclear	2-stage solid fuel rocket	2,500 miles (4,000km)	NATO "Scapegoat"; two upper stages of SS-13; deployed by mobile transporter in eastern USSR.
SS-X-24	ICBM	USSR	L c.68ft 11 in (21m); D c.8ft 2in (2·5m)	Up to 10 MIRV nuclear	Probably 3-stage solid fuel rocket	7,500 miles (12,000km) + (est)	Similar to US Peacekeeper; initial deployment in super-hardened silos.
SS-X-25	ICBM	USSR	L c.59ft 0in (18m); D c.5ft 7in (1·7m)	N/A	Probably 3-stage solid fuel rocket	5,600 miles (9,000km) + (est)	Improved SS-13, similar to US Minuteman; possible multi-shelter base at Plesetsk.
Titan II, LGM-25C	ICBM	USA	L 102 ft 8¼ in (31·3m); D 10ft 0in (3·5m)	c.10mT thermonuclear	Liquid fuel rockets; 99,200lb (45,000kg) or 216,000lb (98,000kg) thrust	9,320 miles (15,000km)	Deactivation of remaining Titan IIs started 1982, scheduled for completion in 1987.

STRATEGIC WEAPONS: SLBMs, SLCMs

L: length   D: diameter   S: span   c: circa   est: estimated
N/A: Not available or not applicable

Weapon	Type	Origin	Dimensions	Warhead	Propulsion	Range	Remarks
Polaris A-3, UGM-27C	SLBM	USA	L 32ft 3¾ in (9·85m); D 6ft 0in (1·83m)	Probably 6x400kT nuclear	Liquid-fuel rocket, 80,000lb (36,000kg) thrust	2,500 miles (4,000km)	To remain operational on 4 Royal Navy Resolution class submarines until early 1990s.
SS-N-4	SLBM	USSR	L c.49ft 2in (15m); D c.5ft 11in (1·8m)	N/A	Probably 2-stage solid fuel rocket	400 miles (650km) est.	NATO "Sark"; tested 1955; installed in some Zulu, Golf and Hotel class submarines; no longer operational.
SS-N-5	SLBM	USSR	L 32ft 10in (10m); D c.5ft 7in (1·5m)	1mT nuclear	2-stage liquid fuel rocket	800 miles (1,300km)	NATO "Serb"; installed in later Golf and Hotel class submarines since 1963; obsolescent.
SS-N-6 Mod 3	SLBM	USSR	L 31ft 8in (9·65m); D c.5ft 5in (1·65m)	2 nuclear MRVs	2-stage liquid fuel rocket	1,860 miles (3,000km)	NATO "Sawfly"; 16 missiles each on 24 Yankee class submarines.
SS-N-8	SLBM	USSR	L 42ft 6in (12·95m); D c.5ft 5in (1·65m)	1x1-2mT or 39 MRV nuclear	2-stage liquid fuel rocket	4,970 miles (8,000km)	Operational aboard 18 Delta I class (12 tubes) and 4 Delta II class (16 tubes).
SS-NX-17	SLBM	USSR	L 36ft 3½ in (11·6m); D 5ft 5in (1·65m)	1mT nuclear	2-stage solid fuel rocket with post-boost vehicle	2,420 miles (3,900km) est.	Tested since 1975; deployed in one Yankee class submarine (12 tubes) since 1977.

AIR WEAPONS: Air-launched Missiles

L: length D: diameter S: span c: circa est: estimated

Weapon	Type (and guidance)	Origin	Dimensions	Warhead	Propulsion/speed	Range (min/max)	Remarks
AS-2	ASM (auto-pilot + active radar homing)	USSR	L c.31ft (9·5m); D c.35½in (90cm); S c.16ft (4·8m)	c.2,200lb (1,000kg) HE	Turbojet; supersonic	132 miles (200km) est; 350 miles (560km) est.	NATO "Kipper"; first seen 1961; single missile carried by Tu-16 "Badger-C".
AS.11	ASM (CLOS)	France	L 47⅝in (121cm); D 6½in (16·4cm); S 19⅝in (50cm)	Delayed impact HE, hollow charge or fragmentation	Solid rocket; 360mph (580km/h)	1,640ft (500m); 9,840ft (3,000m)	Developed in early 1950s from SS.11; widely used; helicopter launched; no longer in production.
Hellfire, AGM-114A	ASM (laser homing)	USA	L 5ft 3¾in (1·62m); D 7in (17·8cm); S 13in (33cm)	c. 20lb (9kg) hollow charge	Solid rocket; subsonic	3·73 miles (6km) max	Developed for US Army AH-64; in service 1985; also to be used on USMC AH-1T helicopters.
Martin Pescador	ASM (command)	Argentina	L 9ft 7¾in (2·94m); D 8⅝in (21·85cm); S 29⅞in (73cm)	88lb (40kg) HE	Solid rocket; supersonic	1·5 miles (2·5m); 5·6 miles (9km)	In service since 1979; helicopter-launched version and improved models under development.
Matra LGBs	Laser-homing bomb	France	Add-on laser seeker + control fins	Standard bomb	Aircraft-launched at speeds up to Mach 0·9	2·5-6·2 miles (4-10km)	In service 1985 with French Air Force; ordered by several countries
MW-1	Submunitions dispenser	Federal Germany	L 17ft 5in (5·3m)	Up to 4,700 submunitions	Charge ejection	8,200 x 1,640ft (2,500 x 500m) max	In production for aircraft of the Federal German Air Force.
Shafrir	AAM (IR homing)	Israel	L 8ft 1¼in (2·47m); D 6¼in (16cm); S 20½in (50cm)	24lb 4oz (11kg) pre-fragmented	Solid rocket; supersonic	3·1 miles (5km) max	Developed from AIM-9 Sidewinder; Mk 2 in service since 1969.

LAND WEAPONS: Air Defence Missiles

L: length D: diameter S: span c: circa est: estimated
N/A: Not available or not applicable

Weapon	Type (and guidance)	Origin	Dimensions	Warhead	Propulsion/speed	Range	Remarks
Bloodhound Mk 2	SAM (SAR)	UK	L 27ft 9in (8·46m); D 21½in (54·6cm); S 9ft 3½in (2·83m)	HE	Ramjet + solid rocket boosters	50 miles (80km) +	In service since 1964 (Mk1 1958) with RAF; also used by Switzerland and Singapore.
Chapparal, MIM-72C	SAM (IR homing)	USA	L 9ft 6½in (2·91m); D 5in (12·7cm); S 24¾in (62·9cm)	M-250 blast fragmentation	Mk 50 rocket; N/A	9,840ft (3,000m) +	Derived from AIM-9 Sidewinder; currently being upgraded with new IR sensors and CM-resistant guidance.
Improved Hawk, MIM-23B	SAM (CWSAR)	USA	L 16ft 8in (5·08m); D 14½in (37cm); S 47¼in (120cm)	120lb (54kg)	Aerojet XM112; Mach 2·5	25 miles (40km)	Upgraded for US Army and USMC with improved reliability and maintainability, plus multiple-target guidance.
Nike Hercules, MIM-14B	SAM (command guided)	SA	L 39ft 8⅜in (12·1m); D 31½in (80cm); S 74in (188cm)	Nuclear or HE	2-stage solid fuel rocket; supersonic	87 miles (140km) +	Operational in US from 1958 (now replaced by Patriot); improvements to units still in service with US allies.
RBS-70 Rayrider	SAM (laser beam-riding)	Sweden	L 4ft 4in (1·32m); D 6¼in (16cm); S 23⅝in (60cm)	2¼lb (1kg) pre-fragmented	Solid booster and sustainer; supersonic	16,400ft (5,000m)	Man-portable, vehicle, turret-mounted and naval variants; RBS70M Nightrider under development.
Redeye, FIM-43A	SAM (IR homing)	USA	L 4ft 0in (1·22m); D 2¾in (7cm); S 5½in (14cm)	Smooth case fragmentation	Solid rocket	c.2 miles (3·3km)	In service since 1964; slow, tail-attack only, vulnerable to IRCM, but produced in large quantities.
SA-3/SA-N-1	SAM (radio command)	USSR	L 22ft 0in (6·7m); D 23½in (60cm) max; S 59in (150cm) max	132lb (60kg) HE	Solid booster and sustainer; Mach 2	6·2-9·3 miles (10-15km)	NATO "Goa"; in service since 1961; vehicle (2, 3 or 4 rounds) or ship (SA-N-1) launched.
SA-4	SAM (radio command)	USSR	L 28ft 10½in (8·8m); D 35½in (90cm); S 90½in (230cm)	220-287lb (100-130kg) HE	4 x solid rocket boosters + ramjet sustain; Mach 2·5	c.43·5 miles (70km)	NATO "Ganef"; deployed (in Warsaw Pact only) since early 1960s.
SA-5	SAM (radio command)	USSR	L 54ft (16·5m); D 39½in (100cm) max; S 156in (396cm)	132lb (60kg) HE	Solid booster and sustainer; Mach 3·5 +	c.186 miles (300km)	NATO "Gammon"; improved version of earlier SA-5 "Griffon"; with SA-3 forms major part of Soviet SAM strength.
SA-6	SAM (radio command + CWSAR homing)	USSR	L 20ft 4in (6·2m); D 13¼in (33·5cm); S 60in (152cm)	176lb (80kg) HE	Integral solid rocket and ramjet; Mach 2·8	c.21·7 miles (35km)	NATO "Gainful"; successful use in 1973 Middle East war, but ineffective against modern ECM.
SA-9	SAM (IR homing)	USSR	L 6ft 6in (2·0m) est; D 4¾in (12cm); S 12in (30cm)	HE	Solid booster and sustainer; Mach 1·5 + est.	5 miles (8km)	NATO "Gaskin"; deployed on BRDM-2 in quad trainable launchers with four reloads in vehicle.
Tan-sam (Type 81)	SAM (autopilot + IR terminal homing)	Japan	L 8ft 10in (2·7m); D 6¼in (16cm); S 23½in (60cm)	HE	Solid rocket; Mach 2·4	6·2 miles (10km)	Quad launcher and fire-control system mounted on 3·5t truck controls two further quad launchers.
Tigercat	SAM (command guided)	UK	L 4ft 10in (1·47m); D 7½in (19cm); S 25½in (65cm)	HE	Dual-thrust solid rocket; c.Mach 0·9	16,400ft (5,000m) +	Derived from ship-mounted Seacat; 3 rounds on wheeled trailer; radar-enhanced version also in service.

LAND WEAPONS: AA Guns (Mobile)

L: length W: width H: height ROF: rate of fire
N/A: Not available or not applicable

Weapon	Type	Origin	Dimensions	Weaponry	Propulsion/speed	Range	Remarks
20mm Twin	Mobile AA gun	Federal Germany	L 16ft 6¼in (5·035m); W 7ft 9in (2·36m); H 6ft 9¾in (2·075m)	2 x 20mm MK20 Rh202	Towed	N/A	Combined ROF 1,200-2,000rds/min; in service with 5 countries.
M3-VDA	Twin SP AA gun	France	L 14ft 7¼in (4·45m); W 7ft 10½in (2·4m)	2 x 20mm	90hp petrol engine, 56mph (90km/h)	6,560ft (2,000m) slant	Based on Panhard M3 wheeled vehicle; in service with several countries.
TCM Mk3	Light AA system	Israel	N/A	2 x 20mm or 25mm	N/A	Dependent on gun mounted	Improved mounting for towed, SP or naval AA guns; in IDF service.
Vulcan M163	SP Air defence system	USA	As M113 APC	6-barrel 20mm M168	As M113 APC	16,400ft (5,000m) radar acquisition	Also towed, wheeled and product improved variants; in US and other service.

LAND WEAPONS: Anti-tank Weapons

L: length D: diameter S: span c: circa est: estimated
N/A: Not available or not applicable

Weapon	Type (and guidance)	Origin	Dimensions	Warhead	Propulsion/ speed	Range (min/max)	Remarks
AT-2 Swatter	ATM (CLOS + IR homing	USSR	L 44in (112cm); D 6in (15cm); S 26in (66cm)	N/A	Probably single-stage solid fuel rocket; 2,460ft/sec (750m/sec)	1,970ft (600m)/ 8,200ft (2,500m)	Triple launcher on BRDM-2; also mounted on some Mi-24 "Hind" helicopters; improved Swatter 2 slightly bigger.
AT-3 Sagger	ATM (CLOS)	USSR	L 34¼in (87cm); D 4¾in (12cm); S 18⅛in (46cm)	6lb (2·7kg)	Solid booster and sustainer; 400ft/sec (120m/sec)	1,640ft (500m)/ 9,840ft (3,000m)	Widely used, man-portable or vehicle-mounted lightweight ATM.
AT-6 Spiral	ATM	USSR	N/A	N/A	N/A	23,000-32,800ft (7-10km) est. max	Initial deployment on Mi-24 "Hind-E" helicopter; possible temporary withdrawal from front-line service.
Carl Gustav M2	84mm RCL	Sweden	L 44½in (113cm)	5lb 8oz (2·5kg) HEAT	N/A	1,300-1,640ft (400-500m) max	Widely used; improved M2-550 fires new HEAT round to 2,300ft (700m) range.
Cobra 2000	ATM (CLOS)	Federal Germany	L 37⅜in (95cm); D 3·93in (10cm); S 18·9in (48cm)	6lb (2·7kg) hollow charge or anti-tank/shrapnel	Solid booster and sustainer; c.280ft/sec (85m/sec)	1,300ft (400m)/ 6,560ft (2,000m)	Produced under licence in Brazil; licences also granted to Italy, Pakistan and Turkey.
Copperhead, M712	CLGP (laser homing)	USA	L 54in (137cm); Calibre 155mm	49·6lb (22·5kg) HESH	Gun-launched; supersonic	1·9 miles (3km)/ 10 miles (16km)	5,000 rounds delivered by early 1984; projected 30,000 total subject to continuing evaluation.
Folgore	RCL rocket launcher	Italy	L 72⅝in (185cm); Calibre 80mm	80mm rocket, hollow charge	Solid rocket; 1,640ft/sec (500m/sec)	164ft (50m)/ 3,280ft (1,000m)	Man-portable, tripod-mounted and vehicle-installed variants; in production.
HOT	ATM (SACLOS)	France/Federal Germany	L 50in (127cm); D 5⅜in (13·6cm); S 12¼in (31cm)	13lb 4oz (6kg) hollow charge	Solid booster and sustainer; 790ft/sec (240m/sec)	250ft (75m)/ 13,120ft (4,000m)	Vehicle-mounted heavy missile; in service in 13 countries.
KAM-3D (Type 64)	ATM (CLOS)	Japan	L 39⅜in (100cm); D 4¾in (12cm); S 23⅝in (60cm)	N/A	Solid booster and sustainer; N/A	1,150ft (350m)/ 5,905ft (1,800m)	Standard Japan Self-defence Force equipment since 1964; vehicle and helicopter installations.
KAM-9L (Type 79)	ATM (SACLOS)	Japan	L 61⅝in (156·5cm); D 6in (15·2cm); S 13in (33·2cm)	4lb 3oz (1·9kg) hollow charge	Solid booster and sustainer; 660ft/sec (200m/sec)	13,120ft (4,000m) max	Similar to US TOW; effective against AFVs and landing craft.
Kun Wu	ATM (CLOS)	Taiwan	N/A	N/A	2-stage solid rocket; N/A	9,840ft (3,000m) est. max	Based on AT-3 Swatter with modified warhead; rear-facing quad launchers in jeeps.
Mamba	ATM (CLOS)	Federal Germany	L 37⅝in (95·5cm); D 4¾in (12cm); S 15¾in (40cm)	6lb (2·7kg) hollow charge	Solid rocket; 460ft/sec (140m/sec)	985ft (300m)/ 6,560ft (2,000m)	Developed from Cobra with single oblique-thrust motor and improved warhead; fibre-optic instead of wire for guidance commands.
Mathogo	ATM (CLOS)	Argentina	L 39¼in (99·8cm); D 4in (10·2cm); S N/A	Hollow charge	Solid booster and sustainer; 295ft/sec (90m/sec)	1,150ft (350m)/ 6,890ft (2,100m)	In production; up to four launchers remote-controlled by single fire control unit.
Picket	ATM (gyro-stabilized to line-of-sight)	Israel	L 30in (76cm); D 3·19in (8·1cm)	HE	Single-stage solid rocket; supersonic	1,640ft (500m) +	In production; 3 folding fins for stabilization; shoulder-launched.
RBS56 Bill	ATM (SACLOS)	Sweden	L 35½in (90cm); D 5⅞in (15cm); S 11⅝in (41cm)	Hollow charge	Solid rocket; 660ft/sec (200m/sec)	490ft (150m), 6,560ft (2,000m)	Designed to fly 3ft 3in (1m) above operator's line of sight, with 30° angled warhead for top attack; test firing completed.
SS.11	ATM (CLOS)	France	L 4¾in (12cm); D 6½in (16·4cm); S 19¾in (50cm)	Hollow charge, anti-personnel or fragmentation	2-stage solid rocket; 377-623ft/sec (115-190m/sec)	1,640ft (500m)/ 9,840ft (3,000m)	In production from 1962; ground, vehicle, ship and aircraft variants; 174,000 delivered by end of production in 1980.
Swingfire	ATM (command control)	UK	L 42⅛in (107cm); D 6¾in (17cm); S 15⅜in (39cm)	Hollow charge	Solid booster and sustainer	490ft (150m)/ 13,120ft (4,000m)	Vehicle-mounted; in service since 1969; pallet and micro-miniaturised versions.

LAND WEAPONS: Battlefield Rockets and Missiles

L: length D: diameter
S: span c: circa est: estimated
N/A: Not available or not applicable

Weapon	Type	Origin	Projectile dimensions	Warhead	Propulsion	Range	Remarks
107mm Type 63	MRL	China	L 32·95in (83·7cm) Calibre 107mm	HE	Solid rocket	5 miles (8km)	12-round truck, trailer and man-portable (in sections) variants; also used by Albania, Vietnam.
140mm MRL	MRL	China	N/A	HE	Solid rocket	6·2 miles (10km) est.	19-round truck-mounted launcher; probably influenced by Soviet designs.
Ching Feng	SSM	Taiwan	L c.23ft (7m); D c.24in (60cm)	HE	Rocket	c.75 miles (120km)	Similar to US Lance; displayed on transporter trailer 1981.
D-10	MRL	Spain	Calibre 300mm	HE	Solid rocket	10·6 miles (17km)	10-round truck-mounted launcher for D3 rockets; in service with Spanish Army.
E-21	MRL	Spain	Calibre 216mm	HE, incendiary or smoke	Solid rocket	9 miles (14·5km)	21-round truck-mounted launcher for E3 rockets; in service with Spanish Army.
FIROS 6	MRL	Italy	Calibre 51mm	HEI, fragmentation, etc.	Solid rocket	3·7 miles (6km)	48-tube vehicle-mounted launcher.
FIROS 25	MRL	Italy	Calibre 122mm	HE etc. or submunitions.	Solid rocket	15·5 miles (25km)	40-tube vehicle-mounted launcher.
Hades	SSM	France	N/A	10-25kt nuclear	Rocket	c.217 miles (350km)	Development started 1983 to replace Pluton in service by 1992.
Honest John, MGR-1B	SSM (unguided)	USA	L 24ft 10in (7·57m); D 30in (76cm); S 54in (1·37m)	Nuclear or HE	Single-stage solid rocket	23 miles (37km)	Operational since early 1950s; replaced in US Army service by Lance.
Kung Feng	MRL	Taiwan	Calibre 127mm	HE	Rocket	N/A	40-tube tracked vehicle (Kung Feng-4) and 45-tube truck (Kung Feng-6) mountings.
LAR 160	MRL	Israel	Calibre 160mm	110lb (50kg) HE or submunitions	Solid rocket	18·6 miles (30km)	Vehicle-mounted launcher up to 50 tubes; operational with IDF.
M-63	MRL	Yugoslavia	Calibre 128mm	HE	Rocket	5·3 miles (8·6km)	32-tube trailer-mounted launcher based on Czech RM-130; in service.

Weapon	Type	Origin	Projectile dimensions	Warhead	Propulsion	Range	Remarks
MAR 290	Artillery rocket	Israel	L 17ft 10½in (5·45m); D 11⅜in (29cm); S 22½in (57cm)	705lb (320kg) HE	Solid rocket	15·5 miles (25km) est.	Based on captured Soviet BM-24; 4 lattice or tube launchers on tank chassis; in service with IDF.
SLAM-Pompero	MRL	Argentina	Calibre 105mm	24lb 4oz (11kg) HE	Solid rocket	7·45 miles (12km)	16-tube vehicle-mounted launcher; production initiated 1979.
SS-1	Battlefield support missile	USSR	L 36ft 11in (11·25m); D 33½in (85cm)	Nuclear or HE	Liquid fuel rocket	186 miles (300km)	NATO "Scud" (data for "Scud-B"); single-round transporter-erector vehicles; in wide service since late 1950s.
SS-21	Battlefield support missile	USSR	L 30ft 11½in (9·44m); D c.18in (46cm)	Nuclear, HE or chemical	Solid rocket	37 miles (60km)	Replacement for Frog-7; deployment delayed since development completed 1976.
SS-22	Battlefield support missile	USSR	N/A	N/A	N/A	c.560 miles (900km)	Replacement for SS-12; no details available.
Type 75	MRL	Japan	Calibre 130mm	HE	Solid rocket	9·3 miles (15km)	30-round launcher on tracked vehicle; in JSDF service since 1975.
Type RWKO14	MRL	Switzerland	Calibre 81mm	HE	Solid rocket	6·2 miles (10km)	2 x 15-tube launchers on wheeled APC.
Valkiri	Artillery rocket	South Africa	Calibre 127mm	Fragmentation	Solid rocket	13·7 miles (22km)	24-tube truck mounted launcher; in South African service since 1981.
YMRL	MRL	Yugoslavia	Calibre 128mm	HE	Solid rocket	11·2 miles (18km)	32-round truck-mounted launcher; in Yugoslav service since 1975.

LAND WEAPONS: Mortars

L: length W: weight
N/A: Not available or not applicable

Weapon	Type	Origin	Dimensions	Projectile	Propulsion/ speed	Range	Remarks
2in ML	Mortar	UK	L 26⅜in (67cm); W 9lb 2oz (4·14kg)	HE, smoke, star, etc.	N/A	1,575ft (480m)	In service since 1930s; obsolescent.
51mm	Mortar	UK	L 29½in (75cm); W 5lb 12oz (2·6kg)	HE, smoke illuminating	N/A	2,460ft (750m)	Replacement for 2in ML; introduced to service 1981.
60mm Commando	Mortar	France	L 26¾in (68cm); W 17lb (7·7kg)	HE, smoke, illuminating, etc.	N/A	3,445ft (1,050m)	Data for Type V; Type A longer and heavier; in service with 20 countries.
60mm Commando	Section mortar	Spain	L 25⅝in (l5cm); W 13lb (5·9kg)	HE, illuminating	N/A	3,510ft (1,070m)	Carried and operated by one man; in service with several countries.
60mm Hotchkiss-Brandt	Mortar	France	L 28½in (72·4cm); W 32lb 10oz (14·8kg)	Various makes and types	N/A	6,725ft (2,050m)	Private-venture; designed to use various standard bombs; in service with several armies.
60mm Model L	Company mortar	Spain	L 25⅝in (65cm); W 24lb (10·9kg)	HE, smoke, illuminating	N/A	6,480ft (1,975m)	In service with several armies.
60mm Soltam	Dual-purpose mortar	Israel	L 29in (74cm); W 35lb 15oz (16·3kg)	HE, frag, smoke, illuminating	N/A	8,370ft (2,550m)	Based on Tampella (Finland) design; data for Standard; also long-range and Commando models.
81mm MO 81-61	Mortar	France	L 45¼in (115cm); W 86lb 14oz (39·4kg)	HE, smoke, etc.	N/A	14,930ft (4,550m)	Data for model C; model L with longer barrel; both in French and other army service.
82mm M-37	Mortar	USSR	L 48⅛in (122·2cm); W 123lb 7oz (56kg)	HE, smoke	N/A	9,840ft (3,000m)	New model has lighter tripod and base-plate; in widespread service.
107mm M30	Rifled mortar	USA	L 60in (152·4cm); W 672lb (305kg)	HE, smoke, gas, illuminating	N/A	22,300ft (6,800m)	No longer in production, but in service with 13 countries.
120mm MO-120-60	Light mortar	France	L 64¼in (163·2cm); W 207lb (94kg)	HE, smoke, illuminating	N/A	21,800ft (6,650m)	Designed for simplicity; 3-man crew; in French and other army service.
120mm MO-120-RT61	Wheeled mortar	France	L 81⅞in (2·08m); W 1,283lb (582kg)	Various, including rocket-assissted	N/A	8 miles (13km) (rocket-assisted)	Complex mortar with gun characteristics; in French and other army service.

LAND WEAPONS: Machine Guns

L: length W: weight

Weapon	Type	Origin	Dimensions	Ammunition	Cyclic rate	Muzzle velocity	Remarks
0·30in Model 1919A4	AFV and company MG	USA	L 41 in (104·4cm); W 31lb (14·06kg)	·30M1 or M2; 250rd belt	400-500rds/min	2,820ft/sec (860m/sec)	Air-cooled, gas-assisted recoil operation; improved M1919A2 for AFV mounting; also in 7·62m x 51 NATO calibre
0·5in M2 HB	Heavy MG	USA	L 65in (165·1cm); W 86lb 4oz (39·1kg)	·50 M2; belt	450-600rds/min	2,815ft/sec (858m/sec)	Air-cooled, recoil-operated; developed early 1930s; ground, AFV and helicopter mountings.
7·62mm FN MAG	GPMG	Belgium	L 49⅝in (126cm); W 23lb 14¾oz (10·85kg)	7·72mm NATO; 50/250rd box	600-1,000rds/min	2,755ft/sec (840m/sec)	Air-cooled, gas-operated; in production since early 1950s; in service with over 75 countries.
7·62mm L4A4 Bren	GPMG	UK	L 45½in (115·6cm); W 23lb 9¾oz (10·71kg)	7·62mm x 51; 30rd box	600rds/min	2,700ft/sec (823m/sec)	Air-cooled, gas-operated; converted from ·303in Bren Mk3; in British and other army service.
7·62mm L7A1	GPMG	UK	L 48½in (123·2cm); W 24lb (10·9kg)	7·62mm x 51; 200rd belt	750-1,000rds/min	2,750ft/sec (838m/sec)	Air-cooled, gas-piston operated; modified FN MAG; in British and some Commonwealth service.
7·62mm RPD	Light MG	USSR	L 40¾in (103·6cm); W 15lb 10½oz (7·1kg)	7·62mm x 39; 100rd belt	700rds/min	2,300ft/sec (700m/sec)	Air-cooled gas-operated; standard Soviet section MG 1950-65; also Chinese Type 56 and North Korean Type 62; obsolescent.
7·62mm RPK	Light MG	USSR	L 41in (104cm); W 15lb 10½oz (7·1kg) with drum	7·62mm x 39; 30/40rd box or 75rd drum	660rds/min	2,400ft/sec (732m/sec)	Air-cooled, gas-operated; Light MG version of AK-47 assault rifle; replacement for RPD.

LAND WEAPONS: Sub-machine Guns

L: length with butt extended W: weight loaded c: circa
N/A: Not available or not applicable

Weapon	Type	Origin	Dimensions	Ammunition	Cyclic rate	Muzzle velocity	Remarks
5·45 AKR	SMG	USSR	L 26⅝in (67·5cm); W N/A	5·45 x 39·5mm; 30rd box	c. 800rds/min	c. 2,625ft/sec (800m/sec)	Derived from AKS-74 rifle; in Soviet service.
9mm L34A1 Sterling	Silenced SMG	UK	L 34in (86·4cm); W 9lb 8oz (4·31kg)	9mm x 19; 34rd box	515-565rds/min	c.985ft/sec (300m/sec)	Blowback operated; silenced version of L2A3; in British and other army service.
9mm MAT 49	SMG	France	L 28⅜in (72cm); W 9lb 3oz (4·17kg)	9mm x 19; 20/32rd box	600rds/min	1,280ft/sec (390m/sec)	Blowback operated; developed late 1940s; in French and other army service.
9mm MP5 SD	Silenced SMG	Federal Germany	L 30¾in (78cm); W 8lb 10¼oz (3·4kg)	9mm x 19, 15/30rd box	800rds/min	935ft/sec (285m/sec)	Delayed blowback operation; silenced version of MP5; in widespread service.
9mm MP5K	SMG	Federal Germany	L 12¾in (32·5cm); W 5lb 8⅞oz (2·52kg)	9mm x 19; 15/30rd box	900rds/min	1,230ft/sec (375m/sec)	Extra short version of MP5 for police and anti-terrorist forces; in wide service.
9mm UZ1	SMG	Israel	L 25⅝in (65cm); W 9lb (4·1kg)	9mm x 19; 25/32rd box	600rds/min	1,310ft/sec (400m/sec)	Blowback operated; in production since 1950; in widespread service.
9mm Viking	SMG	USA	L 23⅝in (60cm); W 7lb 12oz (3·52kg) empty	9mm x 19; 20/36rd box	700-800rds/min	1,310ft/sec (400m/sec)	Blowback operated; in service with several armies.

LAND WEAPONS: Rifles

L: length without bayonet W: weight loaded
N/A: Not available or not applicable

Weapon	Type	Origin	Dimensions	Ammunition	Cyclic rate	Muzzle velocity	Remarks
540 series SIG-Manhurin	Assault rifle	Switzerland	L 37⅜in (95cm); W 8lb 12oz (3·97kg)	5·56mm x 45; 20rd box	650/800rds/min	3,215ft/sec (980m/sec)	Data for SG540; also 7·62mm SG542; in widespread service.
7·62mm AK47	Assault rifle	USSR	L 34¼in (87cm); W 9lb 7½oz (4·3kg)	7·62mm x 39; 30rd box	600rds/min	2,330ft/sec (710m/sec)	Most widely used modern military firearm; superseded in Soviet service by AKM.
7·62mm AKM	Assault rifle	USSR	L 34½in (87·6cm); W 8lb (3·64kg)	7·62mm x 39; 30rd box	600rds/min	2,350ft/sec (715m/sec)	Modernised AK-47; in Warsaw Pact service since 1959; widely exported.
7·62mm SVD	Sniping rifle	USSR	L 48¼in (122·5cm); W 9lb 10½oz (4·385kg)	7·62mm x 54R; 10rd box	Single shot	2,720ft/sec (830m/sec)	Self-loading; standard sniping rifle in Soviet and Warsaw Pact service.
7·62mm FN FAL	Rifle	Belgium	L 42⅞in (109cm); W 9lb 6oz (4·25kg)	7·62mm NATO; 20rd box	650-700rds/min	2,760ft/sec (840m/sec)	In production since early 1950s; popular and in widespread service.
7·62mm L1A1	Self-loading rifle	UK	L 46½in (118cm); W 9lb 12oz (4·42kg)	7·62mm x 51; 10rd box	Single shot	2,760ft/sec (840m/sec)	Modified FN FAL for British Army; no burst capability; to be replaced by IW.
7·62mm Model 58	Assault rifle	Czechoslovakia	L 32¼in (82cm); W 8lb 6¾oz (3·82kg)	7·62mm x 39; 30rd box	800rds/min	2,330ft/sec (710m/sec)	Standard Czechoslovak Army rifle; also sold commercially.
7·62mm Model SP66	Sniping rifle	Federal Germany	N/A	6·62mm x 51; 3rd integral mag.	Single shot	N/A	Bolt action; in Federal German and other army service.
7·62mm NATO M14	Rifle	USA	L 44in (112cm); W 11lb 4oz (5·1kg)	7·62mm x 51; 20rd box	750rds/min	2,800ft/sec (853m(sec)	Developed from M1 Garand; standard US Army rifle since 1959.
7·62mm SSG 69	Sniping rifle	Austria	L 44⅞in (114cm); W 10lb 9oz (4·79kg)	7·62mm x 51; 5rd rotary or 10rd box	Single shot	2,820ft/sec (860m/sec)	Bolt action; in Austrian and other army service.

LAND WEAPONS: Self-propelled Guns

L: length including gun W: width H: height

Weapon	Type	Origin	Dimensions	Weaponry	Propulsion/ speed	Range	Remarks
Abbot	105mm SP gun	UK	L 19ft 2in (5·84m); W 8ft 8in (2·64m); H 8ft 2in (2·49m)	1 x 105mm; 1 x 7·62mm	240bhp multi-fuel engine; 30mph (48km)/h)	240 miles (390km)	Produced 1964-67; in British Army service since 1964; also Indian Army.
M107	175mm SP gun	USA	L 36ft 11in (11·26m); W 10ft 4in (3·15m); H 12ft 1in (3·68m)	1 x 175mm	405hp diesel engine; 35mph (54km/h)	450 miles (725km)	In US Army service since 1963; also in service with allied armies.
Type 75	155m SP howitzer	Japan	L 25ft 6in (7·79m); W 10ft 1in (3·09m)· H 8ft 4in (2·54m)	1 x 155mm; 1 x 12·7mm	450hp diesel engine; 29mph (47km/h)	186 miles (300km)	In JSDF service since 1977; production continuing.

LAND WEAPONS: Tanks

L: length including gun W: weight H: height
N/A: Not available or not applicable

Weapon	Type	Origin	Dimensions	Weaponry	Propulsion/ speed	Range	Remarks
AMX-13	Light tank	France	L 20ft 10in (6·36m); W 8ft 2in (2·5m); H 7ft 7in (2·3m)	1 x 75mm; 1 x 7·5mm/ 7·62mm	250hp petrol engine; 37mph (60km/h)	218 miles (350km)	Developed in late 1940s; in production from 1952; more than 10,000 built; in wide service in serveral versions.
PT-76	Amphibious light tank	USSR	L 25ft 0in (7·625m); W 10ft 4in (3·14m); H 7ft 2in (2·195m)	1 x 76·2mm; 1 x 7·62mm	240hp engine; 27mph (44km/h) road, 6·2mph (10km/h) water	162 miles (260km)	Water-jet propulsion when swimming; mainly used for reconnaissance; many variants.
T-54/55	MBT	USSR	L 29ft 6in (9m); W 10ft 9in (3·27m); H 7ft 10in (2·4m)	1 x 100mm; 1 x 12·7mm; 2 x 7·62mm	520hp diesel engine; 30mph (48km/h)	250 miles (400km)	In service 1950; widely exported; many variants; still in widespread service.
T-62	MBT	USSR	L 30ft 7in (9·33m); W 11ft 0in (3·35m); H 7ft 10in (2·4m)	1 x 115mm; 1 x 12·7mm; 1 x 7·62mm	580hp diesel engine; 28mph (45·5km/h)	280 miles (450km)	First MBT with smooth-bore gun; in Soviet service since 1963; widely exported; many variants.
Type 62	Light tank	China	L 25ft 0in (7·65m); W 9ft 4½in (2·86m); H 8ft 4½in (2·55m)	1 x 85mm; 1 x 12·7mm; 1 x 7·62mm	380hp diesel engine	N/A	Scaled-down version of T-59 (based on Soviet T-54); in production by State Factories.
Type 63	Light tank	China	L 26ft 10in (8·2m); W 11ft 0in (3·35m); H 7ft 2¼in (2·19m)	1 x 85mm; 1 x 12·7mm; 1 x 7·62mm	520hp diesel engine; 25mph (40km/h)	150 miles (240km)	Based on Soviet PT-76 with new turret; in Chinese service since 1960s; exported to some countries.

SEA WEAPONS: Aircraft Carriers

L: length overall B: beam D: draught
N/A: Not available or not applicable

Weapon	Type	Origin	Dimensions	Weaponry	Propulsion/ speed	Range at cruising speed	Remarks
Enterprise	CVN	USA	L 1,123ft (342·3m); B 248ft (75·7m); D 36ft (10·9m)	86 aircraft; 3xSea Sparrow; 3xPhalanx CIWS	4-shaft nuclear, 280,000shp = 30kt	N/A	First USN nuclear-powered aircraft carrier; completed 1961.
Forrestal class	CV	USA	L 1,047ft (319m); B 238ft (72·5m); D 37ft (11·3m)	86 aircraft; 2xBPDMS or Sea Sparrow	4-shaft geared steam turbines, 280,000shp = 33kt	13,800 miles (22,200km)	Four ships: CV-59 *Forrestal*, CV-60 *Saratoga*, CV-61 *Ranger*, CV-62 *Independence*, completed 1952-55.

SEA WEAPONS: Cruisers

L: length overall B: beam D: draught
N/A: Not available or not applicable

Weapon	Type	Origin	Dimensions	Weaponry	Propulsion/ speed	Range at cruising speed	Remarks
Belknap class	CG	USA	L 547ft (166·7m); B 55ft (16·7m); D 29ft (8·7m)	Standard; Asroc; 1 x 5in; 2 Phalanx CIWS	2-shaft geared steam turbines, 85,000shp = 33kt	8,200 miles (13,100km)	9 ships + 1 CGN completed 1964-67.
Kara class	CG	USSR	L 570ft (174m); B 60ft (18m); D 20ft (6m)	8 x SS-N-14; 4 x SA-N-4; 4 x 76mm	2-shaft COGAG, 100,000bhp = 32kt	8,800 miles (14,000km)	7 ships completed 1973-79; all but 2 serve in Black Sea/Mediterranean.
Kresta II class	CG	USSR	L 520ft (158m); B 56ft (17m); D20ft (6m)	8 x SS-N-14; 4 x SA-N-3; 4 x 57mm	2-shaft geared steam turbines, 100,000shp = 34kt	10,500 miles (17,000km)	10 ships completed 1970-78; most with Northern Fleet, others Pacific Fleet.
Moskva class	CHG	USSR	L 625ft (191m); B 112ft (34m); D 25ft (8m)	4 x SA-N-3; 4 x 57mm, 18 "Hormone-A" helos	2-shaft geared steam turbines, 100,000shp = 30kt	9,000 miles (14,500km)	2 ships completed 1967-68; both serve in Black Sea/Mediterranean.

SEA WEAPONS: Destroyers

L: length overall B: beam D: draught
N/A: Not available or not applicable

Weapon	Type	Origin	Dimensions	Weaponry	Propulsion/ speed	Range at cruising speed	Remarks
Charles F Adams class	DDG	USA	L 437ft (133·2m); B 47ft (14·3m); D 22ft (6·7m)	Standard; Asroc; 2 x 5in; 6TT	2-shaft geared steam turbines, 70,000shp = 31·5kt	5,200 miles (8,300km)	23 ships completed 1960-64; design also adopted by Federal German and Australian Navies.
Coontz class	DDG	USA	L 513ft (156·2m); B 53ft (15·9m); D 25ft (7·6m)	Standard; Asroc; 1 x 5in; 6TT	2-shaft geared steam turbines, 85,000shp = 33kt	5,000 miles (8,000km)	10 ships completed 1959-61.
County class	Destroyer	UK	L 520ft 6in (158·7m); B 54ft (16·5m); D 20ft 6in (6·3m)	Exocet; Seaslug; Seacat 2 x 4·5in; 6TT	2-shaft COSAG, 60,000shp = 30kt	4,000 miles (6,500km)	8 ships completed 1962-66; only *Fife* and *Glamorgan* still in RN service; remainder paid off or transferred.
Gearing class	DDE	USA	L 390ft 2in (118·9m); B 40ft 11in (12·47m); D 14ft 4in (4·39m)	Asroc; 4 x 5in; 6TT	2-shaft geared turbines, 60,000shp = 32kt	4,600 miles (7,400km)	93 ships completed 1940s; particulars post FRAM modernisation; many transferred to other navies; none remain in USN.
Hatsuyuki class		Japan	L 426ft 6in (130m); B 44ft 7in (13·6m); D 14ft 1in (4·3m)	Harpon; Sea Sparrow; 1x76mm; 2 Phalanx	2-shaft COGOG, 45,000/13,600shp = 30kt	N/A	12 ships; 7 completed by 1985; remainder scheduled for completion by 1987.
Sheffield class	Destroyer	UK	L 412ft (125·6m); B 47ft (14·3m); D 19ft (5·8m)	Sea Dart; 1x4·5in; Lynx helo; 6TT	2-shaft COGOG, 56,000/8,500shp = 29/18kt	4,000 miles (6,450km)	12 ships completed 1976-84; last 3 42ft (12·8m) longer hull; *Sheffield* and *Coventry* sunk May 1982 off Falklands.
Yamagumo class	Destroyer	Japan	L 374ft (114m); B 38ft 9in (11·8m); D 13ft 1in (4m)	Asroc; 4 x 3in; 6TT	2-shaft diesel, 26,500bhp = 27kt	6,000 miles (9,650km)	6 ships completed 1966-78; last 3 with slightly lengthened hull.

SEA WEAPONS: Frigates

L: length overall B: beam D: draught

Weapon	Type	Origin	Dimensions	Weaponry	Propulsion/ speed	Range at cruising speed	Remarks
Annapolis class	DDH	Canada	L 371 ft (113m); B 42ft (12·8m); D 14ft 4in (4·4m)	2 x 3in; Limbo A/S mortar; Sea King helicopter	2-shaft geared steam turbines, 30,000shp = 28kt	4,600 miles (7,400km)	2 ships completed 1964; Sea Sparrow and other new equipment installed 1982-84.
Baleares class	FF	Spain	L 438ft (133·5m); B 46ft 11in (14·3m); D 15ft 5in (4·7m)	Harpoon; Tartar/Standard; Asroc; 1 x 5in; 6TT	1-shaft geared turbines, 35,000shp = 28kt	4,500 miles (7,250km)	5 ships completed 1973-76; based on US Knox class; to be upgraded to USN standard.
Bremen class	Frigate	Federal Germany	L 428ft 2in (130·5m); B 47ft 3in (14·4m); D 19ft 8in (6m)	Harpoon; Sea Sparrow; Stinger; 1 x 76mm; 4TT	2-shaft CODOG, 50,000/10,400hp = 30/18kt	4,000 miles (6,450km)	6 ships completed 1962-74; modified Netherlands Kortenaer class.
D'Estienne d'Orves class	Frigate	France	L 262ft 6in (80m); B 33ft 10in (10·3m); D 17ft 5in (5·3m)	Exocet; 1x100mm; 4TT	2-shaft diesel, 12,000bhp = 24kt	4,500 miles (7,250km)	17 ships completed 1976-84.
F30 class	Corvette	Spain	L 291ft 4in (88·8m); B 34ft 1in (10·4m); D 12ft 6in (3·8m)	Harpoon; Sea Sparrow; 1 x 3in; 6TT	2-shaft diesel, 26,000shp = 25·5kt	4,000 miles (6,450km)	6 ships completed 1978-82; based on Portuguese Improved Joao Coutinho class.
Garcia class	FF	USA	L 415ft (126·3m); B 44ft (13·5m); D 24ft (7·3m)	Asroc; 2 x 5in; 6TT; SH-2F helicopter	1-shaft geared steam turbine, 20,000shp = 24kt	4,600 miles (7,400km)	10 ships completed 1963-68; 6 similar Brooke class with Tartar.
Grisha class	FFL	USSR	L 236ft 3in (72m); B 32ft 10in (10m); D 12ft 2in (3·7m)	SA-N-4; 2 x 57mm; 30mm Gatling; 4TT	3-shaft CODAG, 40,000shp = 30kt	4,500 miles (7,250km)	55 ships in 3 groups completed by 1985; variations in armament; details for Grisha III class.
Jianghu class	Frigate	China	L 338ft 7in (103·2m); B 33ft 6in (10·2m); D 10ft 2in (3·1m)	Hai Ying 2 (SS-N-2); 2/4 x 3·9in	2-shaft diesel, 24,000shp = 26·5kt	4,000 miles (6,450km)	15 of planned 40 ships completed by 1985; modified Jiandong class with SSMs instead of SAMS.

Weapon	Type	Origin	Dimensions	Weaponry	Propulsion/speed	Range	Remarks
Knox class	FF	USA	L 438ft (133·5m); B 47ft (14·3m); D 25ft (7·6m)	Asroc; Sea Sparrow; 1 x 5in; 4TT; SH-2F	1-shaft geared steam turbine, 35,000shp = 27kt	5,200 miles (8,300km)	46 ships completed 1969-74; most have Asroc replaced by Harpoon.
Leander class	FF	UK	L 372ft (113·4m); B 41ft (12·5m); D 19ft (5·8m)	Exocet; Seacat; 2 x 40mm; 6TT	2-shaft geared steam turbines, 30,000shp = 28kt	4,000 miles (6,450km)	44 ships completed 1963-73; 26 for RN, remainder for other navies; data for Exocet group; also Ikara and Broad-beamed groups.
Maestrale class	ASW frigate	Italy	L 402ft 7in (122·7m); B 42ft 4in (12·9m); D 27ft 7in (8·4m)	Teseo; Aspide; 1 x 5in; 8TT; 2 x AB212 helos	2-shaft CODOG, 50,000/11,000shp = 32/21kt	6,000 miles (9,650km)	8 ships completed 1982-84.
Wielingen class	FF	Belgium	L 349ft (106·4m); B 40ft 4in (12·3m); D 18ft 4in (5·6m)	Exocet; Sea Sparrow; 1 x 3·9in; 2TT	2-shaft CODOG, 28,000/6,000bhp = 29/20kt	6,000 miles (9,650km)	4 ships completed 1978; Goalkeeper CIWS to be fitted.

SEA WEAPONS: Small Combatants

L: length overall   B: beam   D: draught

Weapon	Type	Origin	Dimensions	Weaponry	Propulsion/speed	Range at cruising speed	Remarks
Hauk class	FAC-missile	Norway	L 119ft 9in (36·5m); B 20ft 4in (6·2m); D 6ft 7in (2m)	Penguin; 1 x 40mm; 1 x 20mm; 2TT	2-shaft diesel, 7,000shp = 34kt	440 miles (700km) at 34kt	14 boats completed 1977-80.
Hugin class	FAC-missile	Sweden	L 120ft (36·6m); B 20ft 8in (6·3m); D 5ft 7in (1·7m)	Penguin; 1 x 57mm	2-shaft diesel, 7,200bhp = 35kt	630 miles (1,020km) at 35kt	16 boats completed 1978-82; RB15 SSMs to be fitted.
La Combattante III class	FAC-missile	France	L 184ft 5in 56·2m); B 26ft 3in (8m); D 6ft 11in (2·1m)	Exocet or Penguin; 2 x 76mm; 4 x 30mm; 2TT	4-shaft diesel, 18,000bhp = 36kt	2,000 miles (3,200km)	10 boats completed for Greek Navy 1977-81.
Minister class	FAC-missile	South Africa	L 204ft (62·2m); B 25ft (7·6m); D 8ft (2·4m)	Skorpioen; 2 x 76mm; 2 x 20mm	4-shaft diesel, 12,000shp = 32kt	3,600 miles (5,800km)	10 boats completed 1977-85; first 3 built in Israel, remainder in South Africa; similar to Saar 4 class.
Nanuchka class	Missile corvette	USSR	L 194ft 6in (59·3m); B 41ft 4in (12·6m); D 7ft 10in (2·4m)	SS-N-9; SA-N-4; 2 x 57mm	3-shaft diesel, 24,000shp = 32kt	2,500 miles (4,000km)	17 Nanuchka I and 7 Nanuchka III completed since 1979; Nanuchka III has 1 x 76mm, 1 x 30mm instead of 2 x 57mm.
Osa I/II classes	FAC missile	USSR	L 128ft (39m); B 25ft 7in (7·8m); D 5ft 11in (1·8m)	SS-N-2; SA-N-5 (some); 4 x 30mm	3-shaft diesel, 12,000/15,000shp = 38/40kt (I/II)	800 miles (1,300km)	70 Osa I and 45 Osa II completed 1959-70; many more exported to friendly and allied nations.
P6 class	FAC torpedo	USSR	L 85ft 4in (26m); B 19ft 8in (6m); D 4ft 11in (1·5m)	4 x 25mm; 2TT	4-shaft diesel, 5,000bhp = 43kt	690 miles (1,110km)	500 + completed 1955-59, plus others in China, North Korea; c.200 transferred to other navies; no longer in Soviet service.
Saar 2/3 classes	FAC missile	France	L 147ft 8in (45m); B 23ft (7m); D 8ft 2in (2·5m)	Harpoon; Gabriel II; 3 x 40mm or 1 x 76mm	4-shaft diesel, 13,500bhp = 40 + kt	2,500 miles (4,000km)	6 Saar 2 and 6 Saar 3 built in France to German design for Israel; completed 1968-69.
Sparviero class	Missile hydrofoil	Italy	L 80ft 9in (24·6m); B 39ft 8in (12·1m); D 14ft 5in (4·4m)	Teseo; 1 x 76mm	Gas turbine-driven waterjet, 4,500bhp = 50kt	400 miles (640km)	7 vessels completed 1974-83.
Type 143A	FAC-missile	Federal Germany	L 189ft 4in (24·9m); B 24ft 11in (7·6m); D 8ft 2in (2·5m)	Exocet; 1 x 76mm	4-shaft diesel, 18,000shp = 40kt	2,600 miles (4,200km)	10 boats completed 1982-84; ASMD RAM system to be fitted.
Type 148	FAC-missile	Federal Germany	L 154ft 2in (47m); B 23ft (7m); D 6ft 11in (2·1m)	Exocet; 1 x 76mm; 1 x 40mm	4-shaft diesel, 12,000bhp = 38 kts	600 miles (965km)	20 boats completed 1972-75.

SEA WEAPONS: Submarines (Attack)

L: length overall   B: beam   D: draught
N/A: Not available or not applicable

Weapon	Type	Origin	Dimensions	Weaponry	Propulsion/speed	Range	Remarks
Sturgeon class	SSN	USA	L 292ft (89m); B 32ft (9·65m); D 29ft 6in (8·9m)	Harpoon; Subroc; 4 x 21in TT	1-shaft nuclear, 15,000shp = 30kt	N/A	37 boats completed 1967-75.
Swiftsure class	SSN	UK	L 272ft (82·9m); B 32ft 4in (9·8m); D 27ft (8·2m)	Tigerfish (5 x 21in TT, 20 reloads)	1-shaft nuclear, 15,000shp = 30kt	N/A	6 boats completed 1978-81.

SEA WEAPONS: Submarines (Cruise Missile)

L: length overall   B: beam   D: draught
N/A: Not available or not applicable

Weapon	Type	Origin	Dimensions	Weaponry	Propulsion/speed (surfaced/submerged)	Range	Remarks
Charlie I class	SSGN	USSR	L 308ft (93·9m); B 32ft 6in (9·9m); D 24ft 7in (7·5m)	8 x SS-N-7 tubes; 6TT	1-shaft nuclear, 20,000shp = 28kt	N/A	11 boats completed 1967-72.
Charlie II class	SSGN	USSR	L 337 ft 6in (102·9m); B 32ft 6in (9·9); D 25ft 7in (7·8m)	8 x SS-N-9 tubes; 6TT	1-shaft nuclear, 20,000shp = 25kt	N/A	6 boats completed 1973-80; enlarged version of Charlie I.
Echo II class	SSGN	USSR	L 384ft 8in (117·3m); B 30ft 2in (9·2m); D 25ft 6in (7·8m)	8 x SS-N-3 tubes; 8TT	2-shaft nuclear, 30,000shp = 25kt	N/A	29 boats completed 1963-67; c.7 have SS-N-12 instead of SS-N-3A.
Juliett class	SSG	USSR	L 284ft 5in (86·7m); B 33ft 2in (10·1m); D 23ft (7m)	4 x SS-N-3 tubes; 6TT	2-shaft diesel/electric 8,000/6,000hp = 19/14kt	15,000 miles (24,000km) surfaced	16 boats completed 1961-68.

SEA WEAPONS: Submarines (Patrol)

L: length overall   B: beam   D: draught
N/A: Not available or not applicable

Weapon	Type	Origin	Dimensions	Weaponry	Propulsion/speed (surfaced/submerged)	Range	Remarks
Daphné class	SS	France	L 189ft 8in (57·8m); B 22ft 4in (6·8m); D 15ft 1in (4·6m)	12 x 550mm TT	2-shaft diesel/electric, 1,224/2,600bhp = 13·5/16kt	10,000 miles (16,000km) surfaced	25 boats completed 1964-70, including 14 for export.

Weapon	Type	Origin	Dimensions	Warhead/ weaponry	Propulsion/speed (surfaced/submerged)	Range	Remarks
Foxtrot class	SS	USSR	L 300ft 2in (91·5m) B 26ft 3in (8m); D 20ft (6·1m)	6 x 21in + 4 x 16in TT	3-shaft diesel/electric, 6,000/5,500hp = 18/16kt	20,000 miles (32,000km) surfaced	62 boats completed 1958-71, plus subsequent examples for Cuba (3), India (8), Libya (6).
Oberon class	SS	UK	L 295ft 2in (90m); B 26ft 6in (8·1m); D 18ft (5·5m)	8 x 21in TT	2-shaft diesel/electric, 3,680/6,000hp = 12/17kt	9,000 miles (14,500km) surfaced	21 boats completed 1961-67, including examples for Brazil (3), Canada (3), Chile (2).
Romeo class	SS	USSR	L 251ft 11in (76·8m); B 23ft 11in (7·3m); D 18ft (5·5m)	8 x 21in TT	2-shaft diesel/electric, 4,000/4,000hp = 17/14kt	16,000 miles (26,000km) surfaced	20 + boats completed 1958-61, plus further construction in China (90 +) and North Korea (5); several transfers.
Sauro class	SS	Italy	L 210ft (64m); B 22ft 6in (6·8m); D 18ft 10in (5·7m)	6 x 21in TT	1-shaft diesel/electric, 3,210/3,650hp = 11/20kt	7,000 miles (11,000km) surfaced	6 boats completed since 1980.
Toti class	SS	Italy	L 151ft 6in (46·2m); B 15ft 5in (4·7m); D 13ft 1in (4m)	4 x 21in TT	1-shaft diesel/electric, 2,200/2,200hp = 14/15kt	3,000 miles (4,800km) surfaced	4 boats completed 1968-69.
Type 205	SS	Federal Germany	L 144ft (43·9m); B 14ft 11in (4·5m); D 14ft (4·3m)	8 x 21in TT	1-shaft diesel/electric, 1,200/1,500hp = 10/17kt	4,400 miles (7,000km) surfaced	14 boats completed 1961-69, including 2 for Denmark; 6 remain in German Navy service.
Type 206	SS	Federal Germany	L 159ft 5in (48·6m); B 15ft 1in (4·6m); D 14ft 10in (4·5m)	8 x 21in TT	1-shaft diesel/electric, 1,500/1,800hp = 10/17kt	4,500 miles (7,250km) surfaced	18 boats completed 1973-75, plus 3 IKL 540 class for Israel completed 1977.
Uzushio class	SS	Japan	L 236ft 2in (72m); B 29ft 6in (9m); D 24ft 7in (7·5m)	6 x 21in TT	1-shaft diesel/electric, 3,400/7,200hp = 12/20kt	N/A	7 boats completed 1971-78.
Whiskey class	SS	USSR	L 249ft 4in (76m); B 21ft 4in (6·5m); D 16ft 1in (4·9m)	8 x 21in TT	2-shaft diesel/electric, 4,000/2,700hp = 17/14kt	13,000 miles (21,000km) surfaced	c.260 boats completed 1958-61; c.50 in service; 40 transferred; further construction in China.

SEA WEAPONS: Shipborne AA Guns

Weapon	Type	Origin	Dimensions	Ammunition	Rate of fire	Muzzle velocity	Remarks
Goalkeeper	Air defence system	Netherlands/ USA	Calibre 30mm Weight 14,840lb (6,730kg)	APDS, HEI; 1,190rd linkless feed	4,200 rds/min	3,350ft/sec (1,021m/sec)	GAU-8/A 7-barrel Gatling gun with Signaal radars; adopted by Netherlands and Royal Navy.
TCM 30	Twin 30mm	Israel	L 11ft 8in (3·55m)	Various types; 165rds/gun	1,300 rds/min	3,540ft/sec (1,080m/sec)	Operational aboard Israeli FACs and hydrofoils.

SEA WEAPONS: Shipborne Surface-to-air Missiles

L: length D: diameter S: span
N/A: Not available or not applicable

Weapon	Type (and guidance)	Origin	Dimensions	Warhead	Propulsion/ speed	Range	Remarks
Masurca	SAM (SAR)	France	L 28ft 2½in (8·6m); D 16⅛in (41cm); S 59in (150cm)	HE	Solid booster and sustainer	25 miles (40km) +	Operational aboard *Duquesne, Suffren* and *Colbert.*
Naval Crotale	SAM (radar beam riding + command)	France	L 9ft 7⅜in (2·93m); D 6⅛in (15·6cm); S 21¼in (54cm)	31lb (14kg) fragmentation	Solid rocket Mach 2·3	6·2 miles (10km);	Original 8B model updated to 8S standard to intercept sea-skimming missiles; in French and Saudi Arabian service.
Sea Dart	SAM (SAR)	UK	L 14ft 5¼in (4·4m); D 16½in (42cm); S 35¾in (91cm)	Fragmentation	Solid rocket booster + ramjet sustainer; Mach 3·5	50 miles (80km) +	Lightweight and land versions proposed; in Royal Navy and Argentinian service.
SA-N-1	SAM (command)	USSR	L 22ft (6·7m); D 23⅝in (60cm); S 59in (150cm)	132lb (60kg) HE	Solid booster and sustainer; Mach 2	15·72 miles (25·3km)	NATO "Goa"; naval version of land-based SA-3.
SA-N-5	SAM (IR homing)	USSR	L 4ft 2¾in (1·29m); D N/A; S N/A	5·5lb (2·5kg) fragmentation	Solid booster and sustainer; Mach 1·5	11,800ft (3,600m)	Naval version of SA-7 "Grail"; operational aboard c.170 Soviet Navy ships.
Tartar	SAM (SAR)	USA	L 15ft 1in (4·6m); D 13½in (34·3cm); S 42in (107cm)	HE	Solid rocket; Mach 2	10 miles (16km) +	Operational in US and other navies; to be replaced by Standard.

SEA WEAPONS: Shipborne Surface-to-surface Missiles

L: length D: diameter S: span
N/A: Not available or not applicable

Weapon	Type (and guidance)	Origin	Dimensions	Warhead	Propulsion/ speed	Range	Remarks
Gabriel Mk III	SSM (inertial + active radar homing)	Israel	L 12ft 7⅝in (3·85m); D 13⅜in (34cm); S 43¼in (110cm)	330lb (150kg) HE	Solid rocket; Mach 0·73	37 miles (60km) +	Air-launched Mks I and II in advanced development; in service with IDF and other navies.
HY-2 (CSS-N-2)	SSM	China	As SS-N-2	N/A	N/A	N/A	Based on Soviet SS-N-2; in service with Chinese and Albanian Navies.
Penguin Mk 2	SSM (inertial + IR homing)	Norway	L 9ft 10⅛in (3m); D 11 in (28cn) S 55⅛in (140cm)	265lb (120kg) SAP	Solid rocket; Mach 0·8	18·6 miles (30km)	In service with Norwegian, Greek, Swedish and Turkish Navies; helicopter-launched version ordered by US Navy.
Sea Killer Mk 2	SSM (beam riding + command)	Italy	L 15ft 5in (4·7m); D 8⅛in (20·6cm); S 39⅜in (100cm)	154lb (70kg) SAP	Solid booster and sustainer; 985ft/sec (300m/sec)	15·5 miles (25km) +	In service with Iranian Navy; modified as Marte for Italian Navy helicopters.
SS-N-2	SSM (autopilot or command + IR or radar homing)	USSR	L 20ft 6in (6·25m); D 29½in (75cm); S 108in (275cm)	880-990lb (400-450kg) HE	Solid booster and sustainer	25 miles (40km) or 50 miles (80km)	NATO "Styx"; operational since c.1960; in widespread service; SS-N-2A has bigger warhead, longer range.
SS-N-3	SSM (command + IR or active radar homing)	USSR	N/A	350/800kT nuclear or 1,000lb (450kg) HE	Ramjet or turbojet, transonic	280 miles (450km)	NATO "Shaddock"; operational aboard Soviet Navy cruisers and submarines; also "Sepal" coast defence variant.
SS-NX-22	SSM	USSR	N/A	N/A	N/A	c.75 miles (120km) est	Probably based on SS-N-9; operational aboard Sovremenny (8 launchers), Slava (8) and Tarantul (4) classes.
USN 5in GP	Laser-guided projectile	USA	L 5ft 0⅝in (1·54m); Calibre 5in	30lb (13·6kg) fragmentation/ shaped charge	Gun-launched + solid rocket	15 miles (24km) +	Evaluation completed 1981; in production for US Navy.

Index

238

239

PICTURE CREDITS

Those who supplied photographs for this book are listed below, with page numbers followed by photo position (T = Top, TL or TR = Top Left or Top Right; C = Centre (etc.); B = Bottom (etc.)), except in the case of the US Department of Defense Audio Visual Agencies (DoD) and British Ministry of Defence, whose contributions are listed separately:

US DoD: ½ title; Title; 8-9; 10-11; 12-13; 15 (T); 16-17; 18-19; 20; 22; 24; 25 (BR); 27; 42 (T); 43 (T); 44 (T); 45; 46 (T); 52-53; 58 (C); 59-60; 62 (C); 63 (R, B); 67 (B); 74 (B); 75 (B); 76 (B); 77-79; 80 (T); 81-82; 83 (B); 88; 90; 91 (BL); 92-95; 98; 100; 101 (T, C); 104; 106 (T); 107 (T, C); 109-110; 111 TR); 114; 117 (T, BR); 118 (BR); 120 (CR); 122; 123 (TR); 127 (R); 134-135; 136-137; 140-141; 142 (B); 144 (C, B); 151; 152-153 (T); 154 (BL); 155 (B); 157 (L); 158; 159 (L, R); 161 (T); 162; 164 (T, BL); 166 (B); 167-168; 169 (C, B); 171 (C, B); 173 (C, B); 175 (C, B); 176 (C, B); 182; 183 (T); 184 (T); 187 (T); 188-9; 193-195; 197 (R); 198-199; 201 (CB); 202 (T); 203 (T, C); 204 (TC); 207 (C, B); 208-209; 210 (T); 212 (C, B); 214; 216 (C, B); 217; 218-221; 224.

British MoD: 26 (B); 66 (T); 74 (C); 97 (T); 124; 130 (B); 131 (B); 132; 145 (C, B); 154 (T, BR); 159 (CB); 165 (R); 172; 179 (C, B); 180; 181; 183 (B, R); 186 (B); 196 (B); 197 (L); 200 (C, B); 210 (B); 215 (T, C); 222.

Others: 6-7: Hughes. 14: French Armée de l'Air. 15: B, Rockwell. 17: B, Aerospatiale. 21: ECPA. 22: B, L, DCTN. 25: T, DCTN. 26: T, C, Salamander archives; 32-35: BAe. 36: T, CASA; B, Cessna; 37-41: Dassault-Breguet. 42: B, FNA. 43: B, General Dynamics. 44: C, Hughes. 45-46: C, Grumman. 48: T, Grumman; B, HAL. 49: T, Grumman; B, IAI. 50: IAI. 51: T, Italian Air Force; B, Grumman. 54-55: McDonnell Douglas. 56: T, McDonnell Douglas; B, Pakistan Air Force. 57: T, HAL; B, Salamander Archive. 58: T, Tass; B, Salamander archive. 62: T, Mitsubishi; B, Salamander archive. 63: L, Salamander archive. 65: BAe. 66: C, Hunting Engineering; B, MBB. 67: T, MBB. 68-70: Saab-Scania. 71: T, B, BAe; C, Matra. 72: T, JDW; B, SIAI-Marchetti; B, SOKO. 73: SOKO. 74: T, HAL. 75: T, C, TASS. 76: T, TASS; C, Salamander archive. 80: B, Salamander archive. 82: B, LTV. 83: T, Hsinhua; C, TASS. 84-85: Aerospatiale. 86: Augusta. 87: Bell Helicopter Textron. 89: T, Hughes Helicopters. 91: T, TASS: R (both), MBB. 96: T, Sikorsky; B, Westland. 97: B (two), Westland. 99: T, Israel Air Force; remainder, Grumman. 101: B, Salamander archive. 102: B, BAe. 103: T, Dassault-Breguet; B, Fokker-VFW. 105: Lockheed. 106: C, B, Japan Self-Defense Force. 107: B, Agusta. 108: T, B, Westland; C, Aerospatiale. 111: T, Hughes; B, General Dynamics. 112: T, SNIA; B, BAe. 113: T, Matra; B, Grumman. 115: TL, BAe; BL, Rafael; TR, Saab-Scania; BR, McDonnell Douglas. 116: T, Matra; B, BAe. 117: BL, BAe. 118: T and BL, Aerospatiale. 119: T, Salamander archive; B, Aerospatiale. 120: TL, MBB; TR, Aerospatiale; CL, 1A1; BL, SAMP; BR, Texas Instruments. 123: (RC) Hunting; BL, Matra; R, McDonnell Douglas. 125: Saab-Scania. 127: T, BAe; BL, Westland. 128-129: NATO. 130: T, Panhard. 131: T, C, Panhard. 133: NATO. 134: T, Krauss Maffei. 135: T, Krauss Maffei. 136: T, NATO. 138: C. Foss. 139: Swedish Army. 142: T, C, Thyssen Henschel. 143: T, C, Japan Self-Defense Forces; B, Swedish Army. 144-5: T, Vickers. 146: C. Foss. 147: Royal Ordnance Factory. 148: Panhard. 149: T, Novosti; B, Hagglund and Soner. 150: C. Foss. 152-3: B, Saurer Werke. 155: T, French Army. 156: T, BAe; C, Thomson-CSF; B, Oerlikon. 157: (R) NATO. 160: Ford. 163: T, C. Foss; C, NATO; B, MBB. 164: BR, French Army. 165: L, Hughes. 166: T, C. Foss. 169: T, French Army. 170: T, Saurer-Werke; C, B, FMC. 171: T, French Army. 173: T, TASS. 174: Hughes. 175: Thyssen Henschel. 176: T, NATO. 177: T, Cadillac Gage; 177: C, B, Ray Bonds. 178: T, S. African Army; C, B, AMX. 179: T, Japan Self-Defense Force. 182: BL, NATO; C, TASS. 184: C, B, TASS. 185: T, HeK; C, B, St Etienne arsenal. 186: T, HeK; C, Galil. 187: B, NATO. 190: T, C, DCTN; B, BAe. 191: T, C, MARS; B, Vickers. 192: T, Vickers. 196: T, MARS; 200: T, DCTN; C, B, Vickers. 201: T, DCTN. 202: DCTN. 203: B, Kockums. 204: B (both), Howaldswerke. 205: T, Thyssen Nordseewerke; B (both), Howaldswerke. 206: T, Vickers; B, Japan Maritime Self-Defense Force. 207: T, Royal Netherlands Navy. 211: C, B, DCTN. 212: T, Royal Canadian Navy. 213: Japan Maritime Self-Defense Force. 215: B, Indian Navy. 216: T, NATO. 223: C, B, Bell Aerosystems.